PRINCIPLES OF
REAL ANALYSIS
Third Edition

PRINCIPLES OF REAL ANALYSIS
Third Edition

CHARALAMBOS D. ALIPRANTIS

Departments of Economics and Mathematics
Purdue University

and

OWEN BURKINSHAW

Department of Mathematical Sciences
Indiana University, Purdue University, Indianapolis

ACADEMIC PRESS
An Imprint of Elsevier
San Diego London Boston
New York Sydney Tokyo Toronto

Permissions may be sought directly from Elsevier's Science and Technology Rights Department in
Oxford, UK. Phone: (44) 1865 843830, Fax: (44) 1865 853333, e-mail: permissions@elsevier.co.uk.
You may also complete your request on-line via the Elsevier homepage: http://www.elsevier.com by
selecting "Customer Support" and then "Obtaining Permissions".

ACADEMIC PRESS
An Imprint of Elsevier
525 B Street, Suite 1900, San Diego, California 92101-4495, USA
http://www.apnet.com

ACADEMIC PRESS LIMITED
24–28 Oval Road, London NW1 7DX, UK
http://www.hbuk.co.uk/ap/

Library of Congress Cataloging-in-Publication Data
Aliprantis, Charalambos D.
    Principles of real analysis / Charalambos D. Aliprantis and Owen Burkinshaw.
       p.   cm.
    Includes bibliographical references and index.
    ISBN-13: 978-0-12-050257-8 (acid-free paper) ISBN-10: 0-12-050257-7
    1. Mathematical analysis.   2. Functions of real variables.   I. Burkinshaw, Owen.   II. Title.
QA300.A48   1998
515—dc21   98-3955
                                                                                   CIP

Transferred to Digital Printing 2009

To our wives
Bernadette and Betty,
and to our children
Claire, Dionissi, and Mary.

CONTENTS

PREFACE

This is the third edition of *Principles of Real Analysis*, first published in 1981. The aim of this edition is to accommodate the current needs for the traditional real analysis course that is usually taken by the senior undergraduate or by the first year graduate student in mathematics. This edition differs substantially from the second edition. Each chapter has been greatly improved by incorporating new material and by rearranging the old material. Moreover, a new chapter (Chapter 6) on Hilbert spaces and Fourier analysis has been added.

The subject matter of the book focuses on measure theory and the Lebesgue integral as well as their applications to several functional analytic directions. As in the previous editions, the presentation of measure theory is built upon the notion of a semiring in connection with the classical Carathéodory extension procedure. We believe that this natural approach can be easily understood by the student. An extra bonus of the presentation of measure theory via the semiring approach is the fact that the product of semirings is always a semiring while the product of σ-algebras is a semiring but not a σ-algebra. This simple but important fact demonstrates that the semiring approach is the natural setting for product measures and iterated integrals.

The theory of integration is also studied in connection with partially ordered vector spaces and, in particular, in connection with the theory of vector lattices. The theory of vector lattices provides the natural framework for formalizing and interpreting the basic properties of measures and integrals (such as the Radon–Nikodym theorem, the Lebesgue and Jordan decompositions of a measure, and the Riesz representation theorem). The bibliography at the end of the book includes several books that the reader can consult for further reading and for different approaches to the presentation of measure theory and integration.

In order to supplement the learning effort, we have added many problems (more than 150 for a total of 609) of varying degrees of difficulty. Students who solve a good percentage of these problems will certainly master the material of this book. To indicate to the reader that the development of real analysis was a collective effort by many great scientists from several countries and continents through the ages, we have included brief biographies of all contributors to the subject mentioned in this book.

We take this opportunity to thank colleagues and students from all over the world who sent us numerous comments and corrections to the first two editions. Special thanks are due to our scientific collaborator, Professor Yuri Abramovich, for his comments and constructive criticism during his reading of the manuscript of this edition. The help provided by Professors Achille Basile and Vinchenco Aversa of Università Federico II, Napoli, Italy, during the collection of the biographical data is greatly appreciated. Finally, we thank our students (Anastassia Baxevani, Vladimir Fokin, Hank Hernandez, Igor Kuznetsov, Stavros Muronidis, Mohammad Rahman, and Martin Schlam) of the 1997–98 IUPUI graduate real analysis class who read parts of the manuscript and made many corrections and improvements.

C. D. ALIPRANTIS and O. BURKINSHAW
West Lafayette, Indiana
June, 1998

PRINCIPLES OF REAL ANALYSIS
Third Edition

CHAPTER **1**_____

FUNDAMENTALS OF REAL ANALYSIS

If you are reading this book for the purpose of learning the theory of integration, it is expected that you have a good background in the basic concepts of real analysis. The student who has come this far is assumed to be familiar with set theoretic terminology and the basic properties of real numbers, and to have a good understanding of the properties of continuous functions.

The first section of this chapter covers the fundamentals of set theory. We have kept it to the "minimum amount" of set theory one needs for any modern course in mathematics. The following two sections deal with the real and extended real numbers. Since the basic properties of the real numbers are assumed to be known, the fundamental convergence theorems needed for this book are emphasized. Similarly, the discussion on the extended real numbers is focused on the needed results. The last two sections present a comprehensive treatment of metric spaces.

1. ELEMENTARY SET THEORY

Throughout this book the following commonly used mathematical symbols will be employed:

$\forall$ means "for all" (or "for each");
$\exists$ means "there exists" (or "there is");
$\implies$ means "implies that" (or simply "implies");
$\iff$ means "if and only if."

The basic notions of set theory will be briefly discussed in the first section of this chapter. It is expected that the reader is familiar in one way or another with these concepts. No attempt will be made, however, to develop an axiomatic foundation of set theory. The interested reader can find detailed treatments on the foundation of set theory in references [8], [13], [17], and [20] in the bibliography at the end of this book.

1

The concept of a set plays an important role in every branch of modern mathematics. Although it seems easy, and natural, to define a set as a collection of objects, it has been shown in the past that this definition leads to contradictions. For this reason, in the foundation of set theory the notion of a set is left undefined (like the points and lines in geometry), and is described simply by its properties. In this book we shall mainly work with a number of specific "small" sets (like the Euclidean spaces $\mathbb{R}^n$ and their subsets), and we shall avoid making use of the "big" sets that lead to paradoxes. Therefore, a **set** is considered to be a collection of objects, viewed as a single entity.

Sets will be denoted by capital letters. The objects of a set A are called the **elements** (or the **members** or the **points**) of A. To designate that an object x belongs to a set A, the **membership symbol** $\in$ is used, that is, we write $x \in A$ and read it: x belongs to (or is a member of) A. Similarly, the symbolism $x \notin A$ means that the element x does not belong to A. Braces are also used to denote sets. For instance, the set whose elements are a, b, and c is written as $\{a, b, c\}$. A set having only one element is called a **singleton**.

Two sets A and B are said to be **equal**, in symbols $A = B$, if A and B have precisely the same elements. A set A is called a **subset** of (or that it is included in) a set B, in symbols $A \subseteq B$, if every element of A is also a member of B. Clearly, $A = B$ if and only if $A \subseteq B$ and $B \subseteq A$ both hold. If $A \subseteq B$ and $B \neq A$, then A is called a **proper subset** of B. The set without any elements is called the **empty** (or the **void**) set and is denoted by $\emptyset$. The empty set is a subset of every set.

If A and B are two sets, then we define

i. the **union** $A \cup B$ of A and B to be the set

$$A \cup B = \{x : x \in A \text{ or } x \in B\};$$

ii. the **intersection** $A \cap B$ of A and B to be the set

$$A \cap B = \{x : x \in A \text{ and } x \in B\};$$

iii. the set **difference** $A \setminus B$ of B from A to be the set

$$A \setminus B = \{x : x \in A \text{ and } x \notin B\}.$$

The set $A \setminus B$ is sometimes called the complement of B relative to A. Two sets A and B are called **disjoint** if $A \cap B = \emptyset$.

A number of useful relationships among sets are listed below, and the reader is expected to be able to prove them:

1. $(A \cup B) \cap C = (A \cap C) \cup (B \cap C)$;
2. $(A \cap B) \cup C = (A \cup C) \cap (B \cup C)$;

3. $(A \cup B) \setminus C = (A \setminus C) \cup (B \setminus C)$;
4. $(A \cap B) \setminus C = (A \setminus C) \cap (B \setminus C)$.

The identities (1) and (2) between unions and intersections are referred to as the **distributive laws**.

We remind the reader how one goes about proving the preceding identities by showing (1). Note that an equality between two sets has to be established, and this shall be done by verifying that the two sets contain the same elements. Thus, the argument for (1) goes as follows:

$$x \in (A \cup B) \cap C \iff x \in A \cup B \text{ and } x \in C \iff (x \in A \text{ or } x \in B) \text{ and } x \in C$$
$$\iff x \in A \cap C \text{ or } x \in B \cap C \iff x \in (A \cap C) \cup (B \cap C).$$

Another useful concept is the symmetric difference of two sets. If A and B are sets, then their **symmetric difference** is defined to be the set

$$A \triangle B = (A \setminus B) \cup (B \setminus A).$$

The concepts of union and intersection of two sets can be generalized to unions and intersections of arbitrary families of sets. A **family of sets** is a nonempty set $\mathcal{F}$ whose members are sets by themselves. There is a standard way for denoting a family of sets. If for each element i of a nonempty set I, a subset A_i of a fixed set X is assigned, then $\{A_i\}_{i \in I}$ (or $\{A_i : i \in I\}$ or simply $\{A_i\}$) denotes the family whose members are the sets A_i. The nonempty set I is called the **index set** of the family, and its members are known as indices. Conversely, if $\mathcal{F}$ is a family of sets, then by letting $I = \mathcal{F}$ and $A_i = i$ for each $i \in I$, we can express $\mathcal{F}$ in the form $\{A_i\}_{i \in I}$.

If $\{A_i\}_{i \in I}$ is a family of sets, then the **union** of the family is defined to be the set

$$\bigcup_{i \in I} A_i = \{x : \exists\, i \in I \text{ such that } x \in A_i\},$$

and the **intersection** of the family by

$$\bigcap_{i \in I} A_i = \{x : x \in A_i \text{ for each } i \in I\}.$$

Occasionally, $\bigcup_{i \in I} A_i$ will be denoted by $\bigcup A_i$ and $\bigcap_{i \in I} A_i$ by $\bigcap A_i$. Also, if $I = \mathbb{N} = \{1, 2, \ldots\}$ (the set of natural numbers), then the union and intersection of the family will be denoted by $\bigcup_{n=1}^{\infty} A_n$ and $\bigcap_{n=1}^{\infty} A_n$, respectively. The dummy index n can be replaced, of course, by any other letter.

The **distributive laws** for general families of sets now take the form

$$\left(\bigcup_{i\in I} A_i\right) \cap B = \bigcup_{i\in I}(A_i \cap B) \quad \text{and} \quad \left(\bigcap_{i\in I} A_i\right) \cup B = \bigcap_{i\in I}(A_i \cup B).$$

A family of sets $\{A_i\}_{i\in I}$ is called **pairwise disjoint** if for each pair i and j of distinct indices, the sets A_i and A_j are disjoint, i.e., $A_i \cap A_j = \emptyset$. The set of all subsets of a set A is called the **power set** of A, and is denoted by $\mathcal{P}(A)$. Note that $\emptyset$ and A are members of $\mathcal{P}(A)$. For most of our work in this book, subsets of a fixed set X will be considered (the set X can be thought of as a frame of reference), and all discussions will be considered with respect to the basic set X.

Now, let X be a fixed set. If $P(x)$ is a property (i.e., a well-defined "logical" sentence) involving the elements x of X, then the set of all x for which $P(x)$ is true will be denoted by $\{x \in X: P(x)\}$. For instance, if $X = \{1, 2, \ldots\}$ and $P(x)$ represents the statement "The number $x \in X$ is divisible by 2," then $\{x \in X: P(x)\} = \{2, 4, 6, \ldots\}$.

If A is a subset of X, then its **complement** A^c (relative to X) is the set $A^c = X \setminus A = \{x \in X: x \notin A\}$. It should be obvious that $(A^c)^c = A$, $A \cap A^c = \emptyset$, and $A \cup A^c = X$. Some other properties of the complement operation are stated next (where A and B are assumed to be subsets of X):

5. $A \setminus B = A \cap B^c$;
6. $A \subseteq B$ if and only if $B^c \subseteq A^c$;
7. $(A \cup B)^c = A^c \cap B^c$;
8. $(A \cap B)^c = A^c \cup B^c$.

The identities (7) and (8) are referred to as De Morgan's[1] laws. The generalized De Morgan's laws are going to be very useful, and for this reason we state them as a theorem.

Theorem 1.1 (De Morgan's Laws). *For a family $\{A_i\}_{i\in I}$ of subsets of a set X, the following identities hold*:

$$\left(\bigcup_{i\in I} A_i\right)^c = \bigcap_{i\in I} A_i^c \quad \text{and} \quad \left(\bigcap_{i\in I} A_i\right)^c = \bigcup_{i\in I} A_i^c.$$

---

[1] Augustin De Morgan (1806–1871), a British mathematician. He is well known for his contributions to mathematical logic.

Proof. We establish the validity of the first formula only, and we leave the verification of the other for the reader. Note that

$$x \in \left(\bigcup_{i \in I} A_i \right)^c \iff x \notin \bigcup_{i \in I} A_i \iff x \notin A_i \text{ for all } i \in I$$

$$\iff x \in A_i^c \text{ for all } i \in I \iff x \in \bigcap_{i \in I} A_i^c,$$

and this establishes the first identity. ∎

By a **function** f from a set A to a set B, in symbols $f: A \to B$ (or $A \xrightarrow{f} B$ or even $x \mapsto f(x)$), we mean a specific "rule" that assigns to each element x of A a unique element y in B. The element y is called the **value** of the function f at x (or the **image** of x under f) and is denoted by $f(x)$, that is, $y = f(x)$. The element $y = f(x)$ is also called the **output** of the function when the **input** is x. The set A is called the **domain** of f, and the set $\{y \in B: \exists x \in A \text{ with } y = f(x)\}$ is called the **range** of f. It is tacitly understood that the sets A and B are nonempty.

Two functions $f: A \to B$ and $g: A \to B$ are said to be **equal**, in symbols $f = g$, if $f(x) = g(x)$ holds true for each $x \in A$. A function $f: A \to B$ is called **onto** (or **surjective**) if the range of f is all of B; that is, if for every $y \in B$ there exists (at least one) $x \in A$ such that $y = f(x)$. The function $f: A \to B$ is called **one-to-one** (or **injective**) if $x_1 \neq x_2$ implies $f(x_1) \neq f(x_2)$.

Now, let $f: X \to Y$ be a function. If A is a subset of X, then the **image** $f(A)$ of A under f is the subset of Y defined by

$$f(A) = \{y \in Y: \exists x \in A \text{ such that } y = f(x)\}.$$

Similarly, if B is a subset of Y, then the **inverse image** $f^{-1}(B)$ of B under f is the subset of X defined by $f^{-1}(B) = \{x \in X: f(x) \in B\}$. Regarding images and inverse images of sets, the following relationships hold (we assume that $\{A_i\}_{i \in I}$ is a family of subsets of X and $\{B_i\}_{i \in I}$ a family of subsets of Y):

9. $f(\bigcup_{i \in I} A_i) = \bigcup_{i \in I} f(A_i)$;
10. $f(\bigcap_{i \in I} A_i) \subseteq \bigcap_{i \in I} f(A_i)$;
11. $f^{-1}(\bigcup_{i \in I} B_i) = \bigcup_{i \in I} f^{-1}(B_i)$;
12. $f^{-1}(\bigcap_{i \in I} B_i) = \bigcap_{i \in I} f^{-1}(B_i)$;
13. $f^{-1}(B^c) = (f^{-1}(B))^c$.

Given two functions $f: X \to Y$ and $g: Y \to Z$, their **composition** $g \circ f$ is the function $g \circ f: X \to Z$ defined by $(g \circ f)(x) = g(f(x))$ for each $x \in X$.

If a function $f: X \to Y$ is one-to-one and onto, then for every $y \in Y$ there exists a unique $x \in X$ such that $y = f(x)$; the unique element x is denoted by $f^{-1}(y)$. Thus, in this case, a function $f^{-1}: Y \to X$ can be defined by $f^{-1}(y) = x$, whenever $f(x) = y$. The function f^{-1} is called the **inverse** of f. Note that $(f \circ f^{-1})(y) = y$ for all $y \in Y$ and $(f^{-1} \circ f)(x) = x$ for all $x \in X$. The latter relations are often written as $f \circ f^{-1} = I_Y$ and $f^{-1} \circ f = I_X$, where $I_X: X \to X$ and $I_Y: Y \to Y$ denote the identity functions; that is, $I_X(x) = x$ and $I_Y(y) = y$ for all $x \in X$ and $y \in Y$.

Any function $x: \mathbb{N} \to X$, where $\mathbb{N} = \{1, 2, \ldots\}$ is the set of natural numbers, is called a **sequence** of X. The standard way to denote the value $x(n)$ is by x_n (called the nth **term of the sequence**). We shall denote the sequence x by $\{x_n\}$, and we shall consider it both as a function and as a subset of X. A **subsequence** of a sequence $\{x_n\}$ is a sequence $\{y_n\}$ for which there exists a strictly increasing sequence $\{k_n\}$ of natural numbers (that is, $1 \leq k_1 < k_2 < k_3 < \cdots$) such that $y_n = x_{k_n}$ holds for each n.

If now $\{A_i\}_{i \in I}$ is a family of sets, then the **Cartesian**[2] **product** $\Pi_{i \in I} A_i$ (or ΠA_i) is defined to be the set consisting of all functions $f: I \to \bigcup_{i \in I} A_i$ such that $x_i = f(i) \in A_i$ for each $i \in I$. Such a function is called (for obvious reasons) a **choice function** and quite often is denoted by $(x_i)_{i \in I}$ or simply by (x_i).

If a family of sets consists of two sets, say A and B, then the Cartesian product of the sets A and B is designated by $A \times B$. The members of $A \times B$ are denoted as **ordered pairs**, that is,

$$A \times B = \{(a, b): a \in A \text{ and } b \in B\}.$$

Clearly, $(a, b) = (a_1, b_1)$ if and only if $a = a_1$ and $b = b_1$. Similarly, the Cartesian product of a finite family of sets $\{A_1, \ldots, A_n\}$ is written as $A_1 \times \cdots \times A_n$ and its members are denoted as n-tuples, that is,

$$A_1 \times \cdots \times A_n = \{(a_1, \ldots, a_n): a_i \in A_i \text{ for each } i = 1, \ldots, n\}.$$

Here, again $(a_1, \ldots, a_n) = (b_1, \ldots, b_n)$ if and only if $a_i = b_i$ for $i = 1, \ldots, n$. If $A_1 = A_2 = \cdots = A_n = A$, then it is standard to write $A_1 \times \cdots \times A_n$ as A^n. Similarly, if the family of sets $\{A_i\}_{i \in I}$ satisfies $A_i = A$ for each $i \in I$, then $\Pi_{i \in I} A_i$ is written as A^I, that is, $A^I = \{f \mid f: I \to A\}$.

- *When is the Cartesian product of a family of sets $\{A_i\}_{i \in I}$ nonempty?*

Clearly, if the Cartesian product is nonempty, then each A_i must be nonempty. The following question may, therefore, be asked:

- *If each A_i is nonempty, is then the Cartesian product ΠA_i nonempty?*

Although the answer seems to be affirmative, it is unfortunate that such a statement cannot be proven with the usual axioms of set theory. The affirmative answer

---

[2]René Descartes or Cartesius (1596–1650), an influential French philosopher and mathematician. He is the founder of analytic geometry.

to the last question is known as "the axiom of choice." This axiom will be assumed throughout this book without further explanation. One of its forms is stated below.

Axiom of Choice. *If $\{A_i\}_{i \in I}$ is a nonempty family of sets such that A_i is nonempty for each $i \in I$, then ΠA_i is nonempty.*

A useful equivalent formulation of the axiom of choice is the following:

- *If $\{A_i\}_{i \in I}$ is a nonempty family of pairwise disjoint sets such that $A_i \neq \emptyset$ for each $i \in I$, then there exists a set $E \subseteq \bigcup_{i \in I} A_i$ such that $E \cap A_i$ consists of precisely one element for each $i \in I$.*

For a discussion of the axiom of choice and its history, the reader is referred to [8] and [13].

By a (binary) **relation** on a set X we simply mean a subset $\mathcal{R}$ of $X \times X$. If $(x, y) \in \mathcal{R}$, then x is said to be in the relation $\mathcal{R}$ with y, and this is denoted by $x \mathcal{R} y$. Among the most interesting relations are the equivalence relations. A relation $\mathcal{R}$ on a set X is called an **equivalence relation** if it satisfies the following three properties:

a.　$x \mathcal{R} x$ for each $x \in X$ (*reflexivity*).
b.　If $x \mathcal{R} y$, then $y \mathcal{R} x$ (*symmetry*).
c.　If $x \mathcal{R} y$ and $y \mathcal{R} z$, then $x \mathcal{R} z$ (*transitivity*).

Let $\mathcal{R}$ be an equivalence relation on a set X. Then the **equivalence class** determined by the element $x \in X$ is defined by $[x] = \{y \in X : x \mathcal{R} y\}$. It is easy to observe that any two equivalence classes are either disjoint or else they coincide. Since $x \in [x]$ for each $x \in X$, it follows that $\mathcal{R}$ **partitions** X. That is, there exists a family $\{A_i\}_{i \in I}$ of pairwise disjoint sets (here, the family of equivalence classes) such that $X = \bigcup_{i \in I} A_i$. Conversely, if a family of pairwise disjoint sets $\{A_i\}_{i \in I}$ partitions X (i.e., $X = \bigcup_{i \in I} A_i$), then by letting

$$\mathcal{R} = \{(x, y) \in X \times X : \exists \, i \in I \text{ such that } x \text{ and } y \text{ are in } A_i\},$$

an equivalence relation is defined on X whose equivalence classes are precisely the sets A_i. Thus, the equivalence relations on a set correspond precisely in one-to-one fashion with its partitions.

Another important type of relation is an order relation. A relation, denoted by $\leq$, on a set X is said to be a **partial order** for X (or that X is partially ordered by $\leq$) if it satisfies the following three properties:

α.　$x \leq x$ holds for every $x \in X$ (*reflexivity*).
β.　If $x \leq y$ and $y \leq x$, then $x = y$ (*antisymmetry*).
γ.　If $x \leq y$ and $y \leq z$, then $x \leq z$ (*transitivity*).

An alternative notation for $x \leq y$ is $y \geq x$. A set equipped with an order relation is called a **partially ordered set**.

Now, let X be a partially ordered set. A subset Y of X is said to be a **chain** if for every pair $x, y \in Y$, either $x \leq y$ or else $y \leq x$; a chain is also referred to as a **totally ordered set**. If Y is a subset of X such that $x \leq u$ holds for all $x \in Y$ and some $u \in X$, then u is called an **upper bound** of Y. An element $m \in X$ is called a **maximal** element of X whenever the relation $m \leq x$ implies $x = m$. (Warning: A partially ordered set may contain more than one maximal element.)

The next statement guarantees the existence of maximal elements in certain partially ordered sets. It is known as Zorn's[3] lemma, and it is a powerful tool in analysis.

Zorn's Lemma. *If every chain in a partially ordered set X has an upper bound in X, then X has a maximal element.*

Zorn's lemma (as a statement) is equivalent to the axiom of choice. For a detailed discussion, see [13].

EXERCISES

1. Prove statements 2 through 13 of this section.
2. For two sets A and B show that the following statements are equivalent.

   a. $A \subseteq B$.
   b. $A \cup B = B$.
   c. $A \cap B = A$.

3. Show that $(A \triangle B) \triangle C = A \triangle (B \triangle C)$ hold for every triplet of sets A, B, and C.
4. Give an example of a function $f: X \to Y$ and two subsets A and B of X such that $f(A \cap B) \neq f(A) \cap f(B)$.
5. For a function $f: X \to Y$ show that the following three statements are equivalent.

   a. f is one-to-one.
   b. $f(A \cap B) = f(A) \cap f(B)$ holds for all $A, B \in \mathcal{P}(X)$.
   c. For every pair of subsets A, B of X satisfying $A \cap B = \emptyset$, we have $f(A) \cap f(B) = \emptyset$.

6. Let $f: X \to Y$ be a function. Show that $f(f^{-1}(A)) \subseteq A$ for all $A \subseteq Y$, and $B \subseteq f^{-1}(f(B))$ for all $B \subseteq X$.
7. Show that a function $f: X \to Y$ is onto if and only if $f(f^{-1}(B)) = B$ holds for all $B \subseteq Y$.
8. Assume $X \xrightarrow{f} Y \xrightarrow{g} Z$. If $A \subseteq Z$, show that $(g \circ f)^{-1}(A) = f^{-1}(g^{-1}(A))$.
9. Show that the composition of functions satisfies the associative law. That is, show that if $X \xrightarrow{f} Y \xrightarrow{g} Z \xrightarrow{h} V$, then $(h \circ g) \circ f = h \circ (g \circ f)$.

---

[3] Max Zorn (1906–1993), a German-born American mathematician. He worked in set theory, topology, and algebra.

10. Let $f: X \to Y$. Show that the relation $\mathcal{R}$ on X, defined by $x_1 \mathcal{R} x_2$ if $f(x_1) = f(x_2)$, is an equivalence relation.
11. If X and Y are sets, then show that $\mathcal{P}(X) \cap \mathcal{P}(Y) = \mathcal{P}(X \cap Y)$ and $\mathcal{P}(X) \cup \mathcal{P}(Y) \subseteq \mathcal{P}(X \cup Y)$.

2. COUNTABLE AND UNCOUNTABLE SETS

In this section we shall deal with questions concerning the "size" of a set. Two sets are said to have the "same number of elements" if their elements can be put in one-to-one correspondence with each other. By a one-to-one correspondence between two sets A and B, we mean a function $f: A \to B$ that is one-to-one and onto.

Definition 2.1. *Two sets A and B are said to be **equivalent** (in symbols $A \approx B$) if there exists a function $f: A \to B$ that is one-to-one and onto.*

It is easy to verify that for sets the following properties hold:

1. $A \approx A$.
2. If $A \approx B$, the $B \approx A$.
3. If $A \approx B$ and $B \approx C$, then $A \approx C$.

The dividing line for the sizes of sets is the set of natural numbers $\mathbb{N} = \{1, 2, 3, \ldots\}$. Any subset of $\mathbb{N}$ of the form $\{1, \ldots, n\}$ is called a **segment** of $\mathbb{N}$, and n is called the number of elements of the segment. Clearly, two segments $\{1, \ldots, n\}$ and $\{1, \ldots, m\}$ are equivalent if and only if $n = m$. This shows that a proper subset of a segment cannot be equivalent to the segment.

A set that is equivalent to a segment is called a **finite set**. The empty set is also considered to be finite with zero elements. A set that is not finite is called an **infinite set**.

Definition 2.2. *A set A is called **countable** if it is equivalent to $\mathbb{N}$, that is, if there exists a one-to-one correspondence of $\mathbb{N}$ with the elements of A.*

There is a standard notation for a countable set A. It is usually written as $A = \{a_1, a_2, \ldots\}$, often called an **enumeration** of the set A, and it indicates the one-to-one correspondence of A with the set of natural numbers $\mathbb{N}$.

An infinite set that is not countable is called an **uncountable set**. Our first result compares the infinite sets with the countable ones.

Theorem 2.3. *Every infinite set contains a countable subset.*

Proof. Let A be an infinite set; clearly $A \neq \emptyset$. Pick $a_1 \in A$, and consider the set $A_1 = A \setminus \{a_1\}$. Since A is infinite, A_1 is nonempty. Pick $a_2 \in A_1$, and consider the set $A_2 \setminus \{a_1, a_2\}$. By the same arguments, there exists an element $a_3 \in A_2$. Proceeding in this way, a set $\{a_1, a_2, a_3, \ldots\}$ is obtained that is clearly countable and by construction it is a subset of A. ∎

The following two principles of the natural numbers are needed in order to establish more properties of the countable sets. The first property is known as the "well-ordering principle" of $\mathbb{N}$. Keep in mind that a subset S of $\mathbb{N}$ is said have a **least** (or a **first**) element if there exists $k \in S$ such that $k \leq n$ for each $n \in S$; clearly k is uniquely determined.

THE WELL-ORDERING PRINCIPLE.   *Every nonempty subset of $\mathbb{N}$ has a least element.*

The second property of $\mathbb{N}$ that is needed is known as the "principle of mathematical induction," and is a powerful tool in mathematics.

THE PRINCIPLE OF MATHEMATICAL INDUCTION.   *If a subset S of $\mathbb{N}$ satisfies the properties*

a. $1 \in S$, *and*
b. $n + 1 \in S$ *whenever* $n \in S$,

then $S = \mathbb{N}$.

We now continue our discussion about countable sets.

Theorem 2.4.   *Every subset of a countable set is either finite or else countable.*

Proof.   Let A be a subset of a countable set. Assume that A is not finite; then it will be shown that A is countable. An easy argument shows that one can assume without loss of generality that A is a subset of $\mathbb{N}$.

Now, define a function $f : \mathbb{N} \to A$ inductively as follows: $f(1) = $ the least element of A (that exists by the well-ordering principle); and then if $f(1), \ldots, f(n)$ have been defined, let $f(n + 1)$ be the least element of $A \setminus \{f(1), \ldots, f(n)\}$. (The least element again exists by the well-ordering principle, and the fact that A is not finite.) It is now left to the reader to verify that f is one-to-one and onto. Thus, $A \approx \mathbb{N}$. ∎

It is interesting to observe that (in contrast with the finite sets) an infinite set can be equivalent to some of its proper subsets. To see this, let $X = \{2, 4, 6, \ldots\}$, and note that X is a proper subset of $\mathbb{N}$. Now the function $f : \mathbb{N} \to X$, defined by $f(n) = 2n$ for each n, is one-to-one and onto, so that $X \approx \mathbb{N}$.

Some useful characterizations to ensure that an infinite set is countable are presented next.

Theorem 2.5. *For an infinite set A, the following statements are equivalent:*

  i.   *A is countable.*
  ii.   *There exists a subset B of $\mathbb{N}$ and a function $f: B \to A$ that is onto.*
  iii.   *There exists a function $g: A \to \mathbb{N}$ that is one-to-one.*

Proof. (i) $\Longrightarrow$ (ii) Since A is countable, there exists $f: \mathbb{N} \to A$ that is one-to-one and onto. Thus, (ii) holds with $B = \mathbb{N}$.

(ii) $\Longrightarrow$ (iii) Assume that B is a subset of $\mathbb{N}$ and that $f: B \to A$ is an onto function. Note that since f is onto, $f^{-1}(a) = \{n \in B: f(n) = a\}$ is nonempty for each $a \in A$. Now, define $g: A \to \mathbb{N}$ as follows: $g(a) =$ the least element of $f^{-1}(a)$; this natural number exists by the well-ordering principle. To complete the proof, it must be shown that g is one-to-one. Indeed, if $g(a) = g(b)$ holds, then $a = f(g(a)) = f(g(b)) = b$ also holds, which shows that g is one-to-one.

(iii) $\Longrightarrow$ (i) Assume $g: A \to \mathbb{N}$ to be one-to-one. Then $A \approx g(A)$. Since A is (by hypothesis) infinite, $g(A)$ is an infinite subset of $\mathbb{N}$, and hence, by Theorem 2.4, $g(A) \approx \mathbb{N}$ holds. Therefore, $A \approx \mathbb{N}$ holds, and the proof is complete.   ∎

The next two results are consequences of the preceding theorem. The first one asserts that the countable union of countable sets is countable.

Theorem 2.6. *Let $\{A_1, A_2, \ldots\}$ be a countable family of sets such that each A_i is a countable set. Then $A = \bigcup_{n=1}^{\infty} A_n$ is a countable set.*

Proof. Let $A_n = \{a_1^n, a_2^n, \ldots\}$ for $n = 1, 2, \ldots$, and $A = \bigcup_{n=1}^{\infty} A_n$. Also, let $B = \{2^k \cdot 3^n: k, n \in \mathbb{N}\}$. Now define $f: B \to A$ by $f(2^k \cdot 3^n) = a_k^n$. Then f maps B onto A, and hence A is a countable set by Theorem 2.5.   ∎

The Cartesian product of a finite collection of countable sets is always countable.

Theorem 2.7. *Let $\{A_1, \ldots, A_n\}$ be a finite collection of sets such that each A_i is countable. Then $A_1 \times \cdots \times A_n$ is countable.*

Proof. It suffices to assume that $A_i = \mathbb{N}$ for each i, and hence, $A = \mathbb{N}^n$. Pick n distinct prime numbers, say $p_1, \ldots, p_n$, and define $f: \mathbb{N}^n \to \mathbb{N}$ by $f(k_1, \ldots, k_n) = p_1^{k_1} p_2^{k_2} \cdots p_n^{k_n}$. It should be clear from the fundamental theorem of arithmetic (every natural number has a unique factorization into primes) that f is one-to-one. Hence, by Theorem 2.5, $\mathbb{N}^n$ is countable.   ∎

Our discussion will close with some results concerning the "big" sets. One may ask the question: *How large can a set be?*

To answer this question a definition is needed. Let us write $A \preceq B$ whenever there exists a one-to-one function $f: A \to B$; in other words, $A \preceq B$ if A is equivalent to a subset of B. It should be clear that in this sense one can say that "B has at least as many elements as A."

The relation $\preceq$ satisfies the following properties:

1. $A \preceq A$ for all sets A.
2. If $A \preceq B$ and $B \preceq C$, then $A \preceq C$.
3. If $A \preceq B$ and $B \preceq A$, then $A \approx B$.

Statement (3) is known as the Schröder–Bernstein[4,5] theorem, and is a very important result; for a proof see [13, p. 88], or [28, p. 29].

The next result shows that the power set of a given set has more elements than the set; it is due to G. Cantor.[6]

Theorem 2.8 (Cantor).   *If A is a set, then $A \preceq \mathcal{P}(A)$ and $A \not\approx \mathcal{P}(A)$ hold.*

Proof.   If $A = \emptyset$, then the result is trivial. So, assume that $A \neq \emptyset$. Define $f: A \to \mathcal{P}(A)$ by $f(x) = \{x\}$ for $x \in A$, and note that f is one-to-one. Hence, $A \preceq \mathcal{P}(A)$.

To show that $A \not\approx \mathcal{P}(A)$, assume by way of contradiction that $A \approx \mathcal{P}(A)$. So, there exists $g: A \to \mathcal{P}(A)$ that is one-to-one and onto. Consider the subset B of A defined by $B = \{x \in A: x \notin g(x)\}$, and then pick $a \in A$ such that $g(a) = B$. Now, notice that $a \in B$ if and only if $a \notin B$, which is impossible. This contradiction completes the proof of the theorem.   ∎

Imagine that for every set A a symbol (playing the role of a number) can be assigned that designates the number of elements in the set. Without going into detail, this symbol is called the **cardinal number** of A, and is denoted card A.

It should be clear that if A is a finite set, say $A = \{a_1, \ldots, a_n\}$, then card $A = n$. But what about $\mathbb{N}$? We use the symbol $\aleph_0$ to denote the cardinal number of $\mathbb{N}$. By saying that a set A has cardinal number $\aleph_0$ (in symbols card $A = \aleph_0$), we simply mean that $A \approx \mathbb{N}$. In general, card $A =$ card B means $A \approx B$.

If $\mathfrak{a}$ and $\mathfrak{b}$ are cardinal numbers, then $\mathfrak{a} \leq \mathfrak{b}$ means that there exist two sets A and B such that card $A = \mathfrak{a}$, card $B = \mathfrak{b}$, and $A \preceq B$. Similarly, $\mathfrak{a} < \mathfrak{b}$ among cardinal numbers means that there exist two sets A and B such that card $A = \mathfrak{a}$, card $B = \mathfrak{b}$, $A \preceq B$, and $A \not\approx B$. Note that according to the Schröder–Bernstein theorem, $\mathfrak{a} \leq \mathfrak{b}$ and $\mathfrak{b} \leq \mathfrak{a}$ guarantee $\mathfrak{a} = \mathfrak{b}$.

We mention also a few things about "cardinal arithmetic." By Theorem 2.8, we know that there exists a set with cardinal number greater than $\aleph_0$; this set is $\mathcal{P}(\mathbb{N})$. It can be shown that $\mathcal{P}(\mathbb{N}) \approx \mathbb{R}$ (see Exercise 6 of Section 5), where $\mathbb{R}$ is the

---

[4]Ernst Schröder (1841–1902), a German mathematician. He worked in algebra and mathematical logic.

[5]Felix Bernstein (1878–1956), a German mathematician. He worked in set theory and contributed decisively to the development of the inheritance population genetics.

[6]Georg Cantor (1845–1918), a German mathematician. He was the founder of set theory. He also contributed to the field of classical analysis.

set of all real numbers. The cardinal number of $\mathbb{R}$ is denoted by c, and is called the **cardinality of the continuum**; thus, $\aleph_0 < c$. A cardinal number a satisfying $\aleph_0 \leq a$, is called an **infinite cardinal**.

It is easy to see that if $2 = \{0, 1\}$, then $2^X \approx \mathcal{P}(X)$ for every set X. For this reason it is a custom to denote the cardinal number of $\mathcal{P}(X)$ by $2^{\text{card } X}$. Thus, the following inequalities hold:

$$0 < 1 < 2 < \cdots < n < \cdots < \aleph_0 < 2^{\aleph_0} = c < 2^c < 2^{2^c} < \cdots .$$

The outstanding question dealing with infinite cardinals is the following:

- *If a is an infinite cardinal, is there a cardinal number b such that $a < b < 2^a$?*

The **continuum hypothesis** assumes a negative answer to the above question when $a = \aleph_0$, and the **generalized continuum hypothesis** assumes a negative answer for every infinite cardinal. It has been shown that the continuum hypothesis is "independent" of the ordinary axioms of set theory. There are models of set theory that satisfy the continuum hypothesis and there are models that do not. The interested reader is referred to [8], [13], and [20] for extensive treatments of the cardinal numbers.

EXERCISES

1. Show that the set of all rational numbers is countable.
2. Show that the set of all finite subsets of a countable set is countable.
3. Show that the union of an at-most countable collection of sets, each of which is finite, is an at-most countable set.
4. Let A be an uncountable set and B be a countable subset of A. Show that A is equivalent to $A \setminus B$.
5. Assume that $f: A \rightarrow B$ is a surjective (onto) function between two sets. Establish the following:
   a. card $B \leq$ card A.
   b. If A is countable, then B is at most countable.
6. Show that two nonempty sets A and B are equivalent if and only if there exists a function from A onto B and a function from B onto A.
7. Show that if a finite set X has n elements, then its power set $\mathcal{P}(X)$ has 2^n elements.
8. Show that the set of all sequences with values 0 or 1 is uncountable.
9. If $2 = \{0, 1\}$, then show that $2^X \approx \mathcal{P}(X)$ for every set X.
10. Any complex number that is a root of a (nonzero) polynomial with integer coefficients is called an **algebraic number**. Show that the set of all algebraic numbers is countable.

11. For an arbitrary function $f: \mathbb{R} \to \mathbb{R}$, show that the set

$$A = \left\{ a \in \mathbb{R}: \lim_{x \to a} f(x) \text{ exists and } \lim_{x \to a} f(x) \neq f(a) \right\}$$

is at-most countable.

12. Show that the set of real numbers is uncountable by proving the following:

a. $(0, 1) \approx \mathcal{P}(\mathbb{R})$; and

b. $(0, 1)$ is uncountable.

[HINT: If $(0, 1)$ is countable, then let $\{x_1, x_2, \ldots\}$ be an enumeration of $(0, 1)$. For each n write $x_n = 0.d_{n1}d_{n2} \cdots$ in its decimal expansion, where each d_{ij} is $0, 1, \ldots, 9$. Now consider the real number y of $(0, 1)$ whose decimal expansion $y = 0.y_1 y_2 \cdots$ satisfies $y_n = 1$ if $d_{nn} \neq 1$ and $y_n = 2$ if $d_{nn} = 1$. To obtain a contradiction, show that $y \neq x_n$ for each n.]

13. Using mathematical induction prove the following:

a. If $a \geq -1$, then $(1 + a)^n \geq 1 + na$ for $n = 1, 2, \ldots$ (*Bernoulli's*[7] *inequality*).

b. If $0 < a < 1$, then $1 + 3^n a > (1 + a)^n$ for $n = 1, 2, \ldots$.

c. $\cos(n\pi) = (-1)^n$ for $n = 1, 2, \ldots$.

14. Show that the well-ordering principle implies the principle of mathematical induction.

15. Show that the principle of mathematical induction implies the well-ordering principle.

3. THE REAL NUMBERS

Without any doubt, the most important set for this book will be the set of real numbers $\mathbb{R} = (-\infty, \infty)$. The set of real numbers is also known as the real line. The reason is that by considering a straight line, one can put (in the usual way) the real numbers in one-to-one correspondence with the points of the line. The terms "real line" and "real numbers" will be viewed as identical.

While it is not our purpose to give a complete axiomatic development of the real numbers, it is important to stop and consider exactly what axioms characterize the real numbers. They consist of the field axioms, the order axioms, and the completeness axiom. In algebraic terminology, the set of real numbers is referred to as the one and only "complete ordered field." The name comes from the axiomatic foundation of the real numbers outlined below.

The real numbers are the members of a nonempty set $\mathbb{R}$ equipped with two operations, $+$ and $\cdot$ from $\mathbb{R} \times \mathbb{R}$ into $\mathbb{R}$, called **addition** and **multiplication**, that satisfy the following axioms:

Field Axioms

The letters x, y, and z denote arbitrary real numbers, unless otherwise stated.

---

[7]Jacob (Jacques) Bernoulli (1654–1705), a Swiss mathematician and one of the most prominent members of the famous Bernoulli mathematical family. His major works were in calculus and probability theory.

Axiom 1. $x + y = y + x$ *and* $xy = yx$ *(the commutative laws).*

Axiom 2. $x + (y + z) = (x + y) + z$ *and* $x(yz) = (xy)z$ *(the associative laws).*

Axiom 3. $x(y + z) = xy + xz$ *(the distributive law).*

Axiom 4. *There exists an element* $0 \in \mathbb{R}$ *such that* $x + 0 = x$ *for all* $x \in \mathbb{R}$.

Axiom 5. *For each* $x \in \mathbb{R}$ *there exists an element in* $\mathbb{R}$ *(denoted by* $-x$*) such that* $x + (-x) = 0$.

Axiom 6. *There exists an element* $1 \in \mathbb{R}$ *with* $1 \neq 0$ *satisfying* $1 \cdot x = x$ *for all* $x \in \mathbb{R}$.

Axiom 7. *For each* $x \neq 0$ *there exists an element in* $\mathbb{R}$ *(denoted by* x^{-1}*) satisfying* $xx^{-1} = 1$.

It can be shown that the zero element of Axiom 4 is uniquely determined. Also, it can be established that the element $-x$ given by Axiom 5 is uniquely determined, and that $-x = (-1)x$ holds. In a similar manner, it can be seen that the element x^{-1} of Axiom 7 which satisfies $xx^{-1} = 1$ (where, of course, $x \neq 0$) is uniquely determined.

From the field axioms, one can derive the familiar properties of addition and multiplication. For instance, $0 \cdot x = 0$, $-(-x) = x$, $(-x)(-y) = xy$, $x - y = x + (-y) = -(y - x)$, $(x^{-1})^{-1} = x$. (The reader will find the details in the exercises at the end of this section.)

The next requirement is that $\mathbb{R}$ must be not merely a field but also an "ordered field." This means that $\mathbb{R}$ is equipped with an order relation $\geq$ compatible with the algebraic operations via the following axioms:

Order Axioms

Axiom 8. *For any* $x, y \in \mathbb{R}$, *either* $x \geq y$ *or* $y \geq x$ *holds.*

Axiom 9. *If* $x \geq y$, *then* $x + z \geq y + z$ *holds for each* $z \in \mathbb{R}$.

Axiom 10. *If* $x \geq y$ *and* $z \geq 0$, *then* $xz \geq yz$.

An alternative notation for $x \geq y$ is $y \leq x$. Any number $x \in \mathbb{R}$ satisfying $x > 0$ (i.e., $x \geq 0$ and $x \neq 0$) is called a **positive number** (and likewise, any number x with $x < 0$ is called a **negative number**). From the order axioms, one can derive the ordinary inequality properties of the real numbers. Let us mention one very useful property dealing with inequalities:

- *If* $x + \epsilon \geq y$ *holds for each* $\epsilon > 0$, *then* $x \geq y$ *holds.*

Indeed, if the conclusion is not true, then $y - x > 0$. Let $\epsilon = \frac{1}{2}(y - x) > 0$, and note that our hypothesis implies $\frac{1}{2}(x + y) = x + \frac{1}{2}(y - x) \geq y$. This in turn implies $y - x \leq 0$, which is a contradiction.

The usual way to define the **absolute value** of a real number a is as follows: $|a| = a$ if $a \geq 0$ and $|a| = -a$ if $a < 0$. If $a \vee b$ denotes the larger of the numbers a and b (for instance, $(-1) \vee 2 = 2$ and $1 \vee 1 = 1$), then a moment's thought reveals that $|a| = a \vee (-a)$ for all $a \in \mathbb{R}$. In particular, it follows that $|a| = |-a|$ for each $a \in \mathbb{R}$. The absolute value satisfies the properties:

1. $|a| \geq 0$ for each $a \in \mathbb{R}$, and $|a| = 0$ if and only if $a = 0$;
2. $|ab| = |a| \cdot |b|$ for all $a, b \in \mathbb{R}$; and
3. $|a + b| \leq |a| + |b|$ for all $a, b \in \mathbb{R}$ (*the triangle inequality*).

The non-negative number $|a - b|$ can be viewed geometrically as the distance between the numbers a and b.

The least understood property of the real numbers is the **completeness axiom**, or the **axiom of continuity**. Before stating this axiom, let us recall a few things.

Let A be a nonempty subset of $\mathbb{R}$. An **upper bound** of A is any real number a such that $x \leq a$ for all $x \in A$; similarly, $b \in \mathbb{R}$ is a **lower bound** for A if $b \leq x$ for all $x \in A$. If A has an upper (resp. lower) bound, then A is said to be **bounded from above** (resp. **below**). If A is both, bounded from above and below, then A is called a **bounded set**. A real number is called a **least upper bound** (or a **supremum**) of A if it is an upper bound for A, and it is less than or equal to every other upper bound of A. That is, $x \in \mathbb{R}$ is a least upper bound for A if

i. A is bounded from above by x, and
ii. if A is bounded from above by y, then $x \leq y$.

It should be clear that a given set A can have at most one least upper bound, denoted by sup A. A similar definition is given for the **greatest lower bound** (or **infimum**) of a set A, denoted by inf A. The completeness axiom asserts that every nonempty set bounded from above has a least upper bound and is stated next.

The Completeness Axiom

Axiom 11. *Every nonempty set of real numbers that is bounded from above has a least upper bound.*

From this axiom, it follows easily that every nonempty set of real numbers bounded from below has a greatest lower bound. (If A is nonempty and bounded from below, then the set $B = \{b \in \mathbb{R}: b \leq x \ \forall x \in A\}$ is bounded from above, and so sup B exists. Note that sup $B = $ inf A.) It should be clear also that if a set A has a maximum (resp. a minimum) element, then max $A = $ sup A (resp. min $A = $

inf A). On the other hand, if the supremum of a set exists and sup $A \in A$, then sup A is the maximum element of A. In other words, the supremum of a set generalizes the concept of the maximum element of a set. For more about this axiomatic foundation of the real numbers, see [3] and the exercises at the end of this section.

Regarding the supremum of a set, the following approximation property holds:

Theorem 3.1.   *Assume that the supremum of a subset A of $\mathbb{R}$ exists. Then for every $\epsilon > 0$, there exists some $x \in A$ such that*

$$\sup A - \epsilon < x \leq \sup A.$$

Proof.   If for every $x \in A$ we have $x \leq \sup A - \epsilon$, then $\sup A - \epsilon$ is an upper bound of A, which is less than the least upper bound. But this is impossible. Thus, there exists some $x \in A$ such that $\sup A - \epsilon < x \leq \sup A$.   ∎

Corollary 3.2.   *The set of natural numbers $\mathbb{N}$ is unbounded.*

Proof.   Assume by way of contradiction that $n \leq a$ holds for each $n \in \mathbb{N}$ and some $a \in \mathbb{R}$. Then by the completeness axiom, $s = \sup \mathbb{N}$ exists, and by Theorem 3.1 there exists some $k \in \mathbb{N}$ with $s - 1 < k$. This implies $s < k + 1 \leq s$, which is impossible.   ∎

The next useful property of the real numbers is known as the "Archimedean property." It was used extensively by Archimedes in geometrical proofs but, most likely, it was introduced first by Eudoxus.[8]

Theorem 3.3 (The Archimedean Property).   *If x and y are two positive real numbers, then there exists some natural number n such that $nx > y$.*

Proof.   If $nx \leq y$ holds for each n, then $n \leq y/x$ for each n. That is, $\mathbb{N}$ is bounded from above, contrary to Corollary 3.2.   ∎

An important density property of the rational numbers is described in the next theorem. Recall that a rational number is any real number that can be written as a quotient of two integers.

Theorem 3.4.   *Between any two distinct real numbers there exists a rational number.*

---

[8]Eudoxus of Cnidus (ca 400–347 BC), a Greek scholar of great eminence and influence. He contributed to astronomy, mathematics, geography, and philosophy and helped in the writing of the laws in his home town of Cnidus (Asia Minor). He introduced the "method of exhaustion" for computing areas and volumes and was the first to give a rigorous definition of a real number.

Proof.   It is easy to see that we need consider only positive numbers. So, let $a, b \in \mathbb{R}$ be such that $0 < a < b$.

Consider first the set $A = \{n \in \mathbb{N}: n > \max\{\frac{1}{b-a}, \frac{1}{b}\}\}$. Since $\mathbb{N}$ is not bounded from above, A is nonempty. Fix an element $q \in A$. Clearly, $0 < \frac{1}{q} < b - a$ and $1 < bq$. Now, let $B = \{n \in \mathbb{N}: n < bq\}$. Since $1 \in B$, we see that $B \neq \emptyset$, and clearly B is a finite set. Let $p = \max B$; note that $p \in B$ and $p + 1 \notin B$.

To finish the proof, we shall show that $a < \frac{p}{q} < b$ holds. To this end, note first that by construction $\frac{p}{q} < b$ holds. On the other hand, since $b \leq \frac{p+1}{q}$, we must have

$$a = b - (b - a) < \frac{p+1}{q} - \frac{1}{q} = \frac{p}{q},$$

and we are done.   ∎

With the help of the completeness axiom, we can also establish the existence of "roots" of real numbers.

Theorem 3.5.   *For a real number a and any natural number $n \geq 2$, we have the following:*

1.   *If $a \geq 0$ and n is even, there exists a unique $b \geq 0$ such that $b^n = a$.*
2.   *If $a \in \mathbb{R}$ and n is odd, there exists a unique $b \in \mathbb{R}$ such that $b^n = a$.*

Proof.   We shall establish both cases by assuming $a \geq 0$ and leave the easy details for completing the proof to the reader. If $a = 0$, then clearly $b = 0$, and so we can also suppose that $a > 0$. We shall establish first the uniqueness of b. To see this, assume that $x > 0$ and $y > 0$ satisfy $x^n = y^n = a$. Then

$$0 = x^n - y^n = (x - y)(x^{n-1} + x^{n-2}y + \cdots + xy^{n-2} + y^{n-1}),$$

and since $x^{n-1} + x^{n-2}y + \cdots + xy^{n-2} + y^{n-1} > 0$, we infer that $x - y = 0$, or $x = y$.

For the existence, consider the set $S = \{s \geq 0: s^n \leq a\}$. By the Archimedean property, there exists some $m \in \mathbb{N}$ such that $ma > 1$, or $\frac{1}{m} < a$. So, $0 < (\frac{1}{m})^n < \frac{1}{m} < a$, and thus, S is nonempty and contains positive numbers. On the other hand, if $k \in \mathbb{N}$ satisfies $a \leq k$, then $s \leq k$ for each $s \in S$; otherwise $s > k$ implies $s^n > k \geq a$, a contradiction. Now, by the completeness property, $b = \sup S > 0$ exists in $\mathbb{R}$. We shall complete the proof by proving that $b^n = a$. This will be done by eliminating the possibilities $b^n < a$ and $b^n > a$.

So, assume first that $b^n < a$ is possible. By the binomial theorem, for each $k \in \mathbb{N}$ we have

$$\left(b + \frac{1}{k}\right)^n = \sum_{i=0}^{n} \binom{n}{i} b^{n-i} \frac{1}{k^i} = b^n + \sum_{i=1}^{n} \binom{n}{i} b^{n-i} \frac{1}{k^i} \leq b^n + \frac{r}{k}, \qquad (\star)$$

where $r = \sum_{i=1}^{n} \binom{n}{i} b^{n-i} \geq 0$. By the Archimedean property there exists some $k \in \mathbb{N}$ such that $k(a - b^n) > r$, or $r/k < a - b^n$. But then, it follows from $(\star)$ that for this k we have $(b + \frac{1}{k})^n \leq b^n + (a - b^n) = a$, which implies $b + \frac{1}{k} \in S$, a contradiction.

When $b^n > a$, the situation is similar. As previously, we can use the binomial theorem to get

$$\left(b - \frac{1}{k}\right)^n \geq b^n - \frac{t}{k}, \qquad (\star\star)$$

for some fixed $t \geq 0$ and all $k \in \mathbb{N}$. If we choose again some $k \in \mathbb{N}$ with $\frac{t}{k} < b^n - a$ and $\frac{1}{k} < b$, then it follows from $(\star\star)$ that $(b - \frac{1}{k})^n > a$. But if we pick some $s \in S$ with $0 < b - \frac{1}{k} < s$ (such an s is guaranteed by Theorem 3.1), then we have $s^n > (b - \frac{1}{k})^n > a$, a contradiction. Hence, $b^n = a$ holds. ∎

When feasible, the unique solution b of the equation $b^n = a$ provided by Theorem 3.5 is called the nth-**root** of a and is denoted $\sqrt[n]{a}$ or $a^{\frac{1}{n}}$.

EXERCISES

1. If $a \vee b = \max\{a, b\}$ and $a \wedge b = \min\{a, b\}$, then show that

$$a \vee b = \tfrac{1}{2}(a + b + |a - b|) \quad \text{and} \quad a \wedge b = \tfrac{1}{2}(a + b - |a - b|).$$

2. Show that $||a| - |b|| \leq |a + b| \leq |a| + |b|$ for all $a, b \in \mathbb{R}$.
3. Show that the real numbers $\sqrt{2}$ and $\sqrt{2} + \sqrt{3}$ are irrational numbers.
4. Show that between any two distinct real numbers there is an irrational number.
5. This exercise will introduce (by steps) the familiar process of subtraction in the framework of the axiomatic foundation of real numbers.

   a. Show that the zero element 0 is uniquely determined, i.e., show that if $x + 0^* = x$ for all $x \in \mathbb{R}$ and some $0^* \in \mathbb{R}$, then $0^* = 0$.
   b. Show that the **cancellation law of addition** is valid, i.e., show that $x + a = x + b$ implies $a = b$.
   c. Use the cancellation law of addition to show that $0 \cdot a = 0$ for all $a \in \mathbb{R}$.
   d. Show that for each real number a the real number $-a$ is the unique real number that satisfies the equation $a + x = 0$. (The real number $-a$ is called the **negative** of a.)

  e.   Show that for any two given real numbers a and b, the equation $a + x = b$ has
       a unique solution, namely $x = b + (-a)$. The **subtraction** operation $-$ of $\mathbb{R}$
       is now defined by $a - b = a + (-b)$; the real number $a - b$ is also called the
       difference of b from a.

  f.   For any real numbers a and b show that $-(-a) = a$ and $-(a + b) = -a - b$.

6. This exercise introduces (by steps) the familiar process of division in the framework
   of the axiomatic foundation of real numbers.

  a.   Show that the element 1 is uniquely determined, i.e., show that if $1^* \cdot x = x$ for
       all $x \in \mathbb{R}$ and some $1^* \in \mathbb{R}$, then $1^* = 1$.

  b.   Show that the **cancellation law of multiplication** is valid, i.e., show that $x \cdot a =
       x \cdot b$ with $x \neq 0$ implies $a = b$.

  c.   Show that for each real number $a \neq 0$ the real number a^{-1} is the unique real
       number that satisfies the equation $x \cdot a = 1$. The real number $x = a^{-1}$ is called
       the **inverse** (or the **reciprocal**) of a.

  d.   Show that for any two given real numbers a and b with $a \neq 0$, the equation $ax = b$
       has a unique solution, namely $x = a^{-1}b$. The **division** operation $\div$ (or /) of $\mathbb{R}$
       is now defined by $b \div a = a^{-1}b$; as usual, the real number $b \div a$ is also denoted
       by b/a or $\frac{b}{a}$.

  e.   For any two nonzero $a, b \in \mathbb{R}$ show that $(a^{-1})^{-1} = a$ and $(ab)^{-1} = a^{-1}b^{-1}$.

  f.   Show that $\frac{a}{1} = a$ for each a, $\frac{0}{b} = 0$ for each $b \neq 0$, and $\frac{a}{a} = 1$ for each $a \neq 0$.

7. Establish the following familiar properties of real numbers using the axioms of the
   real numbers together with the properties established in the previous two exercises.

  i.   **The zero product rule:** $ab = 0$ if and only if either $a = 0$ or $b = 0$.

  ii.   **The multiplication rule of signs:** $(-a)b = a(-b) = -(ab)$ and $(-a)(-b) =
        ab$ for all $a, b \in \mathbb{R}$.

  iii. **The multiplication rule for fractions:** For $b, d \neq 0$ and arbitrary real numbers
        a, c we have
        $$\frac{a}{b} \cdot \frac{c}{d} = \frac{ac}{bd}.$$
        In particular, if $\frac{a}{b} \neq 0$, then $(\frac{a}{b})^{-1} = \frac{b}{a}$.

  iv.   **The cancellation law of division:** If $a \neq 0$ and $x \neq 0$, then $\frac{bx}{ax} = \frac{b}{a}$ for each b.

  v.   **The division rule for fractions:** Division by a fraction is the same as multi-
        plication by the reciprocal of the fraction, i.e., whenever the fraction $\frac{a}{b} \div \frac{c}{d}$ is
        defined, we have
        $$\frac{a}{b} \div \frac{c}{d} = \frac{a}{b} \cdot \frac{d}{c} = \frac{ad}{bc}.$$

8. This exercise establishes that there exists essentially one set of real numbers that
   satisfies the eleven axioms stated in this section. To see this, let $\mathbb{R}$ be a set of real
   numbers (i.e., a collection of objects that satisfies all eleven axioms stated in this
   section).

  a.   Show that $1 > 0$.

  b.   A real number a satisfies $a = -a$ if and only if $a = 0$.

c. If $n = 1 + 1 + \cdots + 1$ (where the sum has "n summands" all equal to 1), then show that these elements are all distinct; as usual, we shall call the set $\mathbb{N}$ of all these numbers the natural numbers of $\mathbb{R}$.

d. Let Z consist of $\mathbb{N}$ together with their negative elements and zero; we shall call, of course, Z the set of integers of $\mathbb{R}$. Show that Z consists of distinct elements and that it is closed under addition and multiplication.

e. Define the set Q of rational numbers by $Q = \{\frac{m}{n}: m, n \in Z \text{ and } n \neq 0\}$. Show that Q satisfies itself axioms 1 through 10 and that

$$a = \sup\{r \in Q: r \leq a\} = \inf\{s \in Q: a \leq s\}$$

holds for each $a \in \mathbb{R}$.

f. Now, let $\mathbb{R}'$ be another set of real numbers and let Q' denote its rational numbers. If $1'$ denotes the unit element of $\mathbb{R}'$, then we write $n' = 1' + 1' + \cdots + 1'$ for the sum having "n-summands" all equal to $1'$. Now, define the function $f: Q \to Q'$ by

$$f\left(\frac{m}{n}\right) = \frac{m'}{n'}$$

and extend it to all of $\mathbb{R}$ via the formula

$$f(a) = \sup\{f(r): r \leq a\}.$$

Show that $\mathbb{R}$ and $\mathbb{R}'$ essentially coincide by establishing the following:

i. $a \leq b$ holds in $\mathbb{R}$ if and only if $f(a) \leq f(b)$ holds in $\mathbb{R}'$.

ii. f is one-to-one and onto.

iii. $f(a + b) = f(a) + f(b)$ and $f(ab) = f(a)f(b)$ for all $a, b \in \mathbb{R}$.

9. Consider a two point set $R = \{0, 1\}$ equipped with the following operations:

a. Addition $(+)$: $0 + 0 = 0, 0 + 1 = 1 + 0 = 1$ and $1 + 1 = 0$,

b. Multiplication $(\cdot)$: $0 \cdot 1 = 1 \cdot 0 = 0$ and $1 \cdot 1 = 1$, and

c. Ordering: $0 \geq 0, 1 \geq 1$ and $1 \geq 0$.

Does R with the preceding operations satisfy all eleven axioms defining the real numbers? Explain your answer.

10. Consider the set of rational numbers Q equipped with the usual operations of addition, multiplication, and ordering. Why doesn't Q coincide with the set of real numbers?

11. This exercise establishes the familiar rules of "exponents" based on the axiomatic foundation of real numbers. To avoid unnecessary notation, we shall assume that all real numbers encountered here are positive—and so by Theorem 3.5 all non-negative real numbers have unique roots. As usual, the "integer" powers are defined by

$$a^n = \underbrace{a \cdot a \cdots a}_{n\text{-factors}}, \quad a^0 = 1, \quad a^1 = a, \quad \text{and} \quad a^{-n} = \frac{1}{a^n}.$$

Extending this to rational numbers, for each $m, n \in \mathbb{N}$ we define

$$a^{\frac{m}{n}} = \sqrt[n]{a^m} \quad \text{and} \quad a^{-\frac{m}{n}} = \frac{1}{a^{\frac{m}{n}}} = \frac{1}{\sqrt[n]{a^m}}.$$

Establish the following properties:

a. $a^{\frac{m}{n}} = (\sqrt[n]{a})^m$ for all $m, n \in \mathbb{N}$.
b. If $m, n, p, q \in \mathbb{N}$ satisfy $\frac{m}{n} = \frac{p}{q}$, then $a^{\frac{m}{n}} = a^{\frac{p}{q}}$.
c. If r and s are rational numbers, then:
   i. $a^r a^s = a^{r+s}$ and $\frac{a^r}{a^s} = a^{r-s}$,
   ii. $(ab)^r = a^r b^r$ and $(\frac{a}{b})^r = \frac{a^r}{b^r}$,
   iii. $(a^r)^s = a^{rs}$.

4. SEQUENCES OF REAL NUMBERS

We start with the familiar definition of the convergence of a sequence.

Definition 4.1. *A sequence $\{x_n\}$ of real numbers is said to* **converge** *to $x \in \mathbb{R}$ if for every $\epsilon > 0$ there exists a natural number n_0 (depending on ϵ) such that*

$$|x_n - x| < \epsilon \ \text{for all} \ n > n_0.$$

The real number x is called the **limit** *of the sequence $\{x_n\}$, and we write $x_n \to x$, or $x = \lim_{n\to\infty} x_n$, or simply $x = \lim x_n$.*

We shall say that the terms of a sequence $\{x_n\}$ of a set A satisfy a property (P) **eventually**, if there exists some natural number n_0 such that x_n satisfies the property (P) for all $n > n_0$. In this terminology, a sequence of real numbers $\{x_n\}$ converges to some real number x if and only if for each $\epsilon > 0$ the terms x_n are eventually ϵ-close to x.

It should be clear that if $\lim x_n = x$, then $\lim y_n = x$ for every subsequence $\{y_n\}$ of $\{x_n\}$.

Theorem 4.2. *A sequence of real numbers can have at-most one limit.*

Proof. Assume that a sequence of real numbers $\{x_n\}$ satisfies $x = \lim x_n$ and $y = \lim x_n$. Let $\epsilon > 0$. Then there exists a natural number n_0 such that $|x_n - x| < \epsilon$ and $|x_n - y| < \epsilon$ for all $n > n_0$.

Now, fix $n > n_0$, and use the triangle inequality to get

$$0 \le |x - y| \le |x - x_n| + |x_n - y| < \epsilon + \epsilon = 2\epsilon$$

for all $\epsilon > 0$. This implies $x = y$, and the proof is finished. ∎

Recall that a sequence of real numbers $\{x_n\}$ is said to be bounded if there exists a real number $M > 0$ such that $|x_n| \le M$ for all n. A sequence $\{x_n\}$ of $\mathbb{R}$ is said to be **increasing** if $x_n \le x_{n+1}$ for each n, and **decreasing** if $x_{n+1} \le x_n$ for all n. A **monotone** sequence is either an increasing or a decreasing sequence.

The symbolism $x_n \uparrow x$ means that $\{x_n\}$ is increasing and $x = \sup\{x_n\}$. Similarly, $x_n \downarrow x$ means that $\{x_n\}$ is decreasing with $x = \inf\{x_n\}$. If a sequence $\{x_n\}$ satisfies $x_n = c$ for all n, then it is called a **constant** sequence.

Theorem 4.3. *Every monotone bounded sequence of real numbers is convergent.*

Proof. Assume that $\{x_n\}$ is increasing and bounded. Since $\{x_n\}$ is bounded, it follows from the completeness axiom that $x = \sup\{x_n: n \in \mathbb{N}\}$ exists in $\mathbb{R}$. We claim that $x = \lim x_n$. Indeed, if $\epsilon > 0$ is given, then by Theorem 3.1 there exists n_0 such that $x - \epsilon < x_{n_0} \leq x$. Since $\{x_n\}$ is increasing, it follows that $|x - x_n| = x - x_n \leq x - x_{n_0} < \epsilon$ for all $n > n_0$, and thus, $\lim x_n = x$. The proof for the decreasing case is similar.   ∎

Notice that the preceding theorem implies that an increasing sequence $\{x_n\}$ of real numbers satisfies $x_n \uparrow x$ if and only if $x = \lim x_n$. The basic convergence properties of real sequences are listed below.

1. Every convergent sequence is bounded.
2. If $x_n = c$ for each n, then $\lim x_n = c$.
3. If the three sequences $\{x_n\}$, $\{y_n\}$, and $\{z_n\}$ of $\mathbb{R}$ satisfy $x_n \leq z_n \leq y_n$ for all n, and $\lim x_n = \lim y_n = x$, then $\{z_n\}$ converges and $\lim z_n = x$.

For the next properties assume that $\lim x_n = x$ and $\lim y_n = y$.

4. For each $\alpha, \beta \in \mathbb{R}$ the sequence $\{\alpha x_n + \beta y_n\}$ converges and

$$\lim(\alpha x_n + \beta y_n) = \alpha x + \beta y.$$

5. The sequence $\{x_n y_n\}$ is convergent and $\lim(x_n y_n) = xy$.
6. If $|y_n| \geq \delta > 0$ holds for all n, and some $\delta > 0$, then $\{x_n/y_n\}$ converges and $\lim x_n/y_n = x/y$.
7. If $x_n \geq y_n$ holds for all $n \geq n_0$, then $x \geq y$.

A real number x is said to be a **limit point** (or a **cluster point**) of a sequence of real numbers $\{x_n\}$ if for every $n \in \mathbb{N}$ and $\epsilon > 0$, there exists $k > n$ (depending on ϵ and n) such that $|x_k - x| < \epsilon$.

The limit points of a sequence are characterized as follows:

Theorem 4.4. *Let $\{x_n\}$ be a sequence of real numbers. Then a real number x is a limit point for $\{x_n\}$ if and only if there exists a subsequence $\{x_{k_n}\}$ of $\{x_n\}$ such that $\lim x_{k_n} = x$.*

Proof. Assume that x is a limit point of $\{x_n\}$. Choose a natural number k_1 such that $|x_{k_1} - x| < 1$. Now, inductively, if $k_1, \ldots, k_n$ have been selected, then choose $k_{n+1} > k_n$ such that $|x_{k_{n+1}} - x| < \frac{1}{n+1}$. Thus, we construct a sequence of natural numbers $\{k_n\}$ such that $k_1 < k_2 < \cdots$ and $|x_{k_n} - x| < \frac{1}{n}$ for all n. Clearly, $\{x_{k_n}\}$ is a subsequence of $\{x_n\}$ satisfying $\lim x_{k_n} = x$.

For the converse, assume that a subsequence $\{x_{k_n}\}$ of $\{x_n\}$ satisfies $\lim x_{k_n} = x$. Let $m \in \mathbb{N}$ and $\epsilon > 0$ be fixed. It must be shown that there exists $p > m$ such that $|x_p - x| < \epsilon$. To see this, choose some n_0 such that $|x_{k_n} - x| < \epsilon$ for all $n \geq n_0$. Pick $n > \max\{n_0, m\}$, and let $p = k_n$. Then $p > m$ (since $k_n \geq n$) and $|x_p - x| < \epsilon$, and the proof is finished. ∎

Among the limit points of a sequence, the largest and the smallest ones are of some importance.

Definition 4.5. *Let $\{x_n\}$ be a bounded sequence of $\mathbb{R}$. Then the* **limit superior** *of $\{x_n\}$ is defined by*

$$\limsup x_n = \inf_n \left[\sup_{k \geq n} x_k \right],$$

and the **limit inferior** *of $\{x_n\}$ by*

$$\liminf x_n = \sup_n \left[\inf_{k \geq n} x_k \right].$$

If we write

$$\sup_{k \geq n} x_k = \bigvee_{k=n}^{\infty} x_k \quad \text{and} \quad \inf_{k \geq n} x_k = \bigwedge_{k=n}^{\infty} x_k,$$

then the preceding formulas can be rewritten as follows:

$$\limsup x_n = \bigwedge_{n=1}^{\infty} \left[\bigvee_{k=n}^{\infty} x_k \right] \quad \text{and} \quad \liminf x_n = \bigvee_{n=1}^{\infty} \left[\bigwedge_{k=n}^{\infty} x_k \right].$$

Also, since $\bigvee_{k=n+1}^{\infty} x_k \leq \bigvee_{k=n}^{\infty} x_k$ and $\bigwedge_{k=n}^{\infty} x_k \leq \bigwedge_{k=n+1}^{\infty} x_k$ for each n, it follows that

$$\bigvee_{k=n}^{\infty} x_k \downarrow \limsup x_n \quad \text{and} \quad \bigwedge_{k=n}^{\infty} x_k \uparrow \liminf x_n.$$

Theorem 4.6. *If $\{x_n\}$ is a bounded sequence, then $\liminf x_n$ and $\limsup x_n$ are the smallest and largest limit points of $\{x_n\}$. In particular,*

$$\liminf x_n \leq \limsup x_n.$$

Proof. Let $\{x_n\}$ be a bounded sequence of $\mathbb{R}$. Put $s = \limsup x_n$. We shall show that s is the largest limit point of $\{x_n\}$. The other case can be shown in a similar manner.

We show first that s is a limit point. To this end, let $m \in \mathbb{N}$ and $\epsilon > 0$. Since $\bigvee_{k=n}^{\infty} x_k \downarrow_n s$, there exists $n > m$ such that $s \le \bigvee_{k=n}^{\infty} x_k < s + \epsilon$. This implies the existence of some $k \ge n > m$ such that $s - \epsilon < x_k < s + \epsilon$. Hence, s is a limit point of $\{x_n\}$.

To finish the proof, we show that s is the largest limit point. Let x be a limit point of $\{x_n\}$, and let $\epsilon > 0$. Then for each $n \in \mathbb{N}$, there exists $m > n$ such that $x - \epsilon < x_m < x + \epsilon$. It follows that $x - \epsilon < \bigvee_{k=n}^{\infty} x_k$ for each n, and so, $x - \epsilon \le \bigwedge_{n=1}^{\infty} \bigvee_{k=n}^{\infty} x_k = s$ for each $\epsilon > 0$. Thus, $x \le s$, and the proof is complete. ∎

The next result is known as the Bolzano–Weierstrass[9,10] theorem.

Corollary 4.7 (Bolzano–Weierstrass). *Every bounded sequence of $\mathbb{R}$ has a convergent subsequence.*

Proof. Let $\{x_n\}$ be a bounded sequence. By Theorem 4.6, $\{x_n\}$ has a limit point which by Theorem 4.4 is the limit of a convergent subsequence of $\{x_n\}$. ∎

If $\lim x_n = x$, then x is the only limit point of $\{x_n\}$, and hence, $\limsup x_n = \liminf x_n = x$ holds. The converse of this statement is also true, as the following theorem shows:

Theorem 4.8. *A bounded sequence $\{x_n\}$ of real numbers converges if and only if $\liminf x_n = \limsup x_n = x$. In this case, $\lim x_n = x$.*

Proof. We assume that $\liminf x_n = \limsup x_n = x$ and show that $\lim x_n = x$. The inequalities

$$x_n - x \le \bigvee_{k=n}^{\infty} x_k - \bigwedge_{k=n}^{\infty} x_k \quad \text{and} \quad x - x_n \le \bigvee_{k=n}^{\infty} x_k - \bigwedge_{k=n}^{\infty} x_k$$

---

[9]Bernard Bolzano (1781–1848), a Czech mathematician, philosopher, and theologian. He is well known as one of the early mathematicians who emphasized the need for a foundation of mathematics as well as for rigorous proofs.

[10]Karl Theodor Wilhelm Weierstrass (1815–1897), a German mathematician and one of the most prominent mathematicians of the nineteenth century. He made numerous contributions to analysis and he is remembered for his belief that the "highest aim of science is to achieve general results."

imply that $|x_n - x| \leq \bigvee_{k=n}^{\infty} x_k - \bigwedge_{k=n}^{\infty} x_k$. Since

$$\lim_{n \to \infty} \left[\bigvee_{k=n}^{\infty} x_k - \bigwedge_{k=n}^{\infty} x_k \right] = x - x = 0,$$

it easily follows that $\lim x_n = x$.   ∎

A sequence $\{x_n\}$ in $\mathbb{R}$ is said to be a **Cauchy**[11] **sequence** if for each $\epsilon > 0$ there exists n_0 (depending on ϵ) such that $|x_n - x_m| < \epsilon$ for all $n, m > n_0$.

Clearly, a Cauchy sequence must necessarily be bounded. Also, it should be clear that every convergent sequence is a Cauchy sequence. The converse is also true, and it is expressed by saying that the real numbers form a complete metric space.

Theorem 4.9. *A sequence of real numbers converges if and only if it is a Cauchy sequence.*

Proof. We have only to show that if $\{x_n\}$ is a Cauchy sequence, then $\{x_n\}$ converges in $\mathbb{R}$.

By Corollary 4.7 there exists a subsequence $\{x_{k_n}\}$ of $\{x_n\}$ such that $\lim x_{k_n} = x$. Now, let $\epsilon > 0$. Choose n_0 such that $|x_{k_n} - x| < \epsilon$ and $|x_n - x_m| < \epsilon$ for $n, m > n_0$. Now, if $n > n_0$, then $k_n \geq n > n_0$, and so

$$|x_n - x| \leq |x_n - x_{k_n}| + |x_{k_n} - x| < \epsilon + \epsilon = 2\epsilon.$$

Hence, $\lim x_n = x$.   ∎

Now, let $\{f_n\}$ be a sequence of real-valued functions defined on a nonempty set X. Suppose that there exists a real-valued function g such that $|f_n(x)| \leq g(x)$ for all $x \in X$ and all n. Then for each fixed $x \in X$, the sequence of real numbers $\{f_n(x)\}$ is bounded. Thus, $\limsup f_n(x)$ and $\liminf f_n(x)$ both exist in $\mathbb{R}$. Consequently, $\limsup f_n$ and $\liminf f_n$ of the sequence of functions $\{f_n\}$ can be defined for each $x \in X$ as

$$(\limsup f_n)(x) = \limsup f_n(x) \quad \text{and} \quad (\liminf f_n)(x) = \liminf f_n(x).$$

---

[11] Augustin Louis Cauchy (1789–1875), a great French mathematician. He was a highly creative and prolific researcher who made numerous original contributions to all fields of mathematics of his time and to mathematical physics. His 1821 book *Cours d'analyse* is the first rigorous written treatment of calculus. Because he was on the conservative side during the period of the French revolution he was denied, in many occasions, the academic appointments appropriate for a scientist of his stature.

EXERCISES

1. Show that if $|x| < 1$, then $\lim x^n = 0$.

2. Show that $\lim x_n = x$ holds if and only if every subsequence of $\{x_n\}$ has a subsequence that converges to x.

3. Consider two sequences $\{k_n\}$ and $\{m_n\}$ of strictly increasing natural numbers such that $\{k_1, k_2, \ldots\} \cup \{m_1, m_2, \ldots\} = \mathbb{N}$. Show that a sequence of real numbers $\{x_n\}$ converges in $\mathbb{R}$ if and only if both subsequences $\{x_{k_n}\}$ and $\{x_{m_n}\}$ of $\{x_n\}$ converge in $\mathbb{R}$ and they satisfy $\lim x_{k_n} = \lim x_{m_n}$ (in which case the common limit is also the limit of the sequence).

   In particular, show that a sequence of real numbers $\{x_n\}$ converges in $\mathbb{R}$ if and only if the "even" and "odd" subsequences $\{x_{2n}\}$ and $\{x_{2n-1}\}$ both converge in $\mathbb{R}$ and they satisfy $\lim x_{2n} = \lim x_{2n-1}$.

4. Find the lim sup and lim inf for the sequence $\{(-1)^n\}$.

5. Find the lim sup and lim inf of the sequence $\{x_n\}$ defined by

$$x_1 = \tfrac{1}{3}, \quad x_{2n} = \tfrac{1}{3}x_{2n-1}, \quad \text{and} \quad x_{2n+1} = \tfrac{1}{3} + x_{2n} \quad \text{for } n = 1, 2, \ldots .$$

6. Let $\{x_n\}$ be a bounded sequence. Show that

$$\lim \sup(-x_n) = -\lim \inf x_n \quad \text{and} \quad \lim \inf(-x_n) = -\lim \sup x_n.$$

7. If $\{x_n\}$ and $\{y_n\}$ are two bounded sequences, then show that

   a. $\lim \sup(x_n + y_n) \le \lim \sup x_n + \lim \sup y_n$, and
   b. $\lim \inf(x_n + y_n) \ge \lim \inf x_n + \lim \inf y_n$.

   Moreover, show that if one of the sequences converges, then equality holds in both (a) and (b).

8. Prove that the lim sup and lim inf processes "preserve inequalities." That is, show that if two bounded sequences $\{x_n\}$ and $\{y_n\}$ of real numbers satisfy $x_n \le y_n$ for all $n \ge n_0$, then

$$\lim \inf x_n \le \lim \inf y_n \quad \text{and} \quad \lim \sup x_n \le \lim \sup y_n.$$

9. Show that $\lim \sqrt[n]{n} = 1$ (and conclude from this that $\lim \sqrt[n]{a} = 1$ for each $a > 0$). [HINT: Let $\sqrt{\sqrt[n]{n}} = 1 + x_n$, where $x_n \ge 0$. Using Bernoulli's inequality (see Exercise 13 of Section 2), we see that $\sqrt{n} = (1 + x_n)^n \ge 1 + nx_n > nx_n$, and so $0 < x_n < \frac{1}{\sqrt{n}}$ for each n.]

10. If $\{x_n\}$ is a sequence of strictly positive real numbers, then show that

$$\lim \inf \frac{x_{n+1}}{x_n} \le \lim \inf \sqrt[n]{x_n} \le \lim \sup \sqrt[n]{x_n} \le \lim \sup \frac{x_{n+1}}{x_n}.$$

Conclude from this that if $\lim \frac{x_{n+1}}{x_n}$ exists in $\mathbb{R}$, then $\lim \sqrt[n]{x_n}$ also exists and $\lim \sqrt[n]{x_n} = \lim \frac{x_{n+1}}{x_n}$.

[HINT: Note that $x_n = x_{n_0} \cdot \frac{x_{n_0+1}}{x_{n_0}} \cdot \frac{x_{n_0+2}}{x_{n_0+1}} \cdots \frac{x_n}{x_{n-1}}$ for each $n > n_0$.]

11. The **sequence of averages** of a sequence of real numbers $\{x_n\}$ is the sequence $\{a_n\}$ defined by $a_n = \frac{x_1 + x_2 + \cdots + x_n}{n}$. If $\{x_n\}$ is a bounded sequence of real numbers, then

show that

$$\liminf x_n \leq \liminf a_n \leq \limsup a_n \leq \limsup x_n.$$

In particular, if $x_n \to x$, then show that $a_n \to x$. Does the convergence of $\{a_n\}$ imply the convergence of $\{x_n\}$?

12. For a sequence of real numbers $\{x_n\}$ establish the following:

   a. If $x_{n+1} - x_n \to x$ in $\mathbb{R}$, then $x_n/n \to x$.

   b. If $\{x_n\}$ is bounded and $2x_n \leq x_{n+1} + x_{n-1}$ holds for all $n = 2, 3, \ldots$, then $x_{n+1} - x_n \uparrow 0$.

13. Consider the sequence $\{x_n\}$ defined by $0 < x_1 < 1$ and $x_{n+1} = 1 - \sqrt{1 - x_n}$ for $n = 1, 2, \ldots$. Show that $x_n \downarrow 0$. Also, show that $\frac{x_{n+1}}{x_n} \to \frac{1}{2}$.

14. Show that the sequence $\{x_n\}$ defined by

$$x_n = \left(1 + \frac{1}{n}\right)^n$$

is a convergent sequence.

15. Assume that a sequence $\{x_n\}$ satisfies $|x_{n+1} - x_n| \leq \alpha |x_n - x_{n-1}|$ for $n = 2, 3, \ldots$ and some fixed $0 < \alpha < 1$. Show that $\{x_n\}$ is a convergent sequence.

16. Show that the sequence of real numbers $\{x_n\}$, defined by

$$x_1 = 1 \quad \text{and} \quad x_{n+1} = \frac{1}{3 + x_n} \quad \text{for } n = 1, 2, \ldots,$$

converges and determine its limit.

17. Consider the sequence $\{x_n\}$ of real numbers defined by $x_1 = 1$ and $x_{n+1} = 1 + \frac{1}{1+x_n}$ for $n = 1, 2, \ldots$. Show that $\{x_n\}$ is a convergent sequence and that $\lim x_n = \sqrt{2}$.

18. Define the sequence $\{x_n\}$ by $x_1 = 1$ and

$$x_{n+1} = \frac{1}{2}\left(x_n + \frac{2}{x_n}\right), \quad n = 1, 2, \ldots.$$

Show that $\{x_n\}$ converges and that $\lim x_n = \sqrt{2}$.

19. Define the sequence $x_n = \sum_{k=1}^{n} \frac{1}{k}$ for $n = 1, 2, \ldots$. Show that $\{x_n\}$ does not converge in $\mathbb{R}$.

   [HINT: Show that $x_{2n} - x_n \geq \frac{1}{2}$.]

20. Let $-\infty < a < b < \infty$ and $0 < \lambda < 1$. Define the sequence $\{x_n\}$ by $x_1 = a, x_2 = b$ and

$$x_{n+2} = \lambda x_n + (1 - \lambda)x_{n+1} \quad \text{for } n = 1, 2, \ldots.$$

Show that $\{x_n\}$ converges in $\mathbb{R}$ and find its limit.

21. Let G be a nonempty subset of $\mathbb{R}$ which is a group under addition (i.e., if $x, y \in G$, then $x + y \in G$ and $-x \in G$). Show that between any two distinct real numbers there exists an element of G or else there exists $a \in \mathbb{R}$ such that $G = \{na: n = 0, \pm 1, \pm 2, \ldots\}$.

   [HINT: If $G \neq \{0\}$, let $a = \inf G \cap (0, \infty)$.]

22. Determine the limit points of the sequence $\{\cos n\}$.

   [HINT: Consider the set $G = \{n + 2m\pi: n, m \text{ integers}\}$ and use the previous exercise.]

23. For each n define $f_n: [-1, 1] \to \mathbb{R}$ by $f_n(x) = x^n$. Determine $\limsup f_n$ and $\liminf f_n$.

24. Show that every sequence of real numbers has a monotone subsequence. Use this conclusion to provide an alternate proof of the Bolzano–Weierstrass property of the real numbers: *Every bounded sequence has a convergent subsequence.* (See Corollary 4.7.)

5. THE EXTENDED REAL NUMBERS

The extended real numbers $\mathbb{R}^*$ are the real numbers with two elements adjoint. The two extra elements are denoted by ∞ (or $+\infty$) and $-\infty$, read **plus infinity** and **minus infinity**. Thus, $\mathbb{R}^* = \mathbb{R} \cup \{-\infty, \infty\}$ or, as is customarily written $\mathbb{R}^* = [-\infty, \infty]$.

The algebraic operations for the two infinities are defined as follows:

1. $\infty + \infty = \infty$ and $(-\infty) - \infty = -\infty$;
2. $(\pm\infty) \cdot \infty = \pm\infty$ and $(\pm\infty) \cdot (-\infty) = \mp\infty$;
3. $x + \infty = \infty$ and $x - \infty = -\infty$ for each $x \in \mathbb{R}$;
4. $x \cdot (\pm\infty) = \pm\infty$ if $x > 0$ and $x \cdot (\pm\infty) = \mp\infty$ if $x < 0$.

The expressions $\infty - \infty$ and $-\infty + \infty$ are left (as usual) undefined. In this book we shall agree that

5. $0 \cdot \infty = 0$.

Also, $\mathbb{R}^*$ is ordered, with ∞ the largest element, and $-\infty$ the smallest element. Moreover,

6. $-\infty < x < \infty$ for each $x \in \mathbb{R}$.

In topology, if $\mathbb{R}^*$ is endowed with an appropriate topology, $\mathbb{R}^*$ is referred to as the two-point compactification of $\mathbb{R}$. It turns out that the usual tangent function $\tan: [-\frac{\pi}{2}, \frac{\pi}{2}] \to \mathbb{R}^*$, where, of course, $\tan(-\frac{\pi}{2}) = -\infty$ and $\tan(\frac{\pi}{2}) = \infty$, is a homeomorphism.

One reason for introducing the extended real numbers is that in the theory of measure, one needs to consider sets with infinite measure. Another is that if $\{x_n\}$ is an unbounded sequence of real numbers, then by using Definition 4.5 one can see that $\limsup x_n$ and $\liminf x_n$ exist in $\mathbb{R}^*$ (the values may be plus or minus infinity). Thus, every sequence of real numbers has a limit superior and limit inferior in $\mathbb{R}^*$.

A sequence $\{x_n\}$ of real numbers **converges to** ∞ (denoted $\lim x_n = \infty$) if for every $M > 0$, there exists n_0 (depending on M) such that $x_n > M$ for all $n > n_0$. Similarly, $\lim x_n = -\infty$ means that for every real number $M < 0$, there exists n_0 such that $x_n < M$ for all $n > n_0$.

Theorem 4.3 can now be formulated as follows; the proof is left for the reader.

Theorem 5.1. *Every increasing sequence of real numbers either converges to a real number or to plus infinity.*

Recall that for a sequence $\{x_n\}$ of real numbers, the series $\sum_{n=1}^{\infty} x_n$ is said to be convergent, if the sequence of partial sums $\{\sum_{k=1}^{n} x_k\}$ converges in $\mathbb{R}$. If the sequence of partial sums converges to infinity, then we write $\sum_{n=1}^{\infty} x_n = \infty$ and say that the sum of the series is infinite. The definition of $\sum_{n=1}^{\infty} x_n = -\infty$ is similar. Observe that by Theorem 5.1 every series of non-negative real numbers converges in $\mathbb{R}^*$.

A series $\sum_{n=1}^{\infty} x_n$ of real numbers is said to be **rearrangement invariant** if for every one-to-one and onto function $\sigma: \mathbb{N} \to \mathbb{N}$ (called a **permutation** of $\mathbb{N}$), the series $\sum_{n=1}^{\infty} x_{\sigma_n}$ converges, and moreover $\sum_{n=1}^{\infty} x_n = \sum_{n=1}^{\infty} x_{\sigma_n}$.

Theorem 5.2. *If $\{x_n\}$ is a sequence of non-negative real numbers, then the series $\sum_{n=1}^{\infty} x_n$ is rearrangement invariant.*

Proof. Let $\sigma: \mathbb{N} \to \mathbb{N}$ be a permutation of $\mathbb{N}$. Set $a = \sum_{n=1}^{\infty} x_n$ and $b = \sum_{n=1}^{\infty} x_{\sigma_n}$ (note that both series converge in $\mathbb{R}^*$). To show that $a = b$, it is enough to establish (in view of the symmetry of the situation) that $b \leq a$. The latter is equivalent to showing that $\sum_{m=1}^{n} x_{\sigma_m} \leq a$ holds for all n.

If $n \in \mathbb{N}$, put $k = \max\{\sigma_1, \ldots, \sigma_n\}$ and observe that $\sum_{m=1}^{n} x_{\sigma_m} \leq \sum_{i=1}^{k} x_i \leq a$. The proof is now complete. ∎

In addition to series, **double series** with non-negative terms will appear from time to time in this book. If $\{a_{n,m}\}$ is a double sequence with $0 \leq a_{n,m} \leq \infty$ for each pair n, m, then for each fixed n the series $\sum_{m=1}^{\infty} a_{n,m}$ converges in $\mathbb{R}^*$ (possible to $+\infty$). The double series $\sum_{n=1}^{\infty} \sum_{m=1}^{\infty} a_{n,m}$ is now defined by

$$\sum_{n=1}^{\infty} \sum_{m=1}^{\infty} a_{n,m} = \lim_{k \to \infty} \sum_{n=1}^{k} \left(\sum_{m=1}^{\infty} a_{n,m} \right).$$

Note that the limit to the right always exists in $\mathbb{R}^*$.

The following result on interchanging the order of summation holds:

Theorem 5.3. *If $0 \leq a_{n,m} \leq \infty$ for all m and n, then*

$$\sum_{n=1}^{\infty} \sum_{m=1}^{\infty} a_{n,m} = \sum_{m=1}^{\infty} \sum_{n=1}^{\infty} a_{n,m}.$$

Proof. Set $a = \sum_{n=1}^{\infty} \sum_{m=1}^{\infty} a_{n,m}$ and $b = \sum_{m=1}^{\infty} \sum_{n=1}^{\infty} a_{n,m}$. Note that for each k and p we have

$$\sum_{n=1}^{k} \sum_{m=1}^{p} a_{n,m} = \sum_{m=1}^{p} \sum_{n=1}^{k} a_{n,m} \leq \sum_{m=1}^{p} \left(\sum_{n=1}^{\infty} a_{n,m} \right) \leq \sum_{m=1}^{\infty} \sum_{n=1}^{\infty} a_{n,m} = b,$$

from which it easily follows that $a \leq b$. A similar argument shows that $b \leq a$. Thus, $a = b$, and we are done.   ∎

Theorem 5.4. *Let $0 \leq a_{n,m} \leq \infty$ for all m, n. If $\sigma : \mathbb{N} \to \mathbb{N} \times \mathbb{N}$ is one-to-one and onto, then*

$$\sum_{n=1}^{\infty} a_{\sigma_n} = \sum_{n=1}^{\infty} \sum_{m=1}^{\infty} a_{n,m}.$$

Proof. Let $a = \sum_{n=1}^{\infty} a_{\sigma_n}$ and $b = \sum_{n=1}^{\infty} \sum_{m=1}^{\infty} a_{n,m}$. For each i, let $\sigma_i = (n_i, m_i)$. Then

$$\sum_{i=1}^{k} a_{\sigma_i} = \sum_{i=1}^{k} a_{n_i, m_i} \leq \sum_{i=1}^{k} \left(\sum_{m=1}^{\infty} a_{n_i, m} \right) \leq \sum_{n=1}^{\infty} \left(\sum_{m=1}^{\infty} a_{n,m} \right) = b$$

holds for each k, and so $a \leq b$.

On the other hand, for each k and m, there exists n such that for each $1 \leq i \leq k$ and $1 \leq j \leq m$ there exists $1 \leq r \leq n$ with $\sigma_r = (i, j)$. Thus,

$$\sum_{i=1}^{k} \sum_{j=1}^{m} a_{i,j} \leq \sum_{r=1}^{\infty} a_{\sigma_r} \leq a,$$

which shows that $b \leq a$. Therefore, $a = b$, and the proof is finished.   ∎

The series $\sum_{n=1}^{\infty} a_{\sigma_n}$ is usually called a rearrangement of the double series $\sum_{n=1}^{\infty} \sum_{m=1}^{\infty} a_{n,m}$ into a single series.

EXERCISES

1. Let $\{x_n\}$ be a sequence of $\mathbb{R}^*$. Define a limit point of $\{x_n\}$ in $\mathbb{R}^*$ to be any element x of $\mathbb{R}^*$ for which there exists a subsequence of $\{x_n\}$ that converges to x.

    Show that $\limsup x_n$ and $\liminf x_n$ (use Definition 4.5) are the largest and smallest limit points of $\{x_n\}$ in $\mathbb{R}^*$.

2. Let $\{x_n\}$ be a sequence of positive real numbers such that $\ell = \lim \frac{x_{n+1}}{x_n}$ exists in $\mathbb{R}$. Show that:

    a. if $\ell < 1$, then $\lim x_n = 0$, and
    b. if $\ell > 1$, then $\lim x_n = \infty$.

    [HINT: See the hint of Exercise 10 of Section 4.]

3. Let $0 \leq a_{n,m} \leq \infty$ for all m, n, and let $\sigma : \mathbb{N} \times \mathbb{N} \to \mathbb{N} \times \mathbb{N}$ be one-to-one and onto. Show that

$$\sum_{n=1}^{\infty} \sum_{m=1}^{\infty} a_{n,m} = \sum_{n=1}^{\infty} \sum_{m=1}^{\infty} a_{\sigma(n,m)}.$$

4. Show that

$$\sum_{n=1}^{\infty} \sum_{m=1}^{\infty} \frac{1}{n^2 + m^2} = \infty.$$

5. This exercise describes the *p*-**adic representation** of a real number in $(0, 1)$. We assume that p is a natural number such that $p \geq 2$ and $x \in (0, 1)$.

    a. Divide the interval $[0, 1)$ into the p closed-open intervals $[0, \frac{1}{p})$, $[\frac{1}{p}, \frac{2}{p})$, ..., $[\frac{p-1}{p}, 1)$, and number them consecutively from 0 to $p - 1$. Then x belongs precisely to one of these intervals, say k_1 ($0 \leq k_1 < p$). Next divide the interval $[\frac{k_1}{p}, \frac{k_1+1}{p})$ into p closed-open intervals (of the same length), number them consecutively from 0 to $p - 1$, and let k_2 be the subinterval to which x belongs. Proceeding this way, we construct a sequence $\{k_n\}$ of non-negative integers such that $0 \leq k_n < p$ for each n. Show that

    $$x = \sum_{n=1}^{\infty} \frac{k_n}{p^n}.$$

    b. Apply the same process as in (a) by subdividing each interval now into p open-closed intervals. For example, start with $(0, 1]$ and subdivide it into the open-closed intervals $(0, \frac{1}{p}]$, $(\frac{1}{p}, \frac{2}{p}]$, ..., $(\frac{p-1}{p}, 1]$.

    As in (a), construct a sequence $\{m_n\}$ of non-negative integers such that $0 \leq m_n < p$ for each n. Show that

    $$x = \sum_{n=1}^{\infty} \frac{m_n}{p^n}.$$

    c. Show by an example that the two sequences constructed in (a) and (b) may be different.

    In order to make the *p*-adic representation of a number unique, we shall agree to take the one determined by (a) above. As usual, it will be written as $x = 0.k_1 k_2 \cdots$.

6. Show that $\mathcal{P}(\mathbb{N}) \approx \mathbb{R}$ by establishing the following:

    i.   If A is an infinite set, and $f: A \to B$ is one-to-one such that $B \setminus f(A)$ is at most countable, then show that $A \approx B$.
    ii. Show that the set of real numbers of $(0, 1)$ for which the dyadic (i.e., $p = 2$) representation determined by (a) and (b) of the preceding exercise are different is a countable set.
    iii. For each $x \in (0, 1)$, let $x = 0.k_1 k_2 \cdots$ be the dyadic representation determined by part (a) of the preceding exercise; clearly, each k_i is either 0 or 1. Let $f(x) = \{n \in \mathbb{N}: k_n = 1\}$. Show that $f: (0, 1) \to \mathcal{P}(\mathbb{N})$ is one-to-one such that $\mathcal{P}(\mathbb{N}) \setminus f((0, 1))$ is countable, and conclude from part (i) that $(0, 1) \approx \mathcal{P}(\mathbb{N})$.

7. For a sequence $\{x_n\}$ of real numbers show that the following conditions are equivalent:

    a. The series $\sum_{n=1}^{\infty} x_n$ is rearrangement invariant in $\mathbb{R}$.
    b. For every permutation σ of $\mathbb{N}$ the series $\sum_{n=1}^{\infty} x_{\sigma_n}$ converges in $\mathbb{R}$.
    c. The series $\sum_{n=1}^{\infty} |x_n|$ converges in $\mathbb{R}$.
    d. For every sequence $\{s_n\}$ of $\{-1, 1\}$ the series $\sum_{n=1}^{\infty} s_n x_n$ converges in $\mathbb{R}$.
    e. For every subsequence $\{x_{k_n}\}$ of $\{x_n\}$ the series $\sum_{n=1}^{\infty} x_{k_n}$ converges in $\mathbb{R}$.

f. For every $\epsilon > 0$, there exists an integer k (depending on ϵ) such that for every finite subset S of $\mathbb{N}$ with $\min S \geq k$, we have $|\sum_{n \in S} x_n| < \epsilon$.

(Any series $\sum_{n=1}^{\infty} x_n$ satisfying any one of the above conditions is also referred to as an **unconditionally convergent series**.)

8. A series of the form $\sum_{n=1}^{\infty} (-1)^{n-1} x_n$, where $x_n > 0$ for each n, is called an **alternating series**. Assume that a sequence $\{x_n\}$ of strictly positive real numbers satisfies $x_n \downarrow 0$. Then establish the following:

   a. The alternating series $\sum_{n=1}^{\infty} (-1)^{n-1} x_n$ converges in $\mathbb{R}$.
   b. If $\sum_{n=1}^{\infty} x_n = \infty$, then the alternating series $\sum_{n=1}^{\infty} (-1)^{n-1} x_n$ is not rearrangement invariant.

9. This exercise describes the **integral test** for the convergence of series. Assume that $f : [1, \infty) \to [0, \infty)$ is a decreasing function. We define the sequences $\{\sigma_n\}$ and $\{\tau_n\}$ by

$$\sigma_n = \sum_{k=1}^{n} f(k) \quad \text{and} \quad \tau_n = \int_1^n f(x)\,dx.$$

Establish the following:

   a. $0 \leq \sigma_n - \tau_n \leq f(1)$ for all n.
   b. the sequence $\{\sigma_n - \tau_n\}$ is decreasing—and hence, convergent in $\mathbb{R}$.
   c. Show that the series $\sum_{k=1}^{\infty} f(k)$ converges in $\mathbb{R}$ if and only if the improper Riemann integral $\int_1^{\infty} f(x)\,dx = \lim_{r \to \infty} \int_1^r f(x)\,dx$ exists in $\mathbb{R}$.

10. Use the preceding exercise to show that the series $\sum_{n=1}^{\infty} \frac{1}{n^p}$ does not converge in $\mathbb{R}$ for $0 < p \leq 1$ and converges in $\mathbb{R}$ for all $p > 1$. The following are exercises related to the **harmonic series** $\sum_{n=1}^{\infty} \frac{1}{n}$:

    a. Prove with (at least) three different ways that $\sum_{n=1}^{\infty} \frac{1}{n} = \infty$.
    b. If a computer starting at 12 midnight on December 31, 1939, adds one million terms of the harmonic series every second, what was the value (within an error of 1) of the sum at 12 midnight on December 31, 2005? (Assume that each year has 365 days.)
    c. Show that $\sum_{n=1}^{\infty} \frac{(-1)^{n-1}}{n} = \lim_{n \to \infty} (\frac{1}{n+1} + \frac{1}{n+2} + \cdots + \frac{1}{2n}) = \ln 2$.

11. (**Toeplitz**)[12] Let $\{a_n\}$ be a sequence of positive real numbers (i.e., $a_n > 0$ for each n) and put $b_n = \sum_{i=1}^{n} a_i$. Assume that $b_n \uparrow \sum_{i=1}^{\infty} a_i = \infty$. If $\{x_n\}$ is a sequence of real numbers such that $x_n \to x$ in $\mathbb{R}$, then show that

$$\lim_{n \to \infty} \frac{1}{b_n} \sum_{i=1}^{n} a_i x_i = x.$$

12. (**Kronecker**)[13] Assume that a sequence of positive real numbers $\{b_n\}$ satisfies $0 < b_1 < b_2 < b_3 < \cdots$ and $b_n \uparrow \infty$. If a series $\sum_{n=1}^{\infty} x_n$ of real numbers

---

[12]Otto Toeplitz, (1881–1940), a German mathematician. He made many contributions to analysis—especially to the theory of integral equations and operator theory.

[13]Leopold Kronecker (1823–1891), a German mathematician. He devoted his entire life to attempting to unify (successfully) arithmetic, algebra, and analysis.

converges in $\mathbb{R}$, then show that

$$\lim_{n\to\infty}\frac{1}{b_n}\sum_{i=1}^{n}b_i x_i = 0.$$

In particular, show that if $\{y_n\}$ is a sequence of real numbers such that the series $\sum_{n=1}^{\infty}\frac{y_n}{n}$ converges in $\mathbb{R}$, then $\frac{y_1+\cdots+y_n}{n}\to 0$.

[HINT: Put $b_0 = 0$, $s_0 = 0$, and $s_n = x_1 + \cdots + x_n$ for each $n \geq 1$, and note that $\sum_{i=1}^{n}b_i x_i = \sum_{i=1}^{n}b_i(s_i - s_{i-1}) = b_n s_n - \sum_{i=2}^{n}s_{i-1}(b_i - b_{i-1})$. Now use the preceding exercise.]

6. METRIC SPACES

A **metric** (or a **distance**) d on a nonempty set X is a function $d: X \times X \to \mathbb{R}$ satisfying the three properties:

a. $d(x, y) \geq 0$ for all $x, y \in X$ and $d(x, y) = 0 \iff x = y$;
b. $d(x, y) = d(y, x)$ for all $x, y \in X$;
c. $d(x, y) \leq d(x, z) + d(z, y)$ for all $x, y, z \in X$ (*the triangle inequality*).

The pair (X, d) is called a **metric space**.

In a metric space (X, d) the inequality

$$|d(x, z) - d(y, z)| \leq d(x, y)$$

holds for all points $x, y, z \in X$. Indeed, from the triangle inequality it follows that $d(x, z) \leq d(x, y) + d(y, z)$, and therefore $d(x, z) - d(y, z) \leq d(x, y)$. Interchanging x and y gives $d(y, z) - d(x, z) \leq d(x, y)$, from which the inequality follows.

Here are some examples of metric spaces. The reader should be able to verify by himself that the exhibited functions satisfy the properties of a distance.

Example 6.1. The set of real numbers $\mathbb{R}$ equipped with the distance $d(x, y) = |x - y|$ for all $x, y \in \mathbb{R}$. ∎

Example 6.2. The Euclidean[14] space $\mathbb{R}^n$ equipped with the distance

$$d(x, y) = \left(\sum_{i=1}^{n}(x_i - y_i)^2\right)^{\frac{1}{2}}$$

for $x = (x_1, \ldots, x_n)$ and $y = (y_1, \ldots, y_n)$ in $\mathbb{R}^n$. This distance on $\mathbb{R}^n$ is called the **Euclidean distance**. ∎

Example 6.3. Let X be a nonempty set. Then the function d defined by $d(x, y) = 1$ if $x \neq y$ and $d(x, x) = 0$ is a distance on X. This distance is called the **discrete distance** on X, and X with this distance is called a **discrete metric space**. ∎

---

[14]Euclid (ca 365–300 BC), a famous Greek geometer and the most celebrated mathematician of all time. His name was a synonym for geometry until the twentieth century. Euclid's fame rests upon his classic work on geometry, *The Elements*, (written in thirteen books) that had a profound influence on the human mind and inquiry more than any other work except the Bible.

Example 6.4. Let $X = (0, \infty)$. Then

$$d(x, y) = \left| \frac{1}{x} - \frac{1}{y} \right|$$

for $x, y \in X$ is a distance on X. ∎

If Y is a subset of a metric space (X, d), then Y equipped with the distance d also becomes a metric space.

Now, let us fix a metric space (X, d). If $x \in X$, then the **open ball** at x with radius $r > 0$ is the set $B(x, r) = \{y \in X : d(x, y) < r\}$. The open subsets of X can now be defined in the usual way. A subset A of X is called **open** if for every $x \in A$, there exists some $r > 0$ such that $B(x, r) \subseteq A$.

Every open ball $B(x, r)$ is an open set. Indeed, if $y \in B(x, r)$, then the open ball $B(y, r_1)$, where $r_1 = r - d(x, y) > 0$, satisfies $B(y, r_1) \subseteq B(x, r)$. Reason: $z \in B(y, r_1)$ implies $d(x, z) \leq d(x, y) + d(y, z) < d(x, y) + r_1 = r$, and so $z \in B(x, r)$.

Theorem 6.5. *For a metric space (X, d) the following statements hold:*

i. *X and $\emptyset$ are open sets.*
ii. *Arbitrary unions of open sets are open sets.*
iii. *Finite intersections of open sets are open sets.*

Proof. (i) Obvious.

(ii) Let $\{A_i\}_{i \in I}$ be a family of open subsets of X. Let $x \in \bigcup A_i$. Then there exists some $i \in I$ such that $x \in A_i$. Since A_i is open, there exists $r > 0$ with $B(x, r) \subseteq A_i \subseteq \bigcup A_i$. Hence, $\bigcup A_i$ is open.

(iii) Let $\{A_1, \ldots, A_n\}$ be a finite collection of open sets. If $x \in \bigcap_{i=1}^{n} A_i$, then for each $1 \leq i \leq n$ there exists $r_i > 0$ such that $B(x, r_i) \subseteq A_i$. Put $r = \min\{r_1, \ldots, r_n\}$, and note that $B(x, r) \subseteq \bigcap_{i=1}^{n} A_i$. Hence, $\bigcap_{i=1}^{n} A_i$ is an open set. ∎

A point x is called an **interior point** of a subset A if there exists an open ball $B(x, r)$ such that $B(x, r) \subseteq A$. The set of all interior points of A is denoted by A° and is called the **interior** of A; clearly, $A^\circ \subseteq A$. It is easy to see that A° is the largest open subset of X included in A. Also, note that A is open if and only if $A = A^\circ$.

A subset A of a metric space (X, d) is called **closed** if its complement A^c ($= X \setminus A$) is an open set. The properties of closed sets are stated next.

Theorem 6.6. *For a metric space (X, d) the following statements hold:*

i. *X and $\emptyset$ are closed sets.*
ii. *Arbitrary intersections of closed sets are closed sets.*
iii. *Finite unions of closed sets are closed sets.*

Proof. (i) The result follows from $X^c = \emptyset$, $\emptyset^c = X$, and Theorem 6.5(i).

(ii) Let $\{A_i\}_{i \in I}$ be a family of closed sets. Then by Theorem 6.5 and De Morgan's law, we see that $(\bigcap_{i \in I} A_i)^c = \bigcup_{i \in I} A_i^c$ is open. Thus, $\bigcap_{i \in I} A_i$ is a closed set.

(ii) Combine $(\bigcup_{i=1}^{n} A_i)^c = \bigcap_{i=1}^{n} A_i^c$ with Theorem 6.5(iii). ∎

It should be observed that a set A is open if and only if A^c is closed; and similarly, A is closed if and only if A^c is open. Observe that a set which is not open is not necessarily closed, and vice versa.

A point $x \in X$ is called a **closure point** of a subset A of X if every open ball at x contains (at least) one element of A; that is, $B(x, r) \cap A \neq \emptyset$ for all $r > 0$. The set of all closure points of A is denoted by $\overline{A}$, and is called the **closure** of A; clearly, $A \subseteq \overline{A}$.

Theorem 6.7. *For every subset A of a metric space, $\overline{A}$ is the smallest closed set that includes A.*

Proof. Let A be a subset of a metric space. We show that $\overline{A}$ is closed. Indeed, if $x \notin \overline{A}$, then there exists an open ball $B(x, r)$ such that $B(x, r) \cap A = \emptyset$. If $y \in B(x, r)$, then [since $B(x, r)$ is an open set] there exists $\delta > 0$ such that $B(y, \delta) \subseteq B(x, r)$. Thus, $B(y, \delta) \cap A = \emptyset$, and so $y \notin \overline{A}$. Consequently, $B(x, r) \subseteq (\overline{A})^c$, and so $(\overline{A})^c$ is open, which shows that $\overline{A}$ is closed.

Now, if B is a closed subset such that $A \subseteq B$, then for every $x \in B^c$ there exists an open ball $B(x, r) \subseteq B^c$. Thus, $B(x, r) \cap B = \emptyset$, and in particular, $B(x, r) \cap A = \emptyset$. This shows that no element of B^c is a closure point of A, and therefore $\overline{A} \subseteq B$. ∎

An immediate consequence of the preceding theorem is that a set A is closed if and only if $A = \overline{A}$.

Every set of the form $A = \{x \in X: d(x, a) \leq r\}$, called the **closed ball** at a with radius r, is a closed set. Indeed, assume $d(x, a) > r$, and put $r_1 = d(x, a) - r > 0$. If $d(y, x) < r_1$, then

$$d(a, y) \geq d(a, x) - d(y, x) > d(a, x) - r_1 = r,$$

which shows that A^c is open, and thus, A is closed. Observe that in a discrete metric space, $\overline{B(a, r)}$ may be a proper subset of $\{x \in X: d(x, a) \leq r\}$. However, in the Euclidean space $\mathbb{R}^n$, the closure of every open ball of radius r is the closed ball of radius r. (Why?)

Lemma 6.8. *If A is a subset of a metric space, then $A^o = (\overline{A^c})^c$.*

Proof.　Notice that

$$x \in A^\circ \iff \exists r > 0 \text{ with } B(x, r) \subseteq A$$
$$\iff \exists r > 0 \text{ with } B(x, r) \cap A^c = \emptyset$$
$$\iff x \notin \overline{A^c} \iff x \in (\overline{A^c})^c,$$

and the proof is finished.　　　　　　　　　　　　　　　　■

A point x is called an **accumulation point** of a set A if every open ball $B(x, r)$ contains an element of A distinct from x; that is, $B(x, r) \cap (A \setminus \{x\}) \neq \emptyset$ for each $r > 0$. Note that x need not be an element of A. Clearly, every accumulation point of a set is automatically a closure point of that set. The set of accumulation points of A is called the **derived set** of A, and is denoted by A'. It should be clear that $\overline{A} = A \cup A'$. In particular, it follows that a set is closed if and only if it contains its accumulation points.

A sequence $\{x_n\}$ of a metric space (X, d) is said to be **convergent** to $x \in X$ (in symbols, $\lim x_n = x$, or $x_n \to x$) if $\lim d(x_n, x) = 0$. From the triangle inequality it easily follows that a sequence in a metric space can have at most one limit. (See the proof of Theorem 4.2.)

The next theorem characterizes the closure points of a set in terms of sequences.

Theorem 6.9.　*Let A be a subset of a metric space (X, d). Then a point $x \in X$ belongs to $\overline{A}$ if and only if there exists a sequence $\{x_n\}$ of A such that $\lim x_n = x$.*

Moreover, if x is an accumulation point of A, then there exists a sequence of A with distinct terms that converges to x.

Proof.　Assume that x belongs to the closure of A. For each n pick $x_n \in A$ such that $d(x, x_n) < \frac{1}{n}$. Then $\{x_n\}$ is a sequence of A such that $\lim x_n = x$.

On the other hand, if a sequence $\{x_n\}$ of A satisfies $\lim x_n = x$, then for each $r > 0$ there exists some k such that $d(x, x_n) < r$ for $n > k$. Thus, $B(x, r) \cap A \neq \emptyset$ for each $r > 0$, and so $x \in \overline{A}$.

Next, assume that x is an accumulation point of A. Start by choosing some $x_1 \in A$ such that $x_1 \neq x$ and $d(x, x_1) < 1$. Now, inductively, if $x_1, \ldots, x_n \in (A \setminus \{x\})$ have been chosen, pick $x_{n+1} \in A \setminus \{x\}$ such that $d(x, x_{n+1}) < \min\{\frac{1}{n+1}, d(x, x_n)\}$. Then $\{x_n\}$ is a sequence of A satisfying $x_n \neq x_m$ if $n \neq m$ and $\lim x_n = x$.　　■

A subset A of a metric space (X, d) is called **dense** in X if $\overline{A} = X$. According to Theorem 6.9, a set A is dense in X if and only if for every $x \in X$ there exists a sequence $\{x_n\}$ of A such that $\lim x_n = x$. Also, notice that a set A is dense if and only if $V \cap A \neq \emptyset$ holds for each nonempty open set V.

A point $x \in X$ is called a **boundary point** of a set A if every open ball of x contains points from A and A^c; that is, if $B(x, r) \cap A \neq \emptyset$ and $B(x, r) \cap A^c \neq \emptyset$ for all $r > 0$. The set of all boundary points of a set A is denoted by ∂A and is called the **boundary** of A. By the symmetry of the definition, $\partial A = \partial A^c$ holds for every subset A of X. Also, a simple argument shows that

$$\partial A = \overline{A} \cap \overline{A^c}.$$

We now introduce the concept of continuity.

Definition 6.10. *A function $f : (X, d) \to (Y, \rho)$ between two metric spaces is said to be* **continuous at a point** $a \in X$ *if for every $\epsilon > 0$ there exists $\delta > 0$ (depending on ϵ) such that $\rho(f(x), f(a)) < \epsilon$ whenever $d(x, a) < \delta$.*

The function f is said to be **continuous on** X *(or simply* **continuous***) if f is continuous at every point of X.*

The next theorem presents the most useful characterizations of continuous functions.

Theorem 6.11. *For a function $f : (X, d) \to (Y, \rho)$ between two metric spaces, the following statements are equivalent:*

   i.   *f is continuous on X.*
  ii.   *$f^{-1}(\mathcal{O})$ is an open subset of X whenever $\mathcal{O}$ is an open subset of Y.*
 iii.   *If $\lim x_n = x$ holds in X, then $\lim f(x_n) = f(x)$ holds in Y.*
 iv.   *$f(\overline{A}) \subseteq \overline{f(A)}$ holds for every subset A of X.*
  v.   *$f^{-1}(C)$ is a closed subset of X whenever C is a closed subset of Y.*

Proof. (i) $\Longrightarrow$ (ii) Let $\mathcal{O}$ be an open subset of Y and $a \in f^{-1}(\mathcal{O})$. Since $f(a) \in \mathcal{O}$ and $\mathcal{O}$ is open, there exists some $r > 0$ such that $B(f(a), r) \subseteq \mathcal{O}$. Now by the continuity of f at a, there exists $\delta > 0$ such that $d(x, a) < \delta$ implies $\rho(f(x), f(a)) < r$. But this shows that $B(a, \delta) \subseteq f^{-1}(\mathcal{O})$. Therefore, a is an interior point of $f^{-1}(\mathcal{O})$, and hence, $f^{-1}(\mathcal{O})$ is open.

(ii) $\Longrightarrow$ (iii) Assume $\lim x_n = x$ in X and $r > 0$. Let $V = B(f(x), r)$. By our assumption $f^{-1}(V)$ is an open subset of X, and since x belongs to it, there exists some $\delta > 0$ such that $B(x, \delta) \subseteq f^{-1}(V)$. Pick some k such that $x_n \in B(x, \delta)$ for all $n > k$. Then $f(x_n) \in V$ for all $n > k$, which shows that $\lim f(x_n) = f(x)$.

(iii) $\Longrightarrow$ (iv) Let A be a subset of X. If $y \in f(\overline{A})$, then there exists $x \in \overline{A}$ such that $y = f(x)$. Since $x \in \overline{A}$, there exists (by Theorem 6.9) a sequence $\{x_n\}$ of A such that $\lim x_n = x$. But then, $\{f(x_n)\}$ is a sequence of $f(A)$, and by assumption $\lim f(x_n) = f(x) = y$. By Theorem 6.9 it follows that $y \in \overline{f(A)}$; that is, $f(\overline{A}) \subseteq \overline{f(A)}$.

(iv) $\Longrightarrow$ (v) Let C be a closed subset of Y; clearly, $C = \overline{C}$ holds in Y. Applying our assumption to the set $A = f^{-1}(C)$, we get $f(\overline{A}) \subseteq \overline{f(A)} \subseteq \overline{C} = C$, which shows that $\overline{A} \subseteq f^{-1}(C) = A$. Since $A \subseteq \overline{A}$ is always true, it follows that $A = \overline{A}$, which shows that $A = f^{-1}(C)$ is a closed subset of X.

(v) $\Longrightarrow$ (i) Let $a \in X$ and $\epsilon > 0$. Consider the closed set

$$C = [B(f(a), \epsilon)]^c = \{y \in Y : \rho(f(a), y) \geq \epsilon\}.$$

By hypothesis, $f^{-1}(C)$ is a closed subset of X. Since $a \notin f^{-1}(C)$, there exists some $\delta > 0$ such that $B(a, \delta) \subseteq [f^{-1}(C)]^c$. But then, if $d(x, a) < \delta$, then $\rho(f(x), f(a)) < \epsilon$ holds, so that f is continuous at a. Since a is arbitrary, f is continuous on X. The proof of the theorem is now complete. ∎

From statement (iii) it should be clear that the composition of two continuous functions between metric spaces (whenever it makes sense) must be a continuous function.

Two metric spaces (X, d) and (Y, ρ) are called **homeomorphic** if there exists a one-to-one onto function $f : (X, d) \to (Y, \rho)$ such that f and f^{-1} are both continuous.

Two distances d and ρ on a set X are called **equivalent** if a sequence $\{x_n\}$ of X satisfies $\lim d(x_n, x) = 0$ if and only if $\lim \rho(x_n, x) = 0$. By the preceding theorem, for this to happen it is necessary and sufficient for the identity mapping $I : (X, d) \to (X, \rho)$ to be a homeomorphism. Rephrasing this last statement, d and ρ are equivalent if and only if d and ρ generate the same open sets.

A metric space (X, d) is called **bounded** if there exists a number $M > 0$ such that $d(x, y) \leq M$ for all $x, y \in X$. The **diameter** of a subset A of a metric space (X, d) is defined by

$$d(A) = \sup\{d(x, y) : x, y \in A\}.$$

Thus, (X, d) is bounded if and only if the diameter of X is finite.

If d is a distance on a set X, then the function ρ defined by

$$\rho(x, y) = \frac{d(x, y)}{1 + d(x, y)}$$

is also a distance on X. Moreover, (X, ρ) is bounded and ρ is equivalent to d.

We now turn our attention to complete metric spaces. A sequence $\{x_n\}$ of a metric space (X, d) is called a **Cauchy sequence** if for every $\epsilon > 0$, there exists n_0 (depending on ϵ) such that $d(x_n, x_m) < \epsilon$ for all $n, m > n_0$. Clearly, every convergent sequence is a Cauchy sequence. However, in general, the converse is not true. As an example, let $X = (0, \infty)$ with distance $d(x, y) = |x - y|$, and $x_n = \frac{1}{n}$ for each n. Then $\{x_n\}$ is a Cauchy sequence that does not converge in X.

If a metric space has the property that all of its Cauchy sequences converge (in the space), then the metric space is called a **complete metric space**. Examples of complete metric spaces are provided by the Euclidean spaces $\mathbb{R}^n$ with their Euclidean distances. Recall that according to Theorem 4.9 the real numbers form a complete metric space.

Here is another important example of a complete metric space.

Example 6.12.   Let Ω be a nonempty set. We shall denote by $B(\Omega)$ the set of all real-valued functions defined on Ω that are bounded. That is, a function $f: \Omega \to \mathbb{R}$ belongs to $B(\Omega)$ if and only if there exists a number $M > 0$ (depending on f) such that $|f(\omega)| \leq M$ for all $\omega \in \Omega$. Now for each $f, g \in B(\Omega)$ define

$$D(f, g) = \sup\{|f(\omega) - g(\omega)|: \omega \in \Omega\}.$$

Note that since both f and g are bounded, $D(f, g)$ is a real number. We claim that D is a distance on $B(\Omega)$ and that, in fact, $B(\Omega)$ is complete (with this distance).

To see that D is a distance, only the triangle inequality is verified; the other two properties are trivial. Indeed, if $f, g, h \in B(\Omega)$, then for each $\omega \in \Omega$ we have

$$|f(\omega) - g(\omega)| \leq |f(\omega) - h(\omega)| + |h(\omega) - g(\omega)| \leq D(f, h) + D(h, g),$$

and so $D(f, g) \leq D(f, h) + D(h, g)$.

Next, we shall establish the completeness of $B(\Omega)$. To this end, let $\{f_n\}$ be a Cauchy sequence of $B(\Omega)$. Then, given $\epsilon > 0$ there exists n_0 such that $D(f_n, f_m) < \epsilon$ for all $n, m > n_0$. In particular, note that for each $\omega \in \Omega$, the inequality $|f_n(\omega) - f_m(\omega)| \leq D(f_n, f_m)$ implies that $\{f_n(\omega)\}$ is a Cauchy sequence of real numbers. Thus, $\{f_n(\omega)\}$ converges in $\mathbb{R}$ for each $\omega \in \Omega$; let $f(\omega) = \lim f_n(\omega)$. It easily follows from the inequality $|f_n(\omega) - f_m(\omega)| < \epsilon$ for all $n, m > n_0$ that $|f_n(\omega) - f(\omega)| \leq \epsilon$ for all $n \geq n_0$ and all $\omega \in \Omega$. This last inequality in turn implies that $f \in B(\Omega)$ and $D(f_n, f) \leq \epsilon$ for all $n > n_0$. Hence, $\lim f_n = f$, and so $B(\Omega)$ is complete. ∎

Closed subsets of complete metric spaces are complete metric space in their own right.

Theorem 6.13.   *Let (X, d) be a complete metric space. Then a subset A of X is closed if and only if A (with metric d) is a complete metric space in its own right.*

Proof.   Let A be closed. If $\{x_n\}$ is a Cauchy sequence of A, then $\{x_n\}$ is a Cauchy sequence of X. Since X is complete, there exists $x \in X$ such that $\lim x_n = x$. But since A is closed, $x \in A$. Thus, (A, d) is a complete metric space.

Conversely, assume that (A, d) is a complete metric space. If a sequence $\{x_n\}$ of A satisfies $\lim x_n = x$ in X, then $\{x_n\}$ is a Cauchy sequence of X. But then, $\{x_n\}$

is a Cauchy sequence of A, and hence, it must converge to a unique element of A. This element must be x. Thus, $x \in A$, so that A is a closed subset of X. ∎

The following important result dealing with complete metric spaces is due to G. Cantor. (Keep in mind that the diameter of a set A is defined by $d(A) = \sup\{d(x, y): x, y \in A\}$.)

Theorem 6.14 (Cantor). *Let (X, d) be a complete metric space and let $\{A_n\}$ be a sequence of closed, nonempty subsets of X such that $A_{n+1} \subseteq A_n$ for each n and $\lim d(A_n) = 0$.*

Then the intersection $\bigcap_{n=1}^{\infty} A_n$ consists precisely of one element.

Proof. If $x, y \in \bigcap_{n=1}^{\infty} A_n$, then $x, y \in A_n$ for each n, and hence, $0 \leq d(x, y) \leq d(A_n)$ for each n. Thus, $d(x, y) = 0$, so that $x = y$. This shows that $\bigcap_{n=1}^{\infty} A_n$ contains at most one element.

To show that $\bigcap_{n=1}^{\infty} A_n \neq \emptyset$, proceed as follows. For each n, choose some $x_n \in A_n$. Then, it is easy to see that $d(x_{n+p}, x_n) \leq d(A_n)$ holds for each n and p, and from this it follows that $\{x_n\}$ is a Cauchy sequence of X. Thus, there exists $x \in X$ such that $\lim x_n = x$, and we claim that $x \in \bigcap_{n=1}^{\infty} A_n$. Indeed, since $x_m \in A_n$ for $m \geq n$, we get $x \in \overline{A_n}$ for each n. But since each A_n is closed, $\overline{A_n} = A_n$ holds; therefore, $x \in A_n$ for each n, and we are done. ∎

A subset A of a metric space (X, d) is said to be **nowhere dense** if its closure has an empty interior; that is, if $(\overline{A})^{\circ} = \emptyset$. Since $B^{\circ} = (\overline{B^c})^c$ holds for every subset B, it is easy to see that a subset A is nowhere dense if and only if $(\overline{A})^c$ is dense in X.

A classical nowhere dense subset of the real line is the so-called Cantor set. Because we shall use it later, we pause for a while to describe this set and its properties.

Example 6.15 (The Cantor Set). The Cantor set is a subset of $[0, 1]$ and is constructed as follows:

Let $C_0 = [0, 1]$. Then trisect $[0, 1]$ and remove the middle open interval $(\frac{1}{3}, \frac{2}{3})$. Let $C_1 = [0, \frac{1}{3}] \cup [\frac{2}{3}, 1]$, and note that C_1 is the union of $2^1 = 2$ disjoint closed intervals. Next, trisect each closed interval of C_1 and remove from each one of them the middle open interval. Let C_2 be the set remaining from C_1 after these removals. That is, $C_2 = [0, \frac{1}{9}] \cup [\frac{2}{9}, \frac{1}{3}] \cup [\frac{2}{3}, \frac{7}{9}] \cup [\frac{8}{9}, 1]$; note that C_2 is the union of $2^2 = 4$ disjoint closed intervals.

The (inductive) process of constructing C_{n+1} from C_n should be clear now. Trisect each of the 2^n disjoint closed intervals of C_n, and remove from each one of them the middle open interval. What is left from C_n is then C_{n+1}. Note that C_{n+1} is the union of 2^{n+1} disjoint closed intervals. The graphs of the first few constructions are shown in Figure 1.1.

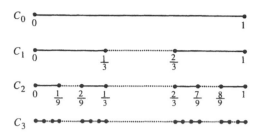

FIGURE 1.1.   The Construction of the Cantor Set

Clearly, $C_{n+1} \subseteq C_n$ holds for all n. The Cantor set of $[0, 1]$ is now defined by $C = \bigcap_{n=1}^{\infty} C_n$. Next, we mention the most interesting properties of the Cantor set C.

1. *The set C is a closed nowhere dense subset of $\mathbb{R}$.*
   Clearly, C is closed as it is an intersection of closed sets. It should also be clear from the above construction that C does not contain any interval and thus, C has empty interior, i.e., is a nowhere dense set.

2. *The total length of the removed intervals from $[0, 1]$ to get C equals one.*
   To see this, note that at the nth step we remove 2^{n-1} open intervals each of which has length 3^{-n}; therefore, a total of $2^{n-1} \cdot 3^{-n}$. Thus, we remove altogether a total length of $\sum_{n=1}^{\infty} 2^{n-1} \cdot 3^{-n} = \frac{1}{2} \sum_{n=1}^{\infty} (\frac{2}{3})^n = 1$.

3. *The set C has cardinality $\mathfrak{c}$, that is, $C \approx \mathbb{R}$.*
   Perhaps the simplest way of proving this is by showing that $C \approx 2^{\mathbb{N}}$, where $2 = \{0, 1\}$. Since $\mathcal{P}(\mathbb{N}) \approx 2^{\mathbb{N}}$ [the mapping $a = \{a_n\} \mapsto g(a) = \{n \in \mathbb{N}: a_n = 1\}$ is one-to-one from $2^{\mathbb{N}}$ onto $\mathcal{P}(\mathbb{N})$] and $\mathcal{P}(\mathbb{N}) \approx \mathbb{R}$ (see Exercise 6 of Section 5), it will follow that $C \approx \mathbb{R}$. The details are included below.

If $x = \{x_n\} \in 2^{\mathbb{N}}$ (i.e., each x_n is either 0 or 1), then let $y_n = 2x_n$ for each n, and define $f(x) = \sum_{n=1}^{\infty} 3^{-n} y_n$. Clearly, each y_n is either 0 or 2. Note first that $f(x) \in C$. Indeed, since $y_1 \neq 1$, we have $f(x) \notin (\frac{1}{3}, \frac{2}{3})$; similarly, since $y_2 \neq 1$, we can see that $f(x) \notin (\frac{1}{9}, \frac{2}{9}) \cup (\frac{7}{9}, \frac{8}{9})$. By induction, we can verify that $f(x)$ does not belong to any of the removed open intervals. Thus, $f(x) \in C$ which shows that $x \mapsto f(x)$ is a mapping from $2^{\mathbb{N}}$ into C.

Now, we claim that $x \mapsto f(x)$ is one-to-one. Indeed, if $a = \{a_n\} \neq b = \{b_n\}$ holds, then let $k = \min\{n \in \mathbb{N}: a_n \neq b_n\}$; assume that $b_k = 1$ and $a_k = 0$. Then, in view of $2 \sum_{n=k+1}^{\infty} 3^{-n} = 3^{-k}$, we have

$$f(b) = 2 \sum_{n=1}^{\infty} b_n 3^{-n} \geq 2 \sum_{n=1}^{k-1} b_n 3^{-n} + 2 \cdot 3^{-k} > 2 \sum_{n=1}^{\infty} a_n 3^{-n} = f(a).$$

Finally, it is not difficult to see that C consists precisely of all numbers of $[0, 1]$ with a triadic representation (see Exercise 5 of Section 5) assuming the values 0 or 2. This implies that $x \mapsto f(x)$ is onto, proving that $C \approx \mathbb{R}$. ∎

A subset Y of a metric space is said to be **meager** (or of **first category**) if there exists a sequence $\{A_n\}$ of nowhere dense subsets such that $Y = \bigcup_{n=1}^{\infty} A_n$. A metric space is called a **Baire**[15] **space** if every nonempty open set is not a meager set. The next result characterizes the Baire spaces.

Theorem 6.16. *For a metric space X the following statements are equivalent*:
1. *X is a Baire space.*
2. *Every countable intersection of open dense sets is also dense.*
3. *If $X = \bigcup_{n=1}^{\infty} F_n$ and each F_n is a closed set, then the open set $\bigcup_{n=1}^{\infty} (F_n)^{\circ}$ is dense.*

Proof. (1) $\Longrightarrow$ (2) We shall use the following property of open dense sets: *If B is an open dense set, then its complement B^c is nowhere dense.* (This follows from Lemma 6.8 by observing that $(\overline{B^c})^{\circ} = (B^c)^{c-c} = (\overline{B})^c = X^c = \emptyset$.)

Assume now that X is a Baire space and let $\{A_n\}$ be a sequence of open dense subsets of X. To show that $A = \bigcap_{n=1}^{\infty} A_n$ is dense in X, we need to show that $A \cap \mathcal{O} \neq \emptyset$ for each nonempty open set $\mathcal{O}$ of X. Suppose, if possible, that $A \cap \mathcal{O} = \emptyset$ for some nonempty open set $\mathcal{O}$. Then $X = (A \cap \mathcal{O})^c = A^c \cup \mathcal{O}^c$, and thus

$$\mathcal{O} = X \cap \mathcal{O} = A^c \cap \mathcal{O} = \left(\bigcap_{n=1}^{\infty} A_n\right)^c \cap \mathcal{O} = \bigcup_{n=1}^{\infty} (A_n^c \cap \mathcal{O}).$$

But this shows that $\mathcal{O}$ is a meager set, and thus, it is empty by hypothesis, contrary to the choice of $\mathcal{O}$. Hence, A is dense in X.

(2) $\Longrightarrow$ (3) Let $\{F_n\}$ be a sequence of closed sets satisfying $X = \bigcup_{n=1}^{\infty} F_n$ and consider the open set $\mathcal{O} = \bigcup_{n=1}^{\infty} (F_n)^{\circ}$. For each n, let $E_n = F_n \setminus (F_n)^{\circ}$ and note that E_n is a nowhere dense closed set. In particular, the set $E = \bigcup_{n=1}^{\infty} E_n$ is a meager set.

Since E_n is a closed nowhere dense set, it easily follows that each $(E_n)^c$ is an open dense set. So, by our hypothesis, $E^c = \bigcap_{n=1}^{\infty} (E_n)^c$ is also a dense set. Now notice that

$$\mathcal{O}^c = X \setminus \mathcal{O} = \bigcup_{n=1}^{\infty} F_n \setminus \bigcup_{n=1}^{\infty} (F_n)^{\circ} \subseteq \bigcup_{n=1}^{\infty} [F_n \setminus (F_n)^{\circ}] = E,$$

and so $E^c \subseteq \mathcal{O}$. Since E^c is dense, we easily conclude that $\mathcal{O}$ is also dense, as desired.

---

[15]René Louis Baire (1874–1932), a French mathematician. Among other topics, he studied extensively the properties of the pointwise limits of sequences of continuous functions.

(3) $\implies$ (1) Let V be a nonempty open set. If V is a meager set, then V can be written as a countable union $V = \bigcup_{n=1}^{\infty} A_n$, where $(\overline{A}_n)^\circ = \emptyset$ for each n. Then

$$X = V^c \cup \overline{A}_1 \cup \overline{A}_2 \cup \overline{A}_3 \cup \cdots$$

is a countable union of closed sets, and so by our hypothesis the open set

$$(V^c)^\circ \cup (\overline{A}_1)^\circ \cup (\overline{A}_2)^\circ \cup (\overline{A}_3)^\circ \cup \cdots = (V^c)^\circ$$

is dense in X. From $(V^c)^\circ \subseteq V^c$, we see that V^c is also dense in X. In particular, we have $V \cap V^c \neq \emptyset$, which is impossible. Hence, V is not a meager set and so X is a Baire space. ∎

The next result is known as Baire's category theorem, and it plays an important role in analysis.

Theorem 6.17 (Baire).   *Every complete metric space is a Baire space.*

Proof.   Let $\{A_n\}$ be a sequence of open and dense subsets of X. According to Theorem 6.16, we have to show that the set $A = \bigcap_{n=1}^{\infty} A_n$ is dense in X. Consequently, if $x \in X$ and $r > 0$ are given, we have to establish that $B(x, r) \cap A \neq \emptyset$. Let $C(a, r) = \{x \in X: d(x, a) \leq r\}$.

Since A_1 is open and dense in X, there exists $x_1 \in X$ and $0 < r_1 \leq 1$ such that $C(x_1, r_1) \subseteq B(x, r) \cap A_1$. Now, inductively, if $x_1, \ldots, x_n$ and $r_1, \ldots, r_n$ have been selected, choose $x_{n+1} \in X$ and $r_{n+1} \leq \frac{1}{n+1}$ such that $C(x_{n+1}, r_{n+1}) \subseteq B(x_n, r_n) \cap A_{n+1}$. Thus, there exists a sequence $\{x_n\}$ of X and a sequence $\{r_n\}$ of real numbers such that $0 < r_n \leq \frac{1}{n}$, and $C(x_{n+1}, r_{n+1}) \subseteq B(x_n, r_n) \cap A_{n+1}$ for each n. Now put $C_n = C(x_n, r_n)$ for each n. Then each C_n is nonempty and closed, and $C_{n+1} \subseteq C_n$ holds for each n. Moreover, $d(C_n) \leq 2r_n \leq \frac{2}{n}$ implies $\lim d(C_n) = 0$. Thus, by Theorem 6.14, there exists $y \in \bigcap_{n=1}^{\infty} C_n$. But then, it easily follows that $y \in B(x, r) \cap A$, and we are done. ∎

A special case of the preceding theorem with many useful applications is stated next.

Theorem 6.18.   *If (X, d) is a complete metric space and $X = \bigcup_{n=1}^{\infty} A_n$, then $(\overline{A}_n)^\circ \neq \emptyset$ for some n.*

A function $f: (X, d) \to (Y, \rho)$ between two metric spaces is called **uniformly continuous** if for every $\epsilon > 0$ there exists some $\delta > 0$ (depending on ϵ) such that $\rho(f(x), f(y)) < \epsilon$ whenever $d(x, y) < \delta$. Clearly, every uniformly continuous function is continuous.

If $X = (0, 1]$ and $Y = \mathbb{R}$, both with the distance $d(x, y) = |x - y|$, then the function $f: X \to Y$, where $f(x) = x^2$, is uniformly continuous, while $g: X \to Y$, with $g(x) = x^{-1}$, is continuous but not uniformly continuous.

For the proof of the next theorem the following simple result is needed:

- *In a metric space (X, d), if $\lim x_n = x$ and $\lim y_n = y$, then*

$$\lim_{n \to \infty} d(x_n, y_n) = d(x, y).$$

To see this, note first that by the triangle inequality we have $|d(x, z) - d(z, y)| \leq d(x, y)$, and then use the chain of inequalities

$$|d(x_n, y_n) - d(x, y)| \leq |d(x_n, y_n) - d(x, y_n)| + |d(x, y_n) - d(x, y)|$$
$$\leq d(x_n, x) + d(y_n, y),$$

to establish the validity of the statement.

Theorem 6.19. *Let A be a subset of a metric space (X, d) and let (Y, ρ) be a complete metric space. If $f: A \to Y$ is a uniformly continuous function, then f has a unique uniformly continuous extension to the closure $\overline{A}$ of A.*

Proof. The uniqueness of the extension should be clear. We only have to establish its existence.

Let $x \in \overline{A}$. Then by Theorem 6.9 there exists a sequence $\{x_n\}$ of A such that $\lim x_n = x$. We shall show that $\{f(x_n)\}$ is a Cauchy sequence of Y. To this end, let $\epsilon > 0$. By the uniform continuity of f, there exists $\delta > 0$ such that $\rho(f(x), f(y)) < \epsilon$ whenever $x, y \in A$ satisfy $d(x, y) < \delta$. Pick n_0 such that $d(x_n, x_m) < \delta$ for all $n, m > n_0$. Thus, $\rho(f(x_n), f(x_m)) < \epsilon$ for all $n, m > n_0$; that is, $\{f(x_n)\}$ is a Cauchy sequence. By the completeness of Y, there exists $y \in Y$ such that $\lim f(x_n) = y$.

If now $\{y_n\}$ is another sequence of A such that $\lim y_n = x$, then by the above $\{f(y_n)\}$ converges in Y. Let $\lim f(y_n) = u$. Now, for each n define $z_{2n} = x_n$ and $z_{2n-1} = y_n$. Note that $\{z_n\}$ is a sequence of A such that $\lim z_n = x$. Thus, $\lim f(z_n)$ exists in Y. In particular, we have $\lim f(z_{2n}) = \lim f(z_{2n-1})$; that is, $y = u$. Therefore, $\lim f(x_n)$ is independent of the choice of the sequence $\{x_n\}$ of A (as long as it converges to x).

Now, define $f^*: \overline{A} \to Y$ by $f^*(x) = \lim f(x_n)$, where $\{x_n\}$ is a sequence of A such that $\lim x_n = x$. Clearly, $f^*(x) = f(x)$ for all $x \in A$. To finish the proof, we show that f^* is uniformly continuous.

If $\epsilon > 0$, choose $\delta > 0$ such that $\rho(f(x), f(y)) < \epsilon$ whenever $x, y \in A$ satisfy $d(x, y) < \delta$. Now if $x, y \in \overline{A}$ satisfy $d(x, y) < \delta$, choose two sequences $\{x_n\}$ and $\{y_n\}$ of A such that $\lim x_n = x$ and $\lim y_n = y$. By the discussion before the theorem, we have $\lim d(x_n, y_n) = d(x, y)$. Pick n_0 such that $d(x_n, y_n) < \delta$ for all $n > n_0$. Then $\rho(f(x_n), f(y_n)) < \epsilon$ for $n > n_0$, and so, by the same remark $\rho(f^*(x), f^*(y)) \leq \epsilon$, which shows that f^* is uniformly continuous on $\overline{A}$. ∎

A function $f: (X, d) \to (Y, \rho)$ between two metric spaces is called an **isometry** if $\rho(f(x), f(y)) = d(x, y)$ holds for all $x, y \in X$. If in addition f is onto, then (X, d) and (Y, ρ) are called **isometric**. Note that two isometric metric spaces are necessarily homeomorphic. Also, observe that every isometry is a uniformly continuous function.

A complete metric space (Y, ρ) is called a **completion** of a metric space (X, d) if there exists an isometry $f: (X, d) \to (Y, \rho)$ such that $f(X)$ is dense in Y. If we think of X and $f(X)$ as identical, then X can be considered as a subset of Y.

Any two completions of a metric space (X, d) must be isometric. Indeed, if (Y_1, ρ_1) and (Y_2, ρ_2) are two completions of (X, d), then there exist two isometries $f: X \to Y_1$ and $g: X \to Y_2$. Thus, g is an isometry from X onto $g(X)$ and f^{-1} is an isometry from $f(X)$ onto X. Then $h = g \circ f^{-1}$ is an isometry from $f(X)$ onto $g(X) \subseteq Y_2$. Since $f(X)$ is dense in Y_1, h is uniformly continuous and Y_2 is complete, it follows from Theorem 6.19 that there exists a uniformly continuous extension h^* of h to all of Y_1. It is now straightforward to show that h^* is an isometry from Y_1 onto Y_2.

Observe that if $f: (X, d) \to (Y, \rho)$ is an isometry and (Y, ρ) is complete, then the closure $\overline{f(X)}$ is a completion of X (since, by Theorem 6.13, $\overline{f(X)}$ is a complete metric space). Next, we shall use this observation to establish that every metric space has a completion.

Theorem 6.20.   *Every metric space has a unique (up to an isometry) completion.*

Proof.   let (X, d) be a metric space. Fix an element $a \in X$. For each $x \in X$, let $f_x: X \to \mathbb{R}$ be defined by $f_x(y) = d(x, y) - d(y, a)$ for each $y \in X$. From the triangle inequality it is easy to see that $|f_x(y)| \leq d(x, a)$ for each $y \in X$, and this shows that f_x is a bounded function for each $x \in X$. That is, $f_x \in B(X)$ for each $x \in X$; see Example 6.12. We have established, therefore, a function $f: X \to B(X)$ by $x \mapsto f_x$. We claim that f is an isometry. (Recall that the distance on $B(X)$ is given by $D(f, g) = \sup\{|f(x) - g(x)|: x \in X\}$.)

Indeed, note first that if $x, z \in X$, then

$$|f_x(y) - f_z(y)| = |d(x, y) - d(y, a) - [d(z, y) - d(y, a)]|$$
$$= |d(x, y) - d(z, y)| \leq d(x, z)$$

holds for all $y \in X$. On the other hand, $|f_x(z) - f_z(z)| = d(x, z)$. Hence,

$$D(f_x, f_z) = \sup\{|f_x(y) - f_z(y)|: y \in X\} = d(x, z).$$

Since $(B(X), D)$ is a complete metric space (see Example 6.12), we see that $(\overline{f(X)}, D)$ is a completion of (X, d).

The uniqueness of the completion (up to an isometry) was established in the discussion before the theorem. The proof is now complete. ∎

EXERCISES

1. For subsets A and B of a metric space (X, d), show that:

   a. $(A \cap B)^{\circ} = A^{\circ} \cap B^{\circ}$.
   b. $A^{\circ} \cup B^{\circ} \subseteq (A \cup B)^{\circ}$.
   c. $\overline{A \cup B} = \overline{A} \cup \overline{B}$.
   d. $\overline{A \cap B} \subseteq \overline{A} \cap \overline{B}$.
   e. If B is open, then $\overline{A} \cap B \subseteq \overline{A \cap B}$.

2. Show that in a Euclidean space $\mathbb{R}^n$ with the Euclidean distance, the closure of any open ball $B(a, r)$ is the closed ball $\{x \in \mathbb{R}^n : d(x, a) \le r\}$. Give an example of a complete metric space for which the corresponding statement is false.

3. If A is a nonempty subset of $\mathbb{R}$, then show that the set

$$B = \{a \in \overline{A}: \text{There exists } \epsilon > 0 \text{ such that } (a, a + \epsilon) \cap A = \emptyset\}$$

   is at most countable.

4. Let $f: (X, d) \to (Y, \rho)$ be a function. Show that f is continuous if and only if $f^{-1}(B^{\circ}) \subseteq (f^{-1}(B))^{\circ}$ for every subset B of Y.

5. Show that the boundary of a closed or open set in a metric space is nowhere dense. Is this statement true for an arbitrary subset?

6. Show that the set of irrational numbers is not a countable union of closed subsets of $\mathbb{R}$.
   [HINT: Use Baire's theorem.]

7. Let (X, d) be a metric space. Show that if $\{x_n\}$ and $\{y_n\}$ are Cauchy sequences of X, then $\{d(x_n, y_n)\}$ converges in $\mathbb{R}$.

8. Show that in a metric space a Cauchy sequence converges if and only if it has a convergent subsequence.

9. Prove that the closed interval $[0, 1]$ is an uncountable set:

   a. by using Cantor's Theorem 6.14, and
   b. by using Theorem 6.18.

10. Let $\{r_1, r_2, \ldots\}$ be an enumeration of all rational numbers in the interval $[0, 1]$ and for each $x \in [0, 1]$ let $A_x = \{n \in \mathbb{N}: r_n \le x\}$. Define the function $f: [0, 1] \to \mathbb{R}$ by the formula

$$f(x) = \sum_{n \in A_x} \frac{1}{2^n}.$$

   Show that f restricted to the set of irrational numbers of $[0, 1]$ is continuous.

11. This exercise concerns connected metric spaces. A metric space (X, d) is said to be **connected** whenever $\emptyset$ and X are the only subsets of X that are simultaneously open and closed. A subset A of a metric space (X, d) is said to be **connected** whenever

(A, d) is itself a connected metric space. Establish the following properties regarding connected metric spaces and connected sets.

a. A metric space (X, d) is connected if and only if every continuous function $f: X \rightarrow \{0, 1\}$ is constant, where the two point set $\{0, 1\}$ is considered to be a metric space under the discrete metric.

b. If in a metric space (X, d) we have $B \subseteq A \subseteq X$, then the set B is a connected subset of (A, d) if and only if B is a connected subset of (X, d).

c. If $f: (X, d) \rightarrow (Y, \rho)$ is a continuous function and A is a connected subset of X, then $f(A)$ is a connected subset of Y.

d. If $\{A_i\}_{i \in I}$ is a family of connected subsets of a metric space such that $\bigcap_{i \in I} A_i \neq \emptyset$, then $\bigcup_{i \in I} A_i$ is likewise a connected set.

e. If A is a subset of a metric space and $a \in A$, then there exists a largest (with respect to inclusion) connected subset C_a of A that contains a. (The connected set C_a is called the **component** of a with respect to A.)

f. If a, b belong to a subset A of a metric space and C_a and C_b are the components of a and b in A, then either $C_a = C_b$ or else $C_a \cap C_b = \emptyset$. Hence, the identity $A = \bigcup_{a \in A} C_a$ shows that A can be written as a disjoint union of connected sets.

g. A nonempty subset of $\mathbb{R}$ with at least two elements is a connected set if and only if it is an interval. Use this and the conclusion of (f) to infer that every open subset of $\mathbb{R}$ can be written as an at-most countable union of disjoint open intervals.

12. Show that $\mathbb{R}^n$ with the Euclidean distance is a connected metric space. Use this conclusion to establish that, if the intersection of two open subsets of $\mathbb{R}^n$ is a proper closed set, then the two open sets must be disjoint.

13. Let C be a nonempty closed subset of $\mathbb{R}$. Show that a function $f: C \rightarrow \mathbb{R}$ is continuous if and only if it can be extended to a continuous real-valued function on $\mathbb{R}$.

14. Show that a metric space is a Baire space if and only if the complement of every meager set is dense.

15. A subset of a metric space is called **co-meager** if its complement is a meager set. For a subset A of a Baire space show that:

a. A is co-meager if and only if it contains a dense G_δ-set.

b. A is meager if and only if it is contained in an F_σ-set whose complement is dense.

7. COMPACTNESS IN METRIC SPACES

We shall discuss in this section the basic properties of compact sets. A family $\{A_i\}_{i \in I}$ of subsets of a set X is said to **cover** a subset A of X if $A \subseteq \bigcup_{i \in I} A_i$. If a subfamily of $\{A_i\}_{i \in I}$ also covers A, then it is called a **subcover**. If (X, d) is a metric space, then any cover of a set consisting of open sets will be called an **open cover** of the set.

Open covers of subsets of the Euclidean spaces $\mathbb{R}^n$ can always be reduced to countable ones, as the next classical result of E. Lindelöf[16] shows.

---

[16]Ernst Leonhard Lindelöf (1870–1940), a Fin mathematician. He contributed to function theory and he is known for his theorems on the existence of solutions to differential equations.

Theorem 7.1 (Lindelöf). *Every open cover of a subset of $\mathbb{R}^n$ can be reduced to an at-most countable subcover.*

Proof. Call a point $a = (a_1, \ldots, a_n) \in \mathbb{R}^n$ *rational* if each component a_i is a rational number. Let A be a subset of $\mathbb{R}^n$ and $\{\mathcal{O}\}_{i \in I}$ be an infinite open cover of A, i.e., $A \subseteq \bigcup_{i \in I} \mathcal{O}_i$.

Now, for each $x \in A$ choose first an index $i_x \in I$ such that $x \in \mathcal{O}_{i_x}$, and then pick a rational point $a_x \in \mathbb{R}^n$ and a rational positive number r_x such that $x \in B(a_x, r_x) \subseteq \mathcal{O}_{i_x}$. Then the collection $\{B(a_x, r_x): x \in A\}$ is an at most countable (why?) open cover of A. Since each $B(a_x, r_x)$ is a subset of some $\mathcal{O}_i$, it easily follows that there exists an at-most countable subcover of $\{\mathcal{O}_i\}_{i \in I}$ for A. ∎

And now we are ready to introduce the important notion of compactness.

Definition 7.2. *Let (X, d) be a metric space. A subset A of X is said to be* **compact** *if every open cover of A can be reduced to a finite subcover. If X is itself a compact set, then (X, d) is referred to as a* **compact metric space**.

The concept of compactness is of fundamental importance in analysis and in mathematics in general. The next result characterizes the compact sets in metric spaces and gives an indication of the usefulness of the compact sets.

Theorem 7.3. *For a subset A of a metric space (X, d) the following statements are equivalent:*

1. *A is a compact set.*
2. *Every infinite subset of A has an accumulation point in A.*
3. *Every sequence in A has a subsequence which converges to a point of A.*

Proof. $(1) \Longrightarrow (2)$ Let S be an infinite subset of the compact set A. Assume by way of contradiction that S has no accumulation point in A.

Thus, for every $x \in A$ there exists some $r_x > 0$ such that $B(x, r_x) \cap (S \setminus \{x\}) = \emptyset$. Note that $B(x, r_x) \cap S \subseteq \{x\}$. Clearly, $A \subseteq \bigcup_{x \in A} B(x, r_x)$ holds, and, in view of the compactness of A, there exist $x_1, \ldots, x_n \in A$ such that $A \subseteq \bigcup_{i=1}^{n} B(x_i, r_{x_i})$. But then,

$$ S = A \cap S \subseteq \bigcup_{i=1}^{n} \left[B(x_i, r_{x_i}) \cap S \right] \subseteq \{x_1, \ldots, x_n\}, $$

which shows that S must be a finite set, contradicting our hypothesis.

$(2) \Longrightarrow (3)$ Let $\{x_n\}$ be a sequence of A. If the sequence $\{x_n\}$ assumes only a finite number of distinct values, then there is nothing to prove—since it must have a constant subsequence.

So, assume that $\{x_n\}$ assumes an infinite number of distinct values. This implies the existence of a subsequence $\{y_n\}$ of $\{x_n\}$ such that $y_n \neq y_m$ holds whenever $n \neq m$. (To see this, set $k_1 = 1$ and proceed inductively as follows: If $k_1 < \cdots < k_n$ have been selected, then choose some $k_{n+1} > k_n$ with $x_{k_{n+1}} \neq x_{k_i}$ for each $1 \leq i \leq n$. Such an integer k_{n+1} must exist since $\{x_n\}$ assumes an infinite number of distinct values. If we let $y_n = x_{k_n}$, then the subsequence $\{y_n\}$ of $\{x_n\}$ has the desired properties.)

Now, the set $\{y_1, y_2, \ldots\}$ is an infinite subset of A, and so by our hypothesis it has an accumulation point in A, say x. Note that we can assume that $y_n \neq x$ for each n; if $y_k = x$ for some k, then replace $\{y_n\}$ by $\{y_{k+n}\}$.

Next, choose m_1 with $d(y_{m_1}, x) < 1$. Now, inductively, if $m_1 < \cdots < m_n$ have been selected, choose m_{n+1} such that

$$d\left(y_{m_{n+1}}, x\right) < \min\left\{\frac{1}{n+1}, d(y_1, x), d(y_2, x), \ldots, d\left(y_{m_n}, x\right)\right\}.$$

Clearly, $m_{n+1} > m_n$ must hold. This shows that $\{y_{m_n}\}$ is a subsequence of $\{y_n\}$, and hence, a subsequence of $\{x_n\}$. In view of $d(y_{m_n}, x) < \frac{1}{n}$, it follows that $\lim y_{m_n} = x$, as required.

    (3) $\Longrightarrow$ (1) Let $A \subseteq \bigcup_{i \in I} \mathcal{O}_i$ be an open cover of A.

Claim I: There exists some $\delta > 0$ such that for each $x \in A$ we have $B(x, \delta) \subseteq \mathcal{O}_i$ for at least one $i \in I$. (Any such number $\delta > 0$ is called a **Lebesgue number** of A for the open cover $\{\mathcal{O}_i\}_{i \in I}$.)

To see this, assume that our claim is false. This means that for each n, there exists some $x_n \in A$ such that $B(x_n, \frac{1}{n}) \cap \mathcal{O}_i^c \neq \emptyset$ holds for each $i \in I$. Let $x \in A$ be the limit of some subsequence of $\{x_n\}$. Pick some $i \in I$ with $x \in \mathcal{O}_i$, and then choose some $r > 0$ with $B(x, r) \subseteq \mathcal{O}_i$. Next, select some n so that $\frac{1}{n} < \frac{r}{2}$ and $d(x, x_n) < \frac{r}{2}$. It follows that $B(x_n, \frac{1}{n}) \subseteq B(x, r) \subseteq \mathcal{O}_i$, contrary to the selection of x_n. This establishes the validity of our claim.

Claim II.   For each $r > 0$, there exist $x_1, \ldots, x_n \in A$ such that

$$A \subseteq \bigcup_{j=1}^{n} B(x_j, r).$$

To see this, let $r > 0$, and assume that the claim is false. Fix some $x_1 \in A$, and then choose $x_2 \in A \setminus B(x_1, r)$. In general, using induction, choose $x_{n+1} \in A \setminus \bigcup_{i=1}^{n} B(x_i, r)$. Clearly, $d(x_n, x_m) \geq r$ holds for $n \neq m$. This implies that no subsequence of $\{x_n\}$ can converge, which contradicts our hypothesis, and our second claim has been established.

To finish the proof, pick a Lebesgue number $\delta > 0$ of A for $\{\mathcal{O}_i\}_{i \in I}$, and then choose $x_1, \ldots, x_n \in A$ such that $A \subseteq \bigcup_{j=1}^{n} B(x_j, \delta)$. Now, for each j pick some

$i_j \in I$ such that $B(x_j, \delta) \subseteq \mathcal{O}_{i_j}$. Thus, $A \subseteq \bigcup_{j=1}^n B(x_j, \delta) \subseteq \bigcup_{j=1}^n \mathcal{O}_{i_j}$ holds. This shows that A is a compact set, and the proof is finished. ∎

The compact subsets of the Euclidean spaces are precisely the closed and bounded subsets. This result (known as the Heine–Borel theorem[17,18]) gave rise to the present definition of compactness. The details follow.

Theorem 7.4 (Heine–Borel). *A subset of a Euclidean space is compact if and only if it is closed and bounded.*

Proof. Let A be a subset of some Euclidean space $\mathbb{R}^k$. Let us denote by d the Euclidean distance of $\mathbb{R}^k$; that is,

$$d(x, y) = \left[\sum_{i=1}^k (x_i - y_i)^2 \right]^{\frac{1}{2}}.$$

Assume that A is a compact set. We shall show first that A is a bounded set. Since $A \subseteq \bigcup_{x \in A} B(x, 1)$, there exists a finite number of points $x_1, \ldots, x_n$ of A such that $A \subseteq \bigcup_{i=1}^n B(x_i, 1)$. Let $M = \max\{d(x_i, x_j): i, j = 1, \ldots, n\}$. If $x, y \in A$, then choose i and j such that $x \in B(x_i, 1)$ and $y \in B(x_j, 1)$. Therefore,

$$d(x, y) \leq d(x, x_i) + d(x_i, x_j) + d(x_j, y) < M + 2 < \infty,$$

so that A is bounded.

Next, we show that A is closed by establishing that A contains its closure points. Let $x \in \overline{A}$. By Theorem 6.9 there exists a sequence $\{x_n\}$ of A such that $x = \lim x_n$. Now, by Theorem 7.3, $\{x_n\}$ has a a subsequence that converges to some point of A. Since every subsequence of $\{x_n\}$ must converge to x, it follows that $x \in A$. That is, $\overline{A} \subseteq A$ holds, so that A is closed.

For the converse, assume that A is a closed and bounded subset of $\mathbb{R}^k$. Pick some $M > 0$ so that $d(x, y) \leq M$ holds for all $x, y \in A$. Fix an element $y \in A$. If $a = (a_1, \ldots, a_k) \in A$, then $|a_i| \leq d(a, 0) \leq d(a, y) + d(y, 0) \leq M + d(y, 0)$ holds for each $1 \leq i \leq k$. Thus, the set of real numbers consisting of the ith coordinates of the elements of A is a bounded set.

Now, let $\{x_n\}$ be a sequence of A. Since the sequence of the first coordinates of $\{x_n\}$ is a bounded sequence of real numbers, it follows from Corollary 4.7 that there exists a subsequence $\{x_n^1\}$ of $\{x_n\}$ such that its first coordinates form a sequence

---

[17]Heinrich Eduard Heine (1821–1881), a German mathematician. He studied several classes of functions and he was the first to formulate and introduce the notion of uniform continuity.

[18]Émile Borel (1871–1958), a French mathematician. He made many contributions to analysis and probability. His 1898 classic book, *Leçons sur la Théorie des Fonctions*, laid the basis of measure theory.

that converges in $\mathbb{R}$. Now, proceed as follows: Choose a subsequence $\{x_n^2\}$ of $\{x_n^1\}$ so that its sequence of second coordinates converges in $\mathbb{R}$, and so on. After k steps, the sequence $\{x_n^k\}$ is a subsequence of $\{x_n\}$ having the property that for each $1 \le i \le k$ the sequence of its ith coordinates form a convergent subsequence. This implies that $\{x_n^k\}$ converges in $\mathbb{R}^k$, and since A is closed, it converges to some point of A. Thus, every sequence of A has a convergent subsequence in A. By Theorem 7.3, A is a compact set, and the proof is finished.   ∎

In general, the Heine–Borel theorem is not valid for arbitrary metric spaces. As in the above proof, if a subset of a metric space is compact, then the set is closed and bounded. However, a closed and bounded subset of a metric space need not be compact. Here is a simple example.

Let X be an infinite set, and let d be the discrete distance on X [that is, $d(x, y) = 1$ if $x \ne y$, and $d(x, x) = 0$]. Note that $B(x, 1) = \{x\}$ for each $x \in X$. Clearly, X is closed and bounded (with respect to this distance), and $X = \bigcup_{x \in X} B(x, 1)$. But this open cover cannot be reduced to a finite subcover. In fact, if X is uncountable, it cannot even be reduced to a countable subcover.

The next result informs us that continuous images of compact sets are also compact sets.

Theorem 7.5. *Let $f: (X, d) \to (Y, \rho)$ be a continuous function, and let A be a compact subset of X. Then $f(A)$ is a compact subset of Y.*

Proof. Let $f(A) \subseteq \bigcup_{i \in I} \mathcal{O}_i$ be an open cover. Then $A \subset \bigcup_{i \in I} f^{-1}(\mathcal{O}_i)$, and by the continuity of f each $f^{-1}(\mathcal{O}_i)$ is an open subset of X. By the compactness of A, there exist indices $i_1, \ldots, i_n$ such that $A \subseteq \bigcup_{m=1}^n f^{-1}(\mathcal{O}_{i_m})$. Thus,

$$f(A) \subseteq f\left(\bigcup_{m=1}^n f^{-1}(\mathcal{O}_{i_m}) \right) = \bigcup_{m=1}^n f(f^{-1}(\mathcal{O}_{i_m})) \subseteq \bigcup_{m=1}^n \mathcal{O}_{i_m},$$

so that $f(A)$ is compact.   ∎

From the last result and Theorem 7.4, it easily follows that if $f: (X, d) \to \mathbb{R}$ is continuous, then f attains a maximum and a minimum value on every compact subset of X.

Every closed subset of a compact metric space (X, d) is compact. Indeed, if C is closed and $\{\mathcal{O}_i\}_{i \in I}$ is an open cover of C, then $X = C^c \cup [\bigcup_{i \in I} \mathcal{O}_i]$ is an open cover of X. Since X is compact, there exist indices $i_1, \ldots, i_n$ such that $X = C^c \cup \mathcal{O}_{i_1} \cup \cdots \cup \mathcal{O}_{i_n}$. But then $C \subseteq \mathcal{O}_{i_1} \cup \cdots \cup \mathcal{O}_{i_n}$, and hence, C is compact.

A function $f: (X, d) \to (Y, \rho)$ is called an **open mapping** if $f(A)$ is open whenever A is open. Similarly, f is called a **closed mapping** if $f(A)$ is closed whenever A is closed.

Theorem 7.6. *Let* (X, d) *be a compact metric space and suppose that* $f : (X, d) \to (Y, \rho)$ *is a continuous function. Then* f *is a closed mapping.*
In particular, if f *is one-to-one and onto, then* f *is a homeomorphism.*

Proof. Let C be a closed subset of X. Then C is a compact subset of X, and by Theorem 7.5 the set $f(C)$ is a compact subset of Y. Hence, $f(C)$ is closed, and so, f is a closed mapping.

If now f is in addition one-to-one and onto, then the relation $(f^{-1})^{-1}(A) = f(A)$ holds for every subset A of X. But then, it follows from the first part and Theorem 6.11(v) that f^{-1} is also continuous, and hence, that f is a homeomorphism. ∎

A continuous function need not be uniformly continuous. However, a continuous function whose domain is compact is always uniformly continuous.

Theorem 7.7. *Let* $f : (X, d) \to (Y, \rho)$ *be a continuous function. If* (X, d) *is compact, then* f *is uniformly continuous.*

Proof. Let $\epsilon > 0$. By the continuity of f, for every x there exists $r_x > 0$ such that $\rho(f(y), f(x)) < \epsilon$ holds whenever $d(x, y) < 2r_x$. Then the collection of open balls $B(x, r_x)$ covers X, and since X is compact, there exists a finite number of points $x_1, \ldots, x_n$ in X such that $X = \bigcup_{i=1}^{n} B(x_i, r_{x_i})$. Let $\delta = \min\{r_{x_1}, \ldots, r_{x_n}\} > 0$.

Now, assume that $x, y \in X$ satisfy $d(x, y) < \delta$. There exists an integer i $(1 \le i \le n)$ such that $d(x, x_i) < r_{x_i}$. Clearly, $\rho(f(x), f(x_i)) < \epsilon$. By the triangle inequality

$$d(y, x_i) \le d(y, x) + d(x, x_i) < \delta + r_{x_i} \le 2r_{x_i}$$

holds. Therefore,

$$\rho(f(x), f(y)) \le \rho(f(x), f(x_i)) + \rho(f(x_i), f(y)) < \epsilon + \epsilon = 2\epsilon.$$

This shows that f is uniformly continuous, and we are finished. ∎

A metric space (X, d) is called **totally bounded** if for each $r > 0$ there exists a finite number of points $x_1, \ldots, x_n$ such that $X = \bigcup_{i=1}^{n} B(x_i, r)$. A compact metric space is totally bounded, but a totally bounded metric space need not be compact; take, for instance, $X = (0, 1)$ with the distance $d(x, y) = |x - y|$.
The next result shows the connection between compactness and completeness.

Theorem 7.8. *A metric space is compact if and only if it is complete and totally bounded.*

Proof. Let (X, d) be a metric space. Assume first that (X, d) is compact. Clearly, (X, d) is totally bounded. By Theorem 7.3, every sequence $\{x_n\}$ has a limit point x in X and hence, if $\{x_n\}$ is a Cauchy sequence, then $\lim x_n = x$. Thus, (X, d) is also complete.

For the converse, assume that (X, d) is complete and totally bounded. According to Theorem 7.3, we must show that every infinite subset of X has an accumulation point. To this end, let A be an infinite subset of X. Let us write (for this proof only) $C(a, r) = \{x \in X : d(a, x) \le r\}$. Since X is totally bounded, there exists a finite subset F of X such that $X = \bigcup_{x \in F} C(x, 1)$. But then, since A is an infinite set, there exists some $x_1 \in F$ such that $A \cap C(x_1, 1)$ is an infinite set. Now, by induction, if $x_1, \ldots, x_n$ have been chosen such that the set $A \cap C(x_1, 1) \cap \cdots \cap C(x_n, \frac{1}{n})$ in an infinite set, then argue as previously shown to select x_{n+1} such that the set

$$A \cap C(x_1, 1) \cap C\left(x_2, \frac{1}{2}\right) \cap \cdots \cap C\left(x_n, \frac{1}{n}\right) \cap C\left(x_{n+1}, \frac{1}{n+1}\right)$$

is infinite. Next, put $E_n = C(x_1, 1) \cap \cdots \cap C(x_n, \frac{1}{n})$ for each n; clearly, each E_n is nonempty and closed. Also, $E_{n+1} \subseteq E_n$ and $d(E_n) \le \frac{2}{n}$ hold for each n.

By Theorem 6.14, there exists $a \in X$ such that $a \in E_n$ for each n. Observe that if $y \in A \cap C(x_1, 1) \cap \cdots \cap C(x_n, \frac{1}{n})$, then $d(a, y) \le d(a, x_n) + d(x_n, y) < \frac{2}{n}$ holds, from which it follows that a is an accumulation point of A. The proof of the theorem is now complete.   ∎

EXERCISES

1. Let $f : (X, d) \to (Y, \rho)$ be a function. Show that f is continuous if and only if f restricted to the compact subsets of X is continuous.
   [HINT: If $\lim x_n = x$, then the set $\{x_n : n \in \mathbb{N}\} \cup \{x\}$ is compact.]
2. A metric space is said to be **separable** if it contains a countable subset that is dense in the space. Show that every compact space (X, d) is separable.
   [HINT: For every n choose a finite subset F_n of X such that $X = \bigcup_{x \in F_n} B(x, \frac{1}{n})$. Now show that $F = \bigcup_{n=1}^{\infty} F_n$ is dense in X.]
3. Show that if (X, d) is a separable metric space (see the preceding exercise for the definition), then card $X \le c$.
4. Let $(X_1, d_1), \ldots, (X_n, d_n)$ be metric spaces, and let $X = X_1 \times \cdots \times X_n$. If $x = (x_1, \ldots, x_n)$ and $y = (y_1, \ldots, y_n)$, define

$$D_1(x, y) = \sum_{m=1}^{n} d_m(x_m, y_m) \quad \text{and} \quad D_2(x, y) = \left(\sum_{m=1}^{n} [d_m(x_m, y_m)]^2 \right)^{\frac{1}{2}}.$$

   a. Show that D_1 and D_2 are distances on X.
   b. Show that D_1 is equivalent to D_2.

c. Show that (X, D_1) is complete if and only if each (X_i, d_i) is complete.

d. Show that (X, D_1) is compact if and only if each (X_i, d_i) is compact.

5. Let $\{(X_n, d_n)\}$ be a sequence of metric spaces, and let $X = \prod_{n=1}^{\infty} X_n$. For each $x = \{x_n\}$ and $y = \{y_n\}$ in X define

$$d(x, y) = \sum_{n=1}^{\infty} \frac{1}{2^n} \cdot \frac{d_n(x_n, y_n)}{1 + d_n(x_n, y_n)}.$$

a. Show that d is a distance on X.

b. Show that (X, d) is a complete metric space if and only if each (X_n, d_n) is complete.

c. Show that (X, d) is a compact metric space if and only if each (X_n, d_n) is compact.

6. A family of sets $\mathcal{F}$ is said to have the **finite intersection property** if every finite intersection of sets of $\mathcal{F}$ is nonempty. Show that a metric space is compact if and only if every family of closed sets with the finite intersection property has a nonempty intersection.

7. Let $f: X \to X$ be a function from a set X into itself. A point $a \in X$ is called a **fixed point** for f if $f(a) = a$. Assume that (X, d) is a compact metric space and $f: X \to X$ satisfies $d(f(x), f(y)) < d(x, y)$ for $x \neq y$. Show that f has a unique fixed point. [HINT: Show that the function $g(x) = d(x, f(x))$ attains its minimum value, that it must be zero.]

8. Let (X, d) be a metric space. A function $f: X \to X$ is called a **contraction** if there exists some $0 < \alpha < 1$ such that $d(f(x), f(y)) \leq \alpha d(x, y)$ for all $x, y \in X$; α is called a *contraction constant*. Show that every contraction f on a complete metric space (X, d) has a unique fixed point; that is, show that there exists a unique point $x \in X$ such that $f(x) = x$. [HINT: Pick $a \in X$, and define the sequence $x_1 = a$, and $x_{n+1} = f(x_n)$ for $n \geq 1$. Show that $\{x_n\}$ is a Cauchy sequence and then that its limit is a fixed point of f.]

9. A property of a metric space is called a **topological property** if it is preserved in a homeomorphic metric space.

a. Show that compactness is a topological property.

b. Show that completeness, boundedness, and total boundedness are not topological properties.

10. Let (X, d) be a metric space. Define the distance of two nonempty subsets A and B of X by

$$d(A, B) = \inf\{d(x, y): x \in A \text{ and } y \in B\}.$$

a. Give an example of two closed sets A and B of some metric space with $A \cap B = \emptyset$ and such that $d(A, B) = 0$.

b. If $A \cap B = \emptyset$, A is closed, and B is compact (and, of course, both are nonempty), then show that $d(A, B) > 0$.

11. Let (X, d) be a compact metric space and $f: X \to X$ an isometry; that is, $d(f(x), f(y)) = d(x, y)$ holds for all $x, y \in X$. Then show that f is onto. Does the conclusion remain true if X is not assumed to be compact?

12. Show that a metric space (X, d) is compact if and only if every continuous real-valued function on X attains a maximum value.

13. This exercise presents a converse of Theorem 7.7. Assume that (X, d) is a metric space such that every real-valued continuous function on X is uniformly continuous.

   a. Show that X is a complete metric space.
   b. Give an example of a noncompact metric space with the above property.
   c. If X has a finite number of isolated points (an element $a \in X$ is said to be an **isolated point** whenever there exists some $r > 0$ such that $B(a, r) \cap (X \setminus \{a\}) = \emptyset$), then show that X is a compact metric space.

14. Consider a function $f: (X, d) \to (Y, \rho)$ between two metric spaces. The graph G of f is the subset of $X \times Y$ defined by

$$G = \{(x, y) \in X \times Y: y = f(x)\}.$$

If (Y, ρ) is a compact metric space, then show that f is continuous if and only if G is a closed subset of $X \times Y$, where $X \times Y$ is considered to be a metric space under the distance $D((x, y), (u, v)) = d(x, u) + \rho(y, v)$; see Exercise 4 of this section. Does the result hold true if (Y, ρ) is not assumed to be compact?

15. A cover $\{V_i\}_{i \in I}$ of a set X is said to be a **pointwise finite cover** whenever each $x \in X$ belongs at most to a finite number of the V_i. Show that a metric space is compact if and only if every pointwise finite open cover of the space contains a finite subcover.

CHAPTER 2

TOPOLOGY AND CONTINUITY

The role of open and closed sets in metric spaces has been discussed previously. Now, the fundamental notion of an open set will be generalized by introducing the concept of a topological space. The properties of open and closed sets will be studied in this setting. This chapter is devoted entirely to topological and function spaces and emphasizes the results needed for this book. For a detailed study of general topology, the reader is referred to the classical book on the subject [18]. For a modern detailed treatment of general topology, you might consult the monograph [22].

The material has been arranged into four sections. The first section discusses the theory of topological spaces. The second section deals with the properties of continuous real-valued functions, and introduces the concepts of vector lattices and function spaces. The notions of pointwise and uniform convergence of sequences are discussed, and their relations to one another are investigated. In the third section of this chapter, we investigate extension and separation properties of real-valued continuous functions. Finally, the fourth section culminates the discussion with a detailed presentation of the classical Stone–Weierstrass approximation theorem.

8. TOPOLOGICAL SPACES

In the previous sections, the fundamental properties of metric spaces were discussed. The open and closed sets played a basic role in that study. In this section the concepts used in a metric space will be generalized by introducing the notion of a topological space. Properties such as closeness, convergence, and continuity will be studied in this setting.

The starting point is the definition of open sets.

Definition 8.1. *Let X be a nonempty set. A collection τ of subsets of X is said to be a* **topology** *on X if τ satisfies the following properties*:

1. $X \in \tau$ *and* $\emptyset \in \tau$.
2. *If U and V belong to τ, then* $U \cap V \in \tau$.
3. *If $\{V_i\}_{i \in I}$ is a family of members of τ, then* $\bigcup_{i \in I} V_i \in \tau$.

If τ is a topology on a set X, then the pair (X, τ) is called a **topological space**. If there is no ambiguity about the topology τ, sometimes for simplicity we shall write X instead of (X, τ). The members of τ are called the **open sets** of X.

Some examples of topological spaces are presented next. It is expected that the reader will be able to verify that the exhibited collections are indeed topologies.

Example 8.2. Let X be a nonempty set. Then $\tau = \{\emptyset, X\}$ is a topology on X, called the **indiscrete topology**. This topology is the smallest possible (with respect to inclusion) topology on X. ∎

Example 8.3. Let X be a nonempty set. Then $\tau = \mathcal{P}(X)$ is a topology on X. Here every subset of X is an open set. This topology is called the **discrete topology**, and it is the "largest" possible topology on X. ∎

Example 8.4. Let (X, d) be a metric space, with the set X uncountable. Let τ denote the collection of all subsets $\mathcal{O}$ of X such that for each $x \in \mathcal{O}$ there exist $r > 0$ and an at-most countable subset A of X (both depending on x) such that $x \notin A$ and $B(x, r) \setminus A \subseteq \mathcal{O}$. Then τ is a topology on X. ∎

Example 8.5. Let (X, d) be a metric space. Then the collection of all open subsets of X satisfies the properties for a topology; see Theorem 65. Thus, every metric space is a topological space. ∎

Example 8.6. Let $X = \mathbf{R}^*$ be the set of extended real numbers. Let τ be the collection of subsets of X defined as follows. A subset $\mathcal{O}$ of X belongs to τ if it satisfies these properties:

1. For each $x \in \mathcal{O} \cap \mathbf{R}$ there exists $r > 0$ (depending on x) such that $(x - r, x + r) \subseteq \mathcal{O}$.
2. If $\infty \in \mathcal{O}$, then there exists $a \in \mathbf{R}$ such that $(a, \infty] \subseteq \mathcal{O}$.
3. If $-\infty \in \mathcal{O}$, then there exists $a \in \mathbf{R}$ such that $[-\infty, a) \subseteq \mathcal{O}$.

Then τ is a topology on $\mathbf{R}^*$. ∎

Example 8.7. Let (X, τ) be a topological space, and Y a subset of X. Define a collection of subsets of Y by

$$\tau_Y = \{V \cap Y : V \in \tau\}.$$

Then (Y, τ_Y) is a topological space. The topology τ_Y is called the topology **induced** by τ on Y, or the **relative topology** of τ on Y. ∎

Unless otherwise stated, for our discussion here, we shall assume that (X, τ) is a fixed topological space. The closed sets are now defined as in the case of metric spaces. A subset A of X is said to be **closed** if its complement A^c is open (i.e., $A^c \in \tau$). By taking complements, the following properties of closed sets follow directly from Definition 8.1:

1. $\emptyset$ and X are closed sets (as well as open sets).
2. Finite unions of closed sets are closed sets.
3. Arbitrary intersections of closed sets are closed sets.

The balls of a metric space are now replaced in this setting by neighborhoods. A **neighborhood** of a point x is any open set containing x. Thus, an open set is a neighborhood for all of its points. In particular, note that a subset A of X is open if and only if every point of $x \in A$ has a neighborhood V_x such that $V_x \subseteq A$. (Indeed, if A has this property, then $A = \bigcup_{x \in A} V_x$ is an open set, since it is a union of open sets.)

The **interior** A° of a subset A of X is defined by

$$A^\circ = \{x \in X : \exists \text{ a neighborhood } V \text{ of } x \text{ such that } V \subseteq A\}.$$

The members of A° are called the **interior points** of A. Clearly, A° is an open set, and $A^\circ \subseteq A$. Moreover, A° is the largest open set that is included in A. It should be also clear that A is open if and only if $A = A^\circ$.

The closure and accumulation points of a set are defined as in metric spaces. A point $x \in X$ is said to be a **closure point** for a set A if every neighborhood V of x contains (at least) one point from A (that is, if $V \cap A \neq \emptyset$ for every neighborhood V of x). The set of all closure points of A is denoted by $\overline{A}$ and is called the **closure** of A. Clearly, $A \subseteq \overline{A}$. As in the proof of Theorem 6.7, it can be shown that $\overline{A}$ is the smallest (with respect to inclusion) closed set that contains A. It follows from this that a set A is closed if and only if $A = \overline{A}$. Some other useful properties that the closure operator satisfies are the following:

a. If $A \subseteq B$, then $\overline{A} \subseteq \overline{B}$.
b. $\overline{A \cup B} = \overline{A} \cup \overline{B}$ for all subsets A and B.
c. $A^\circ = (\overline{A^c})^c$ for every subset A.

A subset A of X is said to be **dense** in X if $\overline{A} = X$.

A point $x \in X$ is said to be an **accumulation point** of a subset A if every neighborhood of x contains a point of A different from x; that is, if $(V \setminus \{x\}) \cap A \neq \emptyset$ holds for each neighborhood V of x. The set of all accumulation points of A is denoted by A' and is called the **derived set** of A. Clearly, $A' \subseteq \overline{A}$. Also, it easily follows that $\overline{A} = A \cup A'$ and from this that a set A is closed if and only if $A' \subseteq A$.

A point $x \in X$ is said to be a **boundary point** of a set A if $V \cap A \neq \emptyset$ and $V \cap A^c \neq \emptyset$ hold for every neighborhood V of x. The set of all boundary point of A is denoted by ∂A and is called the **boundary** of A. Observe that

$$\partial A = \overline{A} \cap \overline{A^c},$$

and hence, ∂A is always a closed set. Also, it is easily seen that $\partial A = \partial A^c$ holds for every subset A of X. Moreover, $\overline{A} = A \cup \partial A$ also holds.

A subset A of X is said to be **nowhere dense** if $(\overline{A})^\circ = \emptyset$. A set A is called a **meager set** (or a set of **first category**) if there exists a sequence $\{A_n\}$ of nowhere dense sets such that $A = \bigcup_{n=1}^{\infty} A_n$. Observe that any subset of a meager set is itself necessarily a meager set.

A sequence $\{x_n\}$ in a topological space (X, τ) is said to **converge** to x (denoted by $\lim x_n = x$) if for every neighborhood V of x there exists an integer k (depending on V) such that $x_n \in V$ for all $n \geq k$. Our first observation is that, in contrast with a metric space, a sequence can have more than one limit. For instance, every sequence in the topological space of Example 8.2 converges to every point of X.

However, in a Hausdorff[1] topological space, every sequence has at most one limit. A topological space (X, τ) is said to be a **Hausdorff space** if for every pair $x, y \in X$ with $x \neq y$ there exist neighborhoods V of x and U of y such that $V \cap U = \emptyset$. That is, a Hausdorff space is a topological space in which any two distinct points can be separated by disjoint neighborhoods. Unless otherwise stated, all topological spaces encountered in this book will be Hausdorff.

Our attention now turns to continuous functions. A function $f: (X, \tau) \to (Y, \tau_1)$ between two topological spaces is said to be **continuous at a point** $a \in X$ if for every neighborhood V of $f(a)$ there exists a neighborhood W of a such that $f(x) \in V$ whenever $x \in W$. If f is continuous at every point of X, then f is called a **continuous function**.

The next result characterizes the continuous functions and is the parallel of Theorem 6.11.

Theorem 8.8.   *For a function $f: (X, \tau) \to (Y, \tau_1)$ between two topological spaces, the following statements are equivalent*:

1. *f is a continuous function.*
2. *$f^{-1}(\mathcal{O})$ is open whenever $\mathcal{O}$ is an open subset of Y.*
3. *$f(\overline{A}) \subseteq \overline{f(A)}$ holds for every subset A of X.*
4. *$f^{-1}(C)$ is a closed subset of X whenever C is a closed subset of Y.*

Proof. $(1) \Longrightarrow (2)$ Let $\mathcal{O}$ be an open subset of Y, and let $a \in f^{-1}(\mathcal{O})$. Then there exists a neighborhood V of a such that $f(x) \in \mathcal{O}$ for all $x \in V$. Thus, $V \subseteq f^{-1}(\mathcal{O})$, and so a is an interior point of $f^{-1}(\mathcal{O})$. Since $a \in f^{-1}(\mathcal{O})$ is arbitrary, it follows that $f^{-1}(\mathcal{O})$ is open.

$(2) \Longrightarrow (3)$ Let $A \subseteq X$ and put $B = \overline{f(A)}$. Since B is closed, B^c is open, and thus, by our hypothesis $f^{-1}(B^c) = [f^{-1}(B)]^c$ is also open. This implies that the set $f^{-1}(B)$ is closed. Now, by virtue of $A \subseteq f^{-1}(B)$ we obtain $\overline{A} \subseteq f^{-1}(B)$. Therefore, $f(\overline{A}) \subseteq B = \overline{f(A)}$, as desired.

$(3) \Longrightarrow (4)$ Let C be a closed subset of Y. Put $A = f^{-1}(C)$. Then, $f(\overline{A}) \subseteq \overline{f(A)} \subseteq \overline{C} = C$ holds, which shows that $\overline{A} \subseteq f^{-1}(C) = A$. Therefore, $A = \overline{A}$, and so $A = f^{-1}(C)$ is a closed set.

$(4) \Longrightarrow (1)$ Let $a \in X$, and let V be a neighborhood of $f(a)$. Since V^c is closed, $f^{-1}(V^c) = [f^{-1}(V)]^c$ is also closed by our assumption, and so $W = f^{-1}(V)$ is an

---

[1]Felix Hausdorff (1868–1942), a German mathematician. His main work was in topology and set theory. He is the founder of general topology and the theory of metric spaces.

open set. Observe now that $a \in W$, and thus, W is a neighborhood of a. Clearly, $x \in W$ implies $f(x) \in V$, so that f is continuous at a. Since a is arbitrary, f is a continuous function. ∎

A countable union of closed sets is not necessarily a closed set, and a countable intersection of open sets need not be an open set. Such sets are nevertheless of importance. A set is called an F_σ-**set** if it is the union of countably many closed sets. Similarly, a set is said to be a G_δ-**set** if it is the intersection of countably many open sets.

The next theorem tells us that the prior classes of sets are in a dual relation.

Theorem 8.9. *A set is an F_σ-set if and only if its complement is a G_δ-set. Similarly, a set is a G_δ-set if and only if its complement is an F_σ-set.*

Proof. Let A be an F_σ-set, so that $A = \bigcup_{n=1}^{\infty} A_n$, where each A_n is closed. Then $A^c = \bigcap_{n=1}^{\infty} A_n^c$ is a countable intersection of open sets, and thus, is a G_δ-set. The proof that the complement of a G_δ-set is an F_σ-set is similar. ∎

The union of a countable collection of F_σ-sets is again an F_σ-set and the countable intersection of G_δ-sets is a G_δ-set. Also, it can be seen easily that a finite union or intersection of F_σ-sets (resp. G_δ-sets) is again an F_σ-set (resp. a G_δ-set).

Now, consider a real-valued function f defined on a topological space (X, τ); that is, $f: X \to \mathbb{R}$. It is possible for f to be discontinuous everywhere. For instance, $f: \mathbb{R} \to \mathbb{R}$ with the value 1 on each irrational and 0 on each rational is discontinuous at every point. It is instructive to examine the set of point where a function is continuous, or the set of points where a function is discontinuous. To do this, we need some preliminary discussion.

Let us denote by $\mathcal{N}_x$ the collection of all neighborhoods of the point x. The **oscillation** $\omega_f(x)$ of f at the point x is the non-negative extended real number defined by

$$\omega_f(x) = \inf_{V \in \mathcal{N}_x} \left\{ \sup_{z,y \in V} |f(z) - f(y)| \right\}.$$

A straightforward verification shows that f is continuous at the point x if and only if $\omega_f(x) = 0$. Rephrasing this last statement, one can see that f is discontinuous at the point x if and only if $\omega_f(x) > 0$. Thus, if $D_n = \{x \in X: \omega_f(x) \geq \frac{1}{n}\}$ and D denotes the set of all points of discontinuity of f, then $D = \bigcup_{n=1}^{\infty} D_n$ holds.

Theorem 8.10. *Let (X, τ) be a topological space, and let $f: X \to \mathbb{R}$. Then the set D of all points of discontinuity of f is an F_σ-set. In particular, the set of points of continuity of f is a G_δ-set.*

Proof. According to the discussion preceding the theorem, we have $D = \bigcup_{n=1}^{\infty} D_n$, where $D_n = \{x \in X : \omega_f(x) \geq \frac{1}{n}\}$. It suffices to show that each D_n is a closed set.

To this end, let $x \notin D_n$. Then $\omega_f(x) < \frac{1}{n}$. It follows from the definition of $\omega_f(x)$ that there exists a neighborhood V of x such that $\sup_{z,y \in V} |f(z) - f(y)| < \frac{1}{n}$. Since V is a neighborhood for each of its members, we easily get that

$$\omega_f(a) \leq \sup_{z,y \in V} |f(z) - f(y)| < \frac{1}{n}$$

for each $a \in V$. Thus, $V \subseteq D_n^c$, which shows that x is an interior point of D_n^c. Since x is arbitrary, D_n^c is open, and hence, D_n is closed. The proof of the theorem is now complete. ∎

We continue our discussion with the introduction of compact sets. Their definition is an abstract version of the one given for metric spaces.

Definition 8.11. *A subset A of a topological space (X, τ) is said to be* **compact** *if every open cover of A can be reduced to a finite subcover.*

In particular, if X itself is a compact set, then (X, τ) is called a **compact topological space.**

It should be obvious from the preceding definition that every finite subset of a topological space is automatically compact. Also, it should be clear that a finite union of compact sets must be a compact set. More properties of the compact sets are included in the next result.

Theorem 8.12. *For a Hausdorff topological space (X, τ) the following statements hold:*

1. *Every compact subset of X is closed.*
2. *If B is a closed subset of a compact set A, then B is compact.*

Proof. (1) Let A be a compact subset of X. We must show that A^c is open. To this end, let $x \in A^c$. Then for each $y \in A$ there exists a neighborhood V_y of y and a neighborhood U_y of x such that $V_y \cap U_y = \emptyset$. Clearly, $A \subseteq \bigcup_{y \in A} V_y$. Since A is compact, there exist $y_1, \ldots, y_n \in A$ such that $A \subseteq \bigcup_{m=1}^{n} V_{y_m}$. Let $\mathcal{O} = \bigcap_{m=1}^{n} U_{y_m}$. Then $\mathcal{O}$ is a neighborhood of x such that $\mathcal{O} \cap A = \emptyset$. Hence, $\mathcal{O} \subseteq A^c$, and so x is an interior point of A^c. Thus, A^c is open, which means that A is closed.

(2) Let $\{\mathcal{O}_i\}_{i \in I}$ be an open cover of B. Then the family of sets $\{B^c\} \cup \{\mathcal{O}_i : i \in I\}$ is an open cover for A. Choose indices $i_1, \ldots, i_n$ such that $A \subseteq B^c \cup \mathcal{O}_{i_1} \cup \cdots \cup \mathcal{O}_{i_n}$. It follows that $B \subseteq \mathcal{O}_{i_1} \cup \cdots \cup \mathcal{O}_{i_n}$ holds, which shows that B is a compact set. (Notice that the proof of part (2) does not require X to be Hausdorff.) ∎

The arguments of the proof of part (1) of the preceding theorem yield also the following separation result.

Theorem 8.13.   *Suppose that A is a compact subset of a Hausdorff topological space and $x \notin A$. Then there exist open sets V and W such that $x \in V, A \subseteq W$, and $V \cap W = \emptyset$ (and hence, also $x \notin \overline{W}$).*

Continuous functions map compact sets to compact sets. The details follow.

Theorem 8.14.   *If $f:(X, \tau) \to (Y, \tau_1)$ is a continuous function and A is a compact subset of X, then $f(A)$ is a compact subset of Y.*

   In particular, every continuous real-valued function on a topological space (X, τ) will attain a maximum and a minimum value on every compact subset of X.

Proof.   For the proof of the first part, repeat the proof of Theorem 7.5. For the second part, let A be a compact subset of X, and let $f: X \to \mathbb{R}$ be continuous. By the preceding, $f(A)$ is a compact subset of $\mathbb{R}$, and hence, by Theorem 7.4, $f(A)$ is closed and bounded. Thus, if $a = \sup\{f(x): x \in A\}$ and $b = \inf\{f(x): x \in A\}$, then since $a, b \in f(A)$, there exist two points $x, y \in A$ such that $a = f(x)$ and $b = f(y)$. The proof of the theorem is now complete.   ∎

Two topological spaces (X, τ) and (Y, τ_1) are called **homeomorphic** if there exists a one-to-one onto function $f:(X, \tau) \to (Y, \tau_1)$ such that f and f^{-1} are both continuous. Any such function is called a **homeomorphism** between (X, τ) and (Y, τ_1).

Theorem 8.15.   *A one-to-one continuous function from a compact topological space onto a Hausdorff topological space is a homeomorphism.*

Proof.   Let $f:(X, \tau) \to (Y, \tau_1)$ be a one-to-one surjective continuous function, where (X, τ) is compact and (Y, τ_1) is Hausdorff. Assume that C is a closed subset of X. By Theorem 8.12(2), C is a compact subset of X, and a glance at Theorem 8.14 guarantees that $f(C)$ is a compact subset of Y. Now, Theorem 8.12(1) implies that $(f^{-1})^{-1}(C) = f(C)$ is a closed set, and this (in view of Theorem 8.8(4)) shows that f^{-1} is also continuous.   ∎

EXERCISES

1.   For any subset A of a topological space show the following:
   a. $A^\circ = (\overline{A^c})^c$.
   b. $\partial A = \overline{A} \setminus A^\circ$.
   c. $(A \setminus A^\circ)^\circ = \emptyset$.

2. If A and B are two subsets of a topological space, then show the following:

   a. $\overline{A \cup B} = \overline{A} \cup \overline{B}$.
   b. $(A \cup B)' = A' \cup B'$.

3. If A is an arbitrary subset of a Hausdorff topological space, then show that its derived set A' is a closed set.

4. Let $X = \mathbb{R}$, and let τ be the topology on X defined in Example 8. In other words, $A \in \tau$ if and only if for each $x \in A$ there exist $\epsilon > 0$ and an at-most countable set B (both depending on x) such that $(x - \epsilon, x + \epsilon) \setminus B \subseteq A$.

   a. Show that τ is a topology on X.
   b. Verify that $0 \in \overline{(0, 1)}$.
   c. Show that there is no sequence $\{x_n\}$ of $(0, 1)$ with $\lim x_n = 0$.

5. If A is a dense subset of a topological space, then show that $\mathcal{O} \subseteq \overline{A \cap \mathcal{O}}$ holds for every open set $\mathcal{O}$. Generalize this conclusion as follows: If A is open, then $A \cap \overline{B} \subseteq \overline{A \cap B}$ for each set B.

6. If $\{\mathcal{O}_i\}_{i \in I}$ is an open cover for a topological space X, then show that a subset A of X is closed if and only if $A \cap \mathcal{O}_i$ is closed in $\mathcal{O}_i$ for each $i \in I$ (where $\mathcal{O}_i$ is considered equipped with the relative topology).

7. If (X, τ) is a Hausdorff topological space, then show the following:

   a. Every finite subset of X is closed.
   b. Every sequence of X converges to at-most one point.

8. For a function $f: (X, \tau) \to (Y, \tau_1)$ show the following:

   a. If τ is the discrete topology, then f is continuous.
   b. If τ is the indiscrete topology and τ_1 is a Hausdorff topology, then f is continuous if and only if f is a constant function.

9. Let f and g be two continuous functions from (X, τ) into a Hausdorff topological space (Y, τ_1). Assume that there exists a dense subset A of X such that $f(x) = g(x)$ for all $x \in A$. Show that $f(x) = g(x)$ holds for all $x \in X$.

10. Let $f: (X, \tau) \to (Y, \tau_1)$ be a function. Show that f is continuous if and only if $f^{-1}(B^\circ) \subseteq [f^{-1}(B)]^\circ$ holds for every subset B of Y.

11. If $f: (X, \tau) \to (Y, \tau_1)$ and $g: (Y, \tau_1) \to (Z, \tau_2)$ are continuous functions, show that their composition $g \circ f: (X, \tau) \to (Z, \tau_2)$ is also continuous.

12. Show that a function $f: X \to \mathbb{R}$, where X is a topological space, is continuous at some $a \in X$ if and only if its oscillation at a is zero, i.e., $\omega_f(a) = 0$.

13. Show that a finite union of nowhere dense sets is again a nowhere dense set. Is this statement true for a countable union of nowhere dense sets?

14. Show that the boundary of an open or closed set is nowhere dense.

15. Let $f: (X, \tau) \to \mathbb{R}$, and let D be the set of all points of X where f is discontinuous. If D^c is dense in X, then show that D is a meager set.

16. Show that there is no function $f: \mathbb{R} \to \mathbb{R}$ having the irrational numbers as the set of its discontinuities.
   [HINT: Use Exercise 6 of Section 6 and Theorem 8.10.]

17. Show that every closed subset of a metric space is a G_δ-set and every open set is an F_σ-set.

18. Let $\mathcal{B}$ be a collection of open sets in a topological space (X, τ). If for each x in an arbitrary open set V there exists some $B \in \mathcal{B}$ with $x \in B \subseteq V$, then $\mathcal{B}$ is called a **base** for τ. In general, a collection $\mathcal{B}$ of subsets of a nonempty set X is said to be a **base** if

    i. $\bigcup_{B \in \mathcal{B}} B = X$, and
    ii. for every pair $A, B \in \mathcal{B}$ and $x \in A \cap B$, there exists some $C \in \mathcal{B}$ with $x \in C \subseteq A \cap B$.

    Show that if $\mathcal{B}$ is a base for a set X, then the collection

    $$\tau = \{V \subseteq X : \forall\, x \in V \text{ there exists } B \in \mathcal{B} \text{ with } x \in B \subseteq V \}$$

    is a topology on X having $\mathcal{B}$ as a base.

19. Let (X, τ) be a topological space, and let $\mathcal{B}$ be a base for the topology τ (see the preceding exercise for the definition). Show that there exists a dense subset A of X such that card $A \le$ card $\mathcal{B}$.

20. Let $f: X \to Y$ be a function. If τ is a topology on X, then the **quotient topology** τ_f determined by f on Y is defined by $\tau_f = \{\mathcal{O} \subseteq Y : f^{-1}(\mathcal{O}) \in \tau\}$.

    a. Show that τ_f is indeed a topology on Y and that $f: (X, \tau) \to (Y, \tau_f)$ is continuous.
    b. If $g: (Y, \tau_f) \to (Z, \tau_1)$ is a function, then show that the composition $g \circ f: (X, \tau) \to (Z, \tau_1)$ is continuous if and only if g is continuous.
    c. Assume that $f: X \to Y$ is onto and that τ^* is a topology on Y such that $f: (X, \tau) \to (Y, \tau^*)$ is an open mapping (i.e., it carries open sets of X onto open sets of Y) and continuous. Show that $\tau^* = \tau_f$.

21. This exercise presents an example of a compact set whose closure is not compact. Start by considering the interval $[0, 1]$ with the topology τ generated by the metric $d(x, y) = |x - y|$. It should be clear that $([0, 1], \tau)$ is a compact topological space. Next put $X = [0, 1] \cup \mathbb{N} = [0, 1] \cup \{2, 3, 4, \ldots\}$, and define

    $$\tau^* = \tau \cup \{ [0, 1] \cup A : A \subseteq \mathbb{N} \}.$$

    a. Show that τ^* is a non-Hausdorff topology on X and that τ^* induces τ on $[0, 1]$.
    b. Show that (X, τ^*) is not a compact topological space.
    c. Show that $[0, 1]$ is a compact subset of (X, τ^*).
    d. Show that $[0, 1]$ is dense in X (and hence, its closure is not compact).
    e. Why doesn't this contradict Theorem 8.12(1)?

22. A topological space (X, τ) is said to be **connected** if a subset of X that is simultaneously closed and open (called a **clopen set**) is either empty or else equal to X.

    a. Show that (X, τ) is connected if and only if the only continuous functions from (X, τ) into $\{0, 1\}$ (with the discrete topology) are the constant ones.
    b. Let $f: (X, \tau) \to (Y, \tau^*)$ be onto and continuous. If (X, τ) is connected, then show that (Y, τ^*) is also connected.

9.　CONTINUOUS REAL-VALUED FUNCTIONS

The properties of topological spaces are the tools for the study of continuous real-valued functions. In this section many important properties of continuous functions will be discussed, and function spaces will be considered. Also, the behavior of the limit of a sequence of continuous functions will be discussed in some detail.

Let X be a topological space. The collection of all continuous real-valued functions on X will be denoted by $C(X)$. The set $C(X)$ is closed under addition and scalar multiplication. That is, if f and g are members of $C(X)$, then the functions $f + g$ and αf defined by

$$(f + g)(x) = f(x) + g(x) \quad \text{and} \quad (\alpha f)(x) = \alpha f(x)$$

for each $x \in X$ and $\alpha \in \mathbb{R}$ belong to $C(X)$. Now it should be clear that $C(X)$ is a vector space.

Also, $C(X)$ has a natural partial ordering defined by $f \geq g$ whenever $f(x) \geq g(x)$ holds for all $x \in X$. Moreover, $C(X)$ is also a *vector lattice*. This means that for every pair $f, g \in C(X)$ the least upper bound $f \vee g$, as well as the greatest lower bound $f \wedge g$, both exist in $C(X)$. They are given by the formulas:

$$(f \vee g)(x) = \max\{f(x), g(x)\} \quad \text{and} \quad (f \wedge g)(x) = \min\{f(x), g(x)\}$$

for each $x \in X$. The absolute value $|f|$ of a function $f \in C(X)$ is defined by $|f| = f \vee (-f)$. That is, $|f|(x) = |f(x)|$ holds for each $x \in X$; clearly, $|f|$ belongs to $C(X)$. Note also that $f \vee g$ and $f \wedge g$ satisfy the identities:

$$f \vee g = \tfrac{1}{2}(f + g + |f - g|) \quad \text{and} \quad f \wedge g = \tfrac{1}{2}(f + g - |f - g|).$$

Observe that the above identities show that $f \vee g$ and $f \wedge g$ are continuous functions because they can be expressed as sums of continuous functions. Since most of the spaces of functions that one encounters are vector lattices, it is appropriate to stop and consider them more closely.

Recall that a relation $\geq$ on a nonempty set X is called an order relation if it satisfies the following properties:

1. $u \geq u$ for all $u \in X$ (*reflexivity*).
2. If $u \geq v$ and $v \geq u$, then $u = v$ (*antisymmetry*).
3. If $u \geq v$ and $v \geq w$, then $u \geq w$ (*transitivity*).

The symbolism $v \leq u$ is an alternate notation for $u \geq v$. Also, $u > v$ (or $v < u$) means $u \geq v$ and $u \neq v$.

An **ordered vector space** is a real vector space E equipped with an order relation satisfying the following two conditions:

4. If $u \geq v$, then $u + w \geq v + w$ for all $w \in E$.
5. If $u \geq v$, then $\alpha u \geq \alpha v$ for all $\alpha \geq 0$.

A vector u in E is called **positive** if $u \geq 0$ holds. The set of all positive vectors is denoted by E^+.

A **vector lattice** E is an ordered vector space with the additional property that for every two vectors $u, v \in E$, the supremum $u \vee v$ and the infimum $u \wedge v$ exist in E. We remind the reader that two vectors $u, v \in E$ have a supremum w in E if $w \geq u$ and $w \geq v$ hold, and whenever z is an upper bound of $\{u, v\}$, then $z \geq w$ holds. Clearly, $w = u \vee v$ is uniquely determined. In other words, $u \vee v$ is the smallest upper bound of the set $\{u, v\}$. The definition of $u \wedge v$ is similar.

If u, v, and w are vectors in a vector lattice E, then the following identities hold:

a. $u \vee v = -[(-u) \wedge (-v)]$;
b. $u \vee v + w = (u + w) \vee (v + w)$;
c. $u \wedge v + w = (u + w) \wedge (v + w)$;
d. $\alpha(u \vee v) = (\alpha u) \vee (\alpha v)$ for each $\alpha \geq 0$.

To indicate how one proves identities in a vector lattice, we shall establish (b). Put $f = u \vee v + w$ and $g = (u + w) \vee (v + w)$. It suffices to show that $f \geq g$ and $g \geq f$ both hold.

Note first that $f = u \vee v + w$ implies $f - w = u \vee v$, and so, $u \leq f - w$ and $v \leq f - w$. Thus, $u + w \leq f$ and $v + w \leq f$ hold, so that $f \geq (u + w) \vee (v + w) = g$. On the other hand, $g = (u + w) \vee (v + w)$ implies $u + w \leq g$ and $v + w \leq g$. Hence, $u \leq g - w$ and $v \leq g - w$, from which it follows that $u \vee v \leq g - w$. Therefore, $f = u \vee v + w \leq g$ also holds.

If E is a vector lattice and $u \in E$, then we define

$$u^+ = u \vee 0, \quad u^- = (-u) \vee 0, \quad \text{and} \quad |u| = u \vee (-u).$$

The element u^+ is called the **positive part**, u^- the **negative part**, and $|u|$ the **absolute value** of u.

Theorem 9.1. *If u is a vector in a vector lattice, then the following identities hold*:

1. $u = u^+ - u^-$,
2. $|u| = u^+ + u^-$, *and*
3. $u^+ \wedge u^- = 0$.

Proof. (1) Applying identity (b) above, we get

$$u^- + u = (-u) \vee 0 + u = 0 \vee u = u^+,$$

from which $u = u^+ - u^-$ follows.

(2) By using (b) and (d) we obtain

$$|u| = u \vee (-u) = (2u) \vee 0 - u = 2(u \vee 0) - u = 2u^+ - u$$
$$= 2u^+ - (u^+ - u^-) = u^+ + u^-.$$

(3) Using (c) and (a), we get

$$u^+ \wedge u^- = (u^+ - u^-) \wedge 0 + u^- = u \wedge 0 + u^-$$
$$= -[(-u) \vee 0] + u^- = -u^- + u^- = 0,$$

and the proof is finished.   ∎

Typical examples of vector lattices are provided by function spaces. A **function space** L is a vector space of real-valued functions defined on some nonempty set X such that the functions $f \vee g$ and $f \wedge g$ belong to L for every pair $f, g \in L$, where

$$(f \vee g)(x) = \max\{f(x), g(x)\} \quad \text{and} \quad (f \wedge g)(x) = \min\{f(x), g(x)\},$$

hold for each $x \in X$.

Note that for every function f in a function space L, the elements f^+, f^-, and $|f|$ of L satisfy

$$f^+(x) = \max\{f(x), 0\}, \quad f^-(x) = \max\{-f(x), 0\}, \quad \text{and} \quad |f|(x) = |f(x)|$$

for each $x \in X$.

Here are some example of function spaces.

1.   The vector space $\mathbb{R}^X$ of all real-valued functions defined on a set X.
2.   The vector space $B(X)$ of all bounded real-valued functions defined on X.
3.   The vector space $C(X)$ of all continuous real-valued functions on X (provided, of course, that X is a topological space).
4.   The vector space $C_b(X)$ of all bounded continuous real-valued functions on a topological space X.

Consider a sequence $\{f_n\}$ of real-valued functions defined on a set X such that $\lim f_n(x)$ exists in $\mathbb{R}$ for each $x \in X$. Then a new function f can be defined by $f(x) = \lim f_n(x)$ for each $x \in X$. If this happens, then the sequence $\{f_n\}$ is said to **converge pointwise** to f (or that f is the **pointwise limit** of $\{f_n\}$) and is written symbolically as $f_n \to f$. In other words, $f_n \to f$ if for each $\epsilon > 0$ and each $x \in X$ there exists some n_0 (depending upon both ϵ and x) such that $|f_n(x) - f(x)| < \epsilon$ for all $n \geq n_0$.

A stronger concept of convergence of a sequence of real-valued functions is that of uniform convergence. A sequence $\{f_n\}$ of real-valued functions is said to **converge uniformly** on X to a function f if for each $\epsilon > 0$ there exists some n_0 (depending only upon ϵ) such that $|f_n(x) - f(x)| < \epsilon$ for all $n \geq n_0$ and all $x \in X$. It should be clear that uniform convergence implies pointwise convergence.

We are interested in determining what properties are possessed by a function that is the "limit" of a sequence of continuous functions. Note first that the pointwise limit of a sequence of continuous functions need not be a continuous function. For an example take $X = [0, 1]$, and let $\{f_n\}$ be the sequence of functions defined by $f_n(x) = x^n$ for each $x \in [0, 1]$. Then each f_n is a continuous function, and $f_n \to f$ holds for the function f defined by $f(x) = 0$ if $x \in [0, 1)$ and $f(1) = 1$. Clearly, f is not continuous. Also, it is easy to see that the convergence is not uniform.

The uniform limit of a sequence of continuous functions is always a continuous function. The details follow.

Theorem 9.2. *Let X be a topological space and let $\{f_n\}$ be a sequence of $C(X)$. If $\{f_n\}$ converges uniformly to f on X, then f is a continuous function.*

Proof. We need to show that f is continuous at every point of X. Therefore, let $a \in X$ and $\epsilon > 0$. Since $\{f_n\}$ converges uniformly to f on X, there exists k such that $|f_k(x) - f(x)| < \epsilon$ for all $x \in X$. On the other hand, since f_k is a continuous function, there exists a neighborhood V of a such that $|f_k(x) - f_k(a)| < \epsilon$ for all $x \in V$. Now note that if $x \in V$, then

$$|f(x) - f(a)| \leq |f(x) - f_k(x)| + |f_k(x) - f_k(a)| + |f_k(a) - f(a)|$$
$$< \epsilon + \epsilon + \epsilon = 3\epsilon,$$

and this shows that f is continuous at a, as desired. ∎

Let X be a nonempty set, and let $B(X)$ denote the collection of all bounded real-valued functions defined on X. Clearly, $B(X)$ is a function space. The **uniform** (or **sup**) **norm** of a function $f \in B(X)$ is defined by

$$\|f\|_\infty = \sup_{x \in X} |f(x)|.$$

The uniform norm satisfies the three characteristic properties of a norm on a vector space, namely:

1. $\|f\|_\infty \geq 0$ for each $f \in B(X)$ and $\|f\|_\infty = 0$ if and only if $f = 0$.
2. $\|\alpha f\|_\infty = |\alpha| \cdot \|f\|_\infty$ for each $f \in B(X)$ and all $\alpha \in \mathbb{R}$.
3. $\|f + g\|_\infty \leq \|f\|_\infty + \|g\|_\infty$ for all $f, g \in B(X)$.

If we set $D(f, g) = \|f - g\|_\infty$ for $f, g \in B(X)$, then D is a distance on $B(X)$ called the **uniform distance** (or the **uniform metric**) such that $(B(X), D)$ is a complete metric space; see Example 6.12. Moreover, by a straightforward verification one can show that a sequence $\{f_n\}$ of $B(X)$ converges to some $f \in B(X)$ with respect to D [i.e., $\lim D(f_n, f) = 0$] if and only if $\{f_n\}$ converges uniformly to f on X. This justifies the name "uniform distance."

Consider now a compact topological space X. Then by Theorem 8.14, every function $f \in C(X)$ is bounded, and hence, $C(X) \subseteq B(X)$. Therefore, $C(X)$ equipped with the uniform distance is a metric space, which is actually complete, as the next result will show.

Theorem 9.3.   *If X is a compact topological space, then $C(X)$ is a complete metric space (with the uniform distance).*

Proof.   Let $\{f_n\}$ be a Cauchy sequence of $C(X)$. Note that the inequality

$$|f_n(x) - f_m(x)| \le \|f_n - f_m\|_\infty \qquad (\star)$$

implies that $\{f_n(x)\}$ is a Cauchy sequence of real numbers for each $x \in X$. Put $f(x) = \lim f_n(x)$ for each $x \in X$. We claim that $f \in C(X)$ and that $\lim \|f_n - f\|_\infty = 0$.

To this end, let $\epsilon > 0$. Choose n_0 such that $\|f_n - f_m\|_\infty < \epsilon$ for $n, m \ge n_0$. It follows from $(\star)$ that $|f_n(x) - f_m(x)| < \epsilon$ holds for all $n, m \ge n_0$ and all $x \in X$. But then $|f_n(x) - f(x)| \le \epsilon$ holds for all $n \ge n_0$ and all $x \in X$. That is, $\{f_n\}$ converges uniformly to f on X. Therefore, by Theorem 9.2, $f \in C(X)$, and clearly, $\lim \|f_n - f\|_\infty = 0$.   ∎

A sequence of real-valued functions $\{f_n\}$ on a set X is said to be **increasing** if $f_n \le f_{n+1}$ holds for all n (and, of course, is called **decreasing** if $f_{n+1} \le f_n$ holds for all n). An increasing or a decreasing sequence of functions is referred to as a **monotone sequence** of functions.

It was observed that pointwise convergence need not imply uniform convergence. The next useful result gives a condition under which pointwise convergence implies uniform convergence, which is a classical result known as Dini's[2] theorem.

Theorem 9.4 (Dini).   *Let X be a compact topological space. If a monotone sequence of $C(X)$ converges pointwise to a continuous function, then it also converges uniformly.*

Proof.   Suppose that the sequence $\{f_n\}$ of $C(X)$ is increasing and convergent pointwise to some $f \in C(X)$; that is, $f_n(x) \uparrow f(x)$ holds for each $x \in X$.

---

[2]Ulisse Dini (1845–1918), an Italian mathematician. He became Professor of mathematics at the University of Pisa at the age of 21. His main contributions were in the the theory of real functions and partial differential equations.

Now, let $\epsilon > 0$. For each n define $\mathcal{O}_n = \{x \in X : f(x) - f_n(x) < \epsilon\}$. Clearly, each $\mathcal{O}_n$ is open. Also, note that since $f_n(x) \leq f_{n+1}(x)$ holds for all $x \in X$, it follows that $\mathcal{O}_n \subseteq \mathcal{O}_{n+1}$ holds for all n. Moreover, since $f_n(x) \uparrow f(x)$ holds for each $x \in X$, we easily get $X = \bigcup_{n=1}^{\infty} \mathcal{O}_n$.

In view of the compactness of X, there exists some k such that $X = \bigcup_{i=1}^{k} \mathcal{O}_i = \mathcal{O}_k$, and so $X = \mathcal{O}_n$ for $n \geq k$. But this merely says that $0 \leq f(x) - f_n(x) < \epsilon$ for all $n \geq k$ and all $x \in X$. In other words, $\{f_n\}$ converges uniformly to f on X.

If $\{f_n\}$ is a decreasing sequence, then apply the above arguments to the sequence $\{-f_n\}$. ∎

Now, consider a sequence $\{f_n\}$ of real-valued functions defined on a set X. For each n let $s_n = \sum_{i=1}^{n} f_i$. We say that the series $\sum_{n=1}^{\infty} f_n(x)$ **converges uniformly** on X if the sequence of partial sums $\{s_n\}$ converges uniformly to a real-valued function defined on X. This is, of course, equivalent to saying that the sequence $\{s_n\}$ is a **uniformly Cauchy sequence** in the sense that for each $\epsilon > 0$ there exists some n_0 such that $|s_n(x) - s_m(x)| < \epsilon$ holds for all $n, m \geq n_0$ and all $x \in X$.

We shall present two useful criteria of uniform convergence of series of functions. The first one is due to K. Weierstrass and is known as the *Weierstrass M-test*.

Theorem 9.5 (Weierstrass' M-test). *Let $\{f_n\}$ be a sequence of real-valued functions defined on a set X and assume that for each n there exists some positive real number a_n such that $|f_n(x)| \leq a_n$ holds for all $x \in X$. If $\sum_{n=1}^{\infty} a_n$ converges, then the series $\sum_{n=1}^{\infty} f_n(x)$ converges uniformly on X.*

Proof. Let $\epsilon > 0$. Pick some n_0 such that $\sum_{k=n}^{m} a_k < \epsilon$ holds for all $n, m \geq n_0$. Now, notice that if $n, m \geq n_0$ and $x \in X$, then

$$|s_n(x) - s_m(x)| = \left| \sum_{k-n}^{m} f_k(x) \right| \leq \sum_{k=n}^{m} |f_k(x)| \leq \sum_{k=n}^{m} a_k < \epsilon.$$

This shows that the sequence $\{s_n\}$ of partial sums is uniformly Cauchy, and hence, a uniformly convergent sequence. ∎

The second criterion of uniform convergence of series is due to J. Dirichlet.[3]

Theorem 9.6 (Dirichlet's test). *Let $\{f_n\}$ and $\{g_n\}$ be two sequences of real-valued functions defined on a set X satisfying the following properties:*

---

[3] Johann Peter Gustav Lejeune Dirichlet (1805–1859), a German mathematician. He worked in number theory, analysis, and mechanics. The modern definition of a function was proposed by him in 1837. Because of his fundamental papers on the convergence of trigonometric series, he is considered to be one of the founders of Fourier analysis. He was the teacher of Riemann.

1.   *There exists a constant $C > 0$ such that $|\sum_{i=1}^{n} f_i(x)| \le C$ holds for all n and all $x \in X$, i.e., the sequence of partial sums of the series $\sum_{n=1}^{\infty} f_n(x)$ is uniformly bounded.*
2.   *For each n we have $g_{n+1}(x) \le g_n(x)$ for all $x \in X$ and $\{g_n\}$ converges uniformly to zero on X.*

Then the series $\sum_{n=1}^{\infty} f_n(x)g_n(x)$ converges uniformly on X.

Proof.   Assume that $\{f_n\}$ and $\{g_n\}$ satisfy the stated conditions. For each n, let $s_n(x) = \sum_{i=1}^{n} f_i(x)g_i(x)$ and $t_n(x) = \sum_{i=1}^{n} f_i(x)$. Clearly,

$$s_n(x) = \sum_{i=1}^{n} t_i(x)[g_i(x) - g_{i+1}(x)] + g_{n+1}(x)t_n(x).$$

Next, fix $\epsilon > 0$ and then choose some n_0 such that $g_n(x) < \epsilon$ holds for all $n \ge n_0$ and all $x \in X$. Now note that if $n > m \ge n_0$ and $x \in X$, then

$$|s_n(x) - s_m(x)| = \left| \sum_{i=m+1}^{n} t_i(x)[g_i(x) - g_{i+1}(x)] + g_{n+1}(x)t_n(x) - g_{m+1}(x)t_m(x) \right|$$

$$\le C \sum_{i=m+1}^{n} [g_i(x) - g_{i+1}(x)] + Cg_{n+1}(x) + Cg_{m+1}(x)$$

$$= 2Cg_{m+1}(x) < 2C\epsilon.$$

This shows that $\{s_n\}$ is a uniformly Cauchy sequence, and hence, the series of functions $\sum_{n=1}^{\infty} f_n(x)g_n(x)$ converges uniformly on X.   ■

We now illustrate the preceding theorems with some examples.

Example 9.7.   Consider the series of functions $\sum_{n=1}^{\infty} \frac{\sin nx}{n^2}$. Since $|\frac{\sin nx}{n^2}| \le \frac{1}{n^2}$ and $\sum_{n=1}^{\infty} \frac{1}{n^2} < \infty$, it follows from the Weierstrass' M-test that the series $\sum_{n=1}^{\infty} \frac{\sin nx}{n^2}$ converges uniformly on $\mathbb{R}$.

Now, consider the series of functions

$$\sum_{n=1}^{\infty} \frac{\cos(2n - 1)x}{n} \quad \text{and} \quad \sum_{n=1}^{\infty} \frac{\sin(2n - 1)x}{n}$$

over the interval $(0, \pi)$. We claim that both series converge uniformly on every closed subinterval $[a, b]$ of $(0, \pi)$.

Notice first that the nth term of each series is dominated by $\frac{1}{n}$. However, since $\sum_{n=1}^{\infty} \frac{1}{n} = \infty$, we cannot apply Weierstrass' M-test to conclude that the series converge uniformly on $[a, b]$. However, we shall reach this conclusion by applying Dirichlet's test. To do this, let $f_n = \cos(2n - 1)x$ and $g_n(x) = \frac{1}{n}$. Clearly, the sequence $\{g_n\}$ converges monotonically and uniformly to zero. On the other hand, from elementary trigonometry, we know that

$$\sum_{k=1}^{n} \cos(2k-1)x = \frac{\sin 2nx}{2\sin x} \quad \text{and} \quad \sum_{k=1}^{n} \sin(2k-1)x = \frac{\sin^2 nx}{\sin x}.$$

For instance, to establish the first identity note that

$$2\sin x \sum_{k=1}^{n} \cos(2k-1)x = \sum_{k=1}^{n} [\sin((2k-1)x + x) - \sin((2k-1)x - x)]$$

$$= \sum_{k=1}^{n} [\sin(2kx) - \sin(2(k-1)x)] = \sin 2nx.$$

Now, if we choose some $\delta > 0$ such that $\sin x > \delta$ for all $x \in [a, b]$, then

$$\left| \sum_{k=1}^{n} \cos(2k-1)x \right| \leq \frac{1}{2\sin x} \leq \frac{1}{2\delta}$$

holds for all n and all $x \in [a, b]$. Hence, Dirichlet's test applies and shows that the series $\sum_{n=1}^{\infty} f_n(x)g_n(x) = \sum_{n=1}^{\infty} \frac{\cos(2n-1)x}{n}$ converges uniformly on $[a, b]$. Similarly, the series $\sum_{n=1}^{\infty} \frac{\sin(2n-1)x}{n}$ converges uniformly on every closed subinterval of $(0, \pi)$. ∎

Assume now that a sequence $\{f_n\}$ of real-valued continuous functions converges pointwise to some function f. Although f is not necessarily a continuous function, nevertheless, something can be said about the points of discontinuity of f. We shall show that the set of all points of discontinuity of f is a meager set. Recall that a set is called a *meager set* if it can be written as a countable union of nowhere dense sets. To do this, we need a lemma.

Lemma 9.8. *Let X be a topological space. Assume that the sequence $\{f_n\}$ of $C(X)$ converges pointwise to some function f. For $\epsilon > 0$ and $n \in \mathbb{N}$, let*

$$V_n(\epsilon) = \{x \in X : |f_n(x) - f(x)| \leq \epsilon\}.$$

Also, put $\mathcal{O}(\epsilon) = \bigcup_{n=1}^{\infty} [V_n(\epsilon)]^\circ$.
Then the set of all points of continuity of f is the G_δ-set $\bigcap_{n=1}^{\infty} \mathcal{O}(\frac{1}{n})$.

Proof. Let $a \in X$ be a point at which f is continuous, and let $\epsilon > 0$. Since $\lim f_n(a) = f(a)$, there exists some k such that $|f_k(a) - f(a)| < \epsilon$. Also, by the continuity of f and f_k at the point a there exists a neighborhood U of a such that $|f(a) - f(x)| < \epsilon$ and $|f_k(x) - f_k(a)| < \epsilon$ for all $x \in U$. Then for $x \in U$ we have

$$|f(x) - f_k(x)| \leq |f(x) - f(a)| + |f(a) - f_k(a)| + |f_k(a) - f_k(x)|$$
$$< \epsilon + \epsilon + \epsilon = 3\epsilon.$$

This implies $U \subseteq [V_k(3\epsilon)]^\circ$, and so $a \in \mathcal{O}(3\epsilon)$ for each $\epsilon > 0$, which shows that $a \in \bigcap_{n=1}^{\infty} \mathcal{O}(\frac{1}{n})$.

For the reverse inclusion, assume that $a \in \bigcap_{n=1}^{\infty} \mathcal{O}(\frac{1}{n})$. To finish the proof, we have to show that f is continuous at a. So, let $\epsilon > 0$. Pick n such that $\frac{1}{n} < \epsilon$. From $a \in \mathcal{O}(\frac{1}{n})$, it follows that there exists some k such that $a \in [V_k(\frac{1}{n})]^\circ$. This in turn means that there exists a neighborhood U of a with $U \subseteq V_k(\frac{1}{n})$; that is, $|f_k(x) - f(x)| \leq \frac{1}{n} < \epsilon$ holds for all $x \in U$. Since f_k is continuous at a, there exists a neighborhood W of a such that $|f_k(x) - f_k(a)| < \epsilon$ for each $x \in W$. Thus, if $x \in U \cap W$, then

$$|f(x) - f(a)| \leq |f(x) - f_k(x)| + |f_k(x) - f_k(a)| + |f_k(a) - f(a)|$$
$$< \epsilon + \epsilon + \epsilon = 3\epsilon,$$

which shows that f is continuous at a.   ∎

We are now ready to prove that the set of discontinuities of a function that is the pointwise limit of a sequence of continuous functions is a meager set.

Theorem 9.9. *Let X be a topological space, and let $\{f_n\}$ be a sequence of $C(X)$. If $\{f_n\}$ converges pointwise to a real-valued function f, then the set D of all points of discontinuity of f is a meager set.*

Proof. According to Lemma 9.8, we have $D = \bigcup_{n=1}^{\infty} [\mathcal{O}(\frac{1}{n})]^c$. To complete the proof, it suffices to show that every set of the form $[\mathcal{O}(\epsilon)]^c$ is a meager set; it will then follow that D is a meager set, since it will be a countable union of meager sets.

So, let $\epsilon > 0$. For each m define

$$F_m(\epsilon) = \bigcap_{i=1}^{\infty} \{x \in X: |f_m(x) - f_{m+i}(x)| \leq \epsilon\},$$

and note that each $F_m(\epsilon)$ is a closed set. Also, since for each $x \in X$ we have $\lim f_n(x) = f(x)$, it is easy to see that $X = \bigcup_{m=1}^{\infty} F_m(\epsilon)$.

Using once more the fact that $\lim f_n(x) = f(x)$ for each $x \in X$, it is easy to see that $F_m(\epsilon) \subseteq V_m(\epsilon)$ holds. Therefore, $[F_m(\epsilon)]^\circ \subseteq [V_m(\epsilon)]^\circ \subseteq \mathcal{O}(\epsilon)$ holds for each m, and so, $\bigcup_{m=1}^{\infty} [F_m(\epsilon)]^{\circ} \subseteq \mathcal{O}(\epsilon)$ also holds. Now, observe that

$$[\mathcal{O}(\epsilon)]^c = X \setminus \mathcal{O}(\epsilon) \subseteq X \setminus \bigcup_{m=1}^{\infty} [F_m(\epsilon)]^{\circ} = \bigcup_{m=1}^{\infty} F_m(\epsilon) \setminus \bigcup_{m=1}^{\infty} [F_m(\epsilon)]^{\circ}$$

$$\subseteq \bigcup_{m=1}^{\infty} [F_m(\epsilon) \setminus [F_m(\epsilon)]^{\circ}] = \bigcup_{m=1}^{\infty} \partial F_m(\epsilon),$$

where the last equality holds since each $F_m(\epsilon)$ is a closed set. Now observe that since each $F_m(\epsilon)$ is closed, its boundary $\partial F_m(\epsilon)$ is a nowhere dense set, and thus, according to the last inclusion $[\mathcal{O}(\epsilon)]^c$ is a meager set, and the proof is finished. ∎

Our next objective is to characterize the (uniform) compact subsets of $C(X)$. To do this we need a definition.

Let X be a topological space, and let S be a subset of $C(X)$. Then the set S is said to be **equicontinuous** at some $x \in X$ if for each $\epsilon > 0$ there exists a neighborhood V of x such that $y \in V$ implies $|f(y) - f(x)| < \epsilon$ for every $f \in S$. If S is equicontinuous at every point of X, then S is called an **equicontinuous set**.

As we have seen before, in a metric space a closed and bounded set need not be compact. However, if X is a compact topological space, then a closed and bounded (with respect to the uniform metric) subset of $C(X)$ is compact if and only if it is equicontinuous. This result is known as the Ascoli–Arzelà[4, 5] theorem and is stated next.

Theorem 9.10 (Ascoli–Arzelà).   *Let X be a compact topological space, and let S be a subset of $C(X)$. Then the following statements are equivalent:*

1.   *S is a compact subset of the metric space $C(X)$ (equipped, of course, with the uniform metric).*
2.   *S is closed, bounded, and equicontinuous.*

Proof.   (1) $\Longrightarrow$ (2) We already know that a compact set is closed and bounded. What remains to be shown is that S is equicontinuous.

To this end, let $\epsilon > 0$. Choose $f_1, \ldots, f_n \in S$ such that $S \subseteq \bigcup_{i=1}^{n} B(f_i, \epsilon)$. If $x \in X$, then pick a neighborhood V_x of x such that $|f_i(y) - f_i(x)| < \epsilon$ holds for all $y \in V_x$ and all $i = 1, \ldots, n$. Now, let $y \in V_x$ and $f \in S$. Choose some i with

---

[4]Guido Ascoli (1887–1957), an Italian mathematician. He contributed to the theory of real functions and differential equations.

[5]Cesare Arzelà (1846–1912), an Italian mathematician. He studied the convergence of sequences of real functions.

$f \in B(f_i, \epsilon)$, and note that

$$|f(y) - f(x)| \leq |f(y) - f_i(y)| + |f_i(y) - f_i(x)| + |f_i(x) - f(x)|$$
$$< \epsilon + \epsilon + \epsilon = 3\epsilon.$$

This shows that S is equicontinuous at x, and since x is arbitrary, S is an equicontinuous set of functions.

(2) $\Longrightarrow$ (1) Let $\{f_n\}$ be a sequence of S. According to Theorem 7.3, it suffices to show that $\{f_n\}$ has a convergent subsequence.

To this end, choose some $M > 0$ satisfying $|f(x)| \leq M$ for all $x \in X$ and $f \in S$. Using the equicontinuity of S and the compactness of X, it is easy to see that for each k there exists a finite subset F_k of X and neighborhoods $\{V_y: y \in F_k\}$ such that $X = \bigcup_{y \in F_k} V_y$ and $|f(x) - f(y)| < \frac{1}{k}$ whenever $x \in V_y$ and $f \in S$.

Let $F = \bigcup_{i=1}^{\infty} F_i$. Clearly, F is at most countable; assume F countable and let $F = \{x_1, x_2, \ldots\}$ be an enumeration of F. Now, since $|f_n(x_1)| \leq M$ holds for all n, there exists a subsequence $\{g_n^1\}$ of $\{f_n\}$ such that $\lim g_n^1(x_1)$ exists in $\mathbb{R}$. Similarly, there exists a subsequence $\{g_n^2\}$ of $\{g_n^1\}$ so that $\lim g_n^2(x_2)$ exists in $\mathbb{R}$. Continuing this way, we can choose (inductively) sequences $\{g_n^i\}$ ($i = 1, 2, \ldots$) such that

a. $\{g_n^1\}$ is a subsequence of $\{f_n\}$,
b. $\{g_n^{i+1}\}$ is a subsequence of $\{g_n^i\}$ for each $i = 1, 2, \ldots$, and
c. $\lim_{n \to \infty} g_n^i(x_i)$ exists in $\mathbb{R}$ for each $i = 1, 2, \ldots$.

Now, consider the diagonal sequence $h_n = g_n^n$, and note that $\{h_n\}$ is a subsequence of $\{f_n\}$ such that $\lim_{n \to \infty} h_n(x_i)$ exists in $\mathbb{R}$ for each i. Moreover, we claim that $\{h_n\}$ is a Cauchy sequence of $C(X)$.

To see this, fix k and then choose some n_0 so that $|h_n(y) - h_m(y)| < \frac{1}{k}$ holds for all $n, m > n_0$ and all $y \in F_k$. Now, if $x \in X$, then pick some $y \in F_k$ such that $x \in V_y$, and note that

$$|h_n(x) - h_m(x)| \leq |h_n(x) - h_n(y)| + |h_n(y) - h_m(y)| + |h_m(y) - h_m(x)|$$
$$< \frac{1}{k} + \frac{1}{k} + \frac{1}{k} = \frac{3}{k}$$

holds for all $n, m > n_0$. That is,

$$\|h_n - h_m\|_\infty = \sup\{|h_n(x) - h_m(x)|: x \in X\} \leq \frac{3}{k}$$

holds for all $n, m > n_0$, so that $\{h_n\}$ is a Cauchy sequence of $C(X)$.

By Theorem 9.3, $\{h_n\}$ converges to some $h \in C(X)$. Since S is closed, $h \in S$, and the proof of the theorem is complete. ∎

Let S be a bounded equicontinuous subset of some $C(X)$-space with X compact. It is not difficult to establish that the (uniform) closure $\overline{S}$ of S is likewise bounded and equicontinuous, and thus by the Ascoli–Arzelà theorem, $\overline{S}$ is a compact subset of $C(X)$. This implies that every sequence of S has a subsequence that converges uniformly. In particular, every bounded equicontinuous sequence has a uniformly convergent subsequence. This last observation is very useful in establishing the existence of solutions to differential equations.

EXERCISES

1. If u, v, and w are vectors in a vector lattice, then establish the following identities:

   a. $u \vee v + u \wedge v = u + v$;

   b. $u - v \vee w = (u - v) \wedge (u - w)$;

   c. $u - v \wedge w = (u - v) \vee (u - w)$;

   d. $\alpha(u \wedge v) = (\alpha u) \wedge (\alpha v)$ if $\alpha \geq 0$;

   e. $|u - v| = u \vee v - u \wedge v$;

   f. $u \vee v = \frac{1}{2}(u + v + |u - v|)$;

   g. $u \wedge v = \frac{1}{2}(u + v - |u - v|)$.

2. If u and v are elements in a vector lattice, then show that:

   a. $|u + v| \vee |u - v| = |u| + |v|$, and

   b. $|u + v| \wedge |u - v| = ||u| - |v||$.

   [HINT: $|u + v| \vee |u - v| = (u + v) \vee (-u - v) \vee (u - v) \vee (-u + v)$

   $= (|u| + v) \vee (|u| - v) = |u| + |v|.$]

3. Show that $|u| \wedge |v| = 0$ holds if and only if $|u + v| = |u - v|$ holds.

4. Show that the vector space consisting of all polynomials (with real coefficients) on $\mathbb{R}$ is not a function space. Prove a similar result for the vector space of all real-valued differentiable functions on $\mathbb{R}$.

5. Let X be a topological space. Consider the collection L of all real-valued functions on X defined by

   $$L = \{f \in \mathbb{R}^X : \exists \{f_n\} \subseteq C(X) \text{ such that } \lim f_n(x) = f(x) \ \forall \, x \in X\}.$$

   Show that L is a function space.

6. Let L be a vector space of real-valued functions defined on a set X. If for every function $f \in L$ the function $|f|$ [defined by $|f|(x) = |f(x)|$ for each $x \in X$] belongs to L, then show that L is a function space.

7. Consider each rational number written in the form $\frac{m}{n}$, where $n > 0$, and m and n are integers without any common factors other than ± 1. Clearly, such a representation is unique. Now, define $f : \mathbb{R} \to \mathbb{R}$ by $f(x) = 0$ if x is irrational and $f(x) = \frac{1}{n}$ if $x = \frac{m}{n}$ as above. Show that f is continuous at every irrational number and discontinuous at every rational number.

8. Let $f: [a, b] \to \mathbb{R}$ be increasing [i.e., $x < y$ implies $f(x) \le f(y)$]. Show that the set of points where f is discontinuous is at-most countable.
    [HINT: If f is discontinuous at c, with $a < c < b$, then choose a rational number r such that $\lim_{x \uparrow c} f(x) < r < \lim_{x \downarrow c} f(x)$.]

9. Give an example of a strictly increasing function $f: [0, 1] \to \mathbb{R}$ which is continuous at every irrational number and discontinuous at every rational number.

10. Recall that a function $f: (X, \tau) \to (Y, \tau_1)$ is called an *open mapping* if $f(V)$ is open whenever V is open. Prove that if $f: \mathbb{R} \to \mathbb{R}$ is a continuous open mapping, then f is a strictly monotone function—and hence, a homeomorphism.

11. Let X be a nonempty set, and for any two functions $f, g \in \mathbb{R}^X$ let
    $$d(f, g) = \sup_{x \in X} \frac{|f(x) - g(x)|}{1 + |f(x) - g(x)|}.$$

    Establish the following:

    a. $(\mathbb{R}^X, d)$ is a metric space.
    b. A sequence $\{f_n\} \subseteq \mathbb{R}^X$ satisfies $d(f_n, f) \to 0$ for some $f \in \mathbb{R}^X$ if and only if $\{f_n\}$ converges uniformly to f.

12. Let $f, f_1, f_2, \ldots$ be real-valued functions defined on a compact metric space (X, d) such that $x_n \to x$ in X implies $f_n(x_n) \to f(x)$ in $\mathbb{R}$. If f is continuous, then show that the sequence of functions $\{f_n\}$ converges uniformly to f.

13. For a sequence $\{f_n\}$ of real-valued functions defined on a topological space X that converges uniformly to a real function f on X, establish the following:

    a. If $x_n \to x$ and f is continuous at x, then $f_n(x_n) \to f(x)$.
    b. If each f_n is continuous at some point $x_0 \in X$, then f is also continuous at the point x_0 and
    $$\lim_{x \to x_0} \lim_{n \to \infty} f_n(x) = \lim_{n \to \infty} \lim_{x \to x_0} f_n(x) = f(x_0).$$

14. Let $f_n: [0, 1] \to \mathbb{R}$ be defined by $f_n(x) = x^n$ for $x \in [0, 1]$. Show that $\{f_n\}$ converges pointwise and find its limit function. Is the convergence uniform?

15. Let $g: [0, 1] \to \mathbb{R}$ be a continuous function with $g(1) = 0$. Show that the sequence of functions $\{f_n\}$ defined by $f_n(x) = x^n g(x)$ for $x \in [0, 1]$, converges uniformly to the constant zero function.

16. Let $\{f_n\}$ be a sequence of continuous real-valued functions defined on $[a, b]$, and let $\{a_n\}$ and $\{b_n\}$ be two sequences of $[a, b]$ such that $\lim a_n = a$ and $\lim b_n = b$. If $\{f_n\}$ converges uniformly to f on $[a, b]$, then show that
    $$\lim_{n \to \infty} \int_{a_n}^{b_n} f_n(x)\, dx = \int_a^b f(x)\, dx.$$

17. Let $\{f_n\}$ be a sequence of continuous real-valued functions on a metric space X such that $\{f_n\}$ converges uniformly to some function f on every compact subset of X. Show that f is a continuous function.

18. Let $\{f_n\}$ and $\{g_n\}$ be two uniformly bounded sequences of real-valued functions on a set X. If both $\{f_n\}$ and $\{g_n\}$ converge uniformly on X, then show that $\{f_n g_n\}$ also converges uniformly on X.

19. Suppose that $\{f_n\}$ is a sequence of monotone real-valued functions defined on $[a, b]$ and not necessarily all increasing or decreasing. Show that if $\{f_n\}$ converges pointwise to a continuous function f on $[a, b]$, then $\{f_n\}$ converges uniformly to f on $[a, b]$.
    [HINT: Use the fact that f must be uniformly continuous on $[a, b]$.]

20. Let X be a topological space and let $\{f_n\}$ be a sequence of real-valued continuous functions defined on X. Suppose that there is a function $f: X \to \mathbb{R}$ such that $f(x) = \lim f_n(x)$ holds for all $x \in X$. Show that f is continuous at a point a if and only if for each $\epsilon > 0$ and each m there exist a neighborhood V of a and some $k > m$ such that $|f(x) - f_k(x)| < \epsilon$ holds for all $x \in V$.

21. Let $\{f_n\}$ be a uniformly bounded sequence of continuous real-valued functions on a closed interval $[a, b]$. Show that the sequence of functions $\{\phi_n\}$, defined by $\phi_n(x) = \int_a^x f_n(t)\, dt$ for each $x \in [a, b]$, contains a uniformly convergent subsequence on $[a, b]$.

22. For each n, let $f_n: \mathbb{R} \to \mathbb{R}$ be a monotone (either increasing or decreasing) function. If there exists a dense subset A of $\mathbb{R}$ such that $\lim f_n(x)$ exists in $\mathbb{R}$ for each $x \in A$, then show that $\lim f_n(x)$ exists in $\mathbb{R}$ at most for all but countably many x.

23. Consider a continuous function $f: [0, \infty) \to \mathbb{R}$. For each n, define the continuous function $f_n: [0, \infty) \to \mathbb{R}$ by $f_n(x) = f(x^n)$. Show that the set of continuous functions $\{f_1, f_2, \ldots\}$ is equicontinuous at $x = 1$ if and only if f is a constant function.

24. Let (X, d) be a compact metric space and let $\mathcal{A}$ be an equicontinuous subset of $C(X)$. Show that $\mathcal{A}$ is uniformly equicontinuous, i.e., show that for each $\epsilon > 0$ there exists some $\delta > 0$ such that $x, y \in X$ and $d(x, y) < \delta$ imply $|f(x) - f(y)| < \epsilon$ for all $f \in \mathcal{A}$.

25. Let X be a connected topological space (see Exercise 22 of Section 8 for the definition) and let $\mathcal{A}$ be an equicontinuous subset of $C(X)$. If for some $x_0 \in X$, the set of real numbers $\{f(x_0): f \in \mathcal{A}\}$ is bounded, then show that $\{f(x): f \in \mathcal{A}\}$ is also bounded for each $x \in X$.

26. Let $\{f_n\}$ be an equicontinuous sequence in $C(X)$, where X is not necessarily compact. If for some function $f: X \to \mathbb{R}$ we have $\lim f_n(x) = f(x)$ for each $x \in X$, then show that $f \in C(X)$.

27. Let X be a compact topological space, and let $\{f_n\}$ be an equicontinuous sequence of $C(X)$. Assume that there exists some $f \in C(X)$ and some dense subset A of X such that $\lim f_n(x) = f(x)$ holds for each $x \in A$. Then show that $\{f_n\}$ converges uniformly to f.

28. Show that for any fixed integer $n > 1$ the set of functions $f \in C[0, 1]$ such that there is some $x \in [0, 1 - \frac{1}{n}]$ for which

$$|f(x + h) - f(x)| \leq nh \quad \text{whenever } 0 < h < \tfrac{1}{n},$$

is nowhere dense in $C[0, 1]$ (with the uniform metric).
    Use the preceding conclusion and Baire's theorem to prove that there exists a continuous real-valued function defined on $[0, 1]$ that is not differentiable at any point of $[0, 1]$.

29. Establish the following result regarding differentiability and uniform convergence. Let $\{f_n\}$ be a sequence of differentiable real-valued functions defined on a bounded

open interval (a, b) such that:

a. for some $x_0 \in (a, b)$ the sequence of real numbers $\{f_n(x_0)\}$ converges in $\mathbb{R}$, and
b. the sequence of derivatives $\{f_n'\}$ converges uniformly to a function $g\colon (a, b) \to \mathbb{R}$.

Then the sequence $\{f_n\}$ converges uniformly to a function $f\colon (a, b) \to \mathbb{R}$ that is differentiable at x_0 and satisfies $f'(x_0) = g(x_0)$.

10. SEPARATION PROPERTIES OF CONTINUOUS FUNCTIONS

It is possible that the only real-valued continuous functions on a topological space are the constant ones. For instance, any indiscrete topological space has this property. Such spaces will be of little interest to us here. In this section, we shall describe a large class of topological spaces with an abundance of continuous real-valued functions. We start our discussion with two notions of separation.

Definition 10.1.   *Two disjoint subsets A and B of a topological space X are*:

a. **separated by open sets**, *if there exist two disjoint open sets V and W satisfying $A \subseteq V$ and $B \subseteq W$, and*
b. **separated by a continuous function**, *if there exists a continuous function $f\colon X \to [0, 1]$ such that $f(a) = 0$ for each $a \in A$ and $f(b) = 1$ for each $b \in B$.*

Lemma 10.2.   *If two disjoint subsets of a topological space are separated by a continuous function, then they are also separated by open sets.*

Proof.   Let A and B be two disjoint subsets of a topological space X and let $f\colon X \to [0, 1]$ be a continuous function such that $f(a) = 0$ for each $a \in A$ and $f(b) = 1$ for each $b \in B$. If $V = \{x \in X\colon f(x) < \frac{1}{2}\}$ and $W = \{x \in X\colon f(x) > \frac{1}{2}\}$, then V and W are two disjoint open sets satisfying $A \subseteq V$ and $B \subseteq W$.   ∎

Topological spaces whose disjoint closed sets can be separated by open sets play an important role in mathematical analysis and they are referred to as *normal spaces*.

Definition 10.3.   *A topological space is said to be **normal** if every pair of disjoint closed sets can be separated by open sets.*

Here are two classes of normal topological spaces.

Lemma 10.4.   *Metric spaces and compact Hausdorff topological spaces are normal spaces.*

Proof.   Let (X, d) be a metric space and let A be a nonempty subset of X. Define the **distance function** $d(\cdot, A): X \rightarrow \mathbb{R}$ of A by

$$d(x, A) = \inf\{d(x, y): y \in A\}.$$

The non-negative number $d(x, A)$ is called the **distance of x from A**. It is easy to see that $|d(x, A) - d(y, A)| \leq d(x, y)$ for all $x, y \in X$, and so, $d(\cdot, A)$ is a uniformly continuous function on X.

Now, if A and B are two nonempty closed disjoint subsets of X, consider the continuous function $f: X \rightarrow \mathbb{R}$ defined by $f(x) = d(x, A) - d(x, B)$. Clearly, the open sets $V = f^{-1}((-\infty, 0))$ and $W = f^{-1}((0, \infty))$ are disjoint sets satisfying $A \subseteq V$ and $B \subseteq W$.

Next, let X be a Hausdorff compact topological space and let A and B be two closed disjoint sets. By Theorem 8.12, both A and B are compact sets. Fix $a \in A$. Then for each $b \in B$ there exist neighborhoods V_b of a and W_b of b such that $V_b \cap W_b = \emptyset$. From $B \subseteq \bigcup_{b \in B} W_b$ and the compactness of B, there exist $b_1, \ldots, b_n \in B$ such that $B \subseteq \bigcup_{i=1}^n W_{b_i} = W_a$. If $V_a = \bigcap_{i=1}^n V_{b_i}$, then V_a is a neighborhood of a satisfying $V_a \cap W_a = \emptyset$. Now from $A \subseteq \bigcup_{a \in A} V_a$ and the compactness of A, we see that there exist $a_1, \ldots, a_k \in A$ such that $A \subseteq \bigcup_{i=1}^k V_{a_i}$. If $V = \bigcup_{i=1}^k V_{a_i}$ and $W = \bigcap_{i=1}^k W_{a_i}$, then V and W are two disjoint open sets satisfying $A \subseteq V$ and $B \subseteq W$.   ■

The next result characterizes the normal topological spaces and is known as Uryson's[6] lemma.

Theorem 10.5 (Uryson's Lemma).   *For a topological space X the following statements are equivalent.*

1.   *X is a normal space.*
2.   *If A is a closed subset and V is an open subset of X satisfying $A \subseteq V$, then there exists an open set W such that $A \subseteq W \subseteq \overline{W} \subseteq V$.*
3.   *Every pair of disjoint closed sets can be separated by a continuous function.*
4.   *If C is a closed subset of X and $f: C \rightarrow [0, 1]$ is a continuous function, then there exists a continuous extension of f to all of X with values in $[0, 1]$.*

Proof.   (1) $\implies$ (2) Assume that A is a closed set and V is an open set such that $A \subseteq V$. Put $C = V^c$ and note that A and C are two disjoint closed sets. So, there exist two disjoint open sets W and U satisfying $A \subseteq W$ and $C \subseteq U$. From $W \cap U = \emptyset$, we easily infer that $\overline{W} \cap U = \emptyset$, and thus $\overline{W} \subseteq U^c \subseteq C^c = V$.

---

[6]Pavel Samuilovich Uryson (1898–1924), a Russian mathematician. Although his scientific activity lasted for only five years—he drowned off the coast of Brittany (France) at the age of 26 while on vacation—he made several important contributions to general topology.

(2) $\implies$ (3) Let A and B be two disjoint closed sets. Put $V = B^c$ and note that $A \subseteq V$. Next, put $r_0 = 0$, $r_1 = 1$, and let $\{r_2, r_3, \ldots\}$ be an enumeration of the rational numbers in the open interval $(0, 1)$. Our hypothesis implies the existence of two open sets V_{r_0} and V_{r_1} such that $A \subseteq V_{r_1} \subseteq \overline{V}_{r_1} \subseteq V_{r_0} \subseteq \overline{V}_{r_0} \subseteq V$.

Now, proceed inductively. Assume that the open sets $V_{r_0}, \ldots, V_{r_n}$ have been chosen so that $r_i < r_j$ implies $A \subseteq V_{r_j} \subseteq \overline{V}_{r_j} \subseteq V_{r_i}$. Observe that there are precisely two rational numbers r_i and r_j among $r_0, r_1, \ldots, r_n$ such that $r_i < r_{n+1} < r_j$ holds, and so no other rational number among $r_0, r_1, \ldots, r_n$ lies in the open interval (r_i, r_j). Clearly, $r_i = \max\{r_k: r_k < r_{n+1} \text{ and } 0 \le k \le n\}$. Similarly, $r_j = \min\{r_k: r_{n+1} < r_k \text{ and } 0 \le k \le n\}$. By our hypothesis, there exists an open set $V_{r_{n+1}}$ such that

$$\overline{V}_{r_j} \subseteq V_{r_{n+1}} \subseteq \overline{V}_{r_{n+1}} \subseteq V_{r_i}.$$

Thus, if $\mathcal{Q}$ denotes the set of all rational numbers, we can construct a collection of open sets $\{V_r: r \in \mathcal{Q} \cap [0, 1]\}$ with the following two properties:

a. $A \subseteq V_r \subseteq V$ for each $r \in \mathcal{Q} \cap [0, 1]$.
b.   If $r, s \in \mathcal{Q} \cap [0, 1]$ satisfy $s > r$, then $\overline{V}_s \subseteq V_r$ holds.

Next, we define the function $f: X \to [0, 1]$ by

$$f(x) = \begin{cases} \sup\{r: x \in V_r\} & \text{if } x \in V_0 \\ 0 & \text{if } x \notin V_0. \end{cases}$$

Clearly, $f(x) = 1$ for all $x \in A$ and $f(x) = 0$ for each $x \in V^c = B$. We claim that f is continuous. To see this, fix $a \in X$ and let $\epsilon > 0$.

Assume first that $0 \le f(a) < 1$ holds. Choose two rational numbers s, t in $[0, 1]$ such that $f(a) < s < t < f(a) + \epsilon$. From $\overline{V}_t \subseteq V_s$ and $f(a) < s$ it follows that $a \notin \overline{V}_t$.

If $f(a) > 0$, choose a rational number $r \in (0, 1)$ such that $f(a) - \epsilon < r < f(a)$ and $a \in V_r$. Put $U = V_r \setminus \overline{V}_t$; clearly, U is a neighborhood of a. Also, note that if $x \in U$, then $r \le f(x) \le t$ holds. Thus, $|f(x) - f(a)| < \epsilon$ holds for all $x \in U$.

If $f(a) = 0$, put $U = X \setminus \overline{V}_t$. Then U is a neighborhood of a, and for $x \in U$ we have $0 \le f(x) \le t$. Therefore, $|f(x) - f(a)| < \epsilon$ holds for all $x \in U$. Hence, in either case f is continuous at a.

Finally, in case $f(a) = 1$, choose a rational number $r \in (0, 1)$ such that $a \in V_r$ and $1 - \epsilon < r$. Clearly, V_r is a neighborhood of a. Also, if $x \in V_r$, then $r \le f(x) \le 1$, from which it follows that $|f(x) - f(a)| < \epsilon$ holds for all $x \in V_r$.

The above arguments show that f is continuous at each point $a \in X$, and hence f is a continuous function separating A and B.

(3) $\implies$ (4) Let C be a (nonempty) closed subset of X and let $f: C \to [0, 1]$ be a continuous function. We shall consider f as a continuous function from C to $[-1, 1]$.

Start by observing that if A and B are two disjoint closed subsets of X and $[\alpha, \beta]$ is an arbitrary closed interval, then there exists a continuous function $\phi: X \to [\alpha, \beta]$ with $\phi = \alpha$ on A and $\phi = \beta$ on B. (Indeed, if $\psi: X \to [0, 1]$ is a continuous function such that $\psi = 0$ on A and $\psi = 1$ on B, then the continuous function $\phi = (\beta - \alpha)\psi + \alpha$ has the desired properties.)

The extension of the function f will be based upon the following property (E): *If $h: C \to [-r, r]$ is a continuous function, then there exists a continuous function $g: X \to \mathbb{R}$ satisfying*

$$|g(x)| \le \frac{1}{3}r \text{ for all } x \in X \quad \text{and} \quad |h(c) - g(c)| \le \frac{2}{3}r \text{ for all } c \in C.$$

To verify property (E), we argue as follows: Let $A = h^{-1}([-r, -\frac{r}{3}])$ and $B = h^{-1}([\frac{r}{3}, r])$. The continuity of h guarantees that the disjoint sets A and B are closed in C. Since C is a closed subset of X, it follows that A and B are also closed subsets of X. So, there exists a continuous function $g: X \to [-\frac{r}{3}, \frac{r}{3}]$ (i.e., $|g(x)| \le \frac{r}{3}$ for each $x \in X$) such that $g = -\frac{r}{3}$ on A and $g = \frac{r}{3}$ on B. Now, let $c \in C$. If $c \in A$, then $-r \le h(c) \le -\frac{r}{3} = g(c)$ and so $|h(c) - g(c)| \le \frac{2}{3}r$. If $c \in B$, then $g(c) = \frac{r}{3} \le h(c) \le r$, and hence, $|h(c) - g(c)| \le \frac{2}{3}r$. Finally, if $c \notin A \cup B$, then $-\frac{r}{3} < h(c) < \frac{r}{3}$, from which it follows that $|h(c) - g(c)| \le \frac{2}{3}r$ holds true in this case too.

Now, we claim that there exists a sequence $\{g_n\}$ of continuous real-valued functions on X such that for each n we have

$$|g_n(x)| \le \frac{1}{3}\left(\frac{2}{3}\right)^{n-1} \text{ for all } x \in X, \qquad (\star)$$

and

$$\left| f(c) - \sum_{i=1}^{n} g_i(c) \right| \le \left(\frac{2}{3}\right)^n \text{ for all } c \in C. \qquad (\star\star)$$

The existence of the sequence $\{g_n\}$ can be established by induction as follows: For $n = 1$, we apply property (E) with $r = 1$ and $h = f$. So, there exists a continuous function $g_1: X \to \mathbb{R}$ satisfying $|g_1(x)| \le \frac{1}{3}$ for each $x \in X$ and $|f(c) - g_1(c)| \le \frac{2}{3}$ for each $c \in C$. Now, for the induction step, assume that $f_1, \ldots, f_n$ have been selected to satisfy $(\star)$ and $(\star\star)$. Applying property (E) with $r = (\frac{2}{3})^n$ and $h = f - \sum_{i=1}^{n} g_i$, we see that there exists some continuous function $g_{n+1}: X \to \mathbb{R}$ satisfying $|g_{n+1}(x)| \le \frac{1}{3}(\frac{2}{3})^n$ for each $x \in X$ and

$$\left| \left[f(c) - \sum_{i=1}^{n} g_i(c) \right] - g_{n+1}(c) \right| = \left| f(c) - \sum_{i=1}^{n+1} g_i(c) \right| \le \left(\frac{2}{3}\right)^{n+1}$$

for all $c \in C$.

From $\sum_{n=1}^{\infty} \frac{1}{3}(\frac{2}{3})^{n-1} = 1$ and $(\star)$, we see that the series $g(x) = \sum_{n=1}^{\infty} g_n(x)$ converges uniformly, and so (by Theorem 9.2) g defines a continuous function from X into $[-1, 1]$. Now, a glance at $(\star\star)$ guarantees that $g(c) = f(c)$ for all $c \in C$. Next, consider the function $|g|: X \to [0, 1]$ and note that $|g|$ is a continuous extension of f to all of X.

(4) $\implies$ (1) Let A and B be two nonempty closed disjoint subsets of X. Then $A \cup B$ is a closed set, and the function $f: A \cup B \to [0, 1]$, defined by $f(a) = 0$ for each $a \in A$ and $f(b) = 1$ for $b \in B$, is continuous. Indeed, if $a \in A$ and $\epsilon > 0$, then B^c is a neighborhood of a and $|f(x) - f(a)| = 0 < \epsilon$ holds for each $x \in (A \cup B) \cap B^c$. This shows that f is continuous at $a \in A$; similarly, f is continuous at every $b \in B$. If g is a continuous extension of f to all of X with values in $[0, 1]$, then g clearly separates A and B. By Lemma 10.2, A and B can be separated by open sets, and so X is a normal space. ∎

Our next result is the celebrated Tietze's[7] extension theorem.

Theorem 10.6 (Tietze's Extension Theorem). *Let C be a closed subset of a normal space X and let $f: C \to \mathbb{R}$ be a continuous function. Then there exists a continuous extension of f to all of X with values in $\mathbb{R}$.*

Proof. Let $f: C \to \mathbb{R}$ be a continuous function, where C is a closed subset of a normal space X. Assume first that $f(x) \geq 0$ for each $x \in C$. Put $h = \frac{f}{1+f}$ and note that $h: C \to [0, 1) \subseteq [0, 1]$ is a continuous function. By Uryson's lemma (Theorem 10.5), there exists a continuous extension $h_0: X \to [0, 1]$ of h.

Next, let $B = h_0^{-1}(\{1\})$ and note that B is closed. Also since $0 \leq h_0(x) < 1$ for each $x \in C$, we see that $B \cap C = \emptyset$. Using Uryson's lemma once more, we see that there exists some continuous function $\phi: X \to [0, 1]$ such that $\phi(c) = 1$ for all $c \in C$ and $\phi(b) = 0$ for all $b \in B$. Now notice that the function $g = \frac{\phi h_0}{1 - \phi h_0}$ is a continuous extension of f to all of X with values in $\mathbb{R}$.

For the general case, assume that $f: C \to \mathbb{R}$ is continuous and write $f = f^+ - f^-$, where $f^+ = f \vee 0: C \to \mathbb{R}$ and $f^- = (-f) \vee 0: C \to \mathbb{R}$ are two continuous non-negative continuous functions. By the first part, there exist two continuous functions $\phi_1, \phi_2: C \to \mathbb{R}$ that extend f^+ and f^-, respectively. Now notice that $\phi = \phi_1 - \phi_1: X \to \mathbb{R}$ is a continuous extension of f. ∎

A topological space (X, τ) is called **locally compact** if every point of X has a neighborhood whose closure is a compact set.

Clearly, every compact topological space is locally compact. By Theorem 7.4 a subset of $\mathbb{R}^n$ is compact if and only if it is closed and bounded. Thus, it follows that a Euclidean space $\mathbb{R}^n$ is not compact but it is locally compact.

---

[7]Heinrich Franz Friedrich Tietze (1880–1964), an Austrian mathematician. He worked in topology and he is well known today for his famous "group transformations."

For locally compact spaces we have the following separation property:

Lemma 10.7. *Let (X, τ) be a Hausdorff locally compact topological space. Assume that V is an open set and A is a compact set such that $A \subseteq V$. Then there exists an open set $\mathcal{O}$ with compact closure such that $A \subseteq \mathcal{O} \subseteq \overline{\mathcal{O}} \subseteq V$.*

Proof. Since each point of A has a neighborhood with compact closure, and since A can be covered by a finite number of these neighborhoods, it easily follows that there exists an open set W with compact closure such that $A \subseteq W$. Replacing W by $W \cap V$ (if necessary), we can assume that $A \subseteq W \subseteq V$ holds.

If $\overline{W} \cap V^c = \emptyset$, then $\mathcal{O} = W$ satisfies $A \subseteq \mathcal{O} \subseteq \overline{\mathcal{O}} \subseteq V$. Otherwise, if $x \in \overline{W} \cap V^c$, then $x \notin A$, and so by Theorem 8.13 there exists an open set U_x such that $A \subseteq U_x$ and $x \notin \overline{U}_x$. Now, observe that the family $\{(\overline{U}_x)^c : x \in \overline{W} \cap V^c\}$ is an open cover of the compact set $\overline{W} \cap V^c$ (its compactness follows from Theorem 8.12). Thus, there exists a finite subset F of $\overline{W} \cap V^c$ such that $\overline{W} \cap V^c \subseteq \bigcup_{x \in F} (\overline{U}_x)^c$. Note that $\left(\bigcap_{x \in F} \overline{U}_x \right) \cap \overline{W} \cap V^c = \emptyset$.

Put $\mathcal{O} = \bigcap_{x \in F}(U_x \cap W)$, and note that $\mathcal{O}$ is an open set such that $A \subseteq \mathcal{O} \subseteq W$. It follows that $\overline{\mathcal{O}} \subseteq \overline{W}$, and consequently,

$$\overline{\mathcal{O}} \cap V^c = \overline{\mathcal{O}} \cap \overline{W} \cap V^c \subseteq \left(\bigcap_{x \in F} \overline{U}_x \right) \cap \overline{W} \cap V^c = \emptyset.$$

Hence, $A \subseteq \mathcal{O} \subseteq \overline{\mathcal{O}} \subseteq V$ holds, and the proof is finished. ∎

The following theorem is a "locally compact version" version of Uryson's lemma.

Theorem 10.8 (Uryson). *Let X be a Hausdorff locally compact topological space, and let A be a compact subset of X. If V is an open set such that $A \subseteq V$, then there exists a continuous function $f : X \to [0, 1]$ such that $f(x) = 1$ for all $x \in A$ and $f(x) = 0$ for all $x \in V^c$.*

Proof. The proof is identical to the proof of the implication (2) $\implies$ (3) of Theorem 10.5. The only difference in the proof is that instead of statement (2) we must invoke Lemma 10.7. ∎

As an application of Theorem 10.8, we shall present a useful result dealing with "partitions of unity."

If $f : X \to \mathbb{R}$ is a function, then the closure of the set $Y = \{x \in X : f(x) \neq 0\}$ is called the **support** of f and is denoted by Supp f. That is, Supp $f = \overline{Y}$. A function is said to have **compact support** if its support is a compact set.

Theorem 10.9. *Let X be a Hausdorff locally compact topological space, and let A be a compact subset of X. If $V_1, \ldots, V_n$ are open sets such that $A \subseteq$*

$\bigcup_{i=1}^{n} V_i$, *then there exist continuous real-valued functions* $f_1, \ldots, f_n$ *on* X *satisfying these properties*:

1. $0 \leq f_i(x) \leq 1$ *holds for all* $x \in X$ *and each* $1 \leq i \leq n$.
2. *Each* f_i *has compact support, and* $\mathrm{Supp}\, f_i \subseteq V_i$.
3. $\sum_{i=1}^{n} f_i(x) = 1$ *holds for all* $x \in A$.

Proof. Let $x \in A$. Then there exists some i ($1 \leq i \leq n$) such that $x \in V_i$. Since $\{x\}$ is a compact set, it follows from Lemma 10.7 that there exists a neighborhood U_x of x with compact closure such that $\overline{U}_x \subseteq V_i$. That is, every $x \in A$ has a neighborhood U_x with compact closure satisfying $\overline{U}_x \subseteq V_i$ for some i.

Let $x_1, \ldots, x_m$ be a finite number of points of A such that $A \subseteq \bigcup_{i=1}^{m} U_{x_i}$. Next, for each i define $\mathcal{O}_i$ to be the union of all those U_{x_j} for which $\overline{U}_{x_j} \subseteq V_i$ holds (if no such U_{x_j} exists, then $\mathcal{O}_i = \emptyset$). Clearly, each $\mathcal{O}_i$ is an open set with compact closure satisfying $\overline{\mathcal{O}_i} \subseteq V_i$. Moreover, $A \subseteq \bigcup_{i=1}^{n} \mathcal{O}_i$ holds. By Lemma 10.7, for each i there exists an open set B_i with compact closure such that $\overline{\mathcal{O}_i} \subseteq B_i \subseteq \overline{B}_i \subseteq V_i$.

By Theorem 10.8, for each i there exists a continuous function $g_i \colon X \to [0, 1]$ such that $g_i(x) = 1$ for each $x \in \overline{\mathcal{O}_i}$ and $g_i(x) = 0$ for each $x \notin B_i$. Also, by the same theorem, there exists a continuous function $h \colon X \to [0, 1]$ such that $h(x) = 1$ for all $x \in A$, and $h(x) = 0$ for all $x \in [\bigcup_{i=1}^{n} \mathcal{O}_i]^c$. Put $g = (1 - h) + \sum_{i=1}^{n} g_i$, and note that g is a continuous function with $g(x) > 0$ for all $x \in X$. Now let $f_i = g_i / g$ for $i = 1, \ldots, n$. We leave it for the reader to verify that $f_1, \ldots, f_n$ satisfy the desired properties.　■

Any collection of functions $f_1, \ldots, f_n$ that satisfies the properties of Theorem 10.9 is referred to as a **partition of unity** for A subordinate to the open cover $\{V_1, \ldots, V_n\}$.

EXERCISES

1. Let (X, d) be a metric space and let A be a nonempty subset of X. The **distance function** of A is the function $d(\cdot, A) \colon X \to \mathbf{R}$ defined by

$$d(x, A) = \inf\{d(x, a) \colon a \in A\}.$$

   Show that $d(x, A) = 0$ if and only if $x \in \overline{A}$.

2. Let (X, d) be a metric space, let A and B be two nonempty disjoint closed sets, and consider the function $f \colon X \to [0, 1]$ defined by $f(x) = \frac{d(x, A)}{d(x, A) + d(x, B)}$. Show that:

   a. f is a continuous function,
   b. $f^{-1}(\{0\}) = A$ and $f^{-1}(\{1\}) = B$, and
   c. if $\inf\{d(a, b) \colon a \in A \text{ and } b \in B\} > 0$, then f is uniformly continuous.

3. Let A and B be two nonempty subsets of a metric space X such that $A \cap \overline{B} = \overline{A} \cap B = \emptyset$. Show that there exist two open disjoint sets U and V such that $A \subseteq U$ and $B \subseteq V$.

4. Show that a closed set of a normal space is itself a normal space.

5. Let X be a normal space and let A and B be two disjoint closed subsets of X. Show that there exist open sets V and W such that $A \subseteq V$, $B \subseteq W$ and $\overline{V} \cap \overline{W} = \emptyset$.

6. Show directly that a topological space is normal if and only if for each closed set A and each open set V with $A \subseteq V$, there exists an open set W such that $A \subseteq W \subseteq \overline{W} \subseteq V$.

7. For a closed subset A of a normal topological space X establish the following:

   a. There exists a continuous function $f: X \to [0, 1]$ satisfying $f^{-1}(\{0\}) = A$ if and only if A is a G_δ-set.

   b. If A is a G_δ-set and B is another closed set satisfying $A \cap B = \emptyset$, then there exists a continuous function $g: X \to [0, 1]$ such that $g^{-1}(\{0\}) = A$ and $g(b) = 1$ for each $b \in B$.

8. Show that a compact subset A of a Hausdorff locally compact topological space is a G_δ-set if and only if there exists a continuous function $f: X \to [0, 1]$ such that $A = f^{-1}(\{0\})$.

9. A topological space X is said to be **perfectly normal** if for every pair of disjoint closed sets A and B there exists a continuous function $f: X \to [0, 1]$ such that $A = f^{-1}(\{0\})$ and $B = f^{-1}(\{1\})$. (Part (b) of Exercise 2 above shows that every metric space is perfectly normal.)

   Show that a Hausdorff normal topological space is perfectly normal if and only if every closed set is a G_δ-set.

10. Show that a nonempty connected normal space is either a singleton or uncountable.

11. Let X be a normal space, let C be a closed subset of X, and let I be a nonempty interval. If $f: C \to I$ is a continuous function, then show that f has a continuous extension to all of X with values in I.

11. THE STONE–WEIERSTRASS APPROXIMATION THEOREM

In this section we shall present some conditions under which a linear subspace of $C(X)$, with X compact, is dense in $C(X)$ with respect to the uniform metric. The main result of this sort is known as the Stone–Weierstrass approximation theorem and is a classical result. Before stating and proving this theorem, we need some preliminary discussion.

A collection L of real-valued functions defined on a set X is said to **separate the points** of X if for every pair of distinct points x and y of X there exists a function $f \in L$ such that $f(x) \neq f(y)$.

Our first result presents a property of vector spaces of functions that separate the points. The constant function $\mathbf{1}$ is the function whose value at every point equals 1.

Lemma 11.1. *Let X be a nonempty set, and let L be a vector space of real-valued functions on X that separate the points of X and contains the constant function $\mathbf{1}$. Then given any two distinct points x and y of X and real numbers α and β, there exists some $f \in L$ such that $f(x) = \alpha$ and $f(y) = \beta$.*

Proof.   Since L separates the points of X there exists $g \in L$ such that $g(x) \neq g(y)$; put $\gamma = g(x) - g(y)$. Then the function $f = \gamma^{-1}[(\alpha - \beta)g + (\beta g(x) - \alpha g(y))\mathbf{1}]$ belongs to L and satisfies $f(x) = \alpha$ and $f(y) = \beta$.   ∎

The next result presents a local approximation property that enables us to approximate a function from above at a given point.

Lemma 11.2.   *Let X be a compact topological space, and let L be a function space of continuous functions that contains the constant function $\mathbf{1}$ and separates the points of X. Then given a function $g \in C(X)$, a point $a \in X$, and $\epsilon > 0$, there exists a function f in L such that*

$$f(a) = g(a) \quad \text{and} \quad f(x) > g(x) - \epsilon \quad \text{for all } x \in X.$$

Proof.   For each $x \in X$, there exists (by Lemma 11.1) a function $f_x \in L$ such that $f_x(a) = g(a)$ and $f_x(x) = g(x)$. Since f_x and g are continuous functions, there exists a neighborhood V_x of x such that $f_x(y) > g(y) - \epsilon$ for all $y \in V_x$.

Since $X = \bigcup_{x \in X} V_x$ and X is compact, there exists a finite number of points $x_1, \ldots, x_n$ of X such that $X = \bigcup_{m=1}^{n} V_{x_m}$. Let $f = f_{x_1} \vee \cdots \vee f_{x_n}$; clearly, $f \in L$ and $f(a) = g(a)$. Also, if $x \in X$, then there exists some m such that $x \in V_{x_m}$. Thus, $f(x) \geq f_{x_m}(x) > g(x) - \epsilon$ holds, which shows that the function f satisfies the required properties.   ∎

The lattice version of the Stone–Weierstrass theorem[8] is presented next.

Theorem 11.3 (Stone–Weierstrass).   *Let X be a compact topological space, and let L be a function space of continuous functions separating the points of X and containing the constant function $\mathbf{1}$. Then L is dense in $C(X)$ with respect to the uniform metric.*

Proof.   Let $g \in C(X)$, and let $\epsilon > 0$. For each $x \in X$, use Lemma 11.2 to choose a continuous function $f_x \in L$ such that $f_x \geq g - \epsilon$ and $f_x(x) = g(x)$. From the inequality $f_x(x) = g(x) < g(x) + \epsilon$ and the continuity of f_x and g at x, it follows that there exists a neighborhood V_x of x such that $f_x(y) < g(y) + \epsilon$ for all $y \in V_x$. Since X is compact, there are points $x_1, \ldots, x_n$ such that $X = \bigcup_{m=1}^{n} V_{x_m}$. Let $f = f_{x_1} \wedge \cdots \wedge f_{x_n}$, and note that $f \in L$.

Also, since $f_{x_m} \geq g - \epsilon$, it easily follows that $f \geq g - \epsilon$. On the other hand, if $x \in X$, then there exists some m such that $x \in V_{x_m}$, and so $f(x) \leq f_{x_m}(x) < g(x) + \epsilon$ holds. Thus, $g(x) - \epsilon \leq f(x) \leq g(x) + \epsilon$ holds for all $x \in X$, and consequently, $\|f - g\|_\infty = \sup\{|f(x) - g(x)|: x \in X\} \leq \epsilon$. The proof of the theorem is now complete.   ∎

---

[8]Marshall Harvey Stone (1903–1989), an American mathematician. He contributed to Boolean algebras, topology, the theory of functions, and functional analysis.

Our next result deals with the uniform approximation of the square-root function by polynomials.

Lemma 11.4.   *There exists a sequence of polynomials that converges uniformly to $\sqrt{x}$ in the interval $[0, 1]$.*

Proof.   Start by defining $P_1(x) = 0$ for all $x \in [0, 1]$, and then, inductively, put

$$P_{n+1}(x) = P_n(x) + \tfrac{1}{2}[x - (P_n(x))^2]$$

for $n \geq 1$. Clearly, $\{P_n\}$ is a sequence of polynomials. We claim that $0 \leq P_n(x) \leq \sqrt{x}$ holds for each n and all $x \in [0, 1]$. The proof of the claim is by induction. For $n = 1$ the claim is trivial. Assume now that $0 \leq P_n(x) \leq \sqrt{x}$ holds for all $x \in [0, 1]$ and some n. Clearly, $0 \leq P_{n+1}(x)$ holds for all $x \in [0, 1]$. Also,

$$\begin{aligned}
\sqrt{x} - P_{n+1}(x) &= \sqrt{x} - P_n(x) - \tfrac{1}{2}[x - (P_n(x))^2] \\
&= [\sqrt{x} - P_n(x)][1 - \tfrac{1}{2}(\sqrt{x} + P_n(x))],
\end{aligned}$$

and the two factors of the last product are nonnegative from our induction hypothesis. Therefore, $P_{n+1}(x) \leq \sqrt{x}$ for all $x \in [0, 1]$.

Now, from the definition of P_{n+1} and the fact that $[P_n(x)]^2 \leq x$ for each $x \in [0, 1]$, it follows that the sequence $\{P_n\}$ is increasing and bounded on $[0, 1]$. Thus, $\{P_n\}$ converges pointwise to some non-negative function f on $[0, 1]$. It easily follows that $[f(x)]^2 = x$ for each $x \in [0, 1]$, and so, $f(x) = \sqrt{x}$.

Finally, since $\sqrt{x}$ is a continuous function and $\{P_n\}$ is increasing, Dini's Theorem 9.4 shows that $\{P_n\}$ converges uniformly to $\sqrt{x}$ on $[0, 1]$. ∎

A vector space $\mathcal{A}$ of real-valued functions on a set X is called an **algebra** of functions whenever the product of any two functions in $\mathcal{A}$ is again in $\mathcal{A}$. Thus, a set $\mathcal{A} \subseteq \mathbb{R}^X$ is an algebra if for every pair $f, g \in \mathcal{A}$ and real numbers α and β we have $\alpha f + \beta g$ and fg in $\mathcal{A}$; where, of course, $(fg)(x) = f(x)g(x)$ for each $x \in X$.

And now we are ready to state and prove the classical Stone–Weierstrass theorem.

Theorem 11.5 (Stone–Weierstrass).   *Let X be a compact topological space, and let $\mathcal{A}$ be an algebra of continuous real-valued functions on X separating the points of X and containing the constant function $\mathbf{1}$. Then $\mathcal{A}$ is dense in $C(X)$ with respect to the uniform metric.*

Proof.   Let $\overline{\mathcal{A}}$ denote the closure of $\mathcal{A}$ in $C(X)$ with respect to the uniform metric. Then $\overline{\mathcal{A}}$ is a closed algebra (why?) containing the constant function $\mathbf{1}$ and separating the points of X. We have to show that $\overline{\mathcal{A}} = C(X)$. By Theorem 11.3, it suffices to show that $\overline{\mathcal{A}}$ is a function space.

To this end, let $f \in \overline{\mathcal{A}}$ with $f \neq 0$. Put $a = \|f\|_\infty = \sup\{|f(x)|: x \in X\} > 0$. Let $\{P_n\}$ be the sequence of polynomials determined by Lemma 11.4 that converges uniformly to $\sqrt{x}$ on $[0, 1]$. Since $\overline{\mathcal{A}}$ is an algebra, the function $g_n = P_n(\frac{f^2}{a^2})$ belongs to $\overline{\mathcal{A}}$ for each n. Moreover, the sequence $\{g_n\}$ converges uniformly on X to $\sqrt{\frac{f^2}{a^2}} = \frac{|f|}{a}$. Thus, $\frac{|f|}{a} \in \overline{\mathcal{A}}$, and so, $|f| \in \overline{\mathcal{A}}$. Therefore, $\overline{\mathcal{A}}$ contains the absolute value of every function of $\overline{\mathcal{A}}$. But then, since

$$f \vee g = \tfrac{1}{2}(f + g + |f - g|) \quad \text{and} \quad f \wedge g = \tfrac{1}{2}(f + g - |f - g|),$$

it follows that $f \vee g$ and $f \wedge g$ belong to $\overline{\mathcal{A}}$ for every pair $f, g \in \overline{\mathcal{A}}$. In other words, $\overline{\mathcal{A}}$ is a function space, and the proof of the theorem is complete. ∎

Since the collection of all polynomials on $\mathbb{R}$ is an algebra of continuous functions that contains the constant function **1** and separates the points of $\mathbb{R}$, the following original result of K. Weierstrass follows immediately from the last theorem.

Corollary 11.6 (Weierstrass).   *Any continuous real-valued function on a compact subset A of $\mathbb{R}$ is the uniform limit on A of a sequence of polynomials.*

EXERCISES

1. Let X be a compact topological space. For a subset L of $C(X)$, let $\overline{L}$ denote the uniform closure of L in $C(X)$. Show the following:

    a. If L is a function space, then so is $\overline{L}$.
    b. If L is an algebra, then so is $\overline{L}$.

2. Let L be the collection of all continuous piecewise linear functions defined on $[0, 1]$. That is, $f \in L$ if and only if $f \in C[0, 1]$ and there exists a finite number of points $0 = x_0 < x_1 < \cdots < x_n = 1$ (depending on f) such that f is linear on each interval $[x_{m-1}, x_m]$. Show that L is a function space but not an algebra. Moreover, show that L is dense in $C[0, 1]$ with respect to the uniform metric.

3. Show that a continuous function $f: (0, 1) \to \mathbb{R}$ is the uniform limit of a sequence of polynomials on $(0, 1)$ if and only if it admits a continuous extension to $[0, 1]$.

4. If f is a continuous function on $[0, 1]$ such that $\int_0^1 x^n f(x)\,dx = 0$ for $n = 0, 1, \ldots$, then show that $f(x) = 0$ for all $x \in [0, 1]$.

5. Show that the algebra generated by the set $\{1, x^2\}$ is dense in $C[0, 1]$ but fails to be dense in $C[-1, 1]$.

6. Let us say that a polynomial is **odd** (resp. **even**) whenever it does not contain any monomial of even (resp. odd) degree.

    Show that a continuous function $f: [0, 1] \to \mathbb{R}$ vanishes at zero (i.e., $f(0) = 0$) if and only if it is the uniform limit of a sequence of odd polynomials on $[0, 1]$.

7. If $f: [0, 1] \to \mathbb{R}$ is a continuous function such that $\int_0^1 f(\sqrt[2n+1]{x})\, dx = 0$ for $n = 0, 1, 2, \ldots$, then show that $f(x) = 0$ for all $x \in [0, 1]$. Does the same conclusion hold true if the interval $[0, 1]$ is replaced by the interval $[-1, 1]$?

8. Assume that a function $f: [0, \infty) \to \mathbb{R}$ is either a polynomial or else a continuous bounded function. Then show that f is identically equal to zero (i.e., show that $f = 0$) if and only if $\int_0^\infty f(x) e^{-nx}\, dx = 0$ for all $n = 1, 2, 3, \ldots$.

9. Show that a continuous bounded function $f: [1, \infty) \to \mathbb{R}$ is identically equal to zero if and only if $\int_1^\infty x^{-n} f(x)\, dx = 0$ for each $n = 8, 9, 10, \ldots$.

10. Let $\mathcal{A}$ be an algebra of continuous real-valued functions defined on a compact topological space X which separates the points of X. Show that the closure $\overline{\mathcal{A}}$ of $\mathcal{A}$ in $C(X)$ with respect to the uniform metric is either all of $C(X)$ or else that there exists some $a \in X$ such that $\overline{\mathcal{A}} = \{f \in C(X): f(a) = 0\}$.
    [HINT: A glance at the proof of Lemma 11.4 shows that the polynomials $P_n(x)$ that approximate $\sqrt{x}$ uniformly on $[0, 1]$ have constant terms zero. This implies that $|f|$ and $\sqrt{|f|}$ both belong to $\overline{\mathcal{A}}$ for each $f \in \overline{\mathcal{A}}$.]

11. Let $\mathcal{A}$ be the vector space generated by the functions $\mathbf{1}$, $\sin x$, $\sin^2 x$, $\sin^3 x$, $\ldots$ defined on $[0, 1]$. That is, $f \in \mathcal{A}$ if and only if there is a nonnegative integer k and real numbers $\alpha_0, \alpha_1, \ldots, \alpha_k$ (all depending on f) such that $f(x) = \sum_{n=0}^{k} \alpha_n \sin^n x$ for each $x \in [0, 1]$. Show that $\mathcal{A}$ is an algebra and that $\mathcal{A}$ is dense in $C[0, 1]$ with respect to the uniform metric.

12. Let X be a compact subset of $\mathbb{R}$. Show that $C(X)$ is a separable metric space (with respect to the uniform metric).

13. Generalize the previous exercise as follows: Show that if (X, d) is a compact metric space, then $C(X)$ is a separable metric space.
    [HINT: Let $\{x_n\}$ be a countable dense subset of X, and let $f_n(x) = d(x, x_n)$. If $\mathcal{A}$ is the algebra generated by the functions $\mathbf{1}$, f_1, f_2, $\ldots$, show that $\overline{\mathcal{A}} = C(X)$.]

14. Let X and Y be two compact metric spaces. Consider the Cartesian product $X \times Y$ equipped with the distance D_1 given in Exercise 4 of Section 7, so that $X \times Y$ is a compact metric space. Show that if $f \in C(X \times Y)$ and $\epsilon > 0$, then there exist functions $\{f_1, \ldots, f_n\} \subseteq C(X)$ and $\{g_1, \ldots, g_n\} \subseteq C(Y)$ such that

$$\left| f(x, y) - \sum_{i=1}^{n} f_i(x) g_i(y) \right| < \epsilon$$

holds for all $(x, y) \in X \times Y$.
    [HINT: Consider the algebra generated in $C(X \times Y)$ by the functions $F(x, y) = f(x)$ and $G(x, y) = g(y)$ for $f \in C(X)$ and $g \in C(Y)$.]

CHAPTER 3

THE THEORY OF MEASURE

At the turn of the nineteenth century it was quite apparent to mathematicians that the properties of continuous functions and Riemann's theory of integration were not rich enough to solve many scientific problems. The inadequacies of the continuous functions led them to search for different classes of functions that would provide solutions to a variety of problems.

Around the beginning of the twentieth century, the theory of measure was originated. At that time, it was realized that to get a better understanding of the structure of functions it was necessary to make a thorough study of the subsets of Euclidean spaces. To study these sets, it became clear that the classical notions of length, area, and volume needed to be generalized. The search for devising ways of assigning a concept of a "measure" to a given set of points has its roots in that period.

E. Borel [4] in 1898 was the first to establish a measure theory on the subsets of the real numbers known today as Borel sets. Soon after (in 1902), H. Lebesgue [21] presented his pioneering work on Lebesgue measure, and a little later (around 1918), C. Carathéodory introduced and studied the properties of outer measures. From then on a rapid development of the theory of measure, which included among its contributors the most prominent mathematicians of the first half of the twentieth century, followed.

This chapter discusses in detail the theory of measure. We start our study by introducing the concept of a semiring of sets and then proceed by studying the properties of measures on semirings. The important notion of the outer measure is introduced and studied next. It is followed by a detailed investigation of the measurable sets and measurable functions. Our attention is then turned to the properties of simple and step functions and to the basic properties of the Lebesgue measure. The chapter culminates with an investigation of convergence in measure and a discussion on abstract measurability properties.

12. SEMIRINGS AND ALGEBRAS OF SETS

In this section the notion of a semiring of sets is introduced and its properties are studied. A semiring of sets is the simplest family of sets for which a measure theory can be built. It turns out that most "reasonable" collections of sets satisfy the semiring properties.

Definition 12.1. *Let X be a nonempty set. A collection S of subsets of X is called a* **semiring** *if it satisfies the following properties:*

1. *The empty set belongs to S; that is $\emptyset \in S$.*
2. *If $A, B \in S$; then $A \cap B \in S$; that is, S is closed under finite intersections.*
3. *The set difference of any two sets of S can be written as a finite union of pairwise disjoint members of S. That is, for every $A, B \in S$; there exist $C_1, \ldots, C_n$ in S (depending on A and B) such that $A \setminus B = \bigcup_{i=1}^n C_i$ and $C_i \cap C_j = \emptyset$ if $i \neq j$.*

Now, let S be a semiring of subsets of X. A subset A of X is called a σ-**set** with respect to S (or simply a σ-set) if there exists a disjoint sequence $\{A_n\}$ of S (i.e., $A_n \cap A_m = \emptyset$ if $n \neq m$) such that $A = \bigcup_{n=1}^\infty A_n$. If $A = \bigcup_{i=1}^\infty A_i$ with $A_1, \ldots, A_n \in S$ and $A_i \cap A_j = \emptyset$ for $i \neq j$, then A is a σ-set. To see this, put $A_i = \emptyset$ for $i > n$. It follows from Definition 12.1 that $A \setminus B$ is a σ-set for every pair A and B in S.

Some basic properties of σ-sets are included in the next theorem.

Theorem 12.2. *For a semiring S, the following statements hold:*

1. *If $A \in S$ and $A_1, \ldots, A_n \in S$, then $A \setminus \bigcup_{i=1}^n A_i$ can be written as a finite union of disjoint sets of S (and hence, it is a σ-set).*
2. *For every sequence $\{A_n\}$ of S, the set $A = \bigcup_{n=1}^\infty A_n$ is a σ-set.*
3. *Countable unions and finite intersections of σ-sets are σ-sets.*

Proof. (1) We use induction on n. For $n = 1$, the statement is true from the definition of the semiring. Now, assume the statement true for some n. Let $A \in S$, and let $A_1, \ldots, A_n, A_{n+1} \in S$. By the induction hypothesis, there exist $B_1, \ldots, B_k \in S$ such that $B = A \setminus \bigcup_{i=1}^n A_i = \bigcup_{i=1}^k B_i$ and $B_i \cap B_j = \emptyset$ if $i \neq j$. Consequently,

$$A \setminus \bigcup_{i=1}^{n+1} A_i = B \setminus A_{n+1} = \bigcup_{i=1}^k (B_i \setminus A_{n+1}).$$

By property (3) of Definition 12.1, each $B \setminus A_{n+1}$ can be written as a finite union of disjoint sets of S. Since $B_i \cap B_j = \emptyset$ if $i \neq j$, it easily follows that $A \setminus \bigcup_{i=1}^{n+1} A_i$ can be written as a finite union of disjoint sets of S. This completes the induction and the proof of (1).

(2) Let $\{A_n\} \subseteq S$. Put $A = \bigcup_{n=1}^\infty A_n$, and then write $A = \bigcup_{n=1}^\infty B_n$ with $B_1 = A_1$ and $B_{n+1} = A_{n+1} \setminus \bigcup_{i=1}^n A_i$ for $n \geq 1$. Observe that $B_i \cap B_j = \emptyset$ if

$i \neq j$, and by statement (1) each B_i is a σ-set. It now follows easily that A is itself a σ-set.

(3) The proof follows from (2), and property (2) of Definition 12.1.   ■

The proof of part (2) of the preceding theorem also guarantees the validity of the following useful result:

Lemma 12.3. *If $\{A_n\}$ is a sequence of sets in a semiring S, then there exists a disjoint sequence $\{C_n\}$ of S such that $\bigcup_{n=1}^{\infty} A_n = \bigcup_{n=1}^{\infty} C_n$ and for each n there exists some k with $C_n \subseteq A_k$.*

Some natural collections of sets happen to satisfy other properties that are stronger than those of a semiring. The "algebra of sets" is such a collection, and its definition follows.

Definition 12.4. *A nonempty collection S of subsets of a set X which is closed under finite intersections and complementation is called an **algebra of sets** (or simply an **algebra**). That is, S is an algebra whenever it satisfies the following properties:*

   i. *If $A, B \in S$, then $A \cap B \in S$.*
   ii. *If $A \subset S$, then $A^c \in S$.*

Three basic properties of an algebra are included in the next theorem.

Theorem 12.5. *For an algebra of sets S, the following statements hold:*
1. $\emptyset, X \in S$.
2. *The algebra S is closed under finite unions and intersections.*
3. *The algebra S is a semiring.*

Proof. (1) Since S is nonempty there exists some $A \in S$. Now, by hypothesis $A^c \in S$, and so, $\emptyset = A \cap A^c \in S$. Moreover, $X = \emptyset^c \in S$.

(2) Let $A, B \in S$. Then $A \cup B = (A^c \cap B^c)^c \in S$, and the rest of the proof can be completed easily by induction.

(3) We have to verify only property (3) of Definition 12.1. But this is obvious in view of the identity $A \setminus B = A \cap B^c$.   ■

We continue by illustrating the notions of semiring and algebra of sets with examples.

Example 12.6. For every nonempty set X, the collection $S = \{\emptyset, X\}$ is an algebra of sets. This is the "smallest" (with respect to inclusion) possible algebra.   ■

Example 12.7. For every nonempty set X, its power set $\mathcal{P}(X)$ (i.e., the collection of all subsets of X) forms an algebra. This is the "largest" possible algebra.   ■

Example 12.8. Let $\mathcal{F}$ be a nonempty pairwise disjoint family of subsets of a set X. Then $S = \mathcal{F} \cup \{\emptyset\}$ is a semiring of subsets of X. To see this, note first that $\emptyset \in S$. Now, if $A, B \in S$, then $A \cap B$ is either empty or equal to A. Likewise, $A \setminus B$ is either empty or equal to A. Thus, $A, B \in S$ implies that $A \cap B$ and $A \setminus B$ both belong to S, and so S is a semiring.   ■

Example 12.9. If $a, b \in \mathbb{R}$, let us write $[a, b) = \emptyset$ if $a \geq b$ and (as usual) $[a, b) = \{x \in \mathbb{R}: a \leq x < b\}$ if $a < b$. Then the collection $S = \{[a, b): a, b \in \mathbb{R}\}$ is a semiring of subsets of $\mathbb{R}$, which is not an algebra (for instance, notice that $[0, 1) \cup [2, 3) \notin S$).   ■

The semiring of the previous example is very important because of its many applications. Its analogue in higher dimensions is presented next.

Example 12.10. Let S denote the collection of all subsets A of $\mathbb{R}^n$ for which there exist intervals $[a_1, b_1), \ldots, [a_n, b_n)$ such that $A = [a_1, b_1) \times \cdots \times [a_n, b_n)$. (If $a_i \geq b_i$ holds for some i, then $[a_i, b_i) = \emptyset$, and so $A = \emptyset$.) Then S is a semiring of subsets of $\mathbb{R}^n$. To see this, note first that only the third property of the semiring definition needs verification; the other two are trivial. The proof is based upon the following identity among sets A, B, C, and D:

$$A \times B \setminus C \times D = [(A \setminus C) \times B] \cup [(A \cap C) \times (B \setminus D)], \qquad (\star)$$

where the sets of the union on the right-hand side are disjoint.

For the proof, use induction on n. For $n = 1$ the result is straightforward. Assume it now true for some n. We have to show that any set of the form

$$[a_1, b_1) \times \cdots \times [a_n, b_n) \times [a_{n+1}, b_{n+1}) \setminus [c_1, d_1) \times \cdots \times [c_n, d_n) \times [c_{n+1}, d_{n+1})$$

can be written as a finite union of disjoint sets from the $(n + 1)$-dimensional collection S. But this can be easily shown by letting

$$A = [a_1, b_1) \times \cdots \times [a_n, b_n), \qquad B = [a_{n+1}, b_{n+1}),$$
$$C = [c_1, d_1) \times \cdots \times [c_n, d_n), \qquad D = [c_{n+1}, d_{n+1})$$

in $(\star)$ and using the induction hypothesis.   ■

An intermediate notion between semirings and algebras is that of a ring of sets. A **ring of sets** (or simply a **ring**) is a nonempty collection of subsets $\mathcal{R}$ of a set X satisfying these properties:

   a.   If $A, B \in \mathcal{R}$, then $A \cup B \in \mathcal{R}$.
   b.   If $A, B \in \mathcal{R}$, then $A \setminus B \in \mathcal{R}$.

Every ring $\mathcal{R}$ contains the empty set. Indeed, since $\mathcal{R}$ is nonempty, there exists $A \in \mathcal{R}$, and so $\emptyset = A \setminus A \in \mathcal{R}$. Clearly, every algebra of sets is a ring of sets. Also, a ring $\mathcal{R}$ is necessarily a semiring. Indeed, if $A, B \in \mathcal{R}$, then the relation $A \cap B = A \setminus (A \setminus B)$ shows that $A \cap B \in \mathcal{R}$.

Another useful concept is that of a σ-algebra of sets.

Definition 12.11. *An algebra $\mathcal{S}$ of subsets of some set X is called a σ-algebra if every union of a countable collection of members of $\mathcal{S}$ is again in $\mathcal{S}$. That is, in addition to $\mathcal{S}$ being an algebra, $\bigcup_{n=1}^{\infty} A_n$ belongs to $\mathcal{S}$ for every sequence $\{A_n\}$ of $\mathcal{S}$.*

By virtue of $\bigcap_{n=1}^{\infty} A_n = (\bigcup_{n=1}^{\infty} A_n^c)^c$, it easily follows that every σ-algebra of sets is also closed under countable intersections.

Every collection of subsets $\mathcal{F}$ of a nonempty set X is contained in a smallest σ-algebra (with respect to the inclusion relation). This σ-algebra is the intersection of all σ-algebras that contain $\mathcal{F}$ (notice that $\mathcal{P}(X)$ is one of them), is called the σ-**algebra generated** by $\mathcal{F}$.

An important σ-algebra of sets is the σ-algebra of all Borel sets of a topological space. Its definition is given next.

Definition 12.12. *The **Borel sets** of a topological space (X, τ) are the members of the σ-algebra generated by the open sets. The σ-algebra of all Borel sets of (X, τ) will be denoted by $\mathcal{B}$.*

EXERCISES

1. If X is a topological space, then show that the collection
$$S = \{C \cap O: C \text{ closed and } O \text{ open}\} = \{C_1 \setminus C_2: C_1, C_2 \text{ closed sets}\}$$
is a semiring of subsets of X.

2. Let S be a semiring of subsets of a set X, and let $Y \subseteq X$. Show that $S_Y = \{Y \cap A: A \in S\}$ is a semiring of Y (called the **restriction semiring** of S to Y).

3. Let S be the collection of all subsets of $[0, 1)$ that can be written as finite unions of subsets of $[0, 1)$ of the form $[a, b)$. Show that S is an algebra of sets but not a σ-algebra.

4. Prove that the σ-sets of the semiring
$$S = \{[a, b): a, b \in \mathbb{R} \text{ and } a \leq b\}$$
form a topology for the real numbers.

5. Let S be a semiring of subsets of a nonempty set X. What additional requirements must be satisfied for S in order to be a base for a topology on X? (For the definition of a base, see Exercise [18] of Section [8].) Prove that if such is the case, then each member of S is both open and closed in this topology.

6. Let A be a fixed subset of a set X. Determine the two σ-algebras of subsets of X generated by

    a. $\{A\}$, and
    b. $\{B: A \subseteq B \subseteq X\}$.

7. Let X be an uncountable set, and let

$$S = \{E \subseteq X: E \text{ or } E^c \text{ is at most countable}\}.$$

Show that S is the σ-algebra generated by the one-point subsets of X.

8. Characterize the metric spaces whose open sets form a σ-algebra.

9. Determine the σ-algebra generated by the nowhere dense subsets of a topological space.

10. Let X be a nonempty set, and let $\mathcal{F}$ be an uncountable collection of subsets of X. Show that any element of the σ-algebra generated by $\mathcal{F}$ belongs to the σ-algebra generated by some countable subcollection of $\mathcal{F}$.

11. Show that every F_σ- and every G_δ-subset of a topological space is a Borel set.

12. Show that every infinite σ-algebra of sets has uncountably many sets.

13. Let (X, τ) be a topological space, let $\mathcal{B}$ be the σ-algebra of its Borel sets, and let Y be an arbitrary subset of X. If Y is considered equipped with the induced topology and $\mathcal{B}_Y$ denotes the σ-algebra of Borel sets of (Y, τ), then show that

$$\mathcal{B}_Y = \{A \cap Y: A \in \mathcal{B}\}.$$

14. Let $A_1, \ldots, A_n$ be sets in some semiring S. Show that there exists a finite number of pairwise disjoint sets $B_1, \ldots, B_m$ of S such that each A_i can be written as a union of sets from the $B_1, \ldots, B_m$.
[HINT: Use induction on n and Theorem 12.2(1).]

13.   MEASURES ON SEMIRINGS

The semirings would not be of importance to us if it were not for the purpose of defining measures on them. The concept of a measure can be thought of as a generalization of the concepts of length and area, and its definition is given next. A real-valued function defined on a family of sets is referred to as a **set function**.

> **Definition 13.1.** *Let S be a semiring of subsets of a set X. A set function $\mu: S \to [0, \infty]$ is called a **measure** on S if it satisfies the following properties:*
>
> 1. *$\mu(\varnothing) = 0$, and*
> 2. *whenever $\{A_n\}$ is a disjoint sequence of S satisfying $\bigcup_{n=1}^{\infty} A_n \in S$, then*
>
> $$\mu\left(\bigcup_{n=1}^{\infty} A_n\right) = \sum_{n=1}^{\infty} \mu(A_n)$$
>
> *holds; that is, μ is σ-**additive**.*

A triplet $(X, \mathcal{S}, \mu)$, where X is a nonempty set, $\mathcal{S}$ is a semiring of subsets of X, and μ is a measure on $\mathcal{S}$ is called a **measure space**.

Theorem 13.2. *For a measure space $(X, \mathcal{S}, \mu)$, the following statements hold:*

1. *If $A_1, \ldots, A_n \in \mathcal{S}$ are pairwise disjoint and $\bigcup_{i=1}^{n} A_i \in \mathcal{S}$, then $\mu(\bigcup_{i=1}^{n} A_i) = \sum_{i=1}^{n} \mu(A_i)$. That is, μ is* **finitely additive**.
2. *If $A, B \in \mathcal{S}$ satisfy $A \subseteq B$, then $\mu(A) \leq \mu(B)$ holds. That is, μ is* **monotone**.

Proof. (1) If $A_1, \ldots, A_n \in \mathcal{S}$ are pairwise disjoint sets such that $\bigcup_{i=1}^{n} A_i \in \mathcal{S}$, let $A_i = \emptyset$ for $i > n$. Then $\{A_i\}$ is a disjoint sequence of $\mathcal{S}$ satisfying $\bigcup_{i=1}^{\infty} A_i = \bigcup_{i=1}^{n} A_i \in \mathcal{S}$. Thus, by the σ-additivity of μ we have

$$\mu\left(\bigcup_{i=1}^{n} A_i\right) = \mu\left(\bigcup_{i=1}^{\infty} A_i\right) = \sum_{i=1}^{\infty} \mu(A_i) = \sum_{i=1}^{n} \mu(A_i),$$

where the last equality holds by virtue of $\mu(\emptyset) = 0$.

(2) Let $A, B \in \mathcal{S}$ satisfy $A \subseteq B$. Choose a finite collection of disjoint sets $C_1, \ldots, C_n$ of $\mathcal{S}$ such that $B \setminus A = \bigcup_{i=1}^{n} C_i$. Then $B = A \cup (B \setminus A) = A \cup C_1 \cup \cdots \cup C_n$ is a finite union of disjoint sets of $\mathcal{S}$. Thus, from part (1) we have

$$\mu(B) = \mu(A) + \mu(C_1) + \cdots + \mu(C_n) \geq \mu(A),$$

and the proof of the theorem is complete. ∎

We continue with some examples of measure spaces.

Example 13.3 (The Counting Measure). Let X be a set, and let $\mathcal{S} = \mathcal{P}(X)$. Define $\mu: \mathcal{S} \to [0, \infty]$ by $\mu(A) = \infty$ if A is an infinite subset of X and $\mu(A) = $ the number of elements of A if A is a finite set. The reader can verify easily that $(X, \mathcal{S}, \mu)$ is a measure space. ∎

The measure of the next example is known as a Dirac[1] measure.

Example 13.4 (The Dirac Measure). Let X be a nonempty set, and let $\mathcal{S} = \mathcal{P}(X)$. Fix an element $a \in X$, and define $\mu: \mathcal{S} \to [0, \infty)$ by $\mu(A) = 0$ if $a \notin A$ and $\mu(A) = 1$ if $a \in A$. It is easy to see that $(X, \mathcal{S}, \mu)$ is a measure space. ∎

Example 13.5. Let $\mathcal{F}$ be a nonempty pairwise disjoint family of subsets of a set X, and let $\mathcal{S} = \mathcal{F} \cup \{\emptyset\}$. In Example 12 we verified that $\mathcal{S}$ is a semiring. Now, for each nonempty set $A \in \mathcal{F}$ fix some $m_A \in [0, \infty]$. Then the set function $\mu: \mathcal{S} \to [0, \infty)$, defined

---

[1] Paul Adrien Maurice Dirac (1902–1984), a famous English theoretical physicist. He won the Nobel prize in Physics at the age of 31 for his pioneering work in quantum theory. This measure was introduced by him in the context of "delta functions."

by $\mu(\emptyset) = 0$ and $\mu(A) = m_A$ if $A \in S$ is nonempty, is a measure. To see this, note that if $A \in S$ can be written as a disjoint union $A = \bigcup_{n=1}^{\infty} A_n$ with $\{A_n\} \subseteq S$, then it must be the case that $A = A_k$ for some k and $A_n = \emptyset$ for $n \neq k$. This easily implies $\mu(A) = \mu(A_k) = \sum_{n=1}^{\infty} \mu(A_n)$ and so μ is σ-additive. ∎

Example 13.6. Assume that the function $f: \mathbb{R} \to \mathbb{R}$ is nondecreasing and left continuous; that is, $\lim_{x \uparrow a} f(x) = f(a)$ holds for each $a \in \mathbb{R}$. Consider the semiring $S = \{[a, b): a, b \in \mathbb{R} \text{ and } a \leq b\}$; see Example 12. Now define $\mu: S \to [0, \infty)$ by $\mu([a, b)) = f(b) - f(a)$ if $a < b$ and $\mu(\emptyset) = 0$. We claim that the set function μ is a measure.

To see that μ is σ-additive, let $a < b$ and let $[a, b) = \bigcup_{n=1}^{\infty} [a_n, b_n)$ with the sequence $\{[a_n, b_n)\}$ disjoint; we can assume that $a_n < b_n$ for each n. Let

$$s = \sum_{n=1}^{\infty} \mu([a_n, b_n)).$$

Rearranging $[a_1, b_1), \ldots, [a_k, b_k)$, we can suppose that

$$a_1 < b_1 \leq a_2 < b_2 \leq \cdots \leq a_k < b_k.$$

Since f is increasing, $\sum_{i=1}^{k}[f(b_i) - f(a_i)] \leq f(b_k) - f(a_1) \leq f(b) - f(a)$, which implies

$$s \leq f(b) - f(a) = \mu([a, b)). \tag{$\star$}$$

For the reverse inequality, let $\delta > 0$ and $0 < \epsilon < b - a$. For each n choose some $c_n < a_n$ satisfying $f(a_n) - f(x) < 2^{-n}\delta$ whenever $c_n < x \leq a_n$. Since $[a, b - \epsilon] \subseteq \bigcup_{n=1}^{\infty}(c_n, b_n)$ and $[a, b - \epsilon]$ is compact, it follows that $[a, b - \epsilon] \subseteq \bigcup_{n=1}^{k}(c_n, b_n)$ must hold for some k. Assume $a_1 = a$. If $b_1 < b - \epsilon$, then by rearranging, we can assume $b_1 \in (c_2, b_2)$. Since $(a_1, b_1) \cap (a_2, b_2) = \emptyset$, it follows that $c_2 < b_1 \leq a_2$. Continuing this process, we obtain $(c_1, b_1), \ldots, (c_m, b_m)$, $1 \leq m \leq k$, with $b - \epsilon \leq b_m$ and, if $m \geq 2$, $c_{i+1} < b_i < a_{i+1}$ for $1 \leq i \leq m - 1$; note that $\sum_{i=1}^{m-1}[f(a_{i+i}) - f(b_i)] \leq \delta$. Consequently,

$$s \geq \sum_{i=1}^{m}[f(b_i) - f(a_i)]$$

$$= f(b_m) - f(a_1) - \sum_{i=1}^{m-1}[f(a_{i+1}) - f(b_i)]$$

$$> f(b - \epsilon) - f(a) - \delta.$$

Since $\delta > 0$ and $0 < \epsilon < b - a$ are arbitrary, the left continuity of f at b implies

$$s \geq f(b) - f(a).$$

The latter combined with $(\star)$ shows that $s = f(b) - f(a)$, so μ is σ-additive. ∎

An important special case of the preceding example is the case when $f(x) = x$ for all $x \in \mathbb{R}$. The resulting measure is called the **Lebesgue measure** on S, and it will be denoted by λ; that is, $\lambda([a, b)) = b - a$. Later, the domain of this measure will be extended to include all the open and closed sets.

Example 13.7. Consider the semiring S of Example 12.10. That is, the semiring S consists of all subsets of $\mathbb{R}^n$ of the form $[a_1, b_1) \times \cdots \times [a_n, b_n)$, with $a_i < b_i$ for $1 \le i \le n$ together with the empty set. Define $\lambda: S \to [0, \infty)$ by $\lambda(\emptyset) = 0$ and

$$\lambda([a_1, b_1) \times \cdots \times [a_n, b_n)) = \prod_{i=1}^{n}(b_i - a_i).$$

Then λ is called the **Lebesgue measure** on S. We postpone the proof of the σ-additivity of λ until Section 18. A proof that requires some additional background will be presented in Theorem 18.1.   ∎

The following theorem characterizes the set functions on semirings that are measures:

Theorem 13.8. *Let S be a semiring, and let $\mu: S \to [0, \infty]$ be a set function. Then μ is a measure on S if and only if μ satisfies the following conditions:*

1. *$\mu(\emptyset) = 0$.*
2. *If $A \in S$ and $A_1, \dots, A_n \in S$ satisfy $\bigcup_{i=1}^{n} A_i \subseteq A$ and $A_i \cap A_j = \emptyset$ for $i \ne j$, then $\sum_{i=1}^{n} \mu(A_i) \le \mu(A)$ holds.*
3. *If $A \in S$ and $\{A_n\} \subseteq S$ satisfy $A \subseteq \bigcup_{n=1}^{\infty} A_n$, then $\mu(A) \le \sum_{n=1}^{\infty} \mu(A_n)$ holds; that is, μ is σ-subadditive.*

Proof. Assume that μ is a measure on S. Then by definition, $\mu(\emptyset) = 0$. For (2) assume that $A \in S$, and that the disjoint sets $A_1, \dots, A_n$ of S satisfy $\bigcup_{i=1}^{n} A_i \subseteq A$. By Theorem 12.2(1), there exist disjoint sets $B_1, \dots, B_m$ of S such that $A \setminus \bigcup_{i=1}^{n} A_i = \bigcup_{i=1}^{m} B_i$. Put $C_1 = A_1, \dots, C_n = A_n$, and $C_{n+i} = B_i$ for $1 \le i \le m$. Then the sets $C_1, \dots, C_{n+m}$ are disjoint and $A = \bigcup_{i=1}^{m+n} C_i$. By the finite additivity property of μ (see Theorem 13.2), we get

$$\mu(A) = \sum_{i=1}^{m+n} \mu(C_i) \ge \sum_{i=1}^{n} \mu(A_i).$$

For the σ-subadditivity of μ, assume that $A \subseteq \bigcup_{n=1}^{\infty} A_n$ holds with $A \in S$ and $\{A_n\} \subseteq S$. Put $B_1 = A_1$ and $B_{n+1} = A_{n+1} \setminus \bigcup_{i=1}^{n} A_i$ for $n \ge 1$. Then $\bigcup_{n=1}^{\infty} B_n = \bigcup_{n=1}^{\infty} A_n$ and $B_n \subseteq A_n$ for each n. Also, the sequence $\{B_n\}$ is disjoint, and by Theorem 12.2(1), for each $n \ge 2$ there exist pairwise disjoint sets $C_1^n, \dots, C_{k_n}^n$ in S such that $B_n = \bigcup_{i=1}^{k_n} C_i^n$. Note that by (2) and $\bigcup_{i=1}^{k_n} C_i^n \subseteq A_n$ for each n, it follows that $\sum_{i=1}^{k_n} \mu(C_i^n) \le \mu(A_n)$. (For $n = 1$, we put $k_1 = 1$ and $C_1^1 = A_1$.)

Now, observe that $A = \bigcup_{n=1}^{\infty}(B_n \cap A) = \bigcup_{n=1}^{\infty} \bigcup_{i=1}^{k_n}(C_i^n \cap A)$, is a disjoint union. So, by the σ-additivity of μ we get

$$\mu(A) = \sum_{n=1}^{\infty} \sum_{i=1}^{k_n} \mu(C_i^n \cap A) \le \sum_{n=1}^{\infty} \sum_{i=1}^{k_n} \mu(C_i^n) \le \sum_{n=1}^{\infty} \mu(A_n).$$

Conversely, if the set function $\mu: S \to [0, \infty]$ satisfies the above three conditions, then μ is σ-additive by combining (2) and (3). Hence, μ is a measure. ∎

We close the section with the definition of a finitely additive measure. A set function $\mu: S \to [0, \infty]$, where S is a semiring, is called a **finitely additive measure** on S if it satisfies these properties:

a. $\mu(\emptyset) = 0$.
b. If $A_1, \ldots, A_n \in S$ are disjoint and $\bigcup_{i=1}^{n} A_i \in S$, then

$$\mu\left(\bigcup_{i=1}^{n} A_i\right) = \sum_{i=1}^{n} \mu(A_i).$$

It easily follows that every finitely additive measure μ is monotone; that is, if $A, B \in S$ satisfy $A \subseteq B$, then $\mu(A) \le \mu(B)$ holds. By Theorem 13.2, every measure is a finitely additive measure, but the converse is not true. See Exercise 7 of this section.

EXERCISES

1. Let $\{a_n\}$ be a sequence of nonnegative real numbers. Set $\mu(\emptyset) = 0$, and for every nonempty subset A of $\mathbb{N}$ put $\mu(A) = \sum_{n \in A} a_n$. Show that $\mu: \mathcal{P}(\mathbb{N}) \to [0, \infty]$ is a measure.

2. Let S be a semiring, and let $\mu: S \to [0, \infty]$ be a set function such that $\mu(A) < \infty$ for some $A \in S$. If μ is σ-additive, then show that μ is a measure.

3. Let X be an uncountable set, and let the σ-algebra

$$S = \{E \subseteq X: E \text{ or } E^c \text{ is at most countable}\};$$

see also Exercise 7 of Section 12. Show that $\mu: S \to [0, \infty)$, defined by $\mu(E) = 0$ if E is at most countable and $\mu(E) = 1$ if E^c is at most countable, is a measure on S.

4. Let X be a nonempty set, and let $f: X \to [0, \infty]$ be a function. Define $\mu: \mathcal{P}(X) \to [0, \infty]$ by $\mu(A) = \sum_{x \in A} f(x)$ if $A \ne \emptyset$ and is at most countable, $\mu(A) = \infty$ if A is uncountable, and $\mu(\emptyset) = 0$. Show that μ is a measure.

5. Let S be a semiring, and let $\mu: S \to [0, \infty]$ be a finitely additive measure. Show that if μ is σ-subadditive, then μ is a measure.

6. Let $\{\mu_n\}$ be an increasing sequence of measures on a semiring S; that is, $\mu_n(A) \le \mu_{n+1}(A)$ holds for all $A \in S$ and all n. Define $\mu: S \to [0, \infty]$ by $\mu(A) = \sup\{\mu_n(A)\}$ for each $A \in S$. Show that μ is a measure.

7. Consider the semiring $S = \{A \subseteq \mathbb{R}: A \text{ is at most countable}\}$, and define the set function $\mu: S \to [0, \infty]$ by $\mu(A) = 0$ if A is finite and $\mu(A) = \infty$ if A is countable. Show that μ is a finitely additive measure that is not a measure.
8. Show that every finitely additive measure is monotone.
9. Consider the set function μ defined in Example 13.6. That is, consider a nondecreasing and left-continuous function $f: \mathbb{R} \to \mathbb{R}$ and then define the set function $\mu: S \to [0, \infty)$ by $\mu([a, b)) = f(b) - f(a)$, where S is the semiring $S = \{[a, b): -\infty < a \le b < \infty\}$. Prove alternately the fact that μ is a measure.

14. OUTER MEASURES AND MEASURABLE SETS

The theory of outer measures will be presented in this section. The concept of an outer measure is due to C. Carathéodory[2] and is defined as follows.

Definition 14.1 (Carathéodory). *A set function $\mu: \mathcal{P}(X) \to [0, \infty]$ defined on the power set $\mathcal{P}(X)$ of some set X is called an **outer measure** if it satisfies these properties:*

1. $\mu(\emptyset) = 0$.
2. $\mu(A) \le \mu(B)$ if $A \subseteq B$; that is, μ is monotone.
3. $\mu(\bigcup_{n=1}^{\infty} A_n) \le \sum_{n=1}^{\infty} \mu(A_n)$ holds for every sequence $\{A_n\}$ of subsets of X; that is, μ is σ-**subadditive**.

An outer measure μ need not be σ-additive on $\mathcal{P}(X)$. However, as we shall see, there always exists a σ-algebra of subsets (called the measurable sets) on which μ is σ-additive. The details will be explained below.

Throughout the rest of this section, μ will denote a fixed outer measure. The next definition describes the measurable sets and is also due to C. Carathéodory.

Definition 14.2 (Carathéodory). *A subset E of X is called **measurable** (more precisely, μ-**measurable**) whenever*

$$\mu(A) = \mu(A \cap E) + \mu(A \cap E^c)$$

holds for all $A \subseteq X$.

Since the σ-subadditivity of μ implies

$$\mu(A) = \mu((A \cap E) \cup (A \cap E^c)) \le \mu(A \cap E) + \mu(A \cap E^c)$$

---

[2]Constantin Carathéodory (1873–1950), a distinguished Greek mathematician. He made many significant contributions to pure and applied mathematics.

for all subsets A and E, it easily follows that a subset E is measurable if and only if

$$\mu(A) \geq \mu(A \cap E) + \mu(A \cap E^c)$$

holds for each subset A of X.

The collection of all measurable sets will be denoted by Λ. That is,

$$\Lambda = \{E \subseteq X: \mu(A) = \mu(A \cap E) + \mu(A \cap E^c) \text{ for all } A \subseteq X\}.$$

If clarity requires μ to be indicated, then we shall write Λ_μ instead of Λ.

The simplest measurable sets are the sets having outer measure zero. Before verifying this, we name these sets.

Definition 14.3. *A set E is called a **null set** if $\mu(E) = 0$.*

It should be clear from the σ-subadditivity property of μ that a countable union of null sets is again a null set. The null sets will play an important role in the theory of integration.

Theorem 14.4. *Every null set is measurable.*

Proof. Let $E \subseteq X$ with $\mu(E) = 0$. Then the monotonicity of μ implies $\mu(A \cap E) = 0$ for each $A \subseteq X$. Consequently, for each subset A of X we have

$$\mu(A) \leq \mu(A \cap E) + \mu(A \cap E^c) = \mu(A \cap E^c) \leq \mu(A),$$

where the first inequality holds by virtue of the σ-subadditivity of μ. Thus, E is measurable. ∎

For more properties of the measurable sets, we need the following:

Lemma 14.5. *Let the sets $E_1, \ldots, E_n$ be disjoint and measurable. Then*

$$\mu\left(\bigcup_{i=1}^{n}(A \cap E_i)\right) = \sum_{i=1}^{n} \mu(A \cap E_i)$$

holds for every subset A of X.

Proof. The proof is by induction on n. Obviously, the result is true for $n = 1$. Assume it now true for some n, and let the sets $E_1, \ldots, E_n, E_{n+1}$ be disjoint and measurable. If $A \subseteq X$, then

$$A \cap \left[\bigcup_{i=1}^{n+1} E_i \right] \cap E_{n+1} = A \cap E_{n+1}$$

$$A \cap \left[\bigcup_{i=1}^{n+1} E_i \right] \cap (E_{n+1})^c = A \cap \left[\bigcup_{i=1}^{n} E_i \right].$$

Therefore, using the measurability of E_{n+1}, we see that

$$\mu \left(\bigcup_{i=1}^{n+1} (A \cap E_i) \right) = \mu \left(A \cap \left[\bigcup_{i=1}^{n+1} E_i \right] \right)$$

$$= \mu \left(A \cap \left(\left[\bigcup_{i=1}^{n+1} E_i \right] \cap E_{n+1} \right) \right) + \mu \left(A \cap \left[\bigcup_{i=1}^{n+1} E_i \right] \cap (E_{n+1})^c \right)$$

$$= \mu(A \cap E_{n+1}) + \mu \left(A \cap \left[\bigcup_{i=1}^{n} E_i \right] \right) = \sum_{i=1}^{n+1} \mu(A \cap E_i),$$

where the last equality holds by the induction hypothesis. The induction is now complete, and the proof is finished. ∎

We are now ready to establish that the collection of all measurable sets is a σ-algebra.

Theorem 14.6. *The collection Λ of all measurable sets is a σ-algebra.*

Proof. It should be clear from the definition of the measurable sets that if $E \in \Lambda$, then $E^c \in \Lambda$; that is, Λ is closed under complementation. Since $\mu(\emptyset) = 0$, we have $\emptyset \in \Lambda$; therefore, $X \in \Lambda$.

Next, we show that if $E_1, E_2 \in \Lambda$, then $E = E_1 \cup E_2 \in \Lambda$. Indeed, note first that $E = E_1 \cup (E_1^c \cap E_2)$, and then that for every subset A of X the relations

$$\mu(A) \leq \mu(A \cap E) + \mu(A \cap E^c)$$
$$\leq \left[\mu(A \cap E_1) + \mu((A \cap E_1^c) \cap E_2) \right] + \mu((A \cap E_1^c) \cap E_2^c)$$
$$= \mu(A \cap E_1) + \left[\mu((A \cap E_1^c) \cap E_2) + \mu((A \cap E_1^c) \cap E_2^c) \right]$$
$$= \mu(A \cap E_1) + \mu(A \cap E_1^c) = \mu(A)$$

imply $E_1 \cup E_2 \in \Lambda$.

It now follows easily that Λ is closed under finite unions and finite intersections. Also, if $E_1, E_2 \in \Lambda$, then $E_1 \setminus E_2 = E_1 \cap E_2^c \in \Lambda$. Thus, Λ is an algebra of sets.

To finish the proof, it remains to be shown that Λ is a σ-algebra of sets. To this end, let $\{E_n\} \subseteq \Lambda$. Put $E = \bigcup_{n=1}^{\infty} E_n$, and define $G_1 = E_1$ and $G_{n+1} = E_{n+1} \setminus \bigcup_{i=1}^{n} E_i$ for $n \geq 1$. Then $\{G_n\} \subseteq \Lambda$, $G_n \cap G_m = \emptyset$ if $n \neq m$, and $E = \bigcup_{n=1}^{\infty} G_n$. Let $F_n = \bigcup_{i=1}^{n} G_i$ for $n \geq 1$, and note that each F_n is a measurable set such that $\bigcup_{n=1}^{\infty} F_n = E$. Now if $A \subseteq X$, then

$$\mu(A) = \mu(A \cap F_n) + \mu\left(A \cap F_n^c\right)$$
$$\geq \mu(A \cap F_n) + \mu(A \cap E^c)$$
$$= \left[\sum_{i=1}^{n} \mu(A \cap G_i)\right] + \mu(A \cap E^c)$$

holds for each n, where the last equality holds by virtue of Lemma 14.5. Hence,

$$\mu(A) \geq \left[\sum_{i=1}^{\infty} \mu(A \cap G_i)\right] + \mu(A \cap E^c) \geq \mu(A \cap E) + \mu(A \cap E^c) \geq \mu(A),$$

and so $E \in \Lambda$. Therefore, Λ is a σ-algebra. ∎

Remarkably, the outer measure μ restricted to Λ is a measure.

Theorem 14.7. *Let μ be an outer measure on X. Then (X, Λ, μ) is a measure space; that is, μ is σ-additive on Λ.*

Proof. Let $\{E_n\}$ be a disjoint sequence of Λ. Put $E = \bigcup_{n=1}^{\infty} E_n$. By the σ-subadditivity property of μ we have

$$\mu(E) \leq \sum_{n=1}^{\infty} \mu(E_n).$$

On the other hand, Lemma 14.5 shows that

$$\sum_{n=1}^{k} \mu(E_n) = \sum_{n=1}^{k} \mu(E \cap E_n) = \mu\left(E \cap \left[\bigcup_{n=1}^{k} E_n\right]\right) \leq \mu(E)$$

holds for every k. Hence, $\sum_{n=1}^{\infty} \mu(E_n) \leq \mu(E)$, so that $\mu(E) = \sum_{n=1}^{\infty} \mu(E_n)$, and the proof is finished. ∎

When μ is restricted to Λ it is often referred to as the measure **induced** by the outer measure μ.

It is useful to know that μ is subtractive on the measurable sets of finite measure.

Theorem 14.8. *Let A and B be measurable sets such that $A \subseteq B$, with $\mu(B) < \infty$. Then $\mu(B \setminus A) = \mu(B) - \mu(A)$ holds.*

Proof. Write $B = A \cup (B \setminus A)$, and then use the additivity of μ to get $\mu(B) = \mu(A) + \mu(B \setminus A)$. Since $\mu(B) < \infty$, it follows that $\mu(B \setminus A) = \mu(B) - \mu(A)$. ∎

We are now ready to describe a process of constructing outer measures. Let $\mathcal{F}$ be a collection of subsets of a set X containing the empty set. Also, let $\mu\colon \mathcal{F} \to [0, \infty]$ be a set function such that $\mu(\emptyset) = 0$. For every subset A of X we define

$$\mu^*(A) = \inf \left\{ \sum_{n=1}^{\infty} \mu(A_n) \colon \{A_n\} \text{ is a sequence of } \mathcal{F} \text{ with } A \subseteq \bigcup_{n=1}^{\infty} A_n \right\}.$$

If there is no sequence $\{A_n\}$ of $\mathcal{F}$ such that $A \subseteq \bigcup_{n=1}^{\infty} A_n$, then we let $\mu^*(A) = \infty$. That is, we adhere to the convention $\inf \emptyset = \infty$.

Theorem 14.9 (Carathéodory). *The set function $\mu^*\colon \mathcal{P}(X) \to [0, \infty]$ is an outer measure (called the **outer measure generated** by the set function $\mu\colon \mathcal{F} \to [0, \infty]$) satisfying*

$$\mu^*(A) \leq \mu(A)$$

for each $A \in \mathcal{F}$.

Proof. Clearly, $\mu^*(A) \geq 0$ holds for every $A \subseteq X$. If $A_n = \emptyset$ for all n, then $0 \leq \mu^*(\emptyset) \leq \sum_{n=1}^{\infty} \mu(\emptyset) = 0$, so that $\mu^*(\emptyset) = 0$.

For the monotonicity of μ^* assume $A \subseteq B$. If $B \subseteq \bigcup_{n=1}^{\infty} A_n$ with $\{A_n\} \subseteq \mathcal{F}$, then $A \subseteq \bigcup_{n=1}^{\infty} A_n$, and so $\mu^*(A) \leq \sum_{n=1}^{\infty} \mu(A_n)$. (If there is no sequence $\{A_n\}$ of $\mathcal{F}$ that covers B, then $\mu^*(B) = \infty$, and $\mu^*(A) \leq \mu^*(B)$ is obvious.) Thus,

$$\mu^*(B) = \inf \left\{ \sum_{n=1}^{\infty} \mu(A_n) \colon \{A_n\} \subseteq \mathcal{F} \text{ and } B \subseteq \bigcup_{n=1}^{\infty} A_n \right\} \geq \mu^*(A).$$

For the σ-subadditivity of μ^*, let $\{A_n\}$ be an arbitrary sequence of subsets of X. If $\sum_{n=1}^{\infty} \mu^*(A_n) = \infty$, then clearly, $\mu^*(\bigcup_{n=1}^{\infty} A_n) \leq \sum_{n=1}^{\infty} \mu^*(A_n)$. Therefore, assume $\sum_{n=1}^{\infty} \mu^*(A_n) < \infty$. Let $\epsilon > 0$. For each i, choose a sequence $\{A_n^i\}$ of $\mathcal{F}$ such that $A_i \subseteq \bigcup_{n=1}^{\infty} A_n^i$, and $\sum_{n=1}^{\infty} \mu(A_n^i) \leq \mu^*(A_i) + 2^{-i}\epsilon$. Then $A_n^i \in \mathcal{F}$ for each i and n, and $\bigcup_{n=1}^{\infty} A_n \subseteq \bigcup_{i=1}^{\infty} \bigcup_{n=1}^{\infty} A_n^i$. Hence,

$$\mu^* \left(\bigcup_{n=1}^{\infty} A_n \right) \leq \sum_{i=1}^{\infty} \sum_{n=1}^{\infty} \mu(A_n^i) \leq \sum_{i=1}^{\infty} [\mu^*(A_i) + 2^{-i}\epsilon] = \sum_{i=1}^{\infty} \mu^*(A_i) + \epsilon$$

for all $\epsilon > 0$, so that $\mu^*(\bigcup_{n=1}^{\infty} A_n) \leq \sum_{n=1}^{\infty} \mu^*(A_n)$.

Since $A = A \cup \emptyset \cup \emptyset \cup \emptyset \cup \cdots$, it easily follows that $\mu^*(A) \leq \mu(A)$ holds true for each $A \in \mathcal{F}$. ∎

It is easy to construct examples where $\mu^*(A) < \mu(A)$ is valid for some $A \in \mathcal{F}$, i.e., μ^* need not be (in general) an extension of μ. For instance, if $X = \{1, 2, 3\}$, $\mathcal{F} = \{\emptyset, \{1\}, \{1, 2\}\}$ and $\mu: \mathcal{F} \to [0, \infty]$ is defined by $\mu(\emptyset) = 0$, $\mu(\{1\}) = 2$ and $\mu(\{1, 2\}) = 1$, then $\mu^*(\{1\}) = 1$. However, as we shall see in the next section, when μ is a measure, the outer measure μ^* is always an extension of μ.

It turns out that the outer measure generated by μ^* coincides with μ^*, i.e., $(\mu^*)^* = \mu^*$. The details follow.

Theorem 14.10. *Let $\mathcal{F}$ be a collection of subsets of a set X containing the empty set and let $\mu: \mathcal{F} \to [0, \infty]$ be a set function satisfying $\mu(\emptyset) = 0$. Assume Φ is also a collection of subsets of X with $\mathcal{F} \subseteq \Phi$ and that $v: \Phi \to [0, \infty]$ denotes the restriction of μ^* to Φ, i.e., $v(A) = \mu^*(A)$ for each $A \in \Phi$. Then the outer measure generated by v coincides with μ^*, i.e., $v^*(A) = \mu^*(A)$ holds for each subset A of X.*

Proof.   Let $A \subseteq X$. We first claim that

$$\mu^*(A) \leq v^*(A).$$

If $v^*(A) = \infty$, then the inequality is obvious. So, assume $v^*(A) < \infty$. In this case, note that if a sequence $\{A_n\}$ of Φ satisfies $A \subseteq \bigcup_{n=1}^{\infty} A_n$, then the σ-subadditivity of μ^* implies $\mu^*(A) \leq \sum_{n=1}^{\infty} \mu^*(A_n) = \sum_{n=1}^{\infty} v(A_n)$. Therefore,

$$\mu^*(A) \leq \inf \left\{ \sum_{n=1}^{\infty} v(A_n): \{A_n\} \subseteq \Phi \text{ and } A \subseteq \bigcup_{n=1}^{\infty} A_n \right\} = v^*(A).$$

Next, we shall show that

$$v^*(A) \leq \mu^*(A)$$

is also true. Again, if $\mu^*(A) = \infty$, then the inequality is obvious. So, assume $\mu^*(A) < \infty$ and let $\epsilon > 0$. Then there exists a sequence $\{A_n\} \subseteq \mathcal{F} \subseteq \Phi$ satisfying $A \subseteq \bigcup_{n=1}^{\infty} A_n$ and $\sum_{n=1}^{\infty} \mu(A_n) < \mu^*(A) + \epsilon$. Now, taking into account (from Theorem 14.9) that $\mu^*(A_n) \leq \mu(A_n)$ holds for each n, we see that

$$v^*(A) \leq \sum_{n=1}^{\infty} v(A_n) = \sum_{n=1}^{\infty} \mu^*(A_n) \leq \sum_{n=1}^{\infty} \mu(A_n) < \mu^*(A) + \epsilon.$$

Since $\epsilon > 0$ is arbitrary, it follows that $\nu^*(A) \leq \mu^*(A)$. Thus, $\nu^*(A) = \mu^*(A)$ holds for all subsets A of X. ∎

Finally, we leave it as an exercise for the reader to prove the following result asserting that the only outer measures are the ones generated by set functions.

Theorem 14.11. *Let $\nu: \mathcal{P}(X) \to [0, \infty]$ be a set function on the power set of a set X. Then ν is an outer measure if and only if there exist a family $\mathcal{F}$ of subsets of X and a set function $\mu: \mathcal{F} \to [0, \infty]$, where $\emptyset \in \mathcal{F}$ and $\mu(\emptyset) = 0$, whose generating outer measure coincides with ν, i.e., $\nu = \mu^*$.*

EXERCISES

Unless otherwise stated, in the exercises below μ is assumed to be an outer measure on some set X.

1. Show that a countable union of null sets is again a null set.
2. If A is a null set, then show that

$$\mu(B) = \mu(A \cup B) = \mu(B \setminus A)$$

   holds for every subset B of X.
3. If a sequence $\{A_n\}$ of subsets of X satisfies $\sum_{n=1}^{\infty} \mu(A_n) < \infty$, then show that the set

$$\{x \in X : \ x \text{ belongs to } A_n \text{ for infinitely many } n\}$$

   is a null set.
4. If E is a measurable subset of X, then show that for every subset A of X the following equality holds:

$$\mu(E \cup A) + \mu(E \cap A) = \mu(E) + \mu(A).$$

5. If A is a nonmeasurable subset of X and E is a measurable set such that $A \subseteq E$, then show that $\mu(E \setminus A) > 0$.
6. Let A be a subset of X, and let $\{E_n\}$ be a disjoint sequence of measurable sets. Show that

$$\mu\left(\bigcup_{n=1}^{\infty} (A \cap E_n)\right) = \sum_{n=1}^{\infty} \mu(A \cap E_n).$$

7. Let $\{A_n\}$ be a sequence of subsets of X. Assume that there exists a disjoint sequence $\{B_n\}$ of measurable sets such that $A_n \subseteq B_n$ holds for each n. Show that

$$\mu\left(\bigcup_{n=1}^{\infty} A_n\right) = \sum_{n=1}^{\infty} \mu(A_n).$$

8. Show that a subset E of X is measurable if and only if for each $\epsilon > 0$ there exists a measurable set F such that $F \subseteq E$, and $\mu(E \setminus F) < \epsilon$.

9. Assume that a subset E of X has the property that for each $\epsilon > 0$, there exists a measurable set F such that $\mu(E \triangle F) < \epsilon$. Show that E is a measurable set.

10. Let $X = \{1, 2, 3\}$, $\mathcal{F} = \{\emptyset, \{1\}, \{1, 2\}\}$ and consider the set function $\mu: \mathcal{F} \rightarrow [0, \infty]$ defined by $\mu(\emptyset) = 0$, $\mu(\{1\}) = 2$ and $\mu(\{1, 2\}) = 1$.

   a. Describe the outer measure μ^* generated by the set function μ.
   b. Describe the σ-algebra of all μ^*-measurable subsets of X (and conclude that the set $\{1\} \in \mathcal{F}$ is not a measurable set).

11. Prove Theorem 14.11.

12. Let $\mathcal{A}$ be the collection of all measurable subsets of X of finite measure. That is, $\mathcal{A} = \{A \in \Lambda: \mu(A) < \infty\}$.

   a. Show that $\mathcal{A}$ is a semiring.
   b. Define a relation $\simeq$ on $\mathcal{A}$ by $A \simeq B$ if $\mu(A \triangle B) = 0$. Show that $\simeq$ is an equivalence relation on $\mathcal{A}$.
   c. Let D denote the set of all equivalence classes of $\mathcal{A}$. For $A \in \mathcal{A}$ let $\dot{A}$ denote the equivalence class of A in D. Now for $\dot{A}, \dot{B} \in D$ define $d(\dot{A}, \dot{B}) = \mu(A \triangle B)$. Show that d is well defined and that (D, d) is a complete metric space. (For this part, see also Exercise [3] of Section [31].)

15. THE OUTER MEASURE GENERATED BY A MEASURE

Throughout this section, $(X, \mathcal{S}, \mu)$ will be a fixed measure space. Our main objective here is to study the remarkable properties of the outer measure μ^* generated by μ. Among other results, we shall establish that the σ-algebra of all μ^*-measurable subsets of X contains the members of the semiring $\mathcal{S}$. It will then become apparent that the semirings are the smallest collections of sets on which a "reasonable measure theory" can be built.

Recall from the previous section that the outer measure $\mu^*: \mathcal{P}(X) \rightarrow [0, \infty]$ generated by μ is defined by

$$\mu^*(A) = \inf \left\{ \sum_{n=1}^{\infty} \mu(A_n): \{A_n\} \text{ is a sequence of } \mathcal{S} \text{ with } A \subseteq \bigcup_{n=1}^{\infty} A_n \right\}.$$

If there is no sequence $\{A_n\}$ of $\mathcal{S}$ such that $A \subseteq \bigcup_{n=1}^{\infty} A_n$, then we let $\mu^*(A) = \infty$. That is, we adhere to the convention $\inf \emptyset = \infty$. Recall also that (as we saw in Theorem 14.9) the set function μ^* is indeed an outer measure.

In general, μ^* need not be a measure, since it may fail to be σ-additive on $\mathcal{P}(X)$. However, if μ^* is restricted to the σ-algebra of all measurable sets, then we already know (by Theorem 14.7) that it is σ-additive.

The next theorem shows that the outer measure μ^* is in fact an extension of μ from $\mathcal{S}$ to $\mathcal{P}(X)$.

Theorem 15.1.   *The outer measure μ^* is an extension of μ. That is, if $A \in S$, then $\mu^*(A) = \mu(A)$ holds.*

Proof.   Let $A \in S$. From Theorem 14.9, we already know that $\mu^*(A) \leq \mu(A)$. Now, let $\{A_n\} \subseteq S$ with $A \subseteq \bigcup_{n=1}^{\infty} A_n$. By the σ-subadditivity of μ (Theorem 13.8), $\mu(A) \leq \sum_{n=1}^{\infty} \mu(A_n)$ holds, from which it follows immediately that $\mu(A) \leq \mu^*(A)$. Hence, $\mu^*(A) = \mu(A)$. ∎

Since the outer measure μ^* generated by a measure is an extension of μ, μ^* is also called the **Carathéodory extension** of μ. From now on the expression "the outer measure of a set A" will also be referred to as "the measure of A."

The measurable sets of an outer measure generated by a measure have a number of useful characterizations. Some of them are included in the next theorem.

Theorem 15.2.   *Let (X, S, μ) be a measure space, and let μ^* be the outer measure generated by μ. For a subset E of X the following statements are equivalent:*

1. *E is measurable with respect to μ^*.*
2. *$\mu(A) - \mu^*(A \cap E) + \mu^*(A \cap E^c)$ holds for all $A \in S$ with $\mu(A) < \infty$.*
3. *$\mu(A) \geq \mu^*(A \cap E) + \mu^*(A \cap E^c)$ holds for all $A \in S$ with $\mu(A) < \infty$.*
4. *$\mu^*(A) \geq \mu^*(A \cap E) + \mu^*(A \cap E^c)$ holds for all $A \subseteq X$.*

Proof. $(1) \Longrightarrow (2)$ and $(2) \Longrightarrow (3)$ are obvious.

$(3) \Longrightarrow (4)$ Let $A \subseteq X$. If $\mu^*(A) = \infty$, then (4) holds trivially. Hence, assume $\mu^*(A) < \infty$. Let $\epsilon > 0$. Choose a sequence $\{A_n\}$ of S with $A \subseteq \bigcup_{n=1}^{\infty} A_n$ and $\sum_{n=1}^{\infty} \mu(A_n) < \mu^*(A) + \epsilon$. It follows that $\mu(A_n) < \infty$ for each n, and so $\mu^*(A_n \cap E) + \mu^*(A_n \cap E^c) \leq \mu(A_n)$ holds for each n, by our hypothesis. Therefore,

$$\mu^*(A \cap E) + \mu^*(A \cap E^c) \leq \mu^*\left(\left[\bigcup_{n=1}^{\infty} A_n\right] \cap E\right) + \mu^*\left(\left[\bigcup_{n=1}^{\infty} A_n\right] \cap E^c\right)$$

$$\leq \sum_{n=1}^{\infty} \mu^*(A_n \cap E) + \sum_{n=1}^{\infty} \mu^*(A_n \cap E^c)$$

$$= \sum_{n=1}^{\infty} [\mu^*(A_n \cap E) + \mu^*(A_n \cap E^c)]$$

$$\leq \sum_{n=1}^{\infty} \mu(A_n) \leq \mu^*(A) + \epsilon$$

for all $\epsilon > 0$, so that $\mu^*(A) \geq \mu^*(A \cap E) + \mu^*(A \cap E^c)$.

(4) $\Longrightarrow$ (1) By the σ-subadditivity property of μ^*, we have

$$\mu^*(A) \leq \mu^*(A \cap E) + \mu^*(A \cap E^c),$$

hence $\mu^*(A) = \mu^*(A \cap E) + \mu^*(A \cap E^c)$ for all $A \subseteq X$. Thus, E is measurable. ∎

A subset E of a measure space (X, S, μ) will be called **measurable** (more precisely μ-measurable) if E is measurable with respect to the outer measure μ^* generated by μ.

Theorem 15.3.   *Every member of S is measurable. That is, $S \subseteq \Lambda$.*

Proof.   Let $E \in S$. We have to show that E is a measurable set. If $A \in S$, then there exist pairwise disjoint sets $B_1, \ldots, B_n$ of S such that $A \cap E^c = A \setminus E = \bigcup_{i=1}^n B_i$. Note that $A \cap E, B_1, \ldots, B_n$ is a disjoint collection of members of S such that $A = (A \cap E) \cup B_1 \cup \cdots \cup B_n$. By the σ-subadditivity property of μ^*, we obtain

$$\mu^*(A \cap E) + \mu^*(A \cap E^c) \leq \mu^*(A \cap E) + \sum_{i=1}^n \mu^*(B_i)$$

$$= \mu(A \cap E) + \sum_{i=1}^n \mu(B_i) = \mu(A),$$

where the last two equalities hold true by virtue of Theorem 15.1 and the σ-additivity of μ on S. By Theorem 15.2, E is measurable, and so, $S \subseteq \Lambda$. ∎

Let us write $A_n \uparrow A$ to mean that the sequence $\{A_n\}$ of subsets of X satisfies $A_n \subseteq A_{n+1}$ for each n and $A = \bigcup_{n=1}^\infty A_n$. Similarly, $A_n \downarrow A$ means $A_{n+1} \subseteq A_n$ for each n and $A = \bigcap_{n=1}^\infty A_n$.

Theorem 15.4.   *For a measure space (X, S, μ) and a sequence of measurable sets $\{E_n\}$, the following statements hold:*

1.   *If $E_n \uparrow E$, then $\mu^*(E_n) \uparrow \mu^*(E)$.*
2.   *If $E_n \downarrow E$ and $\mu^*(E_k) < \infty$ holds for some k, then $\mu^*(E_n) \downarrow \mu^*(E)$.*

Proof.   (1) Let $B_1 = E_1$ and $B_n = E_n \setminus E_{n-1}$ for $n \geq 2$. Then each B_n is measurable, and $B_i \cap B_j = \emptyset$ if $i \neq j$. Also, $E_n = \bigcup_{i=1}^n B_i$ and $E = \bigcup_{i=1}^\infty B_i$. Hence, by Theorem 14.7 we have

$$\mu^*(E) = \sum_{i=1}^\infty \mu^*(B_i) = \lim_{n \to \infty} \sum_{i=1}^n \mu^*(B_i).$$

But $\mu^*(E_n) = \sum_{i=1}^n \mu^*(B_i)$, and thus, $\mu^*(E_n) \uparrow \mu^*(E)$ holds.

(2) Observe that without loss of generality we can assume $\mu^*(E_1) < \infty$. Now, $E_1 \setminus E_n \uparrow E_1 \setminus E$, and so, by part (1), $\lim \mu^*(E_1 \setminus E_n) = \mu^*(E_1 \setminus E)$. Applying Theorem 14.8, we get $\lim[\mu^*(E_1) - \mu^*(E_n)] = \mu^*(E_1) - \mu^*(E)$, from which it follows that $\mu^*(E_n) \downarrow \mu^*(E)$.   ∎

As an application of the results obtained so far, we shall show that the Lebesgue outer measure on $\mathbb{R}$ generalizes the ordinary concept of length. Remember that the Lebesgue measure λ is the measure defined on the semiring $S = \{[a, b): a, b \in \mathbb{R}\}$ by $\lambda([a, b)) = b - a$ for all $a \le b$. It is a custom to call (and we shall do so) the outer measure λ^* generated by λ the **Lebesgue measure** on $\mathbb{R}$.

A subset I of $\mathbb{R}$ is called an interval if for every $x, y \in I$ with $x < y$, we have $[x, y] \subseteq I$. If I has any one of the forms $[a, b)$, $[a, b]$, $(a, b]$, or (a, b) with $-\infty < a < b < \infty$, then I is called a bounded interval, and its length is defined by $|I| = b - a$. If I is unbounded, then its length is said to be infinite, which is written $|I| = \infty$.

Example 15.5.   *Every interval I of $\mathbb{R}$ is Lebesgue measurable and $\lambda^*(I) = |I|$.*

To see this, note that by Theorem 15.3 every set of the form $[a, b)$ with $-\infty < a < b < \infty$ is Lebesgue measurable. Moreover, Theorem 15.1 shows that $\lambda^*([a, b)) = \lambda([a, b)) = b - a = |[a, b)|$.

We shall prove two more cases and leave the rest for the reader. Our first case is a bounded interval of the form $I = [a, b]$. Let $E_n = [a, b + \frac{1}{n})$ for each n. Then each E_n is Lebesgue measurable, and $E_n \downarrow I$ holds. Hence, I is Lebesgue measurable, and by Theorem 15.4(2), we get

$$\lambda^*(I) = \lim_{n \to \infty} \lambda^*\left(\left[a, b + \frac{1}{n}\right)\right) = \lim_{n \to \infty}\left(b + \frac{1}{n} - a\right) = b - a = |I|.$$

The second case is an interval of the form $I = [a, \infty)$. Put $F_n = [a, a + n)$ for each n, and note that $F_n \uparrow I$ holds. Thus, I is Lebesgue measurable, and by Theorem 15.4(1) we have

$$\lambda^*(I) = \lim_{n \to \infty} \lambda^*([a, a + n)) = \lim_{n \to \infty} n = \infty = |I|,$$

as desired.   ∎

Two important classes of measure spaces are introduced next.

Definition 15.6.   *A measure space (X, S, μ) is said to be*

1.   **finite,** *if $\mu^*(X) < \infty$, and*
2.   *σ-**finite,** if there exists a sequence $\{X_n\}$ of subsets of X such that $X = \bigcup_{n=1}^{\infty} X_n$ and $\mu^*(X_n) < \infty$ for each n.*

Clearly, every finite measure space is σ-finite. We also have the following:

Lemma 15.7.   *A measure space (X, S, μ) is σ-finite if and only if there exists a disjoint sequence $\{Y_n\}$ of S such that $X = \bigcup_{n=1}^{\infty} Y_n$ and $\mu(Y_n) < \infty$ for each n.*

Proof.   To see the "only if" part, assume that (X, S, μ) is σ-finite. Choose subsets $X_1, X_2, \ldots$ of X such that $X = \bigcup_{i=1}^{\infty} X_i$ and $\mu^*(X_i) < \infty$ for each i. Now, for each i, choose a sequence $\{A_n^i\}$ of S such that $X_i \subseteq \bigcup_{n=1}^{\infty} A_n^i$ and $\sum_{n=1}^{\infty} \mu(A_n^i) < \mu^*(X_i) + 1$. Then $A_n^i \in S$, $\mu(A_n^i) < \infty$ for each n and i, and $X = \bigcup_{i=1}^{\infty} \bigcup_{n=1}^{\infty} A_n^i$. Now, by applying Theorem 12.2 and Lemma 12.3, our claim can be established.   ∎

The measurable sets of a finite measure space have a simple characterization.

Theorem 15.8.   *Let (X, S, μ) be a finite measure space, and let E be a subset of X. Then E is measurable if and only if*

$$\mu^*(E) + \mu^*(E^c) = \mu^*(X).$$

Proof.   Clearly, if E is measurable, then $\mu^*(E) + \mu^*(E^c) = \mu^*(X)$ holds. For the converse, assume that $\mu^*(E) + \mu^*(E^c) = \mu^*(X)$ holds. Let $A \in S$. The measurability of A applied to E and E^c gives

$$\mu^*(E) = \mu^*(E \cap A) + \mu^*(E \cap A^c),$$
$$\mu^*(E^c) = \mu^*(E^c \cap A) + \mu^*(E^c \cap A^c).$$

By adding the last two equalities, we get

$$\begin{aligned} \mu^*(X) &= \mu^*(E) + \mu^*(E^c) \\ &= \mu^*(E \cap A) + \mu^*(E^c \cap A) + \mu^*(E \cap A^c) + \mu^*(E^c \cap A^c) \\ &\geq \mu^*(A) + \mu^*(A^c) \geq \mu^*(A \cup A^c) = \mu^*(X). \end{aligned}$$

Therefore,

$$\mu^*(A) + \mu^*(A^c) = \mu^*(E \cap A) + \mu^*(E^c \cap A) + \mu^*(E \cap A^c) + \mu^*(E^c \cap A^c).$$

In view of $\mu^*(A^c) \leq \mu^*(E \cap A^c) + \mu^*(E^c \cap A^c)$ and $\mu^*(X) < \infty$, the last equality gives

$$\mu^*(A) \geq \mu^*(A \cap E) + \mu^*(A \cap E^c).$$

This last inequality is the one required by Theorem 15.2 for the measurability of E, and the proof of the theorem is complete.   ∎

Notice that Theorem 15.3 combined with Theorem 12.2 shows that in the definition of the outer measure μ^*, we can consider only disjoint sequences of $\mathcal{S}$. That is, we have the following useful result:

Lemma 15.9. *If* $(X, \mathcal{S}, \mu)$ *is a measure space, then*

$$\mu^*(A) = \inf\left\{ \sum_{n=1}^{\infty} \mu(A_n): \{A_n\} \subseteq \mathcal{S}, \{A_n\} \text{ disjoint, and } A \subseteq \bigcup_{n=1}^{\infty} A_n \right\}$$

$$= \inf\{\mu^*(B): \ B \text{ is a } \sigma\text{-set and } A \subseteq B\}$$

holds true for every subset A of X (with $\inf \emptyset = \infty$*).*

The next result describes the uniqueness of the extension of μ to the σ-algebra of the measurable sets.

Theorem 15.10. *Let* $(X, \mathcal{S}, \mu)$ *be a* σ*-finite measure space,* Σ *a semiring of sets such that* $\mathcal{S} \subseteq \Sigma \subseteq \Lambda$, *and* ν *a measure on* Σ*. If* $\nu = \mu$ *on* $\mathcal{S}$, *then* $\nu = \mu^*$ *on* Σ.

In this case, μ^* *is the one and only extension of* μ *to a measure on* Λ.

Proof. Let ν^* be the outer measure generated by (X, Σ, ν). Now, let $A \subseteq X$, and let $\{A_n\}$ be a sequence of $\mathcal{S}$ such that $A \subseteq \bigcup_{n=1}^{\infty} A_n$. Since μ and ν agree on $\mathcal{S}$, we have

$$\nu^*(A) \leq \sum_{n=1}^{\infty} \nu(A_n) = \sum_{n=1}^{\infty} \mu(A_n).$$

This implies $\nu^*(A) \leq \mu^*(A)$ for all $A \subseteq X$.

Now, assume that $A \in \Sigma$ satisfies $\mu^*(A) < \infty$. We shall show that $\mu^*(A) \leq \nu(A)$ holds, and therefore, $\nu(A) = \mu^*(A)$ will be established in this case. To this end, let $\epsilon > 0$. Choose a disjoint sequence $\{A_n\}$ of $\mathcal{S}$ such that $A \subseteq \bigcup_{n=1}^{\infty} A_n$ and $\sum_{n=1}^{\infty} \mu(A_n) < \mu^*(A) + \epsilon$. Put $B = \bigcup_{n=1}^{\infty} A_n$, and note that $B \in \Lambda_\mu$, $B \in \Lambda_\nu$, and $\mu^*(B) < \mu^*(A) + \epsilon$. Now, observe that $\nu^*(B \setminus A) \leq \mu^*(B \setminus A) = \mu^*(B) - \mu^*(A) < \epsilon$ holds. Therefore,

$$\mu^*(A) \leq \mu^*(B) = \sum_{n=1}^{\infty} \mu(A_n) = \sum_{n=1}^{\infty} \nu(A_n) = \nu^*(B)$$

$$= \nu^*(A) + \nu^*(B \setminus A) < \nu(A) + \epsilon$$

for all $\epsilon > 0$. Hence, $\mu^*(A) \leq \nu(A)$.

For the general case, let $\{X_n\}$ be a disjoint sequence of S covering X such that $\mu^*(X_n) < \infty$ for all n. Now, if $A \in \Sigma$, then by the above $\nu(X_n \cap A) = \mu^*(X_n \cap A)$ holds for all n, and so

$$\mu^*(A) = \mu^*\left(\bigcup_{n=1}^{\infty}[X_n \cap A]\right) = \sum_{n=1}^{\infty}\mu^*(X_n \cap A)$$

$$= \sum_{n=1}^{\infty}\nu(X_n \cap A) = \nu^*\left(\bigcup_{n=1}^{\infty}[X_n \cap A]\right) = \nu(A),$$

and the proof is finished.　　　　　　　　　　　　　　　　　　　　　　　■

A subset A of a measure space (X, S, μ) is said to be σ-**finite** if there exists a sequence $\{A_n\}$ of S such that $A \subseteq \bigcup_{n=1}^{\infty} A_n$ and $\mu(A_n) < \infty$ for each n. Clearly, the sequence $\{A_n\}$ can be chosen to be disjoint.

Repeating verbatim the proof of the last theorem we can establish the following.

- Let (X, S, μ) be a measure space, Σ be a semiring of subsets of X such that $S \subseteq \Sigma \subseteq \Lambda$, and ν be a measure on Σ. If $\nu = \mu$ on S, then $\nu(A) = \mu^*(A)$ holds for every σ-finite set of Σ.

It should be noted that the hypothesis of σ-finiteness in Theorem 15.10 cannot be dropped. As an example, take $X = \mathbb{R}$, $S = \{[a, b): a, b \in \mathbb{R}\}$, and define $\mu: S \to [0, \infty]$ by $\mu(\emptyset) = 0$ and $\mu([a, b)) = \infty$ if $a < b$. Then (X, S, μ) is a measure space that is not σ-finite. Moreover, $\Lambda = P(X)$ and $\mu^*(A) = \infty$ for every nonempty subset A of X. It is easy to see now that the counting measure (see Example 13) is an extension of μ to Λ that is different from μ^*.

We continue with an approximation property of an arbitrary set by a measurable set.

Theorem 15.11. *Let (X, S, μ) be a measure space. If A is a subset of X, then there exists a measurable set E such that $A \subseteq E$ and $\mu^*(E) = \mu^*(A)$.*

In particular, if S is a σ-algebra, then for each $A \subseteq X$ there exists some $E \in S$ with $A \subseteq E$ and $\mu^*(E) = \mu^*(A)$.

Proof. Let $A \subseteq X$. If $\mu^*(A) = \infty$, then the measurable set $E = X$ satisfies $\mu^*(E) = \mu^*(A)$. So, assume $\mu^*(A) < \infty$.

For each i, choose a sequence $\{A_n^i\}$ of the semiring S such that $A \subseteq \bigcup_{n=1}^{\infty} A_n^i$ and $\sum_{n=1}^{\infty}\mu(A_n^i) < \mu^*(A) + \frac{1}{i}$, and let $E_i = \bigcup_{n=1}^{\infty} A_n^i$. Then each E_i is a measurable set such that $A \subseteq E_i$. Put $E = \bigcap_{n=1}^{\infty} E_n$. Then $A \subseteq E$, and E is measurable (and, of course, $E \in S$ if S is a σ-algebra). Also,

$$\mu^*(A) \le \mu^*(E) \le \mu^*(E_i) \le \sum_{n=1}^{\infty} \mu(A_n^i) < \mu^*(A) + \frac{1}{i}$$

for each i implies $\mu^*(E) = \mu^*(A)$. ∎

If A is a subset of X, then any measurable set E satisfying $A \subseteq E$ and $\mu^*(E) = \mu^*(A)$ is called by many authors a **measurable cover** of A.

Is every set a measurable set? Of course, the answer is no. It is easy to construct examples where not every set is measurable. Here is a simple example.

Example 15.12. Let $X = \{1, 2, 3\}$ and $S = \{\emptyset, \{3\}, \{1, 2\}, X\}$; clearly S is a σ-algebra. It should be easy to check that the set function $\mu: S \to [0, \infty)$, defined by

$$\mu(\emptyset) = 0, \quad \mu(\{3\}) = \mu(\{1, 2\}) = 1, \quad \text{and} \quad \mu(X) = 2$$

is a measure.

We claim that the set $E = \{1\}$ is not measurable. To see this, note first that $\mu^*(\{1\}) = \mu^*(\{2\}) = \mu^*(\{1, 2\}) = 1$. Now if $A = \{1, 2\}$, then

$$\mu^*(A \cap E) + \mu^*(A \cap E^c) = \mu^*(\{1\}) + \mu^*(\{2\}) = 1 + 1 > 1 = \mu^*(A),$$

and this shows that E is not measurable. ∎

It is considerably harder to demonstrate the existence of a non-Lebesgue measurable subset of the real line. The following classical example of a non-measurable subset of the real is due to G. Vitali.[3]

Example 15.13 (Vitali). Let λ^* be the outer measure generated on $\mathbb{R}$ by the Lebesgue measure λ. We define a relation $\simeq$ on $[0, 1]$ by saying that $x \simeq y$ whenever $x - y$ is a rational number. An easy verification shows that $\simeq$ is an equivalence relation on $[0, 1]$. Thus, $\simeq$ partitions $[0, 1]$ into equivalence classes. Let E be a subset of $[0, 1]$ intersecting each equivalence class precisely at one point. By the axiom of choice such a set exists. We claim that E is not Lebesgue measurable.

To see this, assume by way of contradiction that E is Lebesgue measurable. Let $r_1, r_2, \ldots$ be an enumeration of the rational numbers of $[-1, 1]$. For each n put

$$E_n = \{r_n + x : x \in E\} = r_n + E,$$

and note that each E_n is Lebesgue measurable. Now, it is not difficult to see that $E_n \cap E_m = \emptyset$ holds if $n \neq m$ and that (since λ^* is translation invariant) $\lambda^*(E_n) = \lambda^*(E)$ for each n. Moreover, $[0, 1] \subseteq \bigcup_{n=1}^{\infty} E_n \subseteq [-1, 2]$. By the σ-additivity of λ^* we get

$$\lambda^*\left(\bigcup_{n=1}^{\infty} E_n \right) = \sum_{n=1}^{\infty} \lambda^*(E_n) = \lim_{n \to \infty} [n\lambda^*(E)] \le \lambda^*([-1, 2]) = 3.$$

---

[3]Giuseppe Vitali (1875–1932), an Italian mathematician. He contributed to the theory of real functions and measure theory.

This implies $\lambda^*(E) = 0$, and hence, $\lambda^*(\bigcup_{n=1}^{\infty} E_n) = 0$. On the other hand, $[0, 1] \subseteq \bigcup_{n=1}^{\infty} E_n$ shows that $1 \leq \lambda^*(\bigcup_{n=1}^{\infty} E_n)$, which is impossible. Therefore, E cannot be a Lebesgue measurable set.   ∎

EXERCISES

1. Let $(X, \mathcal{S}, \mu)$ be a measure space, and let E be a measurable subset of X. Put $\mathcal{S}_E = \{E \cap A: A \in \mathcal{S}\}$, the restriction of $\mathcal{S}$ to E. Show that $(E, \mathcal{S}_E, \mu^*)$ is a measure space.

2. Let $(X, \mathcal{S}, \mu)$ be a measure space. Show that

$$\mu^*(A) = \inf\{\mu^*(B): B \text{ is a } \sigma\text{-set such that } A \subseteq B\}$$

holds for every subset A of X.

3. Complete the details of Example 13.7.

4. Show that every countable subset of $\mathbb{R}$ has Lebesgue measure zero.

5. For a subset A of $\mathbb{R}$ and real numbers a and b, define the set $aA + b = \{ax + b: x \in A\}$. Show that

   a. $\lambda^*(aA + b) = |a|\lambda^*(A)$, and
   b. if A is Lebesgue measurable, then so is $aA + b$.

6. Let $\mathcal{S}$ be a semiring of subsets of a set X, and let $\mu: \mathcal{S} \to [0, \infty]$ be a finitely additive measure that is not a measure. For each $A \subseteq X$ define (as usual)

$$\mu^*(A) = \inf\left\{ \sum_{n=1}^{\infty} \mu(A_n): \{A_n\} \subseteq \mathcal{S} \text{ and } A \subseteq \bigcup_{n=1}^{\infty} A_n \right\}.$$

   Show by a counterexample that it is possible to have $\mu \neq \mu^*$ on $\mathcal{S}$. Why doesn't this contradict Theorem 15.1?
   [HINT: Use Exercise 7 of Section 13.]

7. Let E be an arbitrary measurable subset of a measure space $(X, \mathcal{S}, \mu)$ and consider the measure space $(E, \mathcal{S}_E, \nu)$, where $\mathcal{S}_E = \{E \cap A: A \in \mathcal{S}\}$ and $\nu(E \cap A) = \mu^*(E \cap A)$ (see Exercise 1 of this section). Establish the following properties regarding the measure space $(E, \mathcal{S}_E, \nu)$:

   a. The outer measure ν^* is the restriction of μ^* on E, i.e., $\nu^*(B) = \mu^*(B)$ for each $B \subseteq E$.
   b. The ν-measurable sets of the measure space $(E, \mathcal{S}_E, \nu)$ are precisely the sets of the form $E \cap A$ where A is a μ-measurable subset of X, i.e.,

$$\Lambda_\nu = \{F \subseteq E: F \in \Lambda_\mu\}.$$

8. Show that a subset E of a measure space $(X, \mathcal{S}, \mu)$ is measurable if and only if for each $\epsilon > 0$ there exist a measurable set A_ϵ and two subsets B_ϵ and C_ϵ satisfying

$$E = (A_\epsilon \cup B_\epsilon) \setminus C_\epsilon, \quad \mu^*(B_\epsilon) < \epsilon, \text{ and } \mu^*(C_\epsilon) < \epsilon.$$

9. Let $(X, \mathcal{S}, \mu)$ be a measure space, and let A be a subset of X. Show that if there exists a measurable subset E of X such that $A \subseteq E$, $\mu^*(E) < \infty$, and $\mu^*(E) = \mu^*(A) + \mu^*(E \setminus A)$, then A is measurable.

10. Let A be a subset of $\mathbb{R}$ with $\lambda^*(A) > 0$. Show that there exists a nonmeasurable subset B of $\mathbb{R}$ such that $B \subseteq A$.

11. Give an example of a disjoint sequence $\{E_n\}$ of subsets of some measure space $(X, \mathcal{S}, \mu)$ such that

$$\mu^*\left(\bigcup_{n=1}^{\infty} E_n\right) < \sum_{n=1}^{\infty} \mu^*(E_n).$$

    [HINT: Use the sequence $\{E_n\}$ described in Example 15.13.]

12. Let $(X, \mathcal{S}, \mu)$ be a measure space, and let $\{A_n\}$ be a sequence of subsets of X such that $A_n \subseteq A_{n+1}$ holds for all n. If $A = \bigcup_{n=1}^{\infty} A_n$, then show that $\mu^*(A_n) \uparrow \mu^*(A)$.
    [HINT: Use Theorems 15.11 and 15.4(1).]

13. For subsets of a measure space $(X, \mathcal{S}, \mu)$ let us define the following almost everywhere (a.e.) relations:

    a. $A \subseteq B$ a.e. if $\mu^*(A \setminus B) = 0$;
    b. $A = B$ a.e. if $\mu^*(A \triangle B) = 0$;
    c. $A_n \uparrow A$ a.e. if $A_n \subseteq A_{n+1}$ a.e. for all n and $A = \bigcup_{n=1}^{\infty} A_n$ a.e. (The meaning of $A_n \downarrow A$ a.e. is similar.)

    Generalize Theorem 15.4 by establishing the following properties for a sequence $\{E_n\}$ of measurable sets:

    i. If $E_n \uparrow E$ a.e., then $\mu^*(E_n) \uparrow \mu^*(E)$.
    ii. If $E_n \downarrow E$ a.e. and $\mu^*(E_k) < \infty$ for some k, then $\mu^*(E_n) \downarrow \mu^*(E)$.

    Is (i) true without assuming measurability for the sets E_n?

14. Give an example of a sequence $\{E_n\}$ of measurable sets of some measure space $(X, \mathcal{S}, \mu)$ such that $E_{n+1} \subseteq E_n$ holds for all n and

$$\lim_{n \to \infty} \mu^*(E_n) > \mu^*\left(\bigcap_{n=1}^{\infty} E_n\right).$$

15. For a sequence $\{A_n\}$ of subsets of a set X, define

$$\liminf A_n = \bigcup_{n=1}^{\infty} \bigcap_{i=n}^{\infty} A_i \quad \text{and} \quad \limsup A_n = \bigcap_{n=1}^{\infty} \bigcup_{i=n}^{\infty} A_i.$$

    Now, let $(X, \mathcal{S}, \mu)$ be a measure space and let $\{E_n\}$ be the sequence of measurable sets. Show the following:

    a. $\mu^*(\liminf E_n) \le \liminf \mu^*(E_n)$.
    b. If $\mu^*(\bigcup_{n=1}^{\infty} E_n) < \infty$, then $\mu^*(\limsup E_n) \ge \limsup \mu^*(E_n)$.

16. Give an example of a sequence $\{A_n\}$ of subsets of some measure space $(X, \mathcal{S}, \mu)$ such that $A_{n+1} \subseteq A_n$ for each n, $\mu^*(A_1) < \infty$, and

$$\lim_{n \to \infty} \mu^*(A_n) > \mu^*\left(\bigcap_{n=1}^{\infty} A_n\right).$$

[HINT: If $\{E_n\}$ is the sequence of sets described in Example 15, then let $A_n = \bigcup_{i=n}^{\infty} E_i$.]

17. Let (X, S_1, μ_1) and (X, S_2, μ_2) be two measure spaces. Show that μ_1 and μ_2 generate the same outer measure on X if and only if $\mu_1 = \mu_2^*$ on S_1 and $\mu_2 = \mu_1^*$ on S_2 both hold.

18. Let (X, S, μ) be a measure space. A measurable set A is called an **atom** if $\mu^*(A) > 0$ and for every measurable subset E of A we have either $\mu^*(E) = 0$ or $\mu^*(A \setminus E) = 0$. If (X, S, μ) does not have any atoms, then it is called a **nonatomic measure space**.

    a.   Find the atoms of the measure spaces of Examples 13.3 and 13.4.

    b.   Show that the real line with the Lebesgue measure is a nonatomic measure space.

19. This exercise presents an example of a measure that has infinitely many extensions to a measure on the σ-algebra generated by S. Fix a proper nonempty subset A of a set X (i.e., $A \neq X$) and consider the collection of subsets $S = \{\emptyset, A\}$.

    a.   Show that S is a semiring.

    b.   Show that the set function $\mu: S \to [0, \infty]$ defined by $\mu(\emptyset) = 0$ and $\mu(A) = 1$ is a measure.

    c.   Describe the Carathéodory extension μ^* of μ.

    d.   Determine the σ-algebra of measurable sets Λ_μ.

    e.   Show that μ has uncountably many extensions to a measure on the σ-algebra generated by S. Why doesn't this contradict Theorem 15.10?

16. MEASURABLE FUNCTIONS

An important role in the theory of integration is played by the so-called almost everywhere relations. If μ is an outer measure of X, then a relation involving the elements of X is said to hold **almost everywhere** (or that it holds for **almost all** x) if the set A of all points for which the relation fails to hold is a null set (i.e., $\mu(A) = 0$). For instance, if f and g are real-valued functions defined on X, then $f \leq g$ almost everywhere means that $\mu(\{x \in X: f(x) > g(x)\}) = 0$. Similarly, $\{f_n\}$ converges to f almost everywhere if there is a set A of measure zero such that $\lim f_n(x) = f(x)$ holds for all $x \notin A$.

    If (X, S, μ) is a measure space, then by saying that a relation holds **almost everywhere** (abbreviated a.e.) we mean that the relation holds almost everywhere with respect to the outer measure μ^* generated by μ. The basic almost everywhere relations that will be used in this book are summarized below. We assume that (X, S, μ) is a given measure space and that f and g denote real-valued functions defined on X.

    1. $f = g$ a.e. if $\mu^*(\{x \in X: f(x) \neq g(x)\}) = 0$.

    2. $f \geq g$ a.e. if $\mu^*(\{x \in X: f(x) < g(x)\}) = 0$.

    3. $f_n \to f$ a.e. if $\mu^*(\{x \in X: f_n(x) \nrightarrow f(x)\}) = 0$.

4. $f_n \uparrow f$ a.e. if $f_n \leq f_{n+1}$ a.e. for all n and $f_n \to f$ a.e.
5. $f_n \downarrow f$ a.e. if $f_{n+1} \leq f_n$ a.e. for all n and $f_n \to f$ a.e.

Throughout this section, $(X, \mathcal{S}, \mu)$ will denote a fixed measure space.

Definition 16.1.   *Let $f: X \to \mathbb{R}$ be a function. If $f^{-1}(\mathcal{O})$ is a measurable set for every open subset $\mathcal{O}$ of $\mathbb{R}$, then f is called a* **measurable function**.

Every constant function is measurable. Indeed, if $f(x) = c$ for all $x \in X$ and $\mathcal{O}$ is an open subset of $\mathbb{R}$, then $f^{-1}(\mathcal{O}) = \emptyset$ if $c \notin \mathcal{O}$ and $f^{-1}(\mathcal{O}) = X$ if $c \in \mathcal{O}$.

Recall that the Borel sets of a topological space are the members of the σ-algebra generated by the open sets. The first theorem gives a number of useful characterizations of the measurable functions. (This result is a special case of a more general result regarding measurability of functions; see Theorem 20.6.)

Theorem 16.2.   *For a function $f: X \to \mathbb{R}$, the following statements are equivalent:*

1.   *f is measurable.*
2.   *$f^{-1}((a, b))$ is measurable for each bounded open interval (a, b) of $\mathbb{R}$.*
3.   *$f^{-1}(C)$ is measurable for each closed subset C of $\mathbb{R}$.*
4.   *$f^{-1}([a, \infty))$ is measurable for each $a \in \mathbb{R}$.*
5.   *$f^{-1}((-\infty, a])$ is measurable for each $a \in \mathbb{R}$.*
6.   *$f^{-1}(B)$ is measurable for each Borel subset B of $\mathbb{R}$.*

Proof. $(1) \Longrightarrow (2)$ Obvious.

$(2) \Longrightarrow (3)$ It follows from the relation $f^{-1}(C) = [f^{-1}(C^c)]^c$ and the fact that C^c can be written as a countable union of bounded open intervals.

$(3) \Longrightarrow (4)$ Obvious.

$(4) \Longrightarrow (5)$ Note first that $f^{-1}((-\infty, a)) = [f^{-1}([a, \infty))]^c$ is measurable for each $a \in \mathbb{R}$. The result now follows from the identity

$$f^{-1}((-\infty, a]) = \bigcap_{n=1}^{\infty} f^{-1}\left(\left(-\infty, a + \frac{1}{n}\right)\right).$$

$(5) \Longrightarrow (6)$ Let $\mathcal{A} = \{A \subseteq \mathbb{R}: f^{-1}(A) \text{ is measurable}\}$. An easy verification shows that $\mathcal{A}$ is a σ-algebra of subsets of $\mathbb{R}$ such that $(-\infty, a]$ belongs to $\mathcal{A}$ for each $a \in \mathbb{R}$ by hypothesis. It easily follows that $\mathcal{A}$ contains the open subsets of $\mathbb{R}$, and hence, it also contains the Borel sets. Thus, $f^{-1}(B)$ is measurable for each Borel subset B of $\mathbb{R}$.

$(6) \Longrightarrow (1)$ Obvious.   ■

Two functions that are equal almost everywhere are either both measurable or else both nonmeasurable. The details follow.

Theorem 16.3.   *If f is a measurable function and $g: X \rightarrow \mathbb{R}$ satisfies $f = g$ a.e., then g is a measurable function.*

Proof.   If $A = \{x \in X: f(x) \neq g(x)\}$, then from our hypothesis $\mu^*(A) = 0$, and so, A is measurable. Now, let $\mathcal{O}$ be an open subset of $\mathbb{R}$. Since f is a measurable function, $f^{-1}(\mathcal{O})$ is measurable, and hence, $A^c \cap g^{-1}(\mathcal{O}) = A^c \cap f^{-1}(\mathcal{O})$ is a measurable set. Also, since $A \cap g^{-1}(\mathcal{O})$ has outer measure zero, it is measurable. Hence,

$$g^{-1}(\mathcal{O}) = \left[A \cap g^{-1}(\mathcal{O})\right] \cup \left[A^c \cap g^{-1}(\mathcal{O})\right]$$

is a measurable set, so that g is a measurable function.   ∎

We continue with more properties of measurable functions.

Theorem 16.4.   *If f and g are measurable functions, then the three sets*
a. $\{x \in X: f(x) > g(x)\}$,
b. $\{x \in X: f(x) \geq g(x)\}$, *and*
c. $\{x \in X: f(x) = g(x)\}$
are all measurable.

Proof.   (a) If $r_1, r_2, \ldots$ is an enumeration of the rational numbers of $\mathbb{R}$, then

$$\{x \in X: f(x) > g(x)\} = \bigcup_{n=1}^{\infty} \left[\{x \in X: f(x) > r_n\} \cap \{x \in X: g(x) < r_n\}\right],$$

which is measurable, since it is a countable union of measurable sets.

(b) Note that $\{x \in X: f(x) \geq g(x)\} = \{x \in X: g(x) > f(x)\}^c$, which is measurable by (a).

(c) Observe that

$$\{x \in X: f(x) = g(x)\} = \{x \in X: f(x) \geq g(x)\} \cap \{x \in X: g(x) \geq f(x)\},$$

which is measurable by (b).   ∎

The next result informs us that the usual algebraic combinations of measurable functions again produce measurable functions.

Theorem 16.5.   *For measurable functions f and g the following statements hold:*
1. $f + g$ *is a measurable function.*
2. fg *is a measurable function.*
3. $|f|$, f^+, *and* f^- *are measurable functions.*
4. $f \vee g$ *and* $f \wedge g$ *are measurable functions.*

Proof. (1) Note first that if c is a constant number, then $c - g$ is a measurable function. [Reason: If $a \in \mathbb{R}$, then $\{x \in X: c-g(x) \geq a\} = \{x \in X: g(x) \leq c-a\}$ is a measurable set.] Now, if $a \in \mathbb{R}$, then the set

$$(f + g)^{-1}([a, \infty)) = \{x \in X: f(x) + g(x) \geq a\} = \{x \in X: f(x) \geq a - g(x)\}$$

is measurable by the above observation and Theorem 16.4. Thus, by Theorem 16.2, $f + g$ is a measurable function.

(2) Note first that f^2 is a measurable function. Indeed, if $a \in \mathbb{R}$, then we have $\{x \in X: f^2(x) \leq a\} = \emptyset$ if $a < 0$ and $\{x \in X: f^2(x) \leq a\} = f^{-1}([-\sqrt{a}, \sqrt{a}])$ if $a \geq 0$. Thus, f^2 is a measurable function by Theorem 16.2. Also, if c is a constant number, then cf is measurable. [Reason: If $A = \{x \in X: cf(x) \geq a\}$, then $A = \{x \in X: f(x) \geq a/c\}$ for $c > 0$ and $A = \{x \in X: f(x) \leq a/c\}$ for $c < 0$.] The result now follows from the above observations combined with (1) and the relation

$$fg = \frac{1}{2}[(f + g)^2 - f^2 - g^2].$$

(3) The measurability of $|f|$ follows from the relations

$$\{x \in X: |f(x)| \leq a\} = \emptyset \quad \text{if } a < 0,$$

and

$$\{x \in X: |f(x)| \leq a\} = \{x \in X: f(x) \leq a\} \cap \{x \in X: f(x) \geq -a\} \quad \text{if } a \geq 0.$$

For the measurability of f^+ and f^- use the identities

$$f^+ = \frac{1}{2}(|f| + f) \quad \text{and} \quad f^- = \frac{1}{2}(|f| - f).$$

(4) The identities

$$f \vee g = \frac{1}{2}(f + g + |f - g|) \quad \text{and} \quad f \wedge g = \frac{1}{2}(f + g - |f - g|)$$

show that $f \vee g$ and $f \wedge g$ are measurable functions. ∎

The next result states that the almost everywhere limit of a sequence of measurable functions is again a measurable function.

Theorem 16.6. *For a sequence $\{f_n\}$ of measurable functions, the following statements hold*:

1. *If $f_n \to f$ a.e., then f is a measurable function.*

2. *If $\{f_n(x)\}$ is a bounded sequence for each x, then* $\limsup f_n$ *and* $\liminf f_n$ *are both measurable functions.*

Proof.   (1) Let $A = \{x \in X: \lim f_n(x) = f(x)\}$. Since $f_n \to f$ a.e., it follows that $\mu^*(A^c) = 0$. Thus, A^c is measurable, and hence, A is also a measurable set. Now, let $a \in \mathbb{R}$. Observe that the equality

$$A \cap f^{-1}((a, \infty)) = A \cap \left[\bigcup_{n=1}^{\infty} \bigcap_{i=n}^{\infty} f_i^{-1}\left(\left(a + \frac{1}{n}, \infty\right)\right) \right]$$

and the measurability of each f_i show that $A \cap f^{-1}((a, \infty))$ is a measurable set. Also, $A^c \cap f^{-1}((a, \infty))$ is a measurable set since it is a subset of a set of measure zero. Thus,

$$f^{-1}((a, \infty)) = [A \cap f^{-1}((a, \infty))] \cup [A^c \cap f^{-1}((a, \infty))]$$

is a measurable set. It now easily follows from Theorem 16.2 that f is a measurable function.

(2) Let $\{f_n(x)\}$ be bounded for each x. We shall show that $\limsup f_n$ is a measurable function. The measurability of $\liminf f_n$ will then follow from the identity $\liminf f_n = -\limsup(-f_n)$.

Note first that $\limsup f_n = \bigwedge_{n=1}^{\infty} \bigvee_{k=n}^{\infty} f_k$. Fix a natural number n. Then the function $h_m = f_n \vee f_{n+1} \vee \cdots \vee f_{n+m}$ is measurable for each m [by Theorem 16.5(4)], and since $h_m \uparrow \bigvee_{k=n}^{\infty} f_k = g_n$ (everywhere), it follows from part (1) that each g_n is a measurable function. Since $g_n \downarrow \limsup f_n$ (everywhere), by (1) again it follows that $\limsup f_n$ is a measurable function, and the proof is complete.   ∎

The last two theorems show that the collection of all measurable functions forms a function space containing the limits of its sequences that converge almost everywhere. Moreover, this collection is an algebra under the pointwise product. In other words, the usual operations applied to measurable functions produce measurable functions again.

Therefore, it is natural to ask whether there is a function that fails to be measurable. Clearly, if every subset of X is measurable [i.e., $\Lambda = \mathcal{P}(X)$], then every real-valued function on X is measurable. On the other hand, if there are nonmeasurable sets, then there are also nonmeasurable functions. Indeed, if E is a nonmeasurable set, then the function f defined by $f(x) = 1$ if $x \in E$ and $f(x) = 0$ if $x \notin E$ is not measurable simply because the set $f^{-1}(\{1\}) = E$ is not measurable.

We close this section with a basic theorem connecting almost everywhere convergence and uniform convergence. The result is due to D. F. Egorov.[4]

---

[4]Dimitry Fedorovich Egorov (1869–1931), a Russian mathematician. He is credited with the creation of the famous Moscow school dealing with the theory of functions of one variable.

Theorem 16.7 (Egorov). *Let $\{f_n\}$ be a sequence of measurable functions such that $f_n \to f$ a.e., and let E be a measurable subset of X such that $\mu^*(E) < \infty$. Then for every $\epsilon > 0$, there exists a measurable subset F of E with $\mu^*(F) < \epsilon$, and with $\{f_n\}$ converging uniformly to f on $E \setminus F$.*

Proof. First note that by removing a null set from X, we can assume without loss of generality that $\lim f_n(x) = f(x)$ holds for each $x \in X$. Fix a measurable subset E of X with $\mu^*(E) < \infty$. For each pair of positive integers n and k let

$$E_{n,k} = \{x \in E: |f_m(x) - f(x)| < 2^{-n} \text{ for all } m \geq k\}.$$

Clearly, each $E_{n,k}$ is a measurable subset of X. In addition, $E_{n,k} \subseteq E_{n,k+1}$ holds for all k and n. Since $\lim f_n(x) = f(x)$ for each $x \in X$, it easily follows that

$$E_{n,k} \uparrow_k E \quad \text{for each fixed } n,$$

and so, by Theorem 15.4, we see that

$$\mu^*(E_{n,k}) \uparrow_k \mu^*(E) \quad \text{for each fixed } n.$$

Now, let $\epsilon > 0$. Since $\mu^*(E) < \infty$, for each n there exists a natural number k_n such that $\mu^*(E \setminus E_{n,k_n}) = \mu^*(E) - \mu^*(E_{n,k_n}) < 2^{-n}\epsilon$. Let $F = \bigcup_{n=1}^{\infty}(E \setminus E_{n,k_n})$. Then F is measurable, $F \subseteq E$, and $\mu^*(F) \leq \sum_{n=1}^{\infty}\mu^*(E \setminus E_{n,k_n}) < \epsilon$. Also, if $x \in E \setminus F = \bigcap_{n=1}^{\infty}E_{n,k_n}$, then $|f_m(x) - f(x)| < 2^{-n}$ holds for each $m \geq k_n$. This shows that $\{f_n\}$ converges uniformly to f on $E \setminus F$, and the proof of the theorem is complete. ∎

EXERCISES

1. Let $(X, \mathcal{S}, \mu)$ be a measure space. For a function $f: X \to \mathbb{R}$ show that the following statements are equivalent:

   a. f is a measurable function.
   b. $f^{-1}((-\infty, a))$ is measurable for each $a \in \mathbb{R}$.
   c. $f^{-1}((a, \infty))$ is measurable for each $a \in \mathbb{R}$.

2. Let $(X, \mathcal{S}, \mu)$ be a measure space, and let A be a dense subset of $\mathbb{R}$. Show that a function $f: X \to \mathbb{R}$ is measurable if and only if $\{x \in X: f(x) \geq a\}$ is measurable for each $a \in A$.

3. Give an example of a non-Lebesgue measurable function $f: \mathbb{R} \to \mathbb{R}$ such that $|f|$ is a measurable function and $f^{-1}(\{a\})$ is a measurable set for each $a \in \mathbb{R}$.

4. Show that if $f: \mathbb{R} \to \mathbb{R}$ is continuous a.e., then f is a Lebesgue measurable function.

5. Let $f: \mathbb{R} \to \mathbb{R}$ be a differentiable function. Show that f' is Lebesgue measurable.

6. Let $(X, \mathcal{S}, \mu)$ be a measure space and let $f: X \to \mathbb{R}$ be a measurable function. Show that:

    a. $|f|^p$ is a measurable function for all $p \geq 0$, and

    b.   if $f(x) \neq 0$ for each $x \in X$, then $1/f$ is a measurable function.

7. Let $\{f_n\}$ be a sequence of real-valued measurable functions on a measure space $(X, \mathcal{S}, \mu)$. Then show that the sets

    a. $A = \{x \in X: f_n(x) \to \infty\}$,

    b. $B = \{x \in X: f_n(x) \to -\infty\}$, and

    c. $C = \{x \in X: \lim f_n(x) \text{ exists in } \mathbb{R}\}$

    are all measurable.

8. Let $(X, \mathcal{S}, \mu)$ be a measure space. Assume that $f: X \to \mathbb{R}$ is a measurable function and $g: \mathbb{R} \to \mathbb{R}$ is a continuous function. Show that $g \circ f$ is a measurable function.

9. Let $\mathcal{F}$ be a nonempty family of continuous real-valued functions defined on $\mathbb{R}$. Assume that there exists a function $g: \mathbb{R} \to \mathbb{R}$ such that $f(x) \leq g(x)$ for each $x \in \mathbb{R}$ and all $f \in \mathcal{F}$. Show that the supremum function $h: \mathbb{R} \to \mathbb{R}$, defined by $h(x) = \sup\{f(x): f \in \mathcal{F}\}$, is (Lebesgue) measurable.

10. Show that if $f: X \to \mathbb{R}$ is a measurable function, then either f is constant almost everywhere or else (exclusively) there exists a constant c such that

$$\mu^*(\{x \in X: f(x) > c\}) > 0 \quad \text{and} \quad \mu^*(\{x \in X: f(x) < c\}) > 0.$$

17. SIMPLE AND STEP FUNCTIONS

From this section on, the properties of the characteristic functions are needed. For this reason their properties are listed below, and the reader is expected to be able to verify them. If A is a subset of a set X, then the **characteristic function** χ_A of A is the real-valued function defined on X by $\chi_A(x) = 1$ if $x \in A$ and $\chi_A(x) = 0$ if $x \notin A$. (The characteristic function χ_A is also called by many authors the **indicator function** of A and is also denoted by $\mathbf{1}_A$.)

For subsets A and B of a set X, the following relations hold:

1. $\chi_\emptyset = 0$ and $\chi_X = \mathbf{1}$.

2. $A \subseteq B$ if and only if $\chi_A \leq \chi_B$.

3. $\chi_{A \cap B} = \chi_A \cdot \chi_B = \chi_A \wedge \chi_B$.

4. $\chi_{A \cup B} = \chi_A + \chi_B - \chi_{A \cap B} = \chi_A \vee \chi_B$.

5. $\chi_{A \setminus B} = \chi_A - \chi_{A \cap B}$.

6. If $A = \bigcup_{n=1}^{\infty} A_n$ and $\{A_n\}$ is a disjoint sequence of subsets of X, then

$$\chi_A = \sum_{n=1}^{\infty} \chi_{A_n}.$$

7. $\chi_{A \times B} = \chi_A \cdot \chi_B$. (Here the set B can be taken to be a subset of another set Y.)

Again, throughout this section $(X, \mathcal{S}, \mu)$ is assumed to be a fixed measure space. If a measurable function $\phi: X \to \mathbb{R}$ assumes only a finite number of values, then ϕ

is called a **simple function**. Clearly, finite sums, finite products, and finite suprema and infima of simple functions are again simple functions. In other words, the collection of all simple functions is a function space, which is in addition an algebra of functions.

If ϕ is a simple function assuming the distinct nonzero values $a_1, \ldots, a_n$, then the sets $A_i = \{x \in X: \phi(x) = a_i\}$ are all measurable and pairwise disjoint, and $\phi = \sum_{i=1}^{n} a_i \chi_{A_i}$ holds. This expression is called the **standard representation** of ϕ. Let us agree also that the standard representation of the constant zero function is $\chi_\emptyset$. In general, a simple function ϕ may be represented in more than one way in the form $\phi = \sum_{j=1}^{m} b_j \chi_{B_j}$, where the b_j values are real numbers and the sets B_j are all measurable.

A simple function ϕ is called a **step function** if ϕ has a representation of the form $\phi = \sum_{j=1}^{m} b_j \chi_{B_j}$, where each B_j is a measurable set of finite measure, i.e., $\mu^*(B_j) < \infty$. Clearly, a simple function is a step function if and only if it vanishes outside of a set of finite measure.[5] For this reason, many authors call a step function a simple function with finite support. It should also be clear that the collection of all step functions forms a function space that is in addition an algebra of functions. The step functions will be the "building blocks" for the theory of integration.

Definition 17.1. *Let ϕ be a step function with standard representation $\phi = \sum_{i=1}^{n} a_i \chi_{A_i}$; that is, $a_1, \ldots, a_n$ are the distinct nonzero values of ϕ, and $A_i = \{x \in X: \phi(x) = a_i\}$. Then the* **Lebesgue integral** *of ϕ (or simply, the* **integral** *of ϕ) is defined by*

$$I(\phi) = \sum_{i=1}^{n} a_i \mu^*(A_i).$$

The integral of ϕ is usually denoted by $\int_X \phi \, d\mu$, or simply by $\int \phi \, d\mu$. For the time being we shall use $I(\phi)$ to denote the integral of ϕ, and later on we shall switch to the conventional notation.

Our first theorem describes the linearity property of the integral for step functions. The difficulty of the proof lies in the fact that the standard representation of the sum of two simple functions is not necessarily the sum of their standard representations.

Theorem 17.2 (The Linearity of the Integral). *If ϕ and ψ are step functions, then*

$$I(\alpha\phi + \beta\psi) = \alpha I(\phi) + \beta I(\psi)$$

holds for all $\alpha, \beta \in \mathbb{R}$.

---

[5]Notice that the definition of a step function is more general than the familiar one associated with the Riemann integral. There, a function $\phi: [a, b] \to \mathbb{R}$ is called a step function if there exists some partition $P = \{x_0, x_1, \ldots, x_n\}$ of $[a, b]$ such that ϕ is constant on each open subinterval (x_{i-1}, x_i).

Proof. Clearly, $I(\alpha\phi) = \alpha I(\phi)$ holds for each step function ϕ and $\alpha \in \mathbb{R}$. What needs to be shown, therefore, is that if ϕ and ψ are step functions, then $I(\phi + \psi) = I(\phi) + I(\psi)$ holds—the additivity property of the integral.

Let $\phi = \sum_{i=1}^n a_i \chi_{A_i}$ and $\psi = \sum_{j=1}^m b_j \chi_{B_j}$ be the standard representations of ϕ and ψ. Let $E = (\bigcup_{i=1}^n A_i) \cup (\bigcup_{j=1}^m B_j)$, and note that $\mu^*(E) < \infty$. Next put $A_0 = E \setminus \bigcup_{i=1}^n A_i$, $B_0 = E \setminus \bigcup_{j=1}^m B_j$, and $a_0 = b_0 = 0$. Then the A_i are pairwise disjoint, the B_j are pairwise disjoint, and $\bigcup_{i=0}^n A_i = \bigcup_{j=0}^m B_j = E$.

Now, write $\phi + \psi = \sum_{k=1}^r c_k \chi_{C_k}$ in its standard representation. Since $\phi(x) + \psi(x) \neq 0$ implies either $\phi(x) \neq 0$ or $\psi(x) \neq 0$, it easily follows that $\bigcup_{k=1}^r C_k \subseteq E$. Put $C_0 = E \setminus \bigcup_{k=1}^r C_k$ and $c_0 = 0$. Then the C_k are pairwise disjoint, and

$$C_k = \bigcup_{i=0}^n \bigcup_{j=0}^m (C_k \cap A_i \cap B_j)$$

is a disjoint union for $0 \leq k \leq r$. Thus, $\mu^*(C_k) = \sum_{i=0}^n \sum_{j=0}^m \mu^*(C_k \cap A_i \cap B_j)$. In addition, note that $c_k \mu^*(C_k \cap A_i \cap B_j) = (a_i + b_j)\mu^*(C_k \cap A_i \cap B_j)$ holds for all i, j, and k. Indeed, if $C_k \cap A_i \cap B_j = \emptyset$, then the equality holds trivially, and if $x \in C_k \cap A_i \cap B_j$, then the equality holds by virtue of $c_k = \phi(x) + \psi(x) = a_i + b_j$.

Putting the above together and using the (finite) additivity of μ^* on Λ, we get

$$I(\phi + \psi) = \sum_{k=1}^r c_k \mu^*(C_k) = \sum_{k=0}^r c_k \mu^*(C_k) = \sum_{k=0}^r \sum_{i=0}^n \sum_{j=0}^m c_k \mu^*(C_k \cap A_i \cap B_j)$$

$$= \sum_{i=0}^n \sum_{j=0}^m \sum_{k=0}^r (a_i + b_j)\mu^*(C_k \cap A_i \cap B_j)$$

$$= \sum_{i=0}^n \sum_{j=0}^m (a_i + b_j)\mu^*(A_i \cap B_j)$$

$$= \sum_{i=0}^n a_i \sum_{j=0}^m \mu^*(A_i \cap B_j) + \sum_{j=0}^m b_j \sum_{i=0}^n \mu^*(A_i \cap B_j)$$

$$= \sum_{i=0}^n a_i \mu^*(A_i) + \sum_{j=0}^m b_j \mu^*(B_j)$$

$$= I(\phi) + I(\psi),$$

and the proof is finished.   ∎

From the preceding theorem, it should be clear that if ϕ is a step function having a representation $\phi = \sum_{j=1}^m b_j \chi_{B_j}$, with $b_j \in \mathbb{R}$ and each B_j measurable with finite measure, then $I(\phi) = \sum_{j=1}^{m} b_j \mu^*(B_j)$ holds. In other words, the integral of a step function does not depend upon its particular representation.

We continue with the monotone property of the integral.

Theorem 17.3 (The Monotonicity of the Integral). *For step functions ϕ and ψ, the following statements hold:*

1. *If $\phi \geq 0$ a.e., then $I(\phi) \geq 0$. In particular, if $\phi \geq \psi$ a.e., then $I(\phi) \geq I(\psi)$.*
2. *If $\phi = 0$ a.e., then $I(\phi) = 0$. In particular, if $\phi = \psi$ a.e., then $I(\phi) = I(\psi)$.*

Proof. (1) Let $\phi = \sum_{i=1}^{n} a_i \chi_{A_i}$ be the standard representation of ϕ. Since $\phi \geq 0$ a.e., note that if $a_i < 0$ holds for some i, then necessarily $\mu^*(A_i) = 0$ for that i. Thus, $I(\phi) = \sum_{i=1}^{n} a_i \mu^*(A_i) \geq 0$.

Now, if $\phi \geq \psi$ a.e., then $\phi - \psi \geq 0$ a.e., and so $I(\phi) - I(\psi) = I(\phi - \psi) \geq 0$. Hence, $I(\phi) \geq I(\psi)$.

(2) If $\phi = 0$ a.e., then $\phi \geq 0$ a.e. and $-\phi \geq 0$ a.e. both hold. Thus, by part (1), $I(\phi) \geq 0$ and $-I(\phi) \geq 0$ both hold. Therefore, $I(\phi) = 0$.

Now, if $\phi = \psi$ a.e., then $\phi - \psi = 0$ a.e., and hence, by the preceding case, we obtain $I(\phi) - I(\psi) = I(\phi - \psi) = 0$. That is, $I(\phi) = I(\psi)$. ∎

The next theorem describes a basic continuity property of the integral.

Theorem 17.4 (The Order Continuity of the Integral). *Let $\{\phi_n\}$ be a sequence of step functions. If $\phi_n \downarrow 0$ a.e., then $I(\phi_n) \downarrow 0$.*

In particular, if ϕ is a step function and $\phi_n \uparrow \phi$ a.e., then $I(\phi_n) \uparrow I(\phi)$.

Proof. Assume $\phi_n \downarrow 0$ a.e. Put

$$A_n = \{x \in X : \phi_{n+1} > \phi_n(x)\} \quad \text{and} \quad A_0 = \{x \in X : \phi_n(x) \not\rightarrow 0\}.$$

By assumption, $\mu^*(A_n) = 0$ for $n = 0, 1, 2, \ldots$.

Let $A = \bigcup_{n=0}^{\infty} A_n$. Then $\mu^*(A) = 0$ (use the σ-subadditivity of μ^* to see this), and note that $\phi_n(x) \downarrow 0$ and $\phi_{n+1}(x) \leq \phi_n(x)$ hold for each $x \in A^c$. Also, if $\psi_n = \phi_n \cdot \chi_{A^c}$, then ψ_n is a step function and $\psi_n = \phi_n$ a.e. for each n, $\psi_n(x) \downarrow 0$ for each $x \in X$, and by Theorem 17.3, we have $I(\psi_n) = I(\phi_n)$ for each n. Thus, replacing $\{\phi_n\}$ by $\{\psi_n\}$ if necessary, we can assume without loss of generality that $\phi_n(x) \downarrow 0$ holds for each $x \in X$.

Now, let $\epsilon > 0$. Put $M = \max\{\phi_1(x) : x \in X\}$ and $B = \{x \in X : \phi_1(x) > 0\}$; clearly, $\mu^*(B) < \infty$. For each n, define $E_n = \{x \in X : \phi_n(x) \geq \epsilon\}$. Then $\mu^*(E_1) < \infty$, each E_n is measurable, and $E_n \downarrow \emptyset$ by virtue of $\phi_n(x) \downarrow 0$ for each $x \in X$. By Theorem 15.4, we have $\mu^*(E_n) \downarrow 0$. Pick an integer k such that $\mu^*(E_k) < \epsilon$. Now, if $n \geq k$, then taking into account that $\phi_k(x) < \epsilon$ for each

$x \in B \setminus E_k$, we see that

$$0 \leq \phi_n \leq \phi_k = \phi_k \chi_{E_k} + \phi_k \chi_{B \setminus E_k} \leq M \chi_{E_k} + \epsilon \chi_B,$$

from which it follows that

$$0 \leq I(\phi_n) \leq M\mu^*(E_k) + \epsilon\mu^*(B) < \epsilon[M + \mu^*(B)]$$

for all $n \geq k$. Therefore, $\lim I(\phi_n) = 0$.

Now, if ϕ is a step function and $\phi_n \uparrow \phi$ a.e. holds, then $\phi - \phi_n \downarrow 0$ a.e. also holds, and so, by the previous case, $I(\phi) - I(\phi_n) = I(\phi - \phi_n) \downarrow 0$. Hence, $I(\phi_n) \uparrow I(\phi)$, and the proof is finished. ∎

As an application of the last theorem, we have the following result:

Theorem 17.5.   *Assume that two sequences of step functions $\{\phi_n\}$ and $\{\psi_n\}$ satisfy $\phi_n \uparrow f$ a.e. and $\psi_n \uparrow f$ a.e., where $f: X \to \mathbb{R}^*$ is a given function. Then*

$$\lim_{n \to \infty} I(\phi_n) = \lim_{n \to \infty} I(\psi_n)$$

holds, with the above limits possibly being infinite.

Proof.   Note first that for each fixed m, we have

$$\phi_m \wedge \psi_n \uparrow_n \phi_m \wedge f = \phi_m \text{ a.e.}$$

Therefore, by Theorem 17.4, $\lim_{n \to \infty} I(\phi_m \wedge \psi_n) = I(\phi_m)$.

Now, by virtue of $\phi_m \wedge \psi_n \leq \psi_n$ a.e. for all m and n and the monotonicity of the integral, we get

$$I(\phi_m) = \lim_{n \to \infty} I(\phi_m \wedge \psi_n) \leq \lim_{n \to \infty} I(\psi_n)$$

for all m. Hence, $\lim I(\phi_n) \leq \lim I(\psi_n)$ holds. By the symmetry of the situation, $\lim I(\psi_n) \leq \lim I(\phi_n)$, and so, $\lim I(\phi_n) = \lim I(\psi_n)$. ∎

The next result describes another useful property of step functions.

Theorem 17.6.   *Let $\{\phi_n\}$ be a sequence of step functions. If A is a subset of X such that $\phi_n \uparrow \chi_A$ a.e., then A is a measurable set, and $\lim I(\phi_n) = \mu^*(A)$ holds.*

Proof. The measurability of A follows from Theorem 16.6(1). We can assume (how?) that $\phi_n(x) \uparrow \chi_A(x)$ holds for all $x \in X$.

For each n the set $A_n = \{x \in X: \phi_n(x) > 0\}$ is measurable with finite measure, and $A_n \uparrow A$ holds. It follows that $\chi_{A_n} \uparrow \chi_A$, and thus, by Theorem 17.5, we have

$$\lim_{n \to \infty} I(\phi_n) = \lim_{n \to \infty} I(\chi_{A_n}) = \lim_{n \to \infty} \mu^*(A_n) = \mu^*(A),$$

where the last equality holds by virtue of Theorem 15.4. ∎

We close this section with an important approximation property of measurable functions.

Theorem 17.7. *Let $f: X \to \mathbb{R}$ be a measurable function satisfying $f(x) \geq 0$ for all x. Then there exists a sequence $\{\phi_n\}$ of simple functions such that $0 \leq \phi_n(x) \uparrow f(x)$ holds for all $x \in X$.*

Proof. For each n let $A_n^i = \{x \in X: (i-1)2^{-n} \leq f(x) < i2^{-n}\}$ for $i = 1, 2, \ldots, n2^n$, and note that $A_n^i \cap A_n^j = \emptyset$ if $i \neq j$. Since f is measurable, all the A_n^i are measurable sets.

Now, for each n define $\phi_n = \sum_{i=1}^{n2^n} 2^{-n}(i-1)\chi_{A_n^i}$, and note that $\{\phi_n\}$ is a sequence of simple functions. Also, an easy verification shows that $0 \leq \phi_n(x) \leq \phi_{n+1}(x) \leq f(x)$ holds for all x and all n. Moreover, if x is fixed, then $0 \leq f(x) - \phi_n(x) \leq 2^{-n}$ holds for all sufficiently large n. Thus, $\phi_n(x) \uparrow f(x)$ holds for all x, and the proof of the theorem is finished. ∎

If $f: X \to \mathbb{R}$ is a measurable function such that $f(x) \geq 0$ for almost all x, then there exists a sequence $\{\phi_n\}$ of positive simple functions such that $\phi_n \uparrow f$ a.e. holds. To see this, let $E = \{x \in X: f(x) \geq 0\}$, and then apply Theorem 17.7 to the measurable function $f\chi_E$.

EXERCISES

1. Verify the identities about the characteristic functions at the beginning of Section 17.
2. Let ϕ be a step function and ψ a simple function such that $0 \leq \psi \leq \phi$ a.e. Show that ψ is a step function.
3. Show that if (X, S, μ) is a finite measure space, then every simple function is a step function.
4. Give an alternate proof of the linearity of the integral (Theorem 17.2) based on Exercise 14 of Section 12.
5. Show that $|I(\phi)| \leq I(|\phi|)$ holds for every step function ϕ.
6. Let ϕ be a step function such that $I(|\phi|) = 0$. Show that $\phi = 0$ a.e. holds.

7. Let ϕ be a step function. Let $A = \{x \in X: \phi(x) \neq 0\}$ and $M = \max\{|\phi(x)|: x \in X\}$. Show that $|I(\phi)| \leq M\mu^*(A)$.

8. Let $\{\phi_n\}$ be a sequence of step functions. Show that if ϕ is a step function and $\phi_n \downarrow \phi$ a.e. holds, then $I(\phi_n) \downarrow I(\phi)$ also holds.

9. Let $\{\phi_n\}$ be a sequence of step functions and ϕ a simple function such that $0 \leq \phi_n \uparrow \phi$ a.e. holds. Show that if $\lim I(\phi_n) < \infty$, then ϕ is a step function.

10. Let $(X, \mathcal{S}, \mu)$ be a measure space, and let $f: X \to \mathbb{R}$ be a function. Show that f is a measurable function if and only if there exists a sequence $\{\phi_n\}$ of simple functions such that $\lim \phi_n(x) = f(x)$ holds for all $x \in X$.

11. Let $(X, \mathcal{S}, \mu)$ be a σ-finite measure space, and let $f: X \to \mathbb{R}$ be a measurable function such that $f(x) \geq 0$ for all $x \in X$. Show that there exists a sequence $\{\phi_n\}$ of step functions such that $0 \leq \phi_n \uparrow f(x)$ holds for all $x \in X$.

12. Give a proof of the order continuity of the integral (Theorem 17.4) based on Egorov's Theorem 16.7.

13. Let $(X, \mathcal{S}, \mu)$ be a measure space, and let $f: X \to [0, \infty)$ be a function. Show that f is a measurable function if and only if there exist non-negative constants $c_1, c_2, \ldots$ and measurable sets $E_1, E_2, \ldots$ such that

$$f(x) = \sum_{n=1}^{\infty} c_n \chi_{E_n}(x)$$

holds for each $x \in X$.

14. Let $(X, \mathcal{S}, \mu)$ be a measure space satisfying $\mu^*(X) = 1$, and let $E_1, E_2, \ldots, E_{10}$ be ten measurable sets such that $\mu^*(E_i) = \frac{1}{3}$ holds for each i. Show that four of these sets have an intersection with positive measure. Is the conclusion true for nine measurable sets instead of ten?

15. If $f: X \to [0, 1]$ is a measurable function, then show that either $f = \chi_A$ a.e. for some measurable set A or else (exclusively) there exists a constant $0 < c < \frac{1}{2}$ such that

$$\mu^*(\{x \in X: c < f(x) < 1 - c\}) > 0.$$

16. Let $(X, \mathcal{S}, \mu)$ be a measure space, and let $\phi: X \to \mathbb{R}$ be a simple function having the standard representation $\phi = \sum_{i=1}^{n} a_i \chi_{A_i}$. If $\phi \geq 0$ a.e., then the sum $\sum_{i=1}^{n} a_i \mu^*(A_i)$ makes sense as an extended real number (it may be infinite). Call this extended real number the **Lebesgue integral** of ϕ, and write $I(\phi) = \sum_{i=1}^{n} a_i \mu^*(A_i)$.

   a. If ϕ and ψ are simple functions such that $\phi \geq 0$ a.e. and $\psi \geq 0$ a.e., then show that $I(\phi + \psi) = I(\phi) + I(\psi)$.

   b. If ϕ and ψ are simple functions such that $0 \leq \phi \leq \psi$ a.e., then show that $I(\phi) \leq I(\psi)$.

   c. Show that if $\{\phi_n\}$ and $\{\psi_n\}$ are two sequences of simple functions and $f: X \to \mathbb{R}^*$ such that $0 \leq \phi_n \uparrow f$ a.e. and $0 \leq \psi_n \uparrow f$ a.e., then $\lim I(\phi_n) = \lim I(\psi_n)$ holds (with the limits possibly being infinite).

   d. Assume that $\{\phi_n\}$ is a sequence of simple functions such that $0 \leq \phi_n \uparrow \chi_A$ a.e. holds. Show that $\lim I(\phi_n) = \mu^*(A)$.

   e. Give an example of a sequence $\{\phi_n\}$ of simple functions on some measure space such that $\phi_n \downarrow 0$ (everywhere) and $\lim I(\phi_n) \neq 0$.

17. Let (X, Σ, μ) be a measure space with Σ being a σ-algebra. Let us say that a function $f: X \to \mathbb{R}$ is Σ-*measurable* if $f^{-1}(A) \in \Sigma$ for each open subset A of $\mathbb{R}$. Also, let $\mathcal{M}_\Sigma$ denote the collection of all Σ-measurable functions. Establish the following:

    a. $\mathcal{M}_\Sigma$ is a function space and an algebra of functions.
    b. $\mathcal{M}_\Sigma$ is closed under sequential pointwise limits.
    c. If μ is σ-finite and $f: X \to \mathbb{R}$ is a measurable function, then there exists a Σ-measurable function $g: X \to \mathbb{R}$ such that $f = g$ a.e.

18. THE LEBESGUE MEASURE

The notion of the Lebesgue measure is the natural extension of the concepts of length, area, and volume. In particular, the Lebesgue measure of any geometric figure in $\mathbb{R}^2$ turns out to be its area, while the Lebesgue measure of any geometric solid in $\mathbb{R}^3$ is its volume. Throughout this section, $d(x, y)$ will denote the Euclidean distance of the vectors $x = (x_1, \ldots, x_n)$ and $y = (y_1, \ldots, y_n)$ of $\mathbb{R}^n$. That is, $d(x, y) = \left[\sum_{i=1}^n (x_i - y_i)^2 \right]^{\frac{1}{2}}$.

The objective of this section is to present in some detail the properties of the Lebesgue measure on $\mathbb{R}^n$. For this discussion, S will denote the semiring consisting of the empty set and all sets of the form $A = \prod_{i=1}^n [a_i, b_i)$, where $-\infty < a_i < b_i < \infty$ for each i. This semiring was discussed in Example 12. In Example 13 we have promised the reader to show that the set function $\lambda: S \to [0, \infty)$ defined by $\lambda(\emptyset) = 0$ and $\lambda(\prod_{i=1}^n [a_i, b_i)) = \prod_{i=1}^n (b_i - a_i)$ is σ-additive. We shall do this next.

Theorem 18.1. *The set function* $\lambda: S \to [0, \infty)$ *defined above is a measure, called the* **Lebesgue measure** *on* S.

Proof. The proof is by induction on the dimension of $\mathbb{R}^n$. Let us denote by S_n the semiring S on $\mathbb{R}^n$, and the corresponding set function by λ_n. For $n = 1$ the result has been established in Example 13. Now, assume that λ_n is a measure for some n. We have to show that λ_{n+1} is σ-additive on S_{n+1}.

To this end, let $A \times [a, b) = \bigcup_{i=1}^\infty [A_i \times [a_i, b_i)]$, with $A \in S_n$, $A_i \in S_n$ for each i, and the sequence $\{A_i \times [a_i, b_i)\}$ being pairwise disjoint. It easily follows that $\chi_{A \times [a,b)} = \sum_{i=1}^\infty \chi_{A_i \times [a_i,b_i)}$. Now for $x = (x_1, \ldots, x_n) \in \mathbb{R}^n$ and $t \in \mathbb{R}$, the last equality yields

$$\chi_A(x) \cdot \chi_{[a,b)}(t) = \sum_{i=1}^\infty \chi_{A_i}(x) \cdot \chi_{[a_i,b_i)}(t).$$

Fix $x = (x_1, \ldots, x_n) \in \mathbb{R}^n$, and let $\phi_k(t) = \sum_{i=1}^k \chi_{A_i}(x) \cdot \chi_{[a_i,b_i)}(t)$. Then each ϕ_k is a step function [for $(\mathbb{R}, S_1, \lambda_1)$] such that $\phi_k(t) \uparrow \chi_A(x) \cdot \chi_{[a,b)}(t)$ for each $t \in \mathbb{R}$. By Theorem 17.4 we get $\sum_{i=1}^k (b_i - a_i)\chi_{A_i}(x) \uparrow (b - a)\chi_A(x)$ for each

$x = (x_1, \ldots, x_n) \in \mathbb{R}^n$. Since by our induction hypothesis $(\mathbb{R}^n, S_n, \lambda_n)$ is a measure space, by applying Theorem 17.4 once more, we get

$$\sum_{i=1}^{k} (b_i - a_i)\lambda_n(A_i) \uparrow (b - a)\lambda_n(A).$$

That is,

$$\sum_{i=1}^{\infty} \lambda_{n+1}(A_i \times [a_i, b_i)) = \lambda_{n+1}(A \times [a, b)).$$

Thus, λ_{n+1} is σ-additive and the proof is finished. ∎

An **interval** of $\mathbb{R}^n$ is any set of the form $\prod_{i=1}^{n} I_i$, where each I_i is an interval of $\mathbb{R}$. If each I_i is in addition a bounded open interval of $\mathbb{R}$, then $\prod_{i=1}^{n} I_i$ is called a bounded open interval of $\mathbb{R}^n$. As in Example 15.5, we can show that each interval of $\mathbb{R}^n$ is Lebesgue measurable, and moreover, $\lambda^*(\prod_{i=1}^{n} I_i) = \prod_{i=1}^{n} |I_i|$, where $|I_i|$ denotes the length of I_i.

A useful formula for the outer Lebesgue measure is the following:

$$\lambda^*(A) = \inf\left\{ \sum_{i=1}^{\infty} \lambda^*(I_i) \colon \text{Each } I_i \text{ is a bounded open interval and } A \subseteq \bigcup_{i=1}^{\infty} I_i \right\}$$

for every subset A of $\mathbb{R}^n$.

To prove the formula, note first that if $A = \prod_{i=1}^{n} [a_i, b_i)$ (i.e., $A \in S$), then for each $\epsilon > 0$ the bounded open interval $I_\epsilon = \prod_{i=1}^{n} (a_i - \epsilon, b_i)$ satisfies $A \subseteq I_\epsilon$, and

$$\lambda^*(I_\epsilon) - \lambda(A) = \prod_{i=1}^{n} (b_i - a_i + \epsilon) - \prod_{i=1}^{n} (b_i - a_i) \longrightarrow 0 \quad \text{as } \epsilon \to 0.$$

If $\lambda^*(A) = \infty$, then the formula holds trivially. If $\lambda^*(A) < \infty$, then given $\epsilon > 0$, choose a sequence $\{A_i\}$ of S such that $\sum_{i=1}^{\infty} \lambda(A_i) < \lambda^*(A) + \epsilon$. By the preceding, for each i there exists a bounded open interval I_i such that $A_i \subseteq I_i$ and $\lambda^*(I_i) - \lambda(A_i) < 2^{-i}\epsilon$. Then $A \subseteq \bigcup_{i=1}^{\infty} I_i$ and

$$\lambda^*(A) \leq \sum_{i=1}^{\infty} \lambda^*(I_i) < \sum_{i=1}^{\infty} [\lambda(A_i) + 2^{-i}\epsilon] < \lambda^*(A) + \epsilon,$$

which establishes the validity of the desired formula.

- *From now on, for simplicity, λ^* will be denoted by λ and it will be called the* **Lebesgue measure** *on* $\mathbb{R}^n$.

It should be clear that every finite subset of $\mathbb{R}^n$ has Lebesgue measure zero. Therefore, by the σ-subadditivity of λ, every countable subset of $\mathbb{R}^n$ likewise has Lebesgue measure zero.

Our next result deals with an approximation property of the Lebesgue measurable sets by open sets.

Theorem 18.2. *A subset E of $\mathbb{R}^n$ is Lebesgue measurable if and only if for each $\epsilon > 0$, there exists an open set $\mathcal{O}$ such that $E \subseteq \mathcal{O}$, and $\lambda(\mathcal{O} \setminus E) < \epsilon$.*

Proof. Assume E to be Lebesgue measurable. First, consider the case when $\lambda(E) < \infty$. Given $\epsilon > 0$, choose a sequence $\{I_i\}$ of bounded open intervals such that $E \subseteq \bigcup_{i=1}^{\infty} I_i$ and $\sum_{i=1}^{\infty} \lambda(I_i) < \lambda(E) + \epsilon$. Then $\mathcal{O} = \bigcup_{i=1}^{\infty} I_i$ is an open set such that $E \subseteq \mathcal{O}$ and $\lambda(\mathcal{O}) \leq \sum_{i=1}^{\infty} \lambda(I_i) < \lambda(E) + \epsilon$ hold. Since E is measurable, we get $\lambda(\mathcal{O} \setminus E) = \lambda(\mathcal{O}) - \lambda(E) < \epsilon$.

Now, assume $\lambda(E) = \infty$. For each i let $B_i = \{x \in \mathbb{R}^n : d(x, 0) \leq i\}$ and $E_i = E \cap B_i$. Then $E = \bigcup_{i=1}^{\infty} E_i$, and each E_i is a Lebesgue measurable set such that $\lambda(E_i) < \infty$. By what was shown above, for each i there exists an open set $\mathcal{O}_i$ such that $E_i \subseteq \mathcal{O}_i$ and $\lambda(\mathcal{O}_i \setminus E_i) < 2^{-i}\epsilon$. Put $\mathcal{O} = \bigcup_{i=1}^{\infty} \mathcal{O}_i$. Clearly, $\mathcal{O}$ is an open set, $E \subseteq \mathcal{O}$, and in view of $\mathcal{O} \setminus E \subseteq \bigcup_{i=1}^{\infty}(\mathcal{O}_i \setminus E_i)$, we get $\lambda(\mathcal{O} \setminus E) \leq \sum_{i=1}^{\infty} \lambda(\mathcal{O}_i \setminus E_i) < \epsilon$.

For the converse, assume that for each $\epsilon > 0$ there exists an open set $\mathcal{O}$ such that $E \subseteq \mathcal{O}$ and $\lambda(\mathcal{O} \setminus E) < \epsilon$. For each i choose an open set $\mathcal{O}_i$ such that $E \subseteq \mathcal{O}_i$ and $\lambda(\mathcal{O}_i \setminus E) < i^{-1}$. Put $G = \bigcap_{i=1}^{\infty} \mathcal{O}_i$. Then G is a Lebesgue measurable set such that $E \subseteq G$. Also, $\lambda(G \setminus E) \leq \lambda(\mathcal{O}_i \setminus E) < i^{-1}$ for each i implies that $\lambda(G \setminus E) = 0$, and so $G \setminus E$ is a Lebesgue measurable set. The measurability of E now follows from the identity $E = G \setminus (G \setminus E)$, and the proof is finished. ∎

The next result informs us that every Borel set is Lebesgue measurable. The collection of all Lebesgue measurable sets is denoted (as usual) by Λ.

Theorem 18.3. *Every Borel subset of $\mathbb{R}^n$ is Lebesgue measurable.*

Proof. Let $-\infty < a_i < b_i < \infty$ for $i = 1, \ldots, n$. Choose m such that $a_i + \frac{1}{m} < b_i$ for each $1 \leq i \leq n$. The relation

$$\prod_{i=1}^{n} (a_i, b_i) = \bigcup_{k=m}^{\infty} \left[\prod_{i=1}^{n} \left[a_i + \frac{1}{k}, b_i \right) \right]$$

and the fact that Λ is a σ-algebra imply that $\prod_{i=1}^{n}(a_i, b_i) \in \Lambda$. But then, since every open set can be written as a countable union of such sets, it follows that Λ contains every open set. Therefore, Λ must contain every Borel set since the Borel sets are the members of the σ-algebra generated by the open sets. ∎

It will become apparent in the course of this book that the link connecting measure theory, topology, and analysis is the concept of a regular Borel measure. Its definition follows.

Definition 18.4. *Let (X, τ) be a Hausdorff topological space, and let $\mathcal{B}$ be the σ-algebra of its Borel sets. A measure $\mu: \mathcal{B} \to [0, \infty]$ is called a* **regular Borel measure** *if it satisfies the following properties:*

1. $\mu(K) < \infty$ *for every compact set K.*
2. *If B is a Borel subset of X, then*

$$\mu(B) = \inf\{\mu(\mathcal{O}): \mathcal{O} \text{ open and } B \subseteq \mathcal{O}\}.$$

3. *If $\mathcal{O}$ is an open subset of X, then*

$$\mu(\mathcal{O}) = \sup\{\mu(K): K \text{ compact and } K \subseteq \mathcal{O}\}.$$

A measure μ on the Borel sets of a topological space that satisfies $\mu(K) < \infty$ for each compact set K is referred to as a **Borel measure**.

The next result informs us that property (3) of the definition of a regular Borel measure is true for all Borel sets.

Lemma 18.5. *If μ is a regular Borel measure on a topological space, then*

$$\mu(B) = \sup\{\mu(K): K \text{ compact and } K \subseteq B\}$$

holds for each Borel set B with $\mu(B) < \infty$.

Proof.   Let B be a Borel set with $\mu(B) < \infty$, and let $\epsilon > 0$. Pick an open set V with $B \subseteq V$, and $\mu(V) < \mu(B) + \epsilon$. Similarly, choose an open set W such that $V \setminus B \subseteq W \subseteq V$ and

$$\mu(W) < \mu(V \setminus B) + \epsilon = \mu(V) - \mu(B) + \epsilon < 2\epsilon.$$

Next, choose a compact set C such that $C \subseteq V$ and $\mu(V) < \mu(C) + \epsilon$. Put $K = C \cap W^c$, and note that K is a compact subset of B. Moreover,

$$
\begin{aligned}
0 \leq \mu(B) - \mu(K) &= \mu(B \setminus K) \leq \mu(V \setminus K) = \mu((V \setminus C) \cup W) \\
&\leq [\mu(V) - \mu(C)] + \mu(W) < 3\epsilon
\end{aligned}
$$

holds, and the proof is finished.   ∎

It is important to realize that the Lebesgue measure is a regular Borel measure. The details follow.

Theorem 18.6. *The Lebesgue measure on $\mathbb{R}^n$ is a regular Borel measure.*

Proof. (1) Let K be a compact subset of $\mathbb{R}^n$. Then K is bounded, and so there exists some $A = \prod_{i=1}^{n}[a_i, b_i)$ with $K \subseteq A$. Therefore, $\lambda(K) \leq \lambda(A) < \infty$.

(2) Let B be a Borel set, and let $\epsilon > 0$. By Theorem 18.2 there exists an open set V such that $B \subseteq V$ and $\lambda(V \setminus B) < \epsilon$. Therefore,

$$\lambda(B) \leq \inf\{\lambda(\mathcal{O}): \mathcal{O} \text{ is open and } B \subseteq \mathcal{O}\}$$
$$\leq \lambda(V) = \lambda(V \setminus B) + \lambda(B) \leq \lambda(B) + \epsilon$$

holds for all $\epsilon > 0$, from which the desired equality follows.

(3) Let $\mathcal{O}$ be an open subset of $\mathbb{R}^n$. Pick a sequence $\{K_i\}$ of compact sets with $\mathcal{O} = \bigcup_{i=1}^{\infty} K_i$; for instance, let $\{K_1, K_2, \ldots\}$ be an enumeration of all closed balls with "rational" centers and rational radii that are included in $\mathcal{O}$. Now, for each i let $C_i = \bigcup_{m=1}^{i} K_m$, and note that each C_i is compact and $C_i \uparrow \mathcal{O}$. It follows that $\lambda(C_i) \uparrow \lambda(\mathcal{O})$, and so $\lambda(\mathcal{O}) = \sup\{\lambda(K): K \text{ compact and } K \subseteq \mathcal{O}\}$ holds. The proof of the theorem is now complete. ∎

Later we shall prove a general result (see Theorem [38.4]) that will guarantee that every Borel measure on $\mathbb{R}^n$ is necessarily a regular Borel measure.

If A is a subset of $\mathbb{R}^n$ and $a \in \mathbb{R}^n$, then the set $a + A = \{a + x: x \in A\}$ is called the **translate** of A by the vector a. An alternate notation for $a + A$ is $A + a$. It is easy to see that $\lambda(A) = \lambda(a + A)$ holds for every subset A of $\mathbb{R}^n$ and every a. Also, by using the identities

$$E \cap (a + A) = a + (E - a) \cap A \quad \text{and} \quad E \cap (a + A)^c = a + (E - a) \cap A^c,$$

it can be shown easily that a subset A of $\mathbb{R}^n$ is Lebesgue measurable if and only if $a + A$ is Lebesgue measurable for each $a \in \mathbb{R}^n$.

Let $\mathcal{B}$ denote the σ-algebra of all Borel subsets of $\mathbb{R}^n$. Then $a + A \in \mathcal{B}$ for each $A \in \mathcal{B}$ and each a. Indeed, if $\mathcal{A} = \{A \in \mathcal{B}: a + A \in \mathcal{B} \text{ for all } a \in \mathbb{R}^n\}$, then $\mathcal{A}$ is a σ-algebra of sets containing the open subsets of $\mathbb{R}^n$; hence, $\mathcal{A} = \mathcal{B}$.

A measure $\mu: \mathcal{B} \to [0, \infty]$ is said to be **translation invariant** if $\mu(A) = \mu(a + A)$ holds for all $A \in \mathcal{B}$ and all $a \in \mathbb{R}^n$. The following question is a natural one:

- *What are the translation invariant Borel measures on $\mathbb{R}^n$?*

In answering this question, we shall show that besides a multiplication factor the only translation invariant Borel measure on $\mathbb{R}^n$ is the Lebesgue measure. To establish this, the following result dealing with additive functions is needed. Recall that a function $f: \mathbb{R} \to \mathbb{R}$ is said to be **additive** if $f(x+y) = f(x) + f(y)$ holds for all $x, y \in \mathbb{R}$.

Lemma 18.7.   *Let $f: \mathbb{R} \to \mathbb{R}$ be an additive function. If f is continuous at zero, then there exists a constant c such that $f(x) = cx$ holds for all $x \in \mathbb{R}$.*

   In particular, if $f: \mathbb{R}^n \to \mathbb{R}$ is continuous at zero for each variable and additive for each variable separately, then there exists a constant c such that

$$f(x_1, \ldots, x_n) = cx_1 \cdots x_n$$

holds for all $(x_1, \ldots, x_n) \in \mathbb{R}^n$.

Proof.   If $f: \mathbb{R} \to \mathbb{R}$ is additive, then f satisfies the following properties:

1. $f(0) = 0$ [Reason: $f(0) = f(0+0) = f(0) + f(0)$],
2. $f(-x) = -f(x)$ for each $x \in \mathbb{R}$ [Reason: $f(x) + f(-x) = f(x - x) = f(0) = 0$], and
3. $f(rx) = rf(x)$ for all rational numbers $r \in \mathbb{R}$ and all $x \in \mathbb{R}$.

The proof of (3) goes by steps. If k is a positive integer, then

$$f(kx) = f(x + \cdots + x) = f(x) + \cdots + f(x) = kf(x).$$

So, $nf(\frac{x}{n}) = f(\frac{nx}{n}) = f(x)$, which shows that $f(\frac{1}{n}x) = \frac{1}{n}f(x)$. Thus, if m and n are positive integers, then

$$f\left(\frac{m}{n}x\right) = mf\left(\frac{1}{n}x\right) = \frac{m}{n}f(x).$$

It follows that $f(rx) = rf(x)$ holds for all rational numbers r and all $x \in \mathbb{R}$.

   Since f is continuous at zero, the identity $f(x - y) = f(x) - f(y)$ shows that f is continuous at every point of $\mathbb{R}$. Now, if $x \in \mathbb{R}$, then choose a sequence of rational numbers $\{r_n\}$ such that $\lim r_n = x$. Consequently,

$$f(x) = \lim_{n \to \infty} f(r_n) = \lim_{n \to \infty} r_n f(1) = xf(1) = cx$$

where $c = f(1)$ is a constant, and the result has been proven for the one-dimensional case.

   The general case can be established now by induction. We leave the details as an exercise for the reader.   ∎

We are now ready to prove that, aside from a multiplication factor, the Lebesgue measure is the one and only translation invariant Borel measure on $\mathbb{R}^n$.

Theorem 18.8. *Let μ be a translation invariant Borel measure on $\mathbb{R}^n$. Then there exists a constant c such that $\mu(A) = c\lambda(A)$ holds for every Borel set A of $\mathbb{R}^n$.*

Proof. For each $x \in \mathbb{R}$ let $I_x = \emptyset$ if $x = 0$, $I_x = [0, x)$ if $x > 0$, and $I_x = [x, 0)$ if $x < 0$. Also, for every $x = (x_1, \ldots, x_n) \in \mathbb{R}^n$ let $\operatorname{Sgn} x = 1$ if $\prod_{i=1}^n x_i \geq 0$ and $\operatorname{Sgn} x = -1$ if $\prod_{i=1}^n x_i < 0$.

Assume now that μ is a translation invariant nonzero Borel measure on $\mathbb{R}^n$. Define $f: \mathbb{R}^n \to \mathbb{R}$ by

$$f(x_1, \ldots, x_n) = \operatorname{Sgn} x \cdot \mu\left(\prod_{i=1}^n I_{x_i}\right),$$

where $x = (x_1, \ldots, x_n) \in \mathbb{R}^n$. Then f is additive in each variable separately. For instance, to see that f is additive in the first variable, consider the case when $a > 0$, $b < 0$ with $a + b > 0$. Fix $x_2, \ldots, x_n$ and let $s = \operatorname{Sgn}(x_2, \ldots, x_n)$ and $I = \prod_{i=2}^n I_{x_i}$. Clearly, $[0, a) \setminus [a + b, a) = [0, a + b)$ and $[a + b, a) \times I = (a, 0, 0, \ldots, 0) + [b, 0) \times I$. Since μ is translation invariant,

$$\mu([b, 0) \times I) = \mu([a + b, a) \times I).$$

Consequently,

$$\begin{aligned}
f(a, x_2, \ldots, x_n) + f(b, x_2, \ldots, x_n) &= s[\mu([0, a) \times I) - \mu([b, 0) \times I)] \\
&= s[\mu([0, a) \times I) - \mu([a + b, a) \times I)] \\
&= s\mu(([0, a) \setminus [a + b, a)) \times I) \\
&= s\mu([0, a + b) \times I) \\
&= f(a + b, x_2, \ldots, x_n).
\end{aligned}$$

The other cases can be proven easily in a similar manner.

Next, note that f is continuous at zero for each variable separately. Indeed, if $a_k \uparrow 0$ and $I = \prod_{i=2}^n I_{x_i}$, then $[a_k, 0) \times I \downarrow \emptyset$. Hence, By Theorem 15.4, $\mu([a_k, 0) \times I) \downarrow 0$, which implies $\lim_{k \to \infty} f(a_k, x_2, \ldots, x_n) = 0$. That is, f is left continuous at zero for the first variable. On the other hand, the identity $f(b, x_2, \ldots, x_n) = -f(-b, x_2, \ldots, x_n)$ implies that f is also right continuous for the first variable. Therefore, f is continuous for each variable separately.

By Lemma 18.7, there exists a constant c such that $f(x_1, \ldots, x_n) = cx_1 \cdots x_n$. Since $\mu \neq 0$, it follows that $c > 0$.

Now, let $A = \prod_{i=1}^{n}[a_i, b_i)$, where $-\infty < a_i < b_i < \infty$ for each i. Since $A = (a_1, \ldots, a_n) + \prod_{i=1}^{n}[0, b_i - a_i)$, we get

$$\mu(A) = \mu\left(\prod_{i=1}^{n}[0, b_i - a_i)\right) = f(b_1 - a_1, \ldots, b_n - a_n)$$

$$= c\prod_{i=1}^{n}(b_i - a_i) = c\lambda(A).$$

Thus, $\lambda = c^{-1}\mu$ on $\mathcal{S}$. But then Theorem 15.10 shows that $\lambda = c^{-1}\mu$ holds on $\mathcal{B}$. That is, $\mu(A) = c\lambda(A)$ holds for each $A \in \mathcal{B}$.   ∎

A natural question one may ask at this point is whether every Lebesgue measurable set is a Borel set. The answer is negative. That is, there exist Lebesgue measurable sets that are not Borel sets. One way of establishing this is by a cardinality argument. It can be shown that the σ-algebra of all Borel sets has cardinality $\mathfrak{c}$, while that of the Lebesgue measurable sets is $2^{\mathfrak{c}}$. Hence, the two σ-algebras do not coincide; see [15, p. 133].

Next, we shall present a different proof of the existence of a Lebesgue measurable set that is not a Borel set, the steps of which have some independent interest. For simplicity, the arguments will be given for the one-dimensional case. In fact, the example will be a subset of $[0, 1]$. To do this, we need some preliminary discussion.

The Cantor set C has been described in Example 6.15. It was observed there that the total length of the open intervals removed from $[0, 1]$ to get C is equal to one. It follows from this that $\lambda(C) = 0$. Thus, every subset of C has Lebesgue measure zero, and hence, every subset of C is Lebesgue measurable.

Next, we shall describe the ϵ-Cantor set. The process for constructing it is similar to the one described in Example 6.15. Fix $0 < \epsilon < 1$, and let $\delta = 1 - \epsilon$. Start with $A_0 = [0, 1]$ and remove from the center of A_0 an open interval of length $2^{-1}\delta$. Let A_1 be the remaining set; that is,

$$A_1 = \left[0, \frac{1}{2} - \frac{\delta}{4}\right] \cup \left[\frac{1}{2} + \frac{\delta}{4}, 1\right].$$

Note that A_1 consists of $2^1 = 2$ disjoint closed intervals of the same length and that $\lambda(A_1) = 1 - 2^{-1}\delta$. In the next step, remove from the center of each of the two disjoint closed intervals of A_1 an open interval of length $2^{-3}\delta$. Thus, the "total length" removed from A_1 is $2^{-2}\delta$. Let A_2 be the union of the remaining 2^2 disjoint closed intervals of equal length. Note also that $\lambda(A_2) = \lambda(A_1) - 2^{-2}\delta = 1 - (2^{-1} + 2^{-2})\delta$.

For the general step, suppose that the set A_n has been constructed consisting of the union of 2^n disjoint closed intervals all of the same length and satisfying

$\lambda(A_n) = 1 - (2^{-1} + \cdots + 2^{-n})\delta$. From the center of each of these 2^n closed intervals, delete an open interval of length $2^{-2n-1}\delta$. Let A_{n+1} be the union of the remaining 2^{n+1} disjoint closed intervals of equal length. Since the "total length" removed from A_n is $2^{-n-1}\delta$, it follows that

$$\lambda(A_{n+1}) = 1 - (2^{-1} + \cdots + 2^{-n} + 2^{-n-1})\delta.$$

Clearly, $A_{n+1} \subseteq A_n$ holds for each n, and the ϵ-**Cantor** set is now defined to be the set $C_\epsilon = \bigcap_{n=1}^{\infty} A_n$. Clearly, C_ϵ is a closed set, and nowhere dense in $[0, 1]$. Also,

$$\lambda(C_\epsilon) = \lim_{n\to\infty} \lambda(A_n) = 1 - \left(\sum_{n=1}^{\infty} 2^{-n} \right)\delta = 1 - \delta = \epsilon.$$

Lemma 18.9. *For every $0 < \epsilon < 1$ there exists a real-valued continuous function $f: [0, 1] \to [0, 1]$ such that:*

1. *f is onto,*
2. *f is strictly increasing (and hence, f is one-to-one), and*
3. *f maps the ϵ-Cantor set C_ϵ onto the Cantor set C.*

Proof. Let $\{I_1, I_2, \ldots\}$ be the open intervals removed from $[0, 1]$ in the construction of C_ϵ (counted from left to right, following the construction process inductively). Similarly, let $\{J_1, J_2, \ldots\}$ be the open intervals removed from $[0, 1]$ in order to get the Cantor set C (counted again from left to right). Clearly, $C_\epsilon = [0, 1] \setminus \bigcup_{n=1}^{\infty} I_n$ and $C = [0, 1] \setminus \bigcup_{n=1}^{\infty} J_n$.

Let $I_n = (a_n, b_n)$ and $J_n = (c_n, d_n)$ for each n. Now define $f: [0, 1] \to [0, 1]$ by steps as follows:

1. $f(0) = 0$.
2. For $x \in I_n = (a_n, b_n)$ define

$$f(x) = \frac{d_n - c_n}{b_n - a_n}(x - a_n) + c_n.$$

Note that f maps I_n onto J_n.
3. If $x \neq 0$ and $x \in C_\epsilon$, define

$$f(x) = \sup \left\{ f(y): y < x \text{ and } y \in \bigcup_{n=1}^{\infty} I_n \right\}.$$

A part of the graph for a typical function f is shown in Figure 3.1.

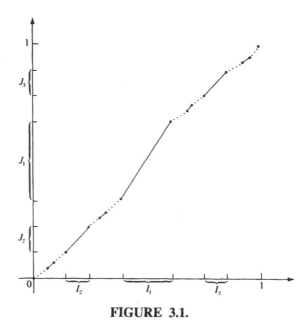

FIGURE 3.1.

We leave the details as an exercise for the reader to verify that f satisfies the properties stated in the lemma. ∎

Since the set C has cardinality $\mathfrak{c}$ (see Example [6.15], it follows from Lemma 18.9 that every ϵ-Cantor set also has cardinality $\mathfrak{c}$.

In view of Example 15, we know that there exist subsets of $[0, 1]$ that are not Lebesgue measurable. The next result describes an interesting property of the non-Lebesgue measurable subsets of $[0, 1]$.

Lemma 18.10. *Let A be a non-Lebesgue measurable subset of $[0, 1]$. Then there exists some $0 < \epsilon < 1$ such that for any Lebesgue measurable subset E of $[0, 1]$ with $\lambda(E) \geq \epsilon$, the set $A \cap E$ is not Lebesgue measurable.*

Proof. Assume by way of contradiction that the conclusion is false. Then for each $0 < \epsilon < 1$ there exists a Lebesgue measurable subset E_ϵ of $[0, 1]$ such that $\lambda(E_\epsilon) \geq \epsilon$ and $A \cap E_\epsilon$ is Lebesgue measurable. Choose a sequence $\{\epsilon_n\}$ of $(0, 1)$ such that $\lim \epsilon_n = 1$.

Put $E = \bigcup_{n=1}^{\infty} E_{\epsilon_n}$, and note that E is a Lebesgue measurable subset of $[0, 1]$. Moreover, $\epsilon_n \leq \lambda(E_{\epsilon_n}) \leq \lambda(E) \leq 1$ holds for all n, and thus, $\lambda(E) = 1$. Now, since $\lambda(A \cap E^c) \leq \lambda([0, 1] \setminus E) = 0$, we have $\lambda(A \cap E^c) = 0$, and so, $A \cap E^c$ is

Lebesgue measurable. But then

$$A = (A \cap E) \cup (A \cap E^c) = \left[\bigcup_{n=1}^{\infty} (A \cap E_{\epsilon_n}) \right] \cup (A \cap E^c)$$

holds and shows that A is a Lebesgue measurable set. However, this is a contradiction, and the proof is complete. ∎

It was observed before that every subset of the Cantor set is Lebesgue measurable (since it has Lebesgue measure zero). We are now ready to establish that there exists a Lebesgue measurable set that is not a Borel set.

Theorem 18.11.   *The Cantor set has a subset that is not a Borel set.*

Proof.   Choose a subset A of $[0, 1]$ that is not Lebesgue measurable. By Lemma 18.10 there exists $0 < \epsilon < 1$ such that $A \cap E$ is not Lebesgue measurable for all Lebesgue measurable subsets E of $[0, 1]$ with $\lambda(E) \geq \epsilon$.

Next, consider the continuous function f of Lemma 18.9 that carries the ϵ-Cantor set C_ϵ onto the Cantor set C. By the continuity of f, we get that f is Lebesgue measurable (where, of course, it is assumed that the Lebesgue measure is restricted to $[0, 1]$). Since $\lambda(C_\epsilon) = \epsilon$, the set $B = A \cap C_\epsilon$ is not Lebesgue measurable. Therefore, the subset $f(B)$ of the Cantor set, which is Lebesgue measurable, cannot be a Borel set—since $B = f^{-1}(f(B))$ is not Lebesgue measurable. The proof of the theorem is now complete. ∎

A useful result (due to H. Steinhaus[6]) regarding Lebesgue measurable sets will be presented next. It will be shown that if A is a Lebesgue measurable subset of $\mathbb{R}^n$ such that $\lambda(A) > 0$, then zero is an interior point of $A - A$. Recall that if A and B are two subsets of $\mathbb{R}^n$, then $A - B = \{a - b\colon a \in A \text{ and } b \in B\}$. The set $A - B$ is called the **algebraic difference** of B from A.

In order to establish this result, we need a lemma.

Lemma 18.12.   *Let E be a subset of $\mathbb{R}^n$ such that $\lambda(E) > 0$. Then for each $0 < \epsilon < 1$, there exists a bounded open interval I of $\mathbb{R}^n$ such that $\epsilon\lambda(I) < \lambda(E \cap I)$.*

Proof.   Fix $0 < \epsilon < 1$, and let E be a subset of $\mathbb{R}^n$ such that $\lambda(E) > 0$. Without loss of generality, we can assume that $\lambda(E) < \infty$. Indeed, if $B_k = \{x \in \mathbb{R}^n\colon d(0, x) \leq k\}$, then $E = \bigcup_{k=1}^{\infty}(E \cap B_k)$, and so $\lambda(E \cap B_k) > 0$ must hold for some k; since $E \cap B_k \subseteq E$, we can replace E by $E \cap B_k$.

---

[6]Hugo Steinhaus (1887–1972), a Polish mathematician. He made many curious and interesting contributions to functional analysis.

Since $0 < \epsilon < 1$, there exists a sequence $\{I_i\}$ of bounded open intervals such that $E \subseteq \bigcup_{i=1}^{\infty} I_i$ and $\epsilon \sum_{i=1}^{\infty} \lambda(I_i) < \lambda(E)$. To complete the proof, we show that there exists some i such that $\epsilon \lambda(I_i) < \lambda(E \cap I_i)$. Indeed, if this is not the case, then $\lambda(E \cap I_i) \leq \epsilon \lambda(I_i)$ holds for all i, and so

$$\lambda(E) = \lambda \left(\bigcup_{i=1}^{\infty} E \cap I_i \right) \leq \sum_{i=1}^{\infty} \lambda(E \cap I_i) \leq \epsilon \sum_{i=1}^{\infty} \lambda(I_i) < \lambda(E),$$

which is impossible. This completes the proof of the lemma.   ∎

Theorem 18.13 (Steinhaus). *If E is a Lebesgue measurable subset of $\mathbb{R}^n$ such that $\lambda(E) > 0$, then the zero vector is an interior point of $E - E$.*

Proof. Assume that E is measurable with $\lambda(E) > 0$. By Lemma 18.12 there exists a bounded open interval $I = \prod_{i=1}^{n}(a_i, b_i)$ such that $\frac{3}{4}\lambda(I) < \lambda(E \cap I)$.

Next, choose $\delta > 0$ such that $I_\delta = \prod_{i=1}^{n}(a_i - \delta, b_i + \delta)$ satisfies $\lambda(I_\delta) < \frac{3}{2}\lambda(I)$, and then consider the open interval $J = (-\delta, \delta) \times \cdots \times (-\delta, \delta) \subseteq \mathbb{R}^n$. To complete the proof, we shall show that $J \subseteq E - E$.

To see this, let $x \in J$. Then

$$(E \cap I) \cup (x + E \cap I) \subseteq I \cup (x + I) \subseteq I_\delta$$

holds. Therefore, $\lambda((E \cap I) \cup (x + E \cap I)) \leq \lambda(I_\delta) < \frac{3}{2}\lambda(I)$, which implies that $(E \cap I) \cap (x + E \cap I) \neq \emptyset$. Indeed, if $(E \cap I) \cap (x + E \cap I) = \emptyset$, then by the additivity of λ, we get $\lambda((E \cap I) \cup (x + E \cap I)) = 2\lambda(E \cap I) > \frac{3}{2}\lambda(I)$ (here we use the measurability of E), which is impossible. Pick a vector $y \in (E \cap I) \cap (x + E \cap I)$. Then $y \in E$, and there exists $z \in E$ such that $x + z = y$. That is, $x = y - z \in E - E$, so that $J \subseteq E - E$, and the proof is complete.   ∎

EXERCISES

1. Let $I = \prod_{i=1}^{n} I_i$ be an interval of $\mathbb{R}^n$. Show that I is Lebesgue measurable and that $\lambda(I) = \prod_{i=1}^{n} |I_i|$, where $|I_i|$ denotes the length of the interval I_i.
2. Let $\mathcal{O}$ be an open subset of $\mathbb{R}$. Show that there exists an at-most countable collection $\{I_\alpha : \alpha \in A\}$ of pairwise disjoint open intervals such that $\mathcal{O} = \bigcup_{\alpha \in A} I_\alpha$. Also, show that $\lambda(\mathcal{O}) = \sum_{\alpha \in A} |I_\alpha|$.
3. Show that the Borel sets of $\mathbb{R}^n$ are precisely the members of the σ-algebra generated by the compact sets.
4. Show that a subset E of $\mathbb{R}^n$ is Lebesgue measurable if and only if for each $\epsilon > 0$ there exists a closed subset F of $\mathbb{R}^n$ such that $F \subseteq E$ and $\lambda(E \setminus F) < \epsilon$.
5. Show that if a subset E of $[0, 1]$ satisfies $\lambda(E) = 1$, then E is dense in $[0, 1]$.
6. If $E \subseteq \mathbb{R}^n$ satisfies $\lambda(E) = 0$, then show that $E^\circ = \emptyset$.

7. Show that if E is a Lebesgue measurable subset of $\mathbb{R}^n$, then there exist an F_σ-set A and a G_δ-set B such that $A \subseteq E \subseteq B$ and $\lambda(B \setminus A) = 0$.

8. Let $\{E_n\}$ be a sequence of nonempty (Lebesgue) measurable subsets of $[0, 1]$ satisfying $\lim \lambda(E_n) = 1$.

    a. Show that for each $0 < \epsilon < 1$ there exists a subsequence $\{E_{k_n}\}$ of $\{E_n\}$ such that $\lambda(\bigcap_{n=1}^{\infty} E_{k_n}) > \epsilon$.

    b. Show that $\bigcap_{k=n}^{\infty} E_k = \emptyset$ is possible for each $n = 1, 2, \ldots$.

9. Assume that a function $f: I \to \mathbb{R}$ defined on a subinterval of $\mathbb{R}$ satisfies a **Lipschitz**[7] **condition**. That is, assume that there exists a constant $C > 0$ such that $|f(x) - f(y)| \le C|x - y|$ holds for all $x, y \in I$. Show that f carries (Lebesgue) null sets to null sets.

    In particular, if a function $f: I \to \mathbb{R}$ defined on a subinterval of $\mathbb{R}$ has a continuous derivative, then show that f carries null sets to null sets.
    [HINT: For the second part note that if J is a closed subinterval of I and $|f'(t)| \le M$ holds for each $t \in J$, then by the Mean Value Theorem we have $|f(x) - f(y)| \le M|x - y|$ for all $x, y \in J$.]

10. Show that the Lebesgue measure of a triangle in $\mathbb{R}^2$ equals its area. Also, determine the Lebesgue measure of a disk in $\mathbb{R}^2$.

11. If μ is a translation invariant Borel measure on $\mathbb{R}^n$, then show that there exists some $c \ge 0$ such that $\mu^*(A) = c\lambda(A)$ for all subsets A of $\mathbb{R}^n$.

12. Show that an arbitrary collection of pairwise disjoint measurable subsets of $\mathbb{R}$, each of which has positive measure, is at most countable.

13. Let G be a proper additive subgroup of $\mathbb{R}^n$ (i.e., $x, y \in G$ imply $x + y \in G$ and $-x \in G$). If G is a measurable set, then show that $\lambda(G) = 0$.
    [HINT: Use Theorem 18.13.]

14. Let $f: \mathbb{R} \to \mathbb{R}$ be additive (i.e., $f(x + y) = f(x) + f(y)$ for all $x, y \in \mathbb{R}$) and Lebesgue measurable. Show that f is continuous (and hence, of the form $f(x) = cx$).
    [HINT: Use the properties of f shown during the proof of Lemma 18.7 to establish that $f^{-1}((0, \epsilon))$ has positive Lebesgue measure for each $\epsilon > 0$, and then use Theorem 18.13 to show that f is continuous at zero.]

15. Show that an arbitrary union of proper intervals of $\mathbb{R}$ is a Lebesgue measurable set.

16. Let C be a closed nowhere dense subset of $\mathbb{R}^n$ such that $\lambda(C) > 0$. Show that the characteristic function χ_C cannot be continuous on the complement of any Lebesgue null set of $\mathbb{R}^n$. Also, show that χ_C will be continuous on the complement of a properly chosen open set whose Lebesgue measure can be made arbitrarily small.

17. Let $f: \mathbb{R}^n \to \mathbb{R}$ be a continuous function. Show that the graph

$$G = \{(x_1, \ldots, x_n, f(x_1, \ldots, x_n)): (x_1, \ldots, x_n) \in \mathbb{R}^n\}$$

of f has $(n + 1)$-dimensional Lebesgue measure zero.
[HINT: If $G_k = \{(x_1, \ldots, x_n, f(x_1, \ldots, x_n)): |x_i| \le k \text{ for } i = 1, \ldots, n\}$, show that $\lambda(G_k) = 0$ and then note that $G_k \uparrow G$.]

---

[7]Rudolf Lipschitz (1832–1903), a German mathematician. His name is associated with the existence and uniqueness of solutions to differential equations.

18. Let X be a Hausdorff topological space, and let μ be a regular Borel measure on X. Show the following:

  a. If A is an arbitrary subset of X, then

$$\mu^*(A) = \inf\{\mu(\mathcal{O}): \mathcal{O} \text{ open and } A \subseteq \mathcal{O}\}.$$

  b. If A is a measurable subset of X with $\mu^*(A) < \infty$, then

$$\mu^*(A) = \sup\{\mu(K): K \text{ compact and } K \subseteq A\}.$$

  c. If μ is σ-finite and A is a measurable subset of X, then

$$\mu^*(A) = \sup\{\mu(K): K \text{ compact and } K \subseteq A\}.$$

19. If A is a measurable subset of $\mathbb{R}$ of positive measure and $0 < \delta < \lambda(A)$, then show that there exists a measurable subset B of A satisfying $\lambda(B) = \delta$.

20. Let E be a Lebesgue measurable subset of $\mathbb{R}$ of finite Lebesgue measure. Show that the function $f_E: \mathbb{R} \to \mathbb{R}$, defined by

$$f_E(x) = \lambda(E \triangle (x + E)),$$

is uniformly continuous.

19. CONVERGENCE IN MEASURE

Let $(X, \mathcal{S}, \mu)$ be a measure space again. The collection of all real-valued μ-measurable functions defined on X will be denoted by $\mathcal{M}$. That is,

$$\mathcal{M} = \{f \in \mathbb{R}^X: f \text{ is } \mu\text{-measurable}\}.$$

It should be clear from Theorem 16.5 that $\mathcal{M}$ is closed under the ordinary algebraic and lattice operations. That is, $\mathcal{M}$ is a function space and an algebra. In the space $\mathcal{M}$, a useful concept of convergence of sequences is defined as follows:

Definition 19.1. *A sequence $\{f_n\}$ of measurable functions converges in measure to a function $f \in \mathcal{M}$, in symbols $f_n \overset{\mu}{\longrightarrow} f$, if for every $\epsilon > 0$ we have*

$$\lim_{n \to \infty} \mu^*(\{x \in X: |f_n(x) - f(x)| \geq \epsilon\}) = 0.$$

The next result summarizes the basic properties of convergence in measure.

Theorem 19.2. *Let $\{f_n\}$ and $\{g_n\}$ be two sequences of measurable functions, and let $f, g \in \mathcal{M}$. Then the following statements hold:*

  1. *If $f_n \overset{\mu}{\longrightarrow} f$ and $g_n \overset{\mu}{\longrightarrow} g$, then $\alpha f_n + \beta g_n \overset{\mu}{\longrightarrow} \alpha f + \beta g$ for all $\alpha, \beta \in \mathbb{R}$.*

  2. *If $f_n \overset{\mu}{\longrightarrow} f$ and $f_n \overset{\mu}{\longrightarrow} g$, then $f = g$ a.e.*

Proof. (1) We can assume $\alpha, \beta \neq 0$. From

$$|\alpha f_n(x) + \beta g_n(x) - [\alpha f(x) + \beta g(x)]| \leq |\alpha||f_n(x) - f(x)| + |\beta||g_n(x) - g(x)|,$$

it follows that

$$\{x \in X: |\alpha f_n(x) + \beta g_n(x) - [\alpha f(x) + \beta g(x)]| \geq \epsilon\}$$
$$\subseteq \left\{x \in X: |f_n(x) - f(x)| \geq \frac{\epsilon}{2|\alpha|}\right\} \cup \left\{x \in X: |g_n(x) - g(x)| \geq \frac{\epsilon}{2|\beta|}\right\}.$$

The conclusion now follows easily.

(2) Let $f_n \xrightarrow{\mu} f$ and $f_n \xrightarrow{\mu} g$. Given $\epsilon > 0$, the triangle inequality implies

$$\{x \in X: |f(x) - g(x)| \geq 2\epsilon\}$$
$$\subseteq \{x \in X: |f_n(x) - f(x)| \geq \epsilon\} \cup \{x \in X: |f_n(x) - g(x)| \geq \epsilon\},$$

from which it easily follows that $\mu^*(\{x \in X: |f(x) - g(x)| \geq 2\epsilon\}) = 0$ for all $\epsilon > 0$. Therefore,

$$\mu^*(\{x \in X: f(x) \neq g(x)\}) = \mu^*\left(\bigcup_{n=1}^{\infty} \left\{x \in X: |f(x) - g(x)| \geq \frac{1}{n}\right\}\right) = 0,$$

so that $f = g$ a.e. holds. ∎

The lattice operations are also continuous with respect to convergence in measure.

Theorem 19.3. *If a sequence $\{f_n\}$ of measurable functions satisfies $f_n \xrightarrow{\mu} f$, then*

$$f_n^+ \xrightarrow{\mu} f^+, \quad f_n^- \xrightarrow{\mu} f^-, \quad \text{and} \quad |f_n| \xrightarrow{\mu} |f|.$$

Proof. The statements easily follow from the standard lattice inequalities $|f_n^+ - f^+| \leq |f_n - f|$, $|f_n^- - f^-| \leq |f_n - f|$, and $||f_n| - |f|| \leq |f_n - f|$. ∎

The convergence in measure is related to the pointwise convergence as follows.

Theorem 19.4. *If a sequence $\{f_n\}$ of measurable functions satisfies $f_n \xrightarrow{\mu} f$ for some $f \in \mathcal{M}$, then there exists a subsequence $\{f_{k_n}\}$ of $\{f_n\}$ such that $f_{k_n} \to f$ a.e.*

Proof. Let $f_n \xrightarrow{\mu} f$. A simple inductive argument shows that there exists a strictly increasing sequence $\{k_n\}$ of natural numbers such that

$$\mu^*\left(\left\{x \in X: |f_k(x) - f(x)| \geq \frac{1}{n}\right\}\right) < 2^{-n}$$

for all $k \geq k_n$. Let $E_n = \{x \in X: |f_{k_n}(x) - f(x)| \geq \frac{1}{n}\}$ for each n, and define the measurable set $E = \bigcap_{m=1}^{\infty} \bigcup_{n=m}^{\infty} E_n$. Then

$$\mu^*(E) \leq \mu^*\left(\bigcup_{n=m}^{\infty} E_n\right) \leq \sum_{n=m}^{\infty} \mu^*(E_n) \leq 2^{-m+1}$$

holds for all m, so that $\mu^*E) = 0$. Also, if $x \notin E$, then there exists some m such that $x \notin \bigcup_{n=m}^{\infty} E_n$, and so, $|f_{k_n}(x) - f(x)| < \frac{1}{n}$ holds for each $n \geq m$. Therefore, $\lim f_{k_n}(x) = f(x)$ for each $x \in E^c$, and so $f_{k_n} \to f$ a.e. holds. ∎

Pointwise convergence does not imply convergence in measure. For instance, let $X = \mathbb{R}$ with the Lebesgue measure λ, and define $f_n = \chi_{[n,n+1]}$ for each n. Clearly, $\lim f_n(x) = 0$ holds for each $x \in \mathbb{R}$. Since

$$\lambda\big(\{x \in X: |f_n(x)| \geq 1\}\big) = \lambda\big([n, n+1]\big) = 1$$

holds for all n, it follows that $\{f_n\}$ does not converge to zero in measure.

However, if the measure space is finite, then pointwise convergence implies convergence in measure.

Theorem 19.5. *Assume $\mu^*(X) < \infty$. If a sequence $\{f_n\}$ of measurable functions satisfies $f_n \to f$ a.e., then $f_n \xrightarrow{\mu} f$ also holds.*

Proof. Fix $\epsilon > 0$ and for each n let $E_n = \{x \in X: |f_n(x) - f(x)| \geq \epsilon\}$. Now let $\delta > 0$. Since $\mu^*(X) < \infty$, it follows from Egorov's Theorem 16.7 that there exists a measurable set A such that $\mu^*(A) < \delta$ and $\{f_n\}$ converges uniformly to f on A^c.

Choose k such that $|f_n(x) - f(x)| < \epsilon$ holds for all $x \in A^c$ and all $n \geq k$. Then $E_n \subseteq A$ holds for all $n \geq k$, and so $\mu^*(E_n) \leq \mu^*(A) < \delta$ for all $n \geq k$. Thus, $\lim \mu^*(E_n) = 0$, and the proof is finished. ∎

It is interesting to observe that there exist sequences of measurable functions that converge in measure, but fail to converge at any points. An example of this type is presented next.

Example 19.6. Consider $[0, 1]$ equipped with the Lebesgue measure λ. For each n subdivide $[0, 1]$ into the n subintervals $[0, \frac{1}{n}], [\frac{1}{n}, \frac{2}{n}], \ldots, [\frac{n-1}{n}, 1]$.

Enumerate all these intervals as follows:

$$\left[0, \frac{1}{2}\right], \left[\frac{1}{2}, 1\right], \left[0, \frac{1}{3}\right], \left[\frac{1}{3}, \frac{2}{3}\right], \left[\frac{2}{3}, 1\right], \left[0, \frac{1}{4}\right], \left[\frac{1}{4}, \frac{2}{4}\right], \left[\frac{2}{4}, \frac{3}{4}\right], \left[\frac{3}{4}, 1\right], \left[0, \frac{1}{5}\right], \cdots .$$

Let f_n be the characteristic function of the nth interval of the above sequence. It is easy to see that $f_n \xrightarrow{\lambda} 0$ holds and that $\{f_n(x)\}$ does not converge to zero for any $x \in [0, 1]$. ∎

EXERCISES

1. Let $\{f_n\}$ be a sequence of measurable functions and let $f: X \rightarrow \mathbb{R}$. Assume that $\lim \mu^*(\{x \in X: |f_n(x) - f(x)| \geq \epsilon\}) = 0$ holds for every $\epsilon > 0$. Show that f is a measurable function.

2. Assume that $\{f_n\} \subseteq \mathcal{M}$ satisfies $f_n \uparrow$ and $f_n \xrightarrow{\mu} f$. Show that $f_n \uparrow f$ a.e. holds.

3. If $\{f_n\} \subseteq \mathcal{M}$ satisfies $f_n \xrightarrow{\mu} f$ and $f_n \geq 0$ a.e. for each n, then show that $f \geq 0$ a.e. holds.

4. Let $\{f_n\} \subseteq \mathcal{M}$ and $\{g_n\} \subseteq \mathcal{M}$ satisfy $f_n \xrightarrow{\mu} f$, $g_n \xrightarrow{\mu} g$, and $f_n = g_n$ a.e. for each n. Show that $f = g$ a.e. holds.

5. Let $(X, \mathcal{S}, \mu)$ be a finite measure space. Assume that two sequences $\{f_n\}$ and $\{g_n\}$ of $\mathcal{M}$ satisfy $f_n \xrightarrow{\mu} f$ and $g_n \xrightarrow{\mu} g$. Show that $f_n g_n \xrightarrow{\mu} fg$. Is this statement true if $\mu^*(X) = \infty$?

6. Show that a sequence of measurable functions $\{f_n\}$ on a finite measure space converges to f in measure if and only if every subsequence of $\{f_n\}$ has in turn a subsequence which converges to f a.e.

7. Define a sequence $\{f_n\}$ of $\mathcal{M}$ to be μ-**Cauchy** whenever for each $\epsilon > 0$ and $\delta > 0$ there exists some k (depending on ϵ and δ) such that $\mu^*(\{x \in X: |f_n(x) - f_m(x)| \geq \epsilon\}) < \delta$ holds for all $n, m \geq k$.

   Show that a sequence $\{f_n\}$ of $\mathcal{M}$ is a μ-Cauchy sequence if and only if there exists a measurable function f such that $f_n \xrightarrow{\mu} f$.

20. ABSTRACT MEASURABILITY

This section can be skipped in the first reading. It deals with collections of sets with stronger properties than those of a semiring. Most of these families were introduced briefly in Section 12. Let us start by recalling the notion of a ring of sets.

Definition 20.1. *A nonempty collection $\mathcal{R}$ of subsets of a set X is said to be a* **ring** *if $A, B \in \mathcal{R}$ imply $A \cup B \in \mathcal{R}$ and $A \setminus B \in \mathcal{R}$.*

It should be obvious that every ring is closed under finite unions. Also, since an arbitrary ring $\mathcal{R}$ is nonempty, it contains a set, say $A \in \mathcal{R}$, and so $\emptyset = A \setminus A \in \mathcal{R}$.

That is, every ring contains the empty set. Also, from the identities

$$A \Delta B = (A \setminus B) \cup (B \setminus A) \quad \text{and} \quad A \cap B = A \setminus (A \setminus B),$$

it easily follows that every ring is closed under symmetric differences and finite intersections.

Recall that a nonempty collection of subsets of a set X is said to be an **algebra** of sets if it is closed under finite intersections and complementation. From the identities

$$A \cup B = (A^c \cap B^c)^c \quad \text{and} \quad A \setminus B = A \cap B^c,$$

it should be obvious that every algebra is a ring. In the converse direction, we have the following result (whose easy proof is left for the reader):

Lemma 20.2.   *A ring $\mathcal{R}$ is an algebra if and only if $X \in \mathcal{R}$.*

A ring $\mathcal{R}$ is called a σ-**ring** if it is closed under countable unions, i.e., if $\{A_n\} \subseteq \mathcal{R}$ implies $\bigcup_{n=1}^{\infty} A_n \in \mathcal{R}$.

Notice that every σ-ring is automatically closed under countable intersections. To see this, let $\{A_n\}$ be a sequence in a σ-ring $\mathcal{R}$ and let $A = \bigcap_{n=1}^{\infty} A_n$. Then $A_1 \setminus A = \bigcup_{n=1}^{\infty} (A_1 \setminus A_n) \in \mathcal{R}$, and so $A = A_1 \setminus (A_1 \setminus A) \in \mathcal{R}$.

The straightforward proof of the next result is also left for the reader.

Lemma 20.3.   *A σ-ring $\mathcal{R}$ is a σ-algebra if and only if $X \in \mathcal{R}$.*

The following scheme indicates the relationships between the various families of sets we have introduced so far:

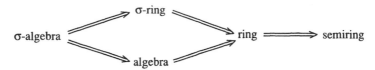

FIGURE 3.2.   The Implication Scheme

In general, no other implication is valid; see Exercise 3 at the end of this section.

Arbitrary intersections of σ-algebras of subsets of a fixed set X are also σ-algebras. Thus, for every nonempty family $\mathcal{F}$ of subsets of a set X there exists a smallest σ-algebra (with respect to inclusion) that includes $\mathcal{F}$. This smallest σ-algebra is called the σ-**algebra generated** by $\mathcal{F}$ and is denoted by $\sigma(\mathcal{F})$. It is the intersection of all σ-algebras that contain $\mathcal{F}$ (notice that $\mathcal{P}(X)$ is one of them).

That is,

$$\sigma(\mathcal{F}) = \bigcap \{\Sigma: \ \Sigma \text{ is a } \sigma\text{-algebra and } \mathcal{F} \subseteq \Sigma\}.$$

If Σ is a σ-algebra and a family of sets $\mathcal{F}$ of Σ (i.e., $\mathcal{F} \subseteq \Sigma$) satisfies $\sigma(\mathcal{F}) = \Sigma$, then $\mathcal{F}$ is called a **family of generators** for Σ. The reader should also notice that for an arbitrary collection of subsets $\mathcal{F}$ of a set X there always exists an algebra, a ring, and a σ-ring generated by $\mathcal{F}$. In this book, we shall make use only of the σ-algebras generated by families of sets.

We now turn our attention to abstract measurability properties of functions. Notice that if $f: X \to Y$ is a function and $\mathcal{A}$ is either an algebra or a σ-algebra (or a ring, or a σ-ring, or a semiring), then

$$f^{-1}(\mathcal{A}) = \{f^{-1}(A): \ A \in \mathcal{A}\}$$

is likewise an algebra or a σ-algebra (or a ring, or a σ-ring, or a semiring).

Lemma 20.4. *If $f: X \to Y$ is a function and $\mathcal{F}$ is a nonempty family of subsets of Y, then*

$$\sigma(f^{-1}(\mathcal{F})) = f^{-1}(\sigma(\mathcal{F})).$$

Proof. Notice first that the identities $f^{-1}(A^c) = [f^{-1}(A)]^c$,

$$f^{-1}\left(\bigcup_{i \in I} A_i\right) = \bigcup_{i \in I} f^{-1}(A_i) \quad \text{and} \quad f^{-1}\left(\bigcap_{i \in I} A_i\right) = \bigcap_{i \in I} f^{-1}(A_i)$$

guarantee that $f^{-1}(\Sigma) = \{f^{-1}(A): \ A \in \Sigma\}$ is a σ-algebra whenever Σ is a σ-algebra.

So, $f^{-1}(\sigma(\mathcal{F}))$ is a σ-algebra satisfying $f^{-1}(\mathcal{F}) \subseteq f^{-1}(\sigma(\mathcal{F}))$. This implies

$$\sigma(f^{-1}(\mathcal{F})) \subseteq f^{-1}(\sigma(\mathcal{F})).$$

For the reverse inclusion, let

$$\mathcal{A} = \{A \in \sigma(\mathcal{F}): \ f^{-1}(A) \in \sigma(f^{-1}(\mathcal{F}))\}.$$

Clearly, $\mathcal{A}$ is a σ-algebra satisfying $\mathcal{F} \subseteq \mathcal{A} \subseteq \sigma(\mathcal{F})$. Hence, $\mathcal{A} = \sigma(\mathcal{F})$. This implies $f^{-1}(\sigma(\mathcal{F})) \subseteq \sigma(f^{-1}(\mathcal{F}))$, and so $f^{-1}(\sigma(\mathcal{F})) = \sigma(f^{-1}(\mathcal{F}))$. ∎

In an abstract setting, a **measurable space** is now a pair (X, Σ), where X is a set and Σ is a σ-algebra of subsets of X. The members of Σ are called the **measurable sets** of X.

Definition 20.5. *A function* $f:(X_1, \Sigma_1) \to (X_2, \Sigma_2)$ *between two measurable spaces is said to be* **measurable** *(more precisely* (Σ_1, Σ_2)-**measurable***) if* $f^{-1}(A) \in \Sigma_1$ *holds for each* $A \in \Sigma_2$, *i.e., if* $f^{-1}(\Sigma_2) \subseteq \Sigma_1$.

A function $f: X \to Y$ *between two topological spaces is said to be* **Borel measurable** *if* $f:(X, \mathcal{B}_X) \to (Y, \mathcal{B}_Y)$ *is measurable, i.e., if* $f^{-1}(B)$ *is a Borel subset of* X *for each Borel subset* B *of* Y.

When (X, Σ) is a measurable space and Y is a topological space, then a function $f: X \to Y$ is called **measurable** if $f:(X, \Sigma) \to (Y, \mathcal{B}_Y)$ is measurable.

Theorem 20.6. *Let* $f:(X_1, \Sigma_1) \to (X_2, \Sigma_2)$ *be a function between two measurable spaces and let* $\mathcal{F}$ *be a family of generators for* Σ_2, *i.e.,* $\sigma(\mathcal{F}) = \Sigma_2$. *Then* f *is measurable if and only if* $f^{-1}(A) \in \Sigma_1$ *for each* $A \in \mathcal{F}$.

Proof. If f is measurable, then clearly $f^{-1}(A) \in \Sigma_1$ holds for each $A \in \mathcal{F}$. On the other hand, if $f^{-1}(A) \in \Sigma_1$ for all $A \in \mathcal{F}$, then it follows from Lemma 20.4 that

$$f^{-1}(\Sigma_2) = f^{-1}(\sigma(\mathcal{F})) = \sigma(f^{-1}(\mathcal{F})) \subseteq \Sigma_1,$$

which shows that f is a measurable function. ∎

We now continue with another important family of sets.

Definition 20.7. *A family of subsets* $\mathcal{D}$ *of a set* X *is said to be a* **Dynkin**[8] **system** *if it satisfies the following three properties*:
1. $X \in \mathcal{D}$,
2. $A, B \in \mathcal{D}$ *and* $A \subseteq B$ *imply* $B \setminus A \in \mathcal{D}$, *and*
3. *whenever a sequence* $\{A_n\} \subseteq \mathcal{D}$ *satisfies* $A_n \uparrow A$, *then* $A \in \mathcal{D}$.

The reader should stop and verify that a Dynkin system is a σ-algebra if and only if it is closed under finite intersections. The Dynkin systems play an important role because of the following result—known as Dynkin's lemma.

Lemma 20.8 (Dynkin's Lemma). *Let* $\mathcal{D}$ *be a Dynkin system and let* $\mathcal{F}$ *be a family of sets such that* $\mathcal{F} \subseteq \mathcal{D}$. *If* $\mathcal{F}$ *is closed under finite intersections, then the* σ-algebra generated by $\mathcal{F}$ is included in $\mathcal{D}$, i.e., $\sigma(\mathcal{F}) \subseteq \mathcal{D}$.

---

[8]Eugene Borisovich Dynkin (1924–), a Russian mathematician. His main research interests are in probability theory. In 1977 he was appointed Professor of mathematics at Cornell University, Ithaca, New York.

Proof. Assume that $\mathcal{F} \subseteq \mathcal{D}$ is nonempty and closed under finite intersections. Let us denote by $\mathcal{D}_0$ the smallest Dynkin system including $\mathcal{F}$ (i.e., $\mathcal{D}_0$ is the intersection of all Dynkin systems that include $\mathcal{F}$). Clearly, $\mathcal{F} \subseteq \mathcal{D}_0 \subseteq \mathcal{D}$. Now let

$$\mathcal{D}_1 = \{A \in \mathcal{D}_0 \colon A \cap C \in \mathcal{D}_0 \text{ for all } C \in \mathcal{F}\}.$$

An easy verification shows that $\mathcal{D}_1$ is itself a Dynkin system satisfying $A \in \mathcal{D}_1$ for all $A \in \mathcal{F}$ (to establish this we must use the assumption that $\mathcal{F}$ is closed under finite intersections). Thus, $\mathcal{F} \subseteq \mathcal{D}_1 \subseteq \mathcal{D}_0$, and so by the definition of $\mathcal{D}_0$, we infer that $\mathcal{D}_1 = \mathcal{D}_0$.

Next, consider the family

$$\mathcal{D}_2 = \{A \in \mathcal{D}_0 \colon A \cap B \in \mathcal{D}_0 \text{ for all } B \in \mathcal{D}_0\}.$$

Again, it is easy to check that $\mathcal{D}_2$ is a Dynkin system such that $\mathcal{F} \subseteq \mathcal{D}_2 \subseteq \mathcal{D}_0$. Thus, $\mathcal{D}_0 = \mathcal{D}_2$, and therefore, $\mathcal{D}_0$ is a Dynkin system which is closed under finite intersections. This easily implies that $\mathcal{D}_0$ is a σ-algebra, and from this we see that $\sigma(\mathcal{F}) = \mathcal{D}_0 \subseteq \mathcal{D}$, as desired. ∎

Here are a few striking consequences of Dynkin's lemma:

Corollary 20.9. *Assume that $\mathcal{D}$ is a Dynkin system and Σ is a σ-algebra such that $\mathcal{D} \subseteq \Sigma$. If $\mathcal{D}$ contains a family of generators of Σ which is closed under finite intersections, then $\mathcal{D} = \Sigma$.*

Corollary 20.10. *Let Σ be a σ-algebra of subsets of a set X and let $\mathcal{F}$ be a family of generators for Σ (i.e., $\sigma(\mathcal{F}) = \Sigma$) which is closed under finite intersections. Then two finite measures μ and ν on Σ coincide (i.e., $\mu = \nu$) if and only if*

1. *$\mu(X) = \nu(X)$, and*
2. *$\mu(F) = \nu(F)$ for each $F \in \mathcal{F}$.*

Proof. Assume that two finite measures μ and ν on Σ satisfy (1) and (2). Let

$$\mathcal{D} = \{A \in \Sigma \colon \mu(A) = \nu(A)\}.$$

Then $X \in \mathcal{D}$ and if $A, B \in \mathcal{D}$ satisfy $A \subseteq B$, then

$$\mu(B \setminus A) = \mu(B) - \mu(A) = \nu(B) - \nu(A) = \nu(B \setminus A),$$

i.e., $B \setminus A \in \mathcal{D}$. Moreover, if a sequence $\{A_n\} \subseteq \mathcal{D}$ satisfies $A_n \uparrow A$, then

$$\mu(A) = \lim_{n \to \infty} \mu(A_n) = \lim_{n \to \infty} v(A_n) = v(A),$$

so that $A \in \mathcal{D}$. Thus, $\mathcal{D}$ is a Dynkin system and $\mathcal{F} \subseteq \mathcal{D}$ holds. By Corollary 20.9, $\mathcal{D} = \Sigma$, and so $\mu = v$.   ■

Corollary 20.11. *Two finite measures defined on the Borel sets of a topological space are equal if and only if they coincide on the open sets.*

We now turn our attention to product semirings. If S_1 and S_2 are two semirings of subsets of X and Y respectively, then the **product semiring** $S_1 \otimes S_2$ is defined by

$$S_1 \otimes S_2 = \{A \times B \colon A \in S_1 \text{ and } B \in S_2\}.$$

Using the identities

$$(A \times B) \cap (A_1 \times B_1) = (A \cap A_1) \times (B \cap B_1)$$
$$(A \times B) \setminus (A_1 \times B_1) = [(A \setminus A_1) \times B] \cup [(A \cap A_1) \times (B \setminus B_1)],$$

it is easy to verify that $S_1 \otimes S_2$ is indeed a semiring of $X \times Y$. It should be noticed that $S_1 \otimes S_2$ *need not* be an algebra even when S_1 and S_2 are both σ-algebras. The **product σ-algebra** generated by S_1 and S_2 is the σ-algebra $\sigma(S_1 \otimes S_2)$ generated by $S_1 \otimes S_2$.

If A is a subset of $X \times Y$ and $x \in X$, then the x-**section** A_x of A is the subset of Y defined by

$$A_x = \{y \in Y \colon (x, y) \in A\}.$$

Similarly, if $y \in Y$, then the y-**section** A^y of A is the subset of X defined by

$$A^y = \{x \in X \colon (x, y) \in A\}.$$

The geometric meanings of the x- and y-sections are shown in Figure 3.3.

Regarding sections, we have the following identities. (The families of sets below are assumed to be families of subsets of $X \times Y$.)

1. $(\bigcup_{i \in I} A_i)_x = \bigcup_{i \in I} (A_i)_x$ and $(\bigcap_{i \in I} A_i)_x = \bigcap_{i \in I} (A_i)_x$.
2. $(\bigcup_{i \in I} A_i)^y = \bigcup_{i \in I} (A_i)^y$ and $(\bigcap_{i \in I} A_i)^y = \bigcap_{i \in I} (A_i)^y$.
3. $(A^c)_x = (A_x)^c$ and $(A^c)^y = (A^y)^c$.

The easy verification of these identities are left for the reader.

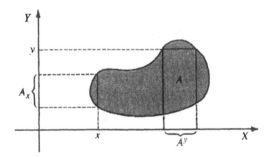

FIGURE 3.3. The Sections of a Set

Lemma 20.12. *Let* (X, Σ_1) *and* (Y, Σ_2) *be measurable spaces. If A belongs to the product σ-algebra, i.e.,* $A \in \sigma(\Sigma_1 \otimes \Sigma_2)$, *then A has "measurable sections."* *That is,* $A_x \in \Sigma_2$ *for each* $x \in X$ *and* $A^y \in \Sigma_1$ *for all* $y \in Y$.

Proof. Let $A \in \sigma(\Sigma_1 \otimes \Sigma_2)$. By the symmetry of the situation, it suffices to show that $A_x \in \Sigma_2$ for each $x \in X$. To this end, fix $x \in X$ and let

$$\Sigma = \{A \in \sigma(\Sigma_1 \otimes \Sigma_2): A_x \in \Sigma_2 \text{ for all } x \in X\}.$$

From $(\emptyset \times \emptyset)_x = \emptyset$ and $(X \times Y)_x = Y$, we see that $\emptyset, X \times Y \in \Sigma$. Also, from the identities listed preceding the lemma, we see that Σ is a σ-algebra. Moreover, if $A \times B \in \Sigma_1 \times \Sigma_2$, then $(A \times B)_x = B \in \Sigma_2$ if $x \in A$ and $(A \times B)_x = \emptyset \in \Sigma_2$ if $x \notin A$, and so $\Sigma_1 \otimes \Sigma_2 \subseteq \Sigma$. This implies $\Sigma = \sigma(\Sigma_1 \otimes \Sigma_2)$, and the proof is finished. ∎

We now come to the notion of joint measurability for functions. Let (X, Σ_1), (Y, Σ_2), and (Z, Σ_3) be three measurable spaces, and let $f: X \times Y \to Z$ be a function. We shall say that:

1. f is **jointly measurable**, if $f: (X \times Y, \sigma(\Sigma_1 \otimes \Sigma_2)) \to (Z, \Sigma_3)$ is measurable,
2. f is **measurable in the variable** x (resp. in y) if for each $y \in Y$ (resp. for each $x \in X$) the function $f(\cdot, y): (X, \Sigma_1) \to (Z, \Sigma_3)$ (resp. the function $f(x, \cdot): (Y, \Sigma_2) \to (Z, \Sigma_3)$) is measurable, and
3. f is **separately measurable** if f is measurable in each variable.

Joint measurability always guarantees separate measurability.

Theorem 20.13. *Let* (X, Σ_1), (Y, Σ_2) *and* (Z, Σ_3) *be three measurable spaces. If a function* $f: X \times Y \to Z$ *is jointly measurable, then it is separately measurable.*

Proof.   Assume $f: X \times Y \rightarrow Z$ is jointly measurable, let $A \in \Sigma_3$ and fix $y \in Y$. Also, let $B = f^{-1}(A) \in \sigma(\Sigma_1 \otimes \Sigma_2)$. Note that

$$f^{-1}(A, y) = \{x \in X: f(x, y) \in A\} = \{x \in X: (x, y) \in B\} = B^y.$$

Now, a glance at Lemma 20.12 guarantees that $B^y \in \Sigma_1$, and consequently, the function $f(\cdot, y): (X, \Sigma_1) \rightarrow (Z, \Sigma_3)$ is measurable. Similarly, it can be shown that the function $f(x, \cdot): (Y, \Sigma_2) \rightarrow (Z, \Sigma_3)$ is measurable for each $x \in X$.   ∎

Does separate measurability imply joint measurability? The answer is in general negative. W. Sierpiński[9] has constructed in [27] a non-Lebesgue measurable subset A of $\mathbb{R}^2$ whose intersection with each line consists of at-most two points; see also [10 p. 130] and [9]. From this property it easily follows that the characteristic function χ_A is separately measurable but not jointly measurable.

However, under certain conditions, separate measurability suffices for the joint measurability. In order to state such a result, we need the notion of a Carathéodory function.

Definition 20.14.   *Let (X, Σ) be a measurable space, and let Y and Z be two topological spaces. A function $f: X \times Y \rightarrow Z$ is said to be a* **Carathéodory function** *if*

1. *for each $x \in X$ the function $f(x, \cdot): Y \rightarrow Z$ is continuous, and*
2. *for each $y \in Y$, the function $f(\cdot, y): (X, \Sigma) \rightarrow (Z, \mathcal{B}_Z)$ (where $\mathcal{B}_Z$ is the σ-algebra of all Borel sets of Z) is measurable.*

Briefly, a function $f(x, y)$ of two variables is a Carathéodory function if f is continuous in y and measurable in x. For the next result, keep in mind that every topological space is considered a measurable space equipped with the σ-algebra of its Borel sets.

Theorem 20.15.   *Let (X, Σ) be a measurable space, Y a separable metric space, and Z a metric space. Then every Carathéodory function $f: X \times Y \rightarrow Z$ is jointly measurable.*

Proof.   Let d denote the metric of Y and ρ the metric of Z. Also, let $\{y_1, y_2, \ldots\}$ be a dense subset of Y. Since the closed sets form a family of generators for $\mathcal{B}_Z$, it suffices to show that $f^{-1}(C) \in \sigma(\Sigma \otimes \mathcal{B}_Y)$. So, let C be a closed subset of Z.

Start by observing that (since $f(x, \cdot)$ is continuous for each $x \in X$), $f(x, y)$ belongs to the closed set C if and only if for each n there exists some y_m with

---

[9]Waclaw Sierpiński (1882–1969), a Polish mathematician. He made many important contributions to set theory, topology, and number theory.

$d(y, y_m) < \frac{1}{n}$ and

$$\rho(f(x, y_m), C) = \inf\{\rho(f(x, y_m), c): c \in C\} < \frac{1}{n}.$$

This implies

$$f^{-1}(C) = \bigcap_{n=1}^{\infty} \bigcup_{m=1}^{\infty} \left[\{x \in X: f(x, y_m) \in B_{\frac{1}{n}}(C)\} \times B\left(y_m, \frac{1}{m}\right)\right], \quad (\star)$$

where $B_{\frac{1}{n}}(C) = \{z \in Z: \rho(z, C) < \frac{1}{n}\}$. The continuity of the function $z \mapsto \rho(z, C)$ guarantees that $B_{\frac{1}{n}}(C)$ is an open set. Since f is measurable in x and $B_{\frac{1}{n}}(C)$ is open (and hence a Borel set), it follows that $\{x \in X: f(x, y_m) \in B_{\frac{1}{n}}(C)\} \in \Sigma$ for each m and n. Now, a glance at $(\star)$ shows that $f^{-1}(C) \in \sigma(\Sigma \otimes \mathcal{B}_Y)$, and thus, f is jointly measurable. ∎

EXERCISES

1. Let $\mathcal{R}$ be a nonempty collection of subsets of a set X. Show that $\mathcal{R}$ is a ring if and only if $\mathcal{R}$ is closed under symmetric differences and finite intersections.
   [HINT: Note that $A\Delta B = (A \setminus B) \cup (B \setminus A)$, $A \cap B = A \setminus (A \setminus B)$, $A \setminus B = A\Delta(A \cap B)$, and $A \cup B = (A\Delta B)\Delta(A \cap B)$.]
2. If $\mathcal{R}$ is a ring, then show that the collection

$$\mathcal{A} = \{A \in \mathcal{R}: \text{Either } A \text{ or } A^c \text{ belongs to } \mathcal{R}\}$$

   is an algebra of sets.
3. In the implication scheme of Figure 3.4 show that no other implication is true by verifying the following regarding an uncountable set X:

   a. The collection of all singleton subsets of X together with the empty set is a semiring but not a ring.
   b. The collection of all finite subsets of X is a ring but is neither an algebra nor a σ-ring.
   c. The collection of all subsets of X that are either finite or have finite complement is an algebra but is neither a σ-algebra nor a σ-ring.
   d. The collection of all at-most countable subsets of X is a σ-ring but not an algebra.

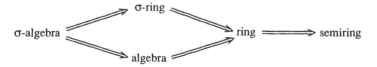

FIGURE 3.4. The Implication Scheme

e.   The collection of all subsets of X that are either at most countable or have at-most a countable complement is a σ-algebra (which is, in fact, the σ-algebra generated by the singletons).

4.   Show that a Dynkin system is a σ-algebra if and only if it is closed under finite intersections.

5.   Give an example of a Dynkin system which is not an algebra.

6.   A **monotone class** of sets is a family $\mathcal{M}$ of subsets of a set X such that if a sequence $\{A_n\}$ of $\mathcal{M}$ satisfies $A_n \uparrow A$ or $A_n \downarrow A$, then $A \in \mathcal{M}$. Establish the following properties regarding monotone classes:

a.   We have the following implication scheme:

$$\sigma\text{-algebra} \implies \text{Dynkin system} \implies \text{monotone class}$$

Give examples to show that no other implication in the above scheme is true.

b.   An algebra is a monotone class if and only if it is a σ-algebra.

c.   The σ-algebra $\sigma(\mathcal{A})$ generated by an algebra $\mathcal{A}$ is the smallest monotone class containing $\mathcal{A}$.

7.   Show that if X and Y are two separable metric spaces, then $\mathcal{B}_{X \times Y} = \mathcal{B}_X \otimes \mathcal{B}_Y$.

8.   Show that the composition function of two measurable functions is measurable.

9.   If (X, Σ) is a measurable space, then show that

a.   the collection of all real-valued measurable functions defined on X is a function space and an algebra of functions, and

b.   any real-valued function on X which is the pointwise limit of a sequence of $(\Sigma, \mathcal{B})$-measurable real-valued functions is itself $(\Sigma, \mathcal{B})$-measurable.

10.   Let (X, Σ) be a measurable space. A Σ-**simple function** is any measurable function $\phi: X \to \mathbb{R}$ which has a finite range, i.e, if ϕ has finite range and its standard representation $\phi = \sum_{i=1}^{n} a_i \chi_{A_i}$ satisfies $A_i \in \Sigma$ for each i.

Show that a function $f: X \to [0, \infty)$ is measurable if and only if there exists a sequence $\{\phi_n\}$ of Σ-simple functions such that $\phi_n(x) \uparrow f(x)$ holds for each $x \in X$.

11.   Use Corollary 20.10 to show that if a measure μ is σ-finite, then μ^* is the one and only extension of μ to a measure on $\sigma(S)$. Give an example of a measure space such that μ^* is not the only extension of μ to a measure on $\sigma(S)$.

12.   Show that the uniform limit of a sequence of measurable functions from a measurable space into a metric space is measurable.

13.   Let $f, g: X \to \mathbb{R}$ be two functions and let $\mathcal{B}$ denote the σ-algebra of all Borel sets of $\mathbb{R}$. Show that there exists a Borel measurable function $h: \mathbb{R} \to \mathbb{R}$ satisfying $f = h \circ g$ if and only if $f^{-1}(\mathcal{B}) \subseteq g^{-1}(\mathcal{B})$ holds.

14.   Let (X, Σ) be a measurable space, Y, Z_1 and Z_2 be separable metric spaces and Ψ a topological space. Assume also that the functions $f_i: X \times Y \to Z_i, (i = 1, 2)$, are Carathéodory functions and $g: Z_1 \times Z_2 \to \Psi$ is Borel measurable. Show that the function $h: X \times Y \to \Psi$, defined by

$$h(x, y) = g(f_1(x, y), f_2(x, y)),$$

is jointly measurable.

15. Let (X, Σ) be a measurable space and (Y, d) a separable metric space. Show that a function $f: X \to Y$ is measurable if and only if the function $x \mapsto d(y, f(x))$, from X into $\mathbb{R}$, is measurable for each fixed $y \in Y$.

16. Let (X, S, μ) be a σ-finite measure space, where S is a σ-algebra. If $f: X \to \mathbb{R}$ is a Λ_μ-measurable function, then show that there exists a S-measurable function $g: X \to \mathbb{R}$ such that $f = g$ a.e.

CHAPTER **4**

THE LEBESGUE INTEGRAL

By definition a Riemann integrable function is bounded, and its domain is a closed interval. These two restrictions make the Riemann integral inadequate to fulfill the requirements of many scientific problems. H. Lebesgue[1] in his classical work [21] introduced a concept of an integral (called today the Lebesgue integral) based on measure theory that generalizes the Riemann integral. It has the advantage of treating at the same time both bounded and unbounded functions and allows their domains to be more general sets. Also, it gives more powerful and useful convergence theorems than the Riemann integral.

In this chapter we study the Lebesgue integral. The concepts of an upper function and its Lebesgue integral are introduced first. Consequently, the Lebesgue integrable functions are introduced as differences of two upper functions, and their properties are studied. The Lebesgue dominated convergence theorem that (under certain conditions) allows us to interchange the processes of limit and integration is proved, and various applications of this powerful result are presented. Next, it is shown that every Riemann integrable function is Lebesgue integrable, and that in this case the two integrals (the Riemann and the Lebesgue) coincide. Furthermore, the relationship between an improper Riemann integral and the Lebesgue integral is obtained. Finally, the chapter culminates with a study of product measures and iterated integrals.

For this chapter, (X, S, μ) will be a fixed measure space, and unless otherwise specified, all properties of the functions will be tacitly referred to in this measure space.

21. UPPER FUNCTIONS

It was mentioned before that the step functions will be the "building blocks" for the Lebesgue integral. Recall that a function ϕ is a step function if and only if there exist a finite collection $\{A_1, \ldots, A_n\}$ of measurable sets with $\mu^*(A_i) < \infty$ for $i = 1, \ldots, n$, and real numbers $a_1, \ldots, a_n$ such that $\phi = \sum_{i=1}^{n} a_i \chi_{A_i}$ holds.

---

[1] Henri Léon Lebesgue (1875–1941), a prominent French mathematician. The founder of the modern theory of integration.

The real number $I(\phi) = \sum_{i=1}^{n} a_i \mu^*(A_i)$ is called the Lebesgue integral of ϕ, and we have already seen (in Section 17) that it is independent of the particular representation of ϕ. From now on, the Lebesgue integral of ϕ will be denoted by its conventional symbol $\int \phi \, d\mu$, or $\int_X \phi \, d\mu$. If clarity requires the variable to be emphasized, then the notation $\int \phi(x) \, d\mu(x)$ will be used. Thus, the integral of the step function is

$$\int \phi \, d\mu = \sum_{i=1}^{n} a_i \mu^*(A_i).$$

The collection of all step functions has both the structure of an algebra and that of a function space.

Theorem 21.1. *The collection of all step functions under the pointwise operations is a function space and an algebra.*

Proof. The proof that the step functions form an algebra is straightforward. To see that this collection is also a function space, note that if the step function ϕ has the standard representation $\phi = \sum_{i=1}^{n} a_i \chi_{A_i}$, then $\phi^+ = \sum_{i=1}^{n} \max\{a_i, 0\} \chi_{A_i}$ holds. Thus, ϕ^+ is a step function, and the conclusion follows. ∎

The basic properties of the Lebesgue integral for step functions were discussed in Section 17. Here we shall deal with the almost everywhere limits of increasing sequences of step functions. Such limits are used to define upper functions.

Definition 21.2. *A function $f : X \to \mathbb{R}$ is called an* **upper function** *if there exists a sequence $\{\phi_n\}$ of step functions such that*

1. $\phi_n \uparrow f$ *a.e., and*
2. $\lim \int \phi_n \, d\mu < \infty$.

Any sequence of step functions that satisfies conditions (1) and (2) of the preceding definition will be referred to as a **generating sequence** for f. Note, in particular, that by Theorem 16.6 every upper function is a measurable function. The collection of all upper functions will be denoted by $\mathcal{U}$. Clearly,

- *every step function is an upper function.*

Also, observe that an upper function *need not* be a positive function.

If f is an upper function with a generating sequence $\{\phi_n\}$, and if $\{\psi_n\}$ is another sequence of step functions such that $\psi_n \uparrow f$ a.e., then it follows from Theorem 17.5 that $\lim \int \phi_n \, d\mu = \lim \int \psi_n \, d\mu$ holds. Thus, $\{\psi_n\}$ is also a generating sequence for f, and therefore, the following definition is well justified.

Definition 21.3. *Let f be an upper function, and let $\{\phi_n\}$ be a sequence of step functions such that $\phi_n \uparrow f$ a.e. holds. Then the* **Lebesgue integral** (*or simply the* **integral**) *of f is defined by*

$$\int f \, d\mu = \lim_{n\to\infty} \int \phi_n \, d\mu.$$

Again, we stress the fact that the value of the Lebesgue integral of an upper function is independent of the generating sequence of step functions. Also, it should be clear that if f is an upper function and g is another function such that $g = f$ a.e., then g is also an upper function and $\int g \, d\mu = \int f \, d\mu$ holds.

The rest of this section is devoted to the properties of upper functions.

Theorem 21.4. *For upper functions f and g, the following statements hold:*
1. *$f + g$ is an upper function and $\int (f + g) \, d\mu = \int f \, d\mu + \int g \, d\mu$.*
2. *αf is an upper function for each $\alpha \geq 0$ and $\int (\alpha f) \, d\mu = \alpha \int f \, d\mu$.*
3. *$f \vee g$ and $f \wedge g$ are upper functions.*

Proof. Choose two generating sequences $\{\phi_n\}$ and $\{\psi_n\}$ for f and g, respectively.

(1) Clearly, $\{\phi_n + \psi_n\}$ is a sequence of step functions, and $\phi_n + \psi_n \uparrow f + g$ a.e. holds. The result now follows by observing that

$$\int (\phi_n + \psi_n) \, d\mu = \int \phi_n \, d\mu + \int \psi_n \, d\mu \uparrow \int f \, d\mu + \int g \, d\mu.$$

(2) Straightforward.

(3) Note that both $\{\phi_n \vee \psi_n\}$ and $\{\phi_n \wedge \psi_n\}$ are sequences of step functions. Now, $\phi_n \wedge \psi_n \uparrow f \wedge g$ a.e. and $\lim \int \phi_n \wedge \psi_n \, d\mu \leq \lim \int \phi_n \, d\mu < \infty$, show that $f \wedge g$ is an upper function.

To see that $f \vee g$ is an upper function, observe first that $\phi_n \vee \psi_n \uparrow f \vee g$ a.e. holds, and then use the identity

$$\phi_n \vee \psi_n = \phi_n + \psi_n - \phi_n \wedge \psi_n$$

to obtain $\int \phi_n \vee \psi_n \, d\mu = \int \phi_n \, d\mu + \int \psi_n \, d\mu - \int \phi_n \wedge \psi_n \, d\mu$. This implies

$$\int \phi_n \vee \psi_n \, d\mu = \int \phi_n \, d\mu \uparrow \int f \, d\mu + \int g \, d\mu - \int f \wedge g \, d\mu < \infty.$$

This finishes the proof of the theorem. ∎

The next theorem states that the integral is a monotone function on $\mathcal{U}$.

Theorem 21.5. *If f and g are upper functions such that $f \geq g$ a.e., then $\int f \, d\mu \geq \int g \, d\mu$ holds. In particular, if $f \in \mathcal{U}$ satisfies $f \geq 0$ a.e., then $\int f \, d\mu \geq 0$.*

Proof. Let $\{\phi_n\}$ and $\{\psi_n\}$ be generating sequences for f and g, respectively. Then $\phi_n \wedge \psi_n \uparrow g$ a.e. holds, and so $\{\phi_n \wedge \psi_n\}$ is also a generating sequence for g. By Theorem 17.3, we have $\int \phi_n \, d\mu \geq \int \phi_n \wedge \psi_n \, d\mu$ for each n. Therefore,

$$\int f \, d\mu = \lim_{n \to \infty} \int \phi_n \, d\mu \geq \lim_{n \to \infty} \int \phi_n \wedge \psi_n \, d\mu = \int g \, d\mu,$$

and the proof is finished.   ∎

It should be noted that if f is an upper function such that $f \geq 0$ a.e., then there exists a sequence of step functions $\{\psi_n\}$ satisfying $\psi_n \geq 0$ a.e. for each n and $\psi_n \uparrow f$ a.e. To see this, notice that if $\phi_n \uparrow f$, then $\phi_n^+ \uparrow f^+ = f$ a.e. holds.

If we take the "upper functions of $\mathcal{U}$," then we get $\mathcal{U}$ again. The details are included in the next theorem.

Theorem 21.6. *Let $f : X \to \mathbb{R}$ be a function. If there exists a sequence $\{f_n\}$ of upper functions such that $f_n \uparrow f$ a.e. and $\lim \int f_n \, d\mu < \infty$, then f is an upper function and $\int f \, d\mu = \lim \int f_n \, d\mu$.*

Proof. For each i choose a sequence $\{\phi_n^i\}$ of step functions such that $\phi_n^i \uparrow_n f_i$ a.e. holds. Now, for each n let $\psi_n = \bigvee_{i=1}^{n} \phi_n^i$, and note that each ψ_n is a step function such that $\psi_n \uparrow f$ a.e. holds. Also, note that $\psi_n \leq f_n$ a.e. for each n, and consequently, $\lim \int \psi_n \, d\mu \leq \lim \int f_n \, d\mu < \infty$ holds by virtue of Theorem 21.5. This shows that f is an upper function.

Now, since for each fixed i we have $\phi_n^i \leq \psi_n$ for all $n \geq i$, it follows that $\int f_i \, d\mu = \lim_{n \to \infty} \int \phi_n^i \, d\mu \leq \lim_{n \to \infty} \int \psi_n \, d\mu$ holds for all i. Therefore,

$$\lim_{n \to \infty} \int f_n \, d\mu = \lim_{n \to \infty} \int \psi_n \, d\mu = \int f \, d\mu,$$

and the proof is complete.   ∎

The integral satisfies an important convergence property for decreasing sequences. It is our familiar **order continuity property** of the integral.

Theorem 21.7. *If $\{f_n\}$ is a sequence of upper functions such that $f_n \downarrow 0$ a.e., then $\lim \int f_n \, d\mu = 0$ holds.*

Proof. Let $\epsilon > 0$. For each n choose a step function ϕ_n such that $0 \leq \phi_n \leq f_n$ a.e. and $\int (f_n - \phi_n) \, d\mu = \int f_n \, d\mu - \int \phi_n \mu < \epsilon 2^{-n}$ (remember that $-\phi_n$ is an upper function). Let $\psi_n = \bigwedge_{i=1}^{n}\phi_i$ for each n. Then $\{\psi_n\}$ is a sequence of step functions satisfying $0 \leq \psi_{n+1} \leq \psi_n$ for each n, and (in view of $\psi_n \leq f_n$ a.e. and $f_n \downarrow 0$ a.e.) we see that $\psi_n \downarrow 0$ a.e. By Theorem 17.4 we have $\lim \int \psi_n \, d\mu = 0$. Pick an integer k such that $\int \psi_n \, d\mu < \epsilon$ for all $n \geq k$. Now, the almost everywhere inequalities

$$0 \leq f_n - \psi_n = f_n - \bigwedge_{i=1}^{n}\phi_i = \bigvee_{i=1}^{n}(f_n - \phi_i) \leq \bigvee_{i=1}^{n}(f_i - \phi_i) \leq \sum_{i=1}^{n}(f_i - \phi_i)$$

imply

$$\int f_n \, d\mu - \int \psi_n \, d\mu \leq \sum_{i=1}^{n}\int (f_i - \phi_i)\,d\mu < \epsilon \left(\sum_{i=1}^{\infty} 2^{-i}\right) = \epsilon.$$

Thus,

$$0 \leq \int f_n \, d\mu < \epsilon + \int \psi_n \, d\mu < 2\epsilon$$

for all $n \geq k$, which shows that $\int f_n \, d\mu \downarrow 0$ holds. ∎

Finally, we mention that $\mathcal{U}$ is not in general a vector space since it fails to be closed under multiplication by negative real numbers. An example of this type is presented in Exercise 2 of this section.

EXERCISES

1. Let L be the collection of all step functions ϕ such that there exist a finite number of sets $A_1, \ldots, A_n$ in S all of finite measure and real numbers $a_1, \ldots, a_n$ such that $\phi = \sum_{i=1}^{n} a_i \chi_{A_i}$. Show that L is a function space. Is L an algebra of functions? [HINT: Use Exercise 14 of Section 12.]

2. Consider the function $f : \mathbb{R} \to \mathbb{R}$ defined by $f(x) = 0$ if $x \notin (0, 1]$, and $f(x) = \sqrt{n}$ if $x \in (\frac{1}{n+1}, \frac{1}{n}]$ for some n. Show that f is an upper function and that $-f$ is not an upper function.
   [HINT: A step function is necessarily bounded.]

3. Compute $\int f \, d\lambda$ for the upper function f of the preceding exercise.

4. Verify that every continuous function $f : [a, b] \to \mathbb{R}$ is an upper function—with respect to the Lebesgue measure on $[a, b]$.

5. Let A be a measurable set, and let f be an upper function. If $\chi_A \leq f$ a.e., then show that $\mu^*(A) < \infty$.

6. Let f be an upper function, and let A be a measurable set of finite measure such that $a \leq f(x) \leq b$ holds for each $x \in A$. Then show that

a. $f\chi_A$ is an upper function, and
b. $a\mu^*(A) \le \int f\chi_A \, d\mu \le b\mu^*(A)$.

7. Let $(X, \mathcal{S}, \mu)$ be a finite measure space, and let f be a positive measurable function. Show that f is an upper function if and only if there exists a real number M such that $\int \phi \, d\mu \le M$ holds for every step function ϕ with $\phi \le f$ a.e. Also, show that if this is the case, then

$$\int f \, d\mu = \sup \left\{ \int \phi \, d\mu \colon \phi \text{ is a step function with } \phi \le f \text{ a.e.} \right\}.$$

8. Show that every monotone function $f : [a, b] \to \mathbb{R}$ is an upper function—with respect to the Lebesgue measure on $[a, b]$.

22. INTEGRABLE FUNCTIONS

It was observed before that the collection $\mathcal{U}$ of all upper functions is not a vector space. However, if we consider the collection of all functions that can be written as an almost everywhere difference of two upper functions, then this set is a function space. The members of this collection are the Lebesgue integrable functions. The details will be explained in the following.

Definition 22.1. *A function $f : X \to \mathbb{R}$ is called* **Lebesgue integrable** (*or simply* **integrable**) *if there exist two upper functions u and v such that $f = u - v$ a.e. holds. The* **Lebesgue integral** (*or simply the* **integral**) *of f is defined by*

$$\int f \, d\mu = \int u \, d\mu - \int v \, d\mu.$$

It should be noted that the value of the integral is independent of the representation of f as a difference of two upper functions. Indeed, if $f = u - v = u_1 - v_1$ a.e. with $u, u_1, v,$ and v_1 all upper functions, then $u + v_1 = u_1 + v$ a.e. holds, and by Theorem 21.4(1) we have $\int u \, d\mu + \int v_1 \, d\mu = \int u_1 \, d\mu + \int v \, d\mu$. Therefore, $\int u \, d\mu - \int v \, d\mu = \int u_1 \, d\mu - \int v_1 \, d\mu$.

An integrable function is necessarily measurable, and every upper function is Lebesgue integrable. Also, it is readily seen that if a function f is Lebesgue integrable and g is another function such that $f = g$ a.e., then g is also Lebesgue integrable and $\int g \, d\mu = \int f \, d\mu$ holds.

Historical Note: The above introduction of the Lebesgue integral is a modification of a method due to P. J. Daniell [5]. Daniell's[2] general approach to integration starts with a function space L on some nonempty set X, together with an "integral" I on L. The function $I : L \to \mathbb{R}$ is said to be an *integral* if

---

[2]P. J. Daniell (1889–1946), a British mathematician. He worked in functional analysis and the theory of integration.

1. $I(\alpha\phi + \beta\psi) = \alpha I(\phi) + \beta I(\psi)$ for all $\alpha, \beta \in \mathbb{R}$ and $\phi, \psi \in L$,
2. $I(\phi) \geq 0$ whenever $\phi \geq 0$, and
3. whenever $\{\phi_n\} \subseteq L$ satisfies $\phi_n(x) \downarrow 0$ for each $x \in X$, then $I(\phi_n) \downarrow 0$.

A function $u : X \to \mathbb{R}$ is called an upper function if there exists a sequence $\{\phi_n\} \subseteq L$ with $\phi_n(x) \uparrow u(x)$ for all $x \in X$ and $\lim I(\phi_n) < \infty$. As in the proof of Theorem 17.5, we can show that $\lim I(\phi_n)$ is independent of the "generating" sequence $\{\phi_n\}$. The real number $I(u) = \lim I(\phi_n)$ is the integral of u. Finally, $f : X \to \mathbb{R}$ is said to be *integrable* if there exist two upper functions u and v with $f = u - v$. The integral of f is then defined by $I(f) = I(u) - I(v)$.

Here our approach to the Lebesgue integral can be considered as a "measure theoretical Daniell method."

The set of integrable functions has all the expected nice properties.

Theorem 22.2. *The collection of all Lebesgue integrable functions is a function space.*

Proof. Let f and g be two integrable functions with representations $f = u - v$ a.e. and $g = u_1 - v_1$ a.e. Then the almost everywhere identities

$$f + g = (u + u_1) - (v + v_1),$$
$$\alpha f = \alpha u - \alpha v \text{ if } \alpha \geq 0,$$
$$\alpha f = [(-\alpha)v] - [(-\alpha)u] \text{ if } \alpha < 0, \text{ and}$$
$$f^+ = (u - v)^1 = u \vee v - v$$

express the above functions as differences of two upper functions. This shows that the collection of all integrable functions is a function space. ∎

Thus, if f is integrable, then $|f|$ is also an integrable function. In particular, it follows from the last theorem that a function f is Lebesgue integrable if and only if f^+ and f^- are both integrable.

The next result describes the linearity property of the integral. Its easy proof follows directly from Definition 22.1 and is left as an exercise for the reader.

Theorem 22.3. *If f and g are two integrable functions, then*

$$\int (\alpha f + \beta g) \, d\mu = \alpha \int f \, d\mu + \beta \int g \, d\mu$$

holds for all $\alpha, \beta \in \mathbb{R}$.

Every positive integrable function is necessarily an upper function.

Theorem 22.4.   *If an integrable function f satisfies $f \geq 0$ a.e., then f is an upper function.*

Proof.   Choose two upper functions u and v such that $f = u - v$ a.e. holds. Since each u and v is the almost everywhere limit of a sequence of step functions, there exists a sequence $\{\psi_n\}$ of step functions such that $\psi_n \to f$ a.e. Since $f \geq 0$ a.e., it follows that $\psi_n^+ \to f$ a.e. also holds.

By Theorem 17.7, there exists a sequence $\{s_n\}$ of simple functions satisfying $0 \leq s_n \uparrow f$ a.e. Now, for each n let $\phi_n = s_n \wedge (\bigvee_{i=1}^{n} \psi_i^+)$. Then $\{\phi_n\}$ is a sequence of step functions such that $0 \leq \phi_n \uparrow f$ a.e. holds. To complete the proof, we show that $\{\int \phi_n \, d\mu\}$ is bounded. Indeed, from $\phi_n + v \leq f + v \leq u$ a.e. and Theorem 21.5, it follows that $\int \phi_n \, d\mu + \int v \, d\mu \leq \int u \, d\mu$, and therefore, $\int \phi_n \, d\mu \leq \int u \, d\mu - \int v \, d\mu < \infty$ holds for all n. The proof of the theorem is now complete.   ∎

If f is an integrable function, then by Theorem 22.4, f^+ and f^- are both upper functions, and so, $f = f^+ - f^-$ is a decomposition of f as a difference of two positive upper functions. In particular,

$$\int f \, d\mu = \int f^+ \, d\mu - \int f^- \, d\mu.$$

(This formula is usually the one used by many authors to define the Lebesgue integral.)

As an application of the preceding theorem, we also have the following useful result.

Theorem 22.5.   *If f is an integrable function, then for every $\epsilon > 0$ the measurable set $\{x \in X : |f(x)| \geq \epsilon\}$ has finite measure.*

Proof.   Fix $\epsilon > 0$, let $A = \{x \in X : |f(x)| \geq \epsilon\}$, and note that $\epsilon \chi_A \leq |f|$ holds. Now, $\frac{1}{\epsilon}|f|$ is an integrable function, in fact, by Theorem 22.4 an upper function. Let $\{\phi_n\}$ be a sequence of step functions such that $\phi_n \uparrow \frac{1}{\epsilon}|f|$ a.e. Then $\{\phi_n \wedge \chi_A\}$ is a sequence of step functions such that $\phi_n \wedge \chi_A \uparrow \chi_A$ a.e. holds. Thus, by Theorem 17.6,

$$\mu^*(A) = \lim_{n \to \infty} \int \phi_n \wedge \chi_A \, d\mu \leq \lim_{n \to \infty} \int \phi_n \, d\mu = \frac{1}{\epsilon} \int |f| \, d\mu < \infty,$$

and the proof is finished.   ∎

Measurable functions "sandwiched" between integrable functions are integrable.

Theorem 22.6. *Let f be a measurable function. If there exist two integrable functions h and g such that $h \leq f \leq g$ a.e., then f is also an integrable function.*

Proof. Writing the given inequality in the form $0 \leq f - h \leq g - h$ a.e., we see that we can assume without loss of generality that $0 \leq f \leq g$ a.e. holds.

By Theorem 22.4, g is an upper function. Pick a sequence $\{\phi_n\}$ of step functions such that $0 \leq \phi_n \uparrow g$ a.e. By Theorem 17.7 there exists a sequence $\{\psi_n\}$ of simple functions such that $0 \leq \psi_n \uparrow f$ a.e. holds. But then $\{\phi_n \wedge \psi_n\}$ is a sequence of step functions such that $\phi_n \wedge \psi_n \uparrow f$ a.e., and $\int \phi_k \wedge \psi_k \, d\mu \leq \lim \int \phi_n \, d\mu = \int g \, d\mu < \infty$ for all k. Hence, $f \in \mathcal{U}$, and so, f is an integrable function. ∎

More properties of the integral are included in the next theorem:

Theorem 22.7. *For integrable functions f and g we have the following:*
1. $\int |f| \, d\mu = 0$ *if and only if* $f = 0$ *a.e.*
2. *If* $f \geq g$ *a.e., then* $\int f \, d\mu \geq \int g \, d\mu$.
3. $|\int f \, d\mu| \leq \int |f| \, d\mu$.

Proof. (1) Clearly, if $f = 0$ a.e., then $\int |f| \, d\mu = 0$ holds. On the other hand, assume that $\int |f| \, d\mu = 0$. Since, by Theorem 22.4, $|f|$ is an upper function, there exists a sequence $\{\phi_n\}$ of step functions such that $0 \leq \phi_n \uparrow |f|$ a.e. holds. By Theorem 21.5, it follows that $\int \phi_n \, d\mu = 0$ for each n, and so, $\phi_n = 0$ a.e. for each n. Thus, $|f| = 0$ a.e., and so $f = 0$ a.e.

(2) Since $f - g \geq 0$ a.e., it follows from Theorem 22.4 that $f - g$ is an upper function. But then, Theorem 21.5 implies that $\int f \, d\mu - \int g \, d\mu = \int (f - g) \, d\mu \geq 0$, so that $\int f \, d\mu \geq \int g \, d\mu$ holds.

(3) The conclusion follows from (2) and the inequality $-|f| \leq f \leq |f|$. ∎

The reader has probably noticed that the functions we have considered so far are real-valued. It is a custom, however, to allow a function to assume infinite values, provided that the set of all points where the function equals $-\infty$ or ∞ is a null set. The reason for this is that neither the integrability character nor the value of the integral of a function changes by altering its values on a null set. Moreover, assigning any value to the sum of two functions at the points where the form $\infty - \infty$ occurs does not affect the integrability and the value of the integral of the sum function (as long as the set of points of all such encounters has measure zero).

If one does not want to deal with functions assuming infinite values (up, of course to null sets), then one may change the infinite values to finite ones (for instance, change all the infinite values to zero) without loosing anything regarding integrability. When an extended real-valued function f is said to **define an integrable function**, it will be meant that f assumes the infinite values (or it is even undefined) on a null set, and that if finite values are assigned to these points, then f becomes an integrable function. To summarize the preceding:

- *Functions that are almost everywhere equal have the same integrability properties and can be considered as identical.*

We continue with a theorem of B. Levi[3] describing a basic monotone property of the integral.

Theorem 22.8 (Levi).   *Assume that a sequence $\{f_n\}$ of integrable functions satisfies $f_n \leq f_{n+1}$ a.e. for each n and $\lim \int f_n \, d\mu < \infty$. Then there exists an integrable function f such that $f_n \uparrow f$ a.e. (and hence, $\int f_n \, d\mu \uparrow \int f \, d\mu$ holds).*

Proof.   Replacing $\{f_n\}$ by $\{f_n - f_1\}$ if necessary, we can assume without loss of generality that $f_n \geq 0$ a.e. holds for each n. Also, an easy argument shows that we can assume that $0 \leq f_n(x) \uparrow$ holds for all $x \in X$. Let $I = \lim \int f_n \, d\mu < \infty$. For each $x \in X$ let $g(x) = \lim f_n(x) \in \mathbb{R}^*$, and consider the set

$$E = \{x \in X : g(x) = \infty\}.$$

Clearly, $E = \bigcap_{i=1}^{\infty} [\bigcup_{n=1}^{\infty} \{x \in X : f_n(x) > i\}]$ holds, and so E is a measurable set. Next, we shall show that $\mu^*(E) = 0$.

By Theorem 22.4, each f_n is an upper function. Thus, for each i there exists a sequence $\{\phi_n^i\}$ of step functions such that $0 \leq \phi_n^i \uparrow_n f_i$ a.e. holds. For each n let $\psi_n = \bigvee_{i=1}^{n} \phi_n^i$, and note that $\{\psi_n\}$ is a sequence of step functions such that $\psi_n \uparrow g$ a.e. and $\lim \int \psi_n \, d\mu = \lim \int f_n \, d\mu = I$. In particular, for each k the sequence of step functions $\{\psi_n \wedge k\chi_E\}$ satisfies $\psi_n \wedge k\chi_E \uparrow k\chi_E$ a.e. From Theorem 17.6, it follows that $\mu^*(E) < \infty$ and $k\mu^*(E) \leq \lim \int \psi_n \, d\mu = I < \infty$ for each k. Hence, $\mu^*(E) = 0$.

Now, define $f : X \to \mathbb{R}$ by $f(x) = g(x)$ if $x \notin E$ and $f(x) = 0$ if $x \in E$. Then $f_n \uparrow f$ a.e. holds, and the result follows from Theorem 21.6. ∎

The series analogue of the preceding theorem is presented next.

Theorem 22.9.   *Let $\{f_n\}$ be a sequence of non-negative integrable functions*

---

[3]Beppo Levi (1875–1961), an Italian mathematician. His main contributions were in algebraic topology, mathematical logic, and analysis.

such that $\sum_{n=1}^{\infty} \int f_n \, d\mu < \infty$. Then $\sum_{n=1}^{\infty} f_n$ defines an integrable function and

$$\int \left(\sum_{n=1}^{\infty} f_n \right) d\mu = \sum_{n=1}^{\infty} \int f_n \, d\mu.$$

Proof. For each n let $g_n = \sum_{i=1}^{n} f_i$, and note that each g_n is an integrable function such that $g_n \uparrow \sum_{i=1}^{\infty} f_i$ a.e. holds. Now, by Levi's theorem, $\sum_{n=1}^{\infty} f_n$ defines an integrable function, and

$$\sum_{n=1}^{\infty} \int f_n \, d\mu = \lim_{n \to \infty} \int g_n \, d\mu = \int \left(\sum_{n=1}^{\infty} f_n \right) d\mu$$

holds.   ∎

The next result is known in the theory of integration as Fatou's[4] lemma.

Theorem 22.10 (Fatou's Lemma). *Let $\{f_n\}$ be a sequence of integrable functions such that $f_n \geq 0$ a.e. for each n and $\liminf \int f_n \, d\mu < \infty$. Then $\liminf f_n$ defines an integrable function, and*

$$\int \liminf f_n \, d\mu \leq \liminf \int f_n \, d\mu.$$

Proof. Without loss of generality, we can suppose that $f_n(x) \geq 0$ holds for all $x \in X$ and all n.

Given n, define $g_n(x) = \inf\{f_i(x) : i \geq n\}$ for each $x \in X$. Then each g_n is a measurable function, and $0 \leq g_n \leq f_n$ holds for all n. Thus, by Theorem 22.6, each g_n is an integrable function. Now, observe that $g_n \uparrow$ and $\lim \int g_n \, d\mu \leq \liminf \int f_n \, d\mu < \infty$ holds. Thus, by Theorem 22.8, there exists an integrable function g such that $g_n \uparrow g$ a.e. holds. It follows that $g = \liminf f_n$ a.e., and therefore, $\liminf f_n$ defines an integrable function. Moreover,

$$\int \liminf f_n \, d\mu = \int g \, d\mu = \lim_{n \to \infty} \int g_n \, d\mu \leq \liminf \int f_n \, d\mu,$$

and the proof is finished.   ∎

We are now in the position to state and prove the Lebesgue dominated convergence theorem—the cornerstone of the theory of integration.

---

[4]Pierre Joseph Louis Fatou (1878–1929), a French mathematician. Besides his work in analysis, he also studied the motion of planets in astronomy.

Theorem 22.11 (The Lebesgue Dominated Convergence Theorem). *Let $\{f_n\}$ be a sequence of integrable functions satisfying $|f_n| \leq g$ a.e. for all n and some fixed integrable function g. If $f_n \to f$ a.e., then f defines an integrable function and*

$$\lim_{n \to \infty} \int f_n \, d\mu = \int \lim_{n \to \infty} f_n \, d\mu = \int f \, d\mu.$$

Proof. Clearly, $|f| \leq g$ a.e. holds, and the integrability of f follows from Theorem 22.6. Observe next that the sequence $\{g - f_n\}$ satisfies the hypotheses of Fatou's lemma and moreover, $\liminf(g - f_n) = g - f$ a.e. Thus,

$$\int g \, d\mu - \int f \, d\mu = \int (g - f) \, d\mu = \int \liminf (g - f_n) \, d\mu$$
$$\leq \liminf \int (g - f_n) \, d\mu = \int g \, d\mu - \limsup \int f_n \, d\mu,$$

and hence,

$$\limsup \int f_n \, d\mu \leq \int f \, d\mu.$$

Similarly, Fatou's lemma applied to the sequence $\{g + f_n\}$ yields

$$\int g \, d\mu + \int f \, d\mu = \int (g + f) \, d\mu = \int \liminf (g + f_n) \, d\mu$$
$$\leq \liminf \int (g + f_n) \, d\mu = \int g \, d\mu + \liminf \int f_n \, d\mu,$$

and so,

$$\int f \, d\mu \leq \liminf \int f_n \, d\mu.$$

Therefore, $\lim \int f_n \, d\mu$ exists in $\mathbb{R}$, and $\lim \int f_n \, d\mu = \int f \, d\mu$ holds. ■

The next theorem characterizes the Lebesgue integrable functions in terms of some given property. It is usually employed to prove that all Lebesgue integrable functions possess a given property.

Theorem 22.12. *Let (X, S, μ) be a measure space and let (P) be a property which may or may not be possessed by an integrable function. Assume that:*

1. *If f and g are integrable functions with property (P), then $f + g$ and αf for each $\alpha \in \mathbb{R}$ also have property (P).*
2. *If f is an integrable function such that for each $\epsilon > 0$ there exists an integrable function g with property (P) satisfying $\int |f - g| \, d\mu < \epsilon$, then f has property (P).*
3. *For each $A \in \mathcal{S}$ with $\mu(A) < \infty$, the characteristic function χ_A has property (P).*

Then every integrable function has property (P).

Proof. Assume first that A is a σ-set with $\mu^*(A) < \infty$. So, there exists a disjoint sequence $\{A_n\}$ of $\mathcal{S}$ such that $A = \bigcup_{n=1}^\infty A_n$. Put $B_n = \bigcup_{k=1}^n A_k$ for each n and note that $B_n \uparrow A$. From $\chi_{B_n} = \sum_{k=1}^n \chi_{A_k}$, (3) and (1), we see that χ_{B_n} has property (P) for each n. Since $\int |\chi_A - \chi_{B_n}| \, d\mu = \mu^*(A) - \mu(B_n) \to 0$, it follows from (2) that χ_A likewise has property (P).

Next, assume that A is an arbitrary measurable set of finite measure and let $\epsilon > 0$. Then there exists a σ-set B of finite measure such that $A \subseteq B$ and $\mu^*(B) < \mu^*(A) + \epsilon$. This implies $\int |\chi_A - \chi_B| \, d\mu = \mu^*(B) - \mu^*(A) < \epsilon$. From the above discussion and (2), we infer that χ_A satisfies property (P).

Now, from (1) we see that every step function satisfies property (P). But then, it follows from (2) that every upper function satisfies property (P). Since every integrable function is the difference of two upper functions, invoking (1) once more, we infer that indeed every integrable function satisfies property (P). ∎

It is easy to see that for every subset E of X, the collection $\mathcal{S}_E = \{E \cap A : A \in \mathcal{S}\}$ of subsets of E (called the restriction of the semiring $\mathcal{S}$ to E) is a semiring of subsets of E. If, in addition, E is a measurable subset of X, then μ^* restricted to $\mathcal{S}_E$ is also a measure. That is, $(E, \mathcal{S}_E, \mu^*)$ is a measure space for every measurable subset E of X. Also, a straightforward verification shows that the measurable subsets of $(E, \mathcal{S}_E, \mu^*)$ are precisely the subsets of E that are measurable subsets of X; see Exercise 7 of Section 15.

If E is a measurable subset of X, then a function $f : E \to \mathbb{R}$ is said to be integrable over E if f is integrable with respect to the measure space $(E, \mathcal{S}_E, \mu^*)$. Of course, the domain of f can be extended to all of X by assigning the values $f(x) = 0$ if $x \notin E$. Then f so defined is an integrable function over X, and in this case $\int_X f \, d\mu = \int_E f \, d\mu$ holds. A function $f : X \to \mathbb{R}$ is said to be **integrable over a measurable subset** E of X if the function $f \chi_E$ is integrable over X, or equivalently, if f restricted to E is integrable with respect to the measure space $(E, \mathcal{S}_E, \mu^*)$. In this case, we shall write $\int f \chi_E \, d\mu = \int_E f \, d\mu$.

The simple proof of the next result is left as an exercise for the reader.

Theorem 22.13. *Every integrable function f is integrable over every measurable subset of X. Moreover,*

$$\int_E f \, d\mu + \int_{E^c} f \, d\mu = \int_X f \, d\mu$$

holds for every measurable subset E of X.

The rest of this section deals with an observation concerning infinite Lebesgue integrals. If $\phi = \sum_{i=1}^n a_i \chi_{A_i}$ is the standard representation of a positive simple function ϕ, then the sum $\sum_{i=1}^n a_i \mu^*(A_i)$ makes sense as an extended non-negative real number. If $\sum_{i=1}^n a_i \mu^*(A_i) = \infty$, then it is a custom to write $\int \phi \, d\mu = \infty$ and say that the Lebesgue integral of ϕ is infinite.

Assume now that $f : X \to \mathbb{R}_+^*$ is a function where there exists a sequence $\{\phi_n\}$ of positive simple functions such that $\phi_n \uparrow f$ a.e. holds. Then $\lim \int \phi_n \, d\mu$ exists as an extended real number, and it can be seen easily that $\lim \int \phi_n \, d\mu$ is independent from the chosen sequence $\{\phi_n\}$. In the case that $\lim \int \phi_n \, d\mu = \infty$, we write $\int f \, d\mu = \infty$ and say that the Lebesgue integral of f is infinite—but we do not call the function integrable! See also Exercise 16 of Section 17. In this sense every positive measurable function f has a Lebesgue integral (finite or infinite) simply because, by Theorem 17.7, there exists a sequence $\{\phi_n\}$ of positive simple function such that $\phi_n \uparrow f$ a.e. holds.

Moreover, if $f : X \to \mathbb{R}^*$ defines a measurable function, then we can write $f = f^+ - f^-$ and (by the above) both integrals $\int f^+ \, d\mu$ and $\int f^- \, d\mu$ exist as (non-negative) extended real numbers. If one of them is a real number, then the integral of f is defined to be the extended real number

$$\int f \, d\mu = \int f^+ \, d\mu - \int f^- \, d\mu.$$

In this manner, we can assign an "integral" to a much larger class of measurable extended real-valued functions.

One advantage of the above extension of the integral is that a number of theorems can be phrased without Lebesgue integrability assumptions on the functions. For instance, Fatou's lemma can be stated as follows:

- *If $\{f_n\}$ is a sequence of measurable functions satisfying $f_n \geq 0$ a.e. for each n, then*

$$\int \liminf f_n \, d\mu \leq \liminf \int f_n \, d\mu$$

holds—where, of course, one or both sides of the inequality may be infinite.

EXERCISES

1. Show by a counterexample that the integrable functions do not form an algebra.
2. Let X be a nonempty set, and let δ be the Dirac measure on X with respect to the point a (see Example 13.4). Show that every function $f : X \to \mathbb{R}$ is integrable and that $\int f \, d\delta = f(a)$.
3. Let μ be the counting measure on $\mathbb{N}$ (see Example 13.3). Show that a function $f : \mathbb{N} \to \mathbb{R}$ is integrable if and only if $\sum_{n=1}^{\infty} |f(n)| < \infty$. Also, show that in this case $\int f \, d\mu = \sum_{n=1}^{\infty} f(n)$.
4. Show that a measurable function f is integrable if and only if $|f|$ is integrable. Give an example of a nonintegrable function whose absolute value is integrable.
5. Let f be an integrable function, and let $\{E_n\}$ be a sequence of disjoint measurable subsets of X. If $E = \bigcup_{n=1}^{\infty} E_n$, then show that

$$\int_E f \, d\mu = \sum_{n=1}^{\infty} \int_{E_n} f \, d\mu.$$

6. Let f be an integrable function. Show that for each $\epsilon > 0$ there exists some $\delta > 0$ (depending on ϵ) such that $|\int_E f \, d\mu| < \epsilon$ holds for all measurable sets with $\mu^*(E) < \delta$. [HINT: Note that $|f| \wedge n \uparrow |f|$.]
7. Show that for every integrable function f the set $\{x \in X : f(x) \neq 0\}$ can be written as a countable union of measurable sets of finite measure—referred to as a σ-**finite set**.
8. Let $f : \mathbb{R} \to \mathbb{R}$ be integrable with respect to the Lebesgue measure. Show that the function $g : [0, \infty) \to \mathbb{R}$ defined by

$$g(t) = \sup \left\{ \int |f(x + y) - f(x)| \, d\lambda(x) : |y| \leq t \right\}$$

for $t \geq 0$ is continuous at $t = 0$. [HINT: Use Theorem 22.12.]
9. Let g be an integrable function and let $\{f_n\}$ be a sequence of integrable functions such that $|f_n| \leq g$ a.e. holds for all n. Show that if $f_n \xrightarrow{\mu} f$, then f is an integrable function and $\lim \int |f_n - f| \, d\mu = 0$ holds. [HINT: Combine Theorem 19.4 with the Lebesgue dominated convergence theorem.]
10. Establish the following generalization of Theorem 22.9: If $\{f_n\}$ is a sequence of integrable functions such that $\sum_{n=1}^{\infty} \int |f_n| \, d\mu < \infty$, then $\sum_{n=1}^{\infty} f_n$ defines an integrable function and

$$\int \left(\sum_{n=1}^{\infty} f_n \right) d\mu = \sum_{n=1}^{\infty} \int f_n \, d\mu.$$

[HINT: By Theorem 22.9, the series $g = \sum_{n=1}^{\infty} |f_n|$ defines an integrable function and $|\sum_{n=1}^{k} f_n| \leq g$ a.e. holds for each k. Now, use the Lebesgue dominated convergence theorem.]
11. Let f be a positive (a.e.) measurable function, and let

$$e_i = \mu^*(\{x \in X : 2^{i-1} < f(x) \leq 2^i\})$$

for each integer i. Show that f is integrable if and only if $\sum_{i=-\infty}^{\infty} 2^i e_i < \infty$.

12. Let $\{f_n\}$ be a sequence of integrable functions such that $0 \le f_{n+1} \le f_n$ a.e. holds for each n. Then show that $f_n \downarrow 0$ a.e. holds if and only if $\int f_n \, d\mu \downarrow 0$.

13. Let f be an integrable function such that $f(x) > 0$ holds for almost all x. If A is a measurable set such that $\int_A f \, d\mu = 0$, then show that $\mu^*(A) = 0$.

14. Let $(X, \mathcal{S}, \mu)$ be a finite measure space and let $f : X \to \mathbb{R}$ be an integrable function satisfying $f(x) > 0$ for almost all x. If $0 < \varepsilon \le \mu^*(X)$, then show that

$$\inf\left\{\int_E f \, d\mu : E \in \Lambda_\mu \text{ and } \mu^*(E) \ge \varepsilon\right\} > 0.$$

15. Let f be a positive integrable function. Define $\nu : \Lambda \to [0, \infty)$ by $\nu(A) = \int_A f \, d\mu$ for each $A \in \Lambda$. Show that:

   a. (X, Λ, ν) is a measure space.
   b. If Λ_ν denotes the σ-algebra of all ν-measurable subsets of X, then $\Lambda \subseteq \Lambda_\nu$. Give an example for which $\Lambda \ne \Lambda_\nu$.
   c. If $\mu^*(\{x \in X : f(x) = 0\}) = 0$, then $\Lambda = \Lambda_\nu$.
   d. If g is an integrable function with respect to the measure space (X, Λ, ν), then fg is integrable with respect to the measure space $(X, \mathcal{S}, \mu)$ and

$$\int g \, d\nu = \int gf \, d\mu.$$

16. (A Change of Variable Formula) Let I be an interval of $\mathbb{R}$, and let $f : I \to \mathbb{R}$ be an integrable function with respect to the Lebesgue measure. For a pair of real numbers a and b with $a \ne 0$, let $J = \{(x - b)/a : x \in I\}$. Show that the function $g : J \to \mathbb{R}$ defined by $g(x) = f(ax+b)$ for $x \in J$ is integrable and that $\int_I f \, d\lambda = |a| \int_J g \, d\lambda$ holds. [HINT: Use Theorem 22.12.]

17. Let $(X, \mathcal{S}, \mu)$ be a finite measure space. For every pair of measurable functions f and g let

$$d(f, g) = \int \frac{|f - g|}{1 + |f - g|} \, d\mu.$$

   a. Show that $(\mathcal{M}, d)$ is a metric space.
   b. Show that a sequence $\{f_n\}$ of measurable functions (i.e., $\{f_n\} \subseteq \mathcal{M}$) satisfies $f_n \xrightarrow{\mu} f$ if and only if $\lim d(f_n, f) = 0$.
   c. Show that $(\mathcal{M}, d)$ is a complete metric space. That is, show that if a sequence $\{f_n\}$ of measurable functions satisfies $d(f_n, f_m) \to 0$ as $n, m \to \infty$, then there exists a measurable function f such that $\lim d(f_n, f) = 0$.

18. Let $f : \mathbb{R} \to \mathbb{R}$ be a Lebesgue integrable function. For each finite interval I let $f_I = \frac{1}{\lambda(I)} \int_I f \, d\lambda$ and $E_I = \{x \in I : f(x) > f_I\}$. Show that

$$\int_I |f - f_I| \, d\lambda = 2 \int_{E_I} (f - f_I) \, d\lambda.$$

19. Let $f : [0, \infty) \to \mathbb{R}$ be a Lebesgue integrable function such that $\int_0^t f(x) \, d\lambda(x) = 0$ for each $t \ge 0$. Show that $f(x) = 0$ holds for almost all x.

20. Let $(X, \mathcal{S}, \mu)$ be a measure space and let $f, f_1, f_2, \ldots$ be non-negative integrable functions satisfying $f_n \to f$ a.e. and $\lim \int f_n \, d\mu = \int f \, d\mu$. If E is a measurable set, then show that

$$\lim_{n \to \infty} \int_E f_n \, d\mu = \int_E f \, d\mu.$$

21. If a Lebesgue integrable function $f : [0, 1] \to \mathbb{R}$ satisfies $\int_0^1 x^{2n} f(x) \, d\lambda(x) = 0$ for each $n = 0, 1, 2, \ldots$, then show that $f = 0$ a.e.
    [HINT: Since the algebra generated by $\{1, x^2\}$ is uniformly dense in $C[0, 1]$, we have $\int_0^1 g(x) f(x) \, dx = 0$ for each $g \in C[0, 1]$.]

22. For each n consider the partition $\{0, 2^{-n}, 2 \cdot 2^{-n}, 3 \cdot 2^{-n}, \ldots, (2^n - 1) \cdot 2^{-n}, 1\}$ of the interval $[0, 1]$ and define the function $r_n : [0, 1] \to \mathbb{R}$ by $r_n(1) = -1$, and

$$r_n(x) = (-1)^{k-1} \text{ for } (k-1)2^{-n} \leq x < k2^{-n} \text{ and each } k = 1, 2, \ldots, 2^n.$$

    a. Draw the graphs of r_1 and r_2.
    b. Show that if $f : [0, 1] \to \mathbb{R}$ is a Lebesgue integrable function, then

$$\lim_{n \to \infty} \int_0^1 r_n(x) f(x) \, d\lambda(x) = 0.$$

    [HINT: For (b) use Theorem 22.12.]

23. Let $\{\epsilon_n\}$ be a sequence of real numbers such that $0 < \epsilon_n < 1$ for each n. Also, let us say that a sequence $\{A_n\}$ of Lebesgue measurable subsets of $[0, 1]$ is *consistent with the sequence* $\{\epsilon_n\}$ if $\lambda(A_n) = \epsilon_n$ for each n. Establish the following properties of $\{\epsilon_n\}$:

    a. The sequence $\{\epsilon_n\}$ converges to zero if and only if there exists a consistent sequence $\{A_n\}$ of measurable subsets of $[0, 1]$ such that $\sum_{n=1}^\infty \chi_{A_n}(x) < \infty$ for almost all x.
    b. The series $\sum_{n=1}^\infty \epsilon_n$ converges in $\mathbb{R}$ if and only if for each consistent sequence $\{A_n\}$ of measurable subsets of $[0, 1]$ we have $\sum_{n=1}^\infty \chi_{A_n}(x) < \infty$ for almost all x.

24. Let $(X, \mathcal{S}, \mu)$ be a finite measure space and let $f : X \to \mathbb{R}$ be a measurable function.

    a. Show that if f^n is integrable for each n and $\lim \int f^n \, d\mu$ exists in $\mathbb{R}$, then $|f(x)| \leq 1$ holds for almost all x.
    b. If f^n is integrable for each n, then show that $\int f^n \, d\mu = c$ (a constant) for $n = 1, 2, \ldots$ if and only if $f = \chi_A$ for some measurable subset A of X.

23. THE RIEMANN INTEGRAL AS A LEBESGUE INTEGRAL

It will be shown in this section that the Lebesgue integral is a generalization of the Riemann[5] integral. We start by reviewing the definition of the Riemann integral.

---

[5]Georg Friedrich Bernhard Riemann (1826–1866), a German mathematician, one of the greatest mathematicians of all time. Although his life was short, his contributions left the impact of a real genius. He made path-breaking contributions to the theory of complex functions, space geometry, and mathematical physics.

For simplicity, the details will be given for functions of one variable, and at the end it will be indicated how to carry out the same results for functions of several variables. Unless otherwise specified, throughout our discussion, f will be a fixed bounded real-valued function on a closed interval $[a, b]$ of $\mathbb{R}$.

A collection of points $P = \{x_0, x_1, \ldots, x_n\}$ is called a **partition** of $[a, b]$ if

$$a = x_0 < x_1 < \cdots < x_n = b$$

holds. Every partition $P = \{x_0, x_1, \ldots, x_n\}$ divides $[a, b]$ into the n closed subintervals

$$[x_0, x_1], [x_1, x_2], \ldots, [x_{n-1}, x_n].$$

The length of the largest of these subintervals is called the **mesh** of P and is denoted by $|P|$; that is, $|P| = \max\{x_i - x_{i-1} : i = 1, \ldots, n\}$. A partition P is said to be **finer** than another partition Q if $Q \subseteq P$ holds. If P and Q are partitions, then $P \cup Q$ is also a partition that is finer than both P and Q.

For a partition $P = \{x_0, x_1, \ldots, x_n\}$ of $[a, b]$, let

$$m_i = \inf\{f(x) : x \in [x_{i-1}, x_i]\} \quad \text{and} \quad M_i = \sup\{f(x) : x \in [x_{i-1}, x_i]\}$$

for each $i = 1, \ldots, n$. Then the **lower sum** $S_*(f, P)$ of f corresponding to the partition P is defined by

$$S_*(f, P) = \sum_{i=1}^{n} m_i(x_i - x_{i-1}),$$

and similarly, the **upper sum** $S^*(f, P)$ of f by

$$S^*(f, P) = \sum_{i=1}^{n} M_i(x_i - x_{i-1}).$$

Clearly, $S_*(f, P) \leq S^*(f, P)$ holds for every partition P of $[a, b]$.

Lemma 23.1. *If a partition P is finer than another partition Q (i.e., $Q \subseteq P$), then*

$$S_*(f, Q) \leq S_*(f, P) \quad and \quad S^*(f, P) \leq S^*(f, Q).$$

Proof. We show that $S_*(f, Q) \leq S_*(f, P)$ holds. The other inequality can be proven in a similar manner.

To establish the inequality, it is enough to assume that P has only one more point than Q, say t. So, let $Q = \{x_0, x_1, \ldots, x_n\}$ and $P = Q \cup \{t\}$. Then there exists some k $(1 \le k \le n)$ such that $x_{k-1} < t < x_k$, and thus, $P = \{x_0, x_1, \ldots, x_{k-1}, t, x_k, \ldots, x_n\}$. Let $c_1 = \inf\{f(x) : x \in [x_{k-1}, t]\}$ and $c_2 = \inf\{f(x) : x \in [t, x_k]\}$. Observe that both $m_k \le c_1$ and $m_k \le c_2$ hold. Therefore,

$$S_*(f, Q) = \sum_{i=1}^{n} m_i(x_i - x_{i-1}) = \sum_{i \ne k} m_i(x_i - x_{i-1}) + m_k(x_k - x_{k-1})$$

$$\le \sum_{i \ne k} m_i(x_i - x_{i-1}) + c_1(t - x_{k-1}) + c_2(x_k - t)$$

$$= S_*(f, P)$$

holds, and the proof is finished.   ∎

Lemma 23.2.   *For every pair of partitions P and Q, we have*

$$S_*(f, P) \le S^*(f, Q).$$

Proof.   From Lemma 23.1, it follows that

$$S_*(f, P) \le S_*(f, P \cup Q) \le S^*(f, P \cup Q) \le S^*(f, Q),$$

as claimed.   ∎

The preceding lemma states that every upper sum is an upper bound for the collection of all lower sums of f, and similarly, every lower sum is a lower bound for the collection of all upper sums.

Thus, if the **lower Riemann integral** of f is defined by

$$I_*(f) = \sup\{S_*(f, P) : P \text{ is a partition of } [a, b]\},$$

and the **upper Riemann integral** of f by

$$I^*(f) = \inf\{S^*(f, P) : P \text{ is a partition of } [a, b]\},$$

then

$$S_*(f, P) \le I_*(f) \le I^*(f) \le S^*(f, Q)$$

holds for every pair of partitions P and Q of $[a, b]$.

Definition 23.3. *A bounded function $f : [a, b] \to \mathbb{R}$ is called* **Riemann inte-grable** *if $I_*(f) = I^*(f)$. In this case, the common value is called the* **Riemann integral** *of f and is denoted by the classical symbol $\int_a^b f(x)\,dx$.*

Historical Note: Riemann's definition of the integral is a generalization of Eudoxus' method of exhaustion as was used by Archimedes[6] in his computation of the area of a circle. Interestingly, the value of the limit of the areas of the inscribed (or circumscribed) polygons that were employed by Archimedes to compute the area of the circle, was also called by him the integral ($\tau\grave{o}\ \pi\grave{\alpha}\nu$). It should be historically correct to call the Riemann integral the Eudoxus–Archimedes integral or the Eudoxus–Archimedes–Riemann integral (or even the Archimedes–Riemann integral).

A characterization for the Riemann integrability of a function, known as *Riemann's criterion*, is presented next.

Theorem 23.4 (Riemann's Criterion). *A bounded function $f : [a, b] \to \mathbb{R}$ is Riemann integrable if and only if for every $\epsilon > 0$ there exists a partition P of $[a, b]$ such that $S^*(f, P) - S_*(f, P) < \epsilon$ holds.*

Proof. Assume that f is Riemann integrable and let $\epsilon > 0$. Then let $I = \int_a^b f(x)\,dx$. Then there exist two partitions P_1 and P_2 of $[a, b]$ such that $I - S_*(f, P_1) < \epsilon$ and $S^*(f, P_2) - I < \epsilon$. Then (by Lemma 23.1), the partition $P = P_1 \cup P_2$ satisfies

$$
\begin{aligned}
S^*(f, P) - S_*(f, P) &\le S^*(f, P_2) - S_*(f, P_1) \\
&= [S^*(f, P_2) - I] + [I - S_*(f, P_1)] \\
&< \epsilon + \epsilon = 2\epsilon.
\end{aligned}
$$

---

[6]Archimedes of Syracuse (287–212 BC), a Greek mathematician and inventor. He was the most celebrated mathematician of antiquity and perhaps the best mathematician of all times. He used Eudoxus' method of exhaustion to compute areas and volumes very successfully. In his classic work, *The Measurement of the Circle*, he established that a circle has the same area as a right triangle having one leg equal to the radius of the circle and the other equal to the circumference of the circle, and that the volume of a sphere is four times the volume of a right cone with radius and height equal to the radius of the sphere.

Conversely, if the condition is satisfied, then since

$$0 \leq I^*(f) - I_*(f) \leq S^*(f, P) - S_*(f, P)$$

holds for every partition P of $[a, b]$, we have $0 \leq I^*(f) - I_*(f) < \epsilon$ for all $\epsilon > 0$. Hence, $I^*(f) - I_*(f) = 0$, or $I_*(f) = I^*(f)$, which shows that f is Riemann integrable. ∎

Now, let $P = \{x_0, x_1, \ldots, x_n\}$ be a partition of $[a, b]$. A collection of points $T = \{t_1, \ldots, t_n\}$ is said to be a **selection of points** for P if $x_{i-1} \leq t_i \leq x_i$ holds for $i = 1, \ldots, n$. We write

$$R_f(P, T) = \sum_{i=1}^{n} f(t_i)(x_i - x_{i-1}),$$

and call it (as usual) a **Riemann sum** associated with the partition P.

The following theorem of J.-C. Darboux[7] presents some powerful approximation formulas for the Riemann integral. The theorem can be viewed as an abstract formulation of Eudoxus' exhaustion method.

Theorem 23.5 (Darboux). *Let $f : [a, b] \to \mathbb{R}$ be Riemann integrable, and let $\{P_n\}$ be a sequence of partitions of $[a, b]$ such that $\lim |P_n| = 0$. Then*

$$\lim_{n \to \infty} S_*(f, P_n) = \lim_{n \to \infty} S^*(f, P_n) = \int_a^b f(x)\, dx.$$

In particular, if a sequence of partitions $\{P_n\}$ satisfies $\lim |P_n| = 0$ and T_n is a selection of points for P_n, then

$$\lim_{n \to \infty} R_f(P_n, T_n) = \int_a^b f(x)\, dx.$$

Proof. Choose a constant $c > 0$ such that $|f(x)| < c$ holds for all $x \in [a, b]$. Let $\epsilon > 0$. By Theorem 23.4, there exists a partition $P = \{x_0, x_1, \ldots, x_m\}$ of $[a, b]$ such that $S^*(f, P) - S_*(f, P) < \epsilon$. Choose n_0 such that

$$|P_n| < \frac{\epsilon}{2cm} \quad \text{and} \quad |P_n| < \min\{x_1 - x_0, x_2 - x_1, \ldots, x_m - x_{m-1}\}$$

---

[7]Jean-Gastin Darboux (1842–1917), a French mathematician. He was a geometer who used his geometric intuition to solve various problems in analysis and differential equations.

for all $n \geq n_0$. Fix $n \geq n_0$, and let $P_n = \{t_0, t_1, \ldots, t_k\}$. Put

$$
\begin{aligned}
M_j^P &= \sup\{f(x) : x \in [x_{j-1}, x_j]\} \quad \text{for } j = 1, 2, \ldots, m, \\
M_i &= \sup\{f(x) : x \in [t_{i-1}, t_i]\} \quad \text{for } i = 1, \ldots, k.
\end{aligned}
$$

The definitions of m_j^P and m_i are analogous (replace the sups by infs). Then

$$
0 \leq \int_a^b f(x)\, dx - S_*(f, P_n) \leq S^*(f, P_n) - S_*(f, P_n)
$$

$$
= \sum_{i=1}^k (M_i - m_i)(t_i - t_{i-1}) = V + W,
$$

where V is the sum of the terms $(M_i - m_i)(t_i - t_{i-1})$ for which $[t_{i-1}, t_i]$ lies entirely in some subinterval of the partition P, and W is the sum of the remaining terms. The sums V and W are estimated separately.

We estimate V first. Note that $V = \Sigma_1 + \cdots + \Sigma_m$, where each Σ_j is the sum of the terms $(M_i - m_i)(t_i - t_{i-1})$ for which $[t_{i-1}, t_i] \subseteq [x_{j-1}, x_j]$ holds. But if $[t_{i-1}, t_i] \subseteq [x_{j-1}, x_j]$, then $M_i - m_i \leq M_j^P - m_j^P$ holds. Also, the sum of the lengths of those subintervals of the partition P_n that lie inside $[x_{j-1}, x_j]$ never exceeds $x_j - x_{j-1}$. Thus, $\Sigma_j \leq (M_j^P - m_j^P)(x_j - x_{j-1})$ holds, and hence

$$
V \leq \sum_{j=1}^m \left(M_j^P - m_j^P \right)(x_j - x_{j-1}) = S^*(f, P) - S_*(f, P) < \epsilon.
$$

Now, we estimate W. Let $(M_i - m_i)(t_i - t_{i-1})$ be a term of the sum W. Since $|P_n| < \min\{x_j - x_{j-1} : j = 1, \ldots, m\}$, there exists exactly one j with $1 \leq j \leq n$ such that $x_{j-1} < t_{i-1} < x_j < t_i < x_{j+1}$. Thus, the sum W has at most m terms, and since

$$
(M_i - m_i)(t_i - t_{i-1}) \leq 2c|P_n| < 2c \cdot \frac{\epsilon}{2cm} = \frac{\epsilon}{m}
$$

it follows that $W < m(\epsilon/m) = \epsilon$. Thus,

$$
0 \leq \int_a^b f(x)\, dx - S_*(f, P_n) \leq V + W < \epsilon + \epsilon = 2\epsilon
$$

for all $n \geq n_0$. That is, $\lim S_*(f, P_n) = \int_a^b f(x)\, dx$.

Similarly, $0 \leq S^*(f, P_n) - \int_a^b f(x)\, dx \leq V + W < 2\epsilon$ holds for all $n \geq n_0$, and so, $\lim S^*(f, P_n) = \int_a^b f(x)\, dx$. The last part follows immediately from the inequalities

$$S_*(f, P_n) \le R_f(P_n, T_n) \le S^*(f, P_n).$$

The proof of the theorem is now complete.   ∎

We are now ready to establish that the Lebesgue integral is a generalization of the Riemann integral. Here, "f is Lebesgue integrable" means that f is integrable with respect to the Lebesgue measure.

Theorem 23.6.   *Every Riemann integrable function $f : [a, b] \to \mathbb{R}$ is Lebesgue integrable, and in this case the two integrals coincide. That is,*

$$\int f \, d\lambda = \int_a^b f(x) \, dx.$$

Proof.   For each n let $P_n = \{x_0, x_1, \ldots, x_{2^n}\}$ be the partition that divides $[a, b]$ into 2^n subintervals all of the same length $(b - a)2^{-n}$; that is, $x_i = a + i(b - a)2^{-n}$. Let

$$\phi_n = \sum_{i=1}^{2^n} m_i \chi_{[x_{i-1}, x_i)} \quad \text{and} \quad \psi_n = \sum_{i=1}^{2^n} M_i \chi_{[x_{i-1}, x_i)},$$

where $m_i = \inf\{f(x) : x \in [x_{i-1}, x_i]\}$ and $M_i = \sup\{f(x) : x \in [x_{i-1}, x_i]\}$. Clearly, $\{\phi_n\}$ and $\{\psi_n\}$ are two sequences of step functions satisfying the propeties $\phi_n(x) \uparrow \le f(x) \le \psi_n(x) \downarrow$ for each $x \in [a, b)$.

Now, if $\phi_n(x) \uparrow g(x)$ and $\psi_n(x) \downarrow h(x)$, then by Theorem 22.6, both functions g and h are Lebesgue integrable and $g(x) \le f(x) \le h(x)$ holds for all $x \in [a, b)$. Also, by definition, $\int \phi_n \, d\lambda = S_*(f, P_n)$ and $\int \psi_n \, d\lambda = S^*(f, P_n)$. Therefore, since $\psi_n(x) - \phi_n(x) \downarrow h(x) - g(x) \ge 0$, it follows that

$$0 \le \int (h - g) \, d\lambda = \lim_{n \to \infty} \int (\psi_n - \phi_n) \, d\lambda = \lim_{n \to \infty} \int \psi_n \, d\lambda - \lim_{n \to \infty} \int \phi_n \, d\lambda$$

$$= \lim_{n \to \infty} S^*(f, P_n) - \lim_{n \to \infty} S_*(f, P_n) = 0,$$

where the last equality holds true by virtue of Theorem 23.5. This implies $h - g = 0$ a.e., and hence, $h = g = f$ a.e. holds. In particular, $\phi_n \uparrow f$ a.e. and $\psi_n \downarrow f$ a.e. hold, which show that f is Lebesgue integrable—in fact, an upper function. Finally,

$$\int f \, d\lambda = \lim_{n \to \infty} \int \phi_n \, d\lambda = \lim_{n \to \infty} S_*(f, P_n) = \int_a^b f(x) \, dx.$$

and the proof of the theorem is finished.   ∎

The next theorem, due to H. Lebesgue and G. Vitali, characterizes the Riemann integrable functions in terms of their discontinuities. (The almost everywhere relations are considered with respect to the Lebesgue measure.)

Theorem 23.7 (Lebesgue–Vitali). *A bounded function* $f : [a, b] \to \mathbb{R}$ *is Riemann integrable if and only if it is continuous almost everywhere.*

Proof. For each n, let P_n, ϕ_n, and ψ_n be as they were introduced in the proof of Theorem 23.6.

Assume first that f is Riemann integrable. Then a glance at the proof of Theorem 23.6 guarantees the existence of a (Lebesgue) null subset A of $[a, b]$ such that $\phi_n(x) \uparrow f(x)$ and $\psi_n(x) \downarrow f(x)$ for all $x \notin A$. Clearly, $D = A \cup (\bigcup_{n=1}^{\infty} P_n)$ has Lebesgue measure zero, and we claim that f is continuous on $[a, b] \setminus D$.

To see this, let $s \in [a, b] \setminus D$ and $\epsilon > 0$. Pick some n with $f(s) - \phi_n(s) < \epsilon$ and $\psi_n(s) - f(s) < \epsilon$. Then there exists some subinterval $[x_{i-1}, x_i]$ of the partition P_n such that $s \in (x_{i-1}, x_i)$. Clearly, $\phi_n(s) = m_i$ and $\psi_n(s) = M_i$. Therefore, if $x \in (x_{i-1}, x_i)$, then

$$-\epsilon < m_i - f(s) \leq f(x) - f(s) \leq M_i - f(s) < \epsilon.$$

Since (x_{i-1}, x_i) is a neighborhood of s, the last inequality shows that f is continuous at s. This establishes that f is continuous almost everywhere.

For the converse, assume that f is continuous almost everywhere. Let $s \neq b$ be a point of continuity of f. If $\epsilon > 0$ is given, then choose some $\delta > 0$ such that

$$f(s) - \epsilon < f(x) < f(s) + \epsilon \qquad\qquad (\star)$$

for all $x \in [a, b]$ with $|x - s| < \delta$. Pick some k so that $|P_k| < \delta$. Then for some subinterval $[x_{i-1}, x_i]$ of P_k, we must have $s \in [x_{i-1}, x_i)$. In particular, $|x - s| < \delta$ must hold for all $x \in [x_{i-1}, x_i]$, and so, from $(\star)$ we get

$$f(s) - \epsilon \leq m_i = \phi_k(s) < f(s) + \epsilon.$$

Since $\phi_n(s) \uparrow$, it easily follows that $\phi_n(s) \uparrow f(s)$. Similarly, $\psi_n(s) \downarrow f(s)$ holds.

Since f is continuous almost everywhere, we conclude that $\phi_n \uparrow f$ a.e. and $\psi_n \downarrow f$ a.e. both hold. This shows that f is Lebesgue integrable. Moreover, by the Lebesgue dominated convergence theorem we have

$$S_*(f, P_n) = \int \phi_n \, d\lambda \uparrow \int f \, d\lambda \quad \text{and} \quad S^*(f, P_n) = \int \psi_n \, d\lambda \downarrow \int f \, d\lambda.$$

Thus, $\lim[S^*(f, P_n) - S_*(f, P_n)] = 0$, and so, by Theorem 23.3, the function f is Riemann integrable. The proof of the theorem is now complete. ∎

An immediate consequence of Theorem 23.7 is the following:

Theorem 23.8.   *The collection of all Riemann integrable functions on a closed interval is a function space and an algebra of functions.*

It is easy to present examples of bounded Lebesgue integrable functions that are not Riemann integrable. Here is an example:

Example 23.9.   Let $f : [0, 1] \to \mathbb{R}$ be defined by $f(x) = 0$ if x is a rational number and $f(x) = 1$ if x is irrational (in other words, f is the characteristic function of the irrationals of $[0, 1]$). Then f is discontinuous at every point of $[0, 1]$, and thus, by Theorem 23.7, f is not Riemann integrable. On the other hand, $f = \mathbf{1}$ a.e. holds (since the set of rational numbers has Lebesgue measure zero), and so, f is Lebesgue integrable. Also, note that $\int f \, d\lambda = 1$ holds. ∎

It follows from Theorem 23.7 that if a function $f : [a, b] \to \mathbb{R}$ is Riemann integrable, then f restricted to any closed subinterval of $[a, b]$ is also Riemann integrable there. Moreover, by the same theorem, if two functions f and g are Riemann integrable on $[a, c]$ and $[c, b]$, then the function $h : [a, b] \to \mathbb{R}$, defined by $h(x) = f(x)$ if $x \in [a, c]$, and $h(x) = g(x)$ if $x \in (c, b]$, is Riemann integrable.

Clearly, by Theorem 23.7, every continuous function on a closed interval is Riemann integrable. To compute the Riemann (and hence, the Lebesgue) integral of a continuous function, one usually uses the fundamental theorem of calculus, one form of which is stated next. Since any "reasonable" calculus book contains a proof of this important result, its proof is omitted. (See also Exercise 6 at the end of the section.) The fundamental theorem of calculus is due to I. Newton[8] and independently to G. Leibniz.[9]

Theorem 23.10 (The Fundamental Theorem of Calculus).   *For a continuous function $f : [a, b] \to \mathbb{R}$ we have the following:*

---

[8]Isaac Newton (1642–1727), a great British mathematician, physicist, astronomer, and philosopher. He discovered the law of gravity and was one of the founders of calculus. His pioneering original contributions to mathematics and science revolutionized the modern scientific approach.

[9]Gottfried Wilhelm Leibniz (1646–1716), a prominent German mathematician and philosopher. He was a person with "universal" scientific interests. Besides his philosophical and metaphysical contributions, he contributed substantially to mathematics, mathematical logic, and physics. Together with Newton, he is considered the founder of calculus.

1. If $A : [a, b] \to \mathbb{R}$ is an **area function** of f (i.e., $A(x) = \int_c^x f(t)\, dt$ holds for all $x \in [a, b]$ and some fixed $c \in [a, b]$), then A is an antiderivative of f. That is, $A'(x) = f(x)$ holds for each $x \in [a, b]$.

2. If $F : [a, b] \to \mathbb{R}$ is an antiderivative of f, i.e., $F'(x) = f(x)$ holds for each $x \in [a, b]$, then

$$\int_a^b f(x)\, dx = F(b) - F(a).$$

In a conventional way, the integral $\int_b^a f(x)\, dx$ is defined to be $-\int_a^b f(x)\, dx$; that is, $\int_b^a f(x)\, dx = -\int_a^b f(x)\, dx$. Also, $\int_a^a f(x)\, dx$ is defined to be zero. By doing so, the useful identity

$$\int_c^d f(x)\, dx = \int_c^e f(x)\, dx + \int_e^d f(x)\, dx$$

holds regardless of the ordering between the points c, d, and e of $[a, b]$.

We now indicate how to extend the above results to functions of several variables. In the general case, the interval $[a, b]$ is replaced by a **cell** $J = [a_1, b_1] \times \cdots \times [a_n, b_n]$, and its Lebesgue measure is $\lambda(J) = \prod_{i=1}^n (b_i - a_i)$. A **partition** P of J is a set of points of the form $P = P_1 \times \cdots \times P_n$, where P_i is a partition of $[a_i, b_i]$ for each $i = 1, \dots, n$. Clearly, any partition P divides J into a finite number of subcells.

Now, if $f : J \to \mathbb{R}$ is a bounded function and the partition P divides J into the subcells $J_1, \dots, J_k$, then we define again the numbers

$$m_i = \inf\{ f(x_1, \dots, x_n) : (x_1, \dots, x_n) \in J_i \},$$
$$M_i = \sup\{ f(x_1, \dots, x_n) : (x_1, \dots, x_n) \in J_i \}.$$

The lower and upper sums corresponding to the partition P are defined as before by the formulas

$$S_*(f, P) = \sum_{i=1}^k m_i \lambda(J_i) \quad \text{and} \quad S^*(f, P) = \sum_{i=1}^k M_i \lambda(J_i),$$

respectively.

The **lower Riemann integral** of f is defined (as before) by

$$I_*(f) = \sup\{ S_*(f, P) : P \text{ is a partition of } J \},$$

and the **upper Riemann integral** of f by

$$I^*(f) = \inf\{S^*(f, P) : P \text{ is a partition of } J\}.$$

As in the one-dimensional case,

$$-\infty < I_*(f) \leq I^*(f) < \infty$$

holds. The function f is called **Riemann integrable** if $I_*(f) = I^*(f)$. This common number is called the **Riemann integral** of f and is denoted by

$$\int_{a_1}^{b_1} \int_{a_2}^{b_2} \cdots \int_{a_n}^{b_n} f(x_1, x_2, \ldots, x_n)\, dx_1 dx_2 \cdots dx_n.$$

All the results given in this section are valid in this general setting. Their proofs parallel the ones presented here, and for this reason we leave them as an exercise for the reader.

EXERCISES

1. Let $f : [a, b] \to \mathbb{R}$ be Riemann integrable. Show that f is Riemann integrable on every closed subinterval of $[a, b]$. Also, show that

$$\int_c^d f(x)\, dx = \int_c^e f(x)\, dx + \int_e^d f(x)\, dx$$

holds for every three points c, d, and e of $[a, b]$.

2. Let $f : [a, b] \to \mathbb{R}$ be Riemann integrable. Then show that

$$\int_a^b f(x)\, dx = \lim_{n \to \infty} \frac{b - a}{n} \sum_{i=1}^{n} f\left(a + \frac{i(b - a)}{n}\right).$$

3. Let $\{f_n\}$ be a sequence of Riemann integrable functions on $[a, b]$ such that $\{f_n\}$ converges uniformly to a function f. Show that f is Riemann integrable and that

$$\lim_{n \to \infty} \int_a^b f_n(x)\, dx = \int_a^b f(x)\, dx.$$

4. For each n, let $f_n : [0, 1] \to \mathbb{R}$ be defined by $f_n(x) = \frac{nx^{n-1}}{1+x}$ for all $x \in [0, 1]$. Then show that $\lim \int_0^1 f_n(x)\, dx = \frac{1}{2}$.
   [HINT: Use integration by parts.]

5. Let $f : [a, b] \to \mathbb{R}$ be an increasing function. Show that f is Riemann integrable.
   [HINT: Verify that f satisfies Riemann's criterion.]

6. (*The Fundamental Theorem of Calculus*) If $f : [a, b] \to \mathbb{R}$ is a Riemann integrable function, define its **area function** $A : [a, b] \to \mathbb{R}$ by $A(x) = \int_a^x f(t)\, dt$ for each $x \in [a, b]$. Show that

    a. A is a uniformly continuous function.

    b.   If f is continuous at some point c of $[a, b]$, then A is differentiable at c and $A'(c) = f(c)$ holds.

    c.   Give an example of a Riemann integrable function f whose area function A is differentiable and satisfies $A' \neq f$.

    [HINT: For part (c) use the function defined in Exercise 7 of Section 9.]

7.   (**Arzelà**) Let $\{f_n\}$ be a sequence of Riemann integrable functions on $[a, b]$ such that $\lim f_n(x) = f(x)$ holds for each $x \in [a, b]$ and f is Riemann integrable. Also, assume that there exists a constant M such that $|f_n(x)| \leq M$ holds for all $x \in [a, b]$ and all n. Show that

$$\lim_{n \to \infty} \int_a^b f_n(x)\,dx = \int_a^b f(x)\,dx.$$

8.   Determine the lower and upper Riemann integrals for the function of Example 23.9.

9.   Let C be the Cantor set (see Example 6.15). Show that χ_C is Riemann integrable over $[0, 1]$, and that $\int_0^1 \chi_C\,dx = 0$.

10.   Let $0 < \epsilon < 1$, and consider the ϵ-Cantor set C_ϵ of $[0, 1]$. Show that χ_{C_ϵ} is not Riemann integrable over $[0, 1]$. Also, determine $I_*(\chi_{C_\epsilon})$ and $I^*(\chi_{C_\epsilon})$.
    [HINT: Show that the set of all discontinuities of χ_{C_ϵ} is C_ϵ.]

11.   Give a proof of the Riemann integrability of a continuous function based upon its uniform continuity (Theorem 7.7).

12.   Establish the familiar change of variable formula for the Riemann integral of continuous functions: *If* $[a, b] \xrightarrow{g} [c, d] \xrightarrow{f} \mathbb{R}$ *are continuous functions with* g *continuously differentiable (i.e.,* g *has a continuous derivative), then*

$$\int_a^b f(g(x))g'(x)\,dx = \int_{g(a)}^{g(b)} f(u)\,du.$$

13.   Let $f : [0, \infty) \to \mathbb{R}$ be a continuous function such that $\lim_{x \to \infty} f(x) = \delta$. Show that $\lim_{n \to \infty} \int_0^a f(nx)\,dx = a\delta$ for each $a > 0$.

14.   Let $f : [0, \infty) \to \mathbb{R}$ be a continuous function such that $f(x + 1) = f(x)$ for all $x \geq 0$. If $g : [0, 1] \to \mathbb{R}$ is an arbitrary continuous function, then show that

$$\lim_{n \to \infty} \int_0^1 g(x)f(nx)\,dx = \left(\int_0^1 g(x)\,dx \right) \cdot \left(\int_0^1 f(x)\,dx \right).$$

15.   Let $f : [0, 1] \to [0, \infty)$ be Riemann integrable on every closed subinterval of $(0, 1]$. Show that f is Lebesgue integrable over $[0, 1]$ if and only if $\lim_{\epsilon \downarrow 0} \int_\epsilon^1 f(x)\,dx$ exists in $\mathbb{R}$. Also, show that if this is the case, then $\int f\,d\lambda = \lim_{\epsilon \downarrow 0} \int_\epsilon^1 f(x)\,dx$.

16.   As an application of the preceding exercise, show that the function $f : [0, 1] \to \mathbb{R}$ defined by $f(x) = x^p$ if $x \in (0, 1]$ and $f(0) = 0$ is Lebesgue integrable if and only if $p > -1$. Also, show that if f is Lebesgue integrable, then

$$\int f\,d\lambda = \frac{1}{1 + p}.$$

17.   Let $f : [0, 1] \to \mathbb{R}$ be a function and define $g : [0, 1] \to \mathbb{R}$ by $g(x) = e^{f(x)}$.

a. Show that if f is measurable (or Borel measurable), then so is g.
b. If f is Lebesgue integrable, is then g necessarily Lebesgue integrable?
c. Give an example of an essentially unbounded function f which is continuous on $(0, 1]$ such that f^n is Lebesgue integrable for each $n = 1, 2, \ldots$. (A function f is "essentially unbounded," if for each $M > 0$ the set $\{x \in [0, 1] : f(x)| > M\}$ has positive measure.)

[HINT: For (b) consider the function $f(x) = x^{-\frac{1}{2}}$.]

18. Let $f : [0, 1] \to \mathbb{R}$ be Lebesgue integrable. Assume that f is differentiable at $x = 0$ and $f(0) = 0$. Show that the function $g : [0, 1] \to \mathbb{R}$ defined by $g(x) = x^{-\frac{3}{2}} f(x)$ for $x \in (0, 1]$ and $g(0) = 0$ is Lebesgue integrable.

19. Let $f : [a, b] \times [c, d] \to \mathbb{R}$ be a continuous function. Show that the Riemann integral of f can be computed with two iterated integrations. That is, show that

$$\int_a^b \int_c^d f(x, y)\,dx\,dy = \int_a^b \left[\int_c^d f(x, y)\,dy \right] dx = \int_c^d \left[\int_a^b f(x, y)\,dx \right] dy.$$

Generalize this to a continuous function of n variables.

20. Assume that $f : [a, b] \to \mathbb{R}$ and $g : [a, b] \to \mathbb{R}$ are two continuous functions such that $f(x) \le g(x)$ for $x \in [a, b]$. Let $A = \{(x, y) \in \mathbb{R}^2 : x \in [a, b]$ and $f(x) \le y \le g(x)\}$.

a. Show that A is a closed set—and hence, a measurable subset of $\mathbb{R}^2$.
b. If $h : A \to \mathbb{R}$ is a continuous function, then show that h is Lebesgue integrable over A and that

$$\int_A h\,d\lambda = \int_a^b \left[\int_{f(x)}^{g(x)} h(x, y)\,dy \right] dx.$$

21. Let $f : [a, b] \to \mathbb{R}$ be a differentiable function—with one-sided derivatives at the end-points. If the derivative f' is uniformly bounded on $[a, b]$, then show that f' is Lebesgue integrable and that

$$\int_{[a,b]} f'\,d\lambda = f(b) - f(a).$$

22. Let $f, g : [a, b] \to \mathbb{R}$ be two Lebesgue integrable functions satisfying

$$\int_a^x f(t)\,d\lambda(t) \le \int_a^x g(t)\,d\lambda(t)$$

for each $x \in [a, b]$. If $\phi : [a, b] \to \mathbb{R}$ is a non-negative decreasing function, then show that the functions ϕf and ϕg are both Lebesgue integrable over $[a, b]$ and that they satisfy

$$\int_a^x \phi(t) f(t)\,d\lambda(t) \le \int_a^x \phi(t) g(t)\,d\lambda(t)$$

for all $x \in [a, b]$.

[HINT: Prove it first for a decreasing function of the form $\phi = \sum_{i=1}^k c_i \chi_{[a_{i-1}, a_i)}$, where $a_0 < a_1 < \cdots < a_k$ is a partition of $[a, b]$, and then use the fact that there exists a sequence $\{\phi_n\}$ of such step functions satisfying $\phi_n(t) \uparrow \phi(t)$ for almost all t in $[a, b]$; see Exercise 8 of Section 21.]

24. APPLICATIONS OF THE LEBESGUE INTEGRAL

If a function $f : [a, \infty) \rightarrow \mathbb{R}$ (where, of course, $a \in \mathbb{R}$) is Riemann integrable on every closed subinterval of $[a, \infty)$, then its **improper Riemann integral** is defined by

$$\int_a^\infty f(x)\,dx = \lim_{r \to \infty} \int_a^r f(x)\,dx,$$

provided that the limit on the right-hand side exists in $\mathbb{R}$. The existence of the prior limit is also expressed by saying that the (improper Riemann) integral $\int_a^\infty f(x)\,dx$ exists. Similarly, if $f : (-\infty, a] \rightarrow \mathbb{R}$ is Riemann integrable on every closed subinterval of $(-\infty, a]$, then $\int_{-\infty}^a f(x)\,dx$ is defined by

$$\int_{-\infty}^a f(x)\,dx = \lim_{r \to -\infty} \int_r^a f(x)\,dx$$

whenever the limit exists in $\mathbb{R}$. It should be clear that if $\int_a^\infty f(x)\,dx$ exists, then $\int_b^\infty f(x)\,dx$ also exists for each $b > a$ and

$$\int_a^\infty f(x)\,dx = \int_a^b f(x)\,dx + \int_b^\infty f(x)\,dx.$$

Theorem 24.1. *Assume that $f : [a, \infty) \rightarrow \mathbb{R}$ is Riemann integrable on every closed subinterval of $[a, \infty)$. Then $\int_a^\infty f(x)\,dx$ exists if and only if for every $\epsilon > 0$ there exists some $M > 0$ (depending on ϵ) such that $|\int_s^t f(x)\,dx| < \epsilon$ for all $s, t \geq M$.*

Proof. Assume that $I = \int_a^\infty f(x)\,dx$ exists. Pick a real number $M > 0$ such that $|I - \int_a^r f(x)\,dx| < \epsilon$ holds for all $r \geq M$. If $s, t \geq M$, then

$$\left| \int_s^t f(x)\,dx \right| = \left| \int_a^t f(x)\,dx - \int_a^s f(x)\,dx \right|$$

$$\leq \left| I - \int_a^t f(x)\,dx \right| + \left| I - \int_a^s f(x)\,dx \right| < 2\epsilon.$$

Conversely, assume that the condition is satisfied. If $\{a_n\}$ is a sequence of $[a, \infty)$ such that $\lim a_n = \infty$, then it is easy to see that the sequence $\{\int_a^{a_n} f(x)\,dx\}$ is

Cauchy. Thus, $A = \lim \int_a^{a_n} f(x)\,dx$ exists in $\mathbb{R}$. Now, let $\{b_n\}$ be another sequence of $[a, \infty)$ with $\lim b_n = \infty$; let $B = \lim \int_a^{b_n} f(x)\,dx$. From the inequality

$$|A - B| \leq \left| A - \int_a^{a_n} f(x)\,dx \right| + \left| \int_{a_n}^{b_n} f(x)\,dx \right| + \left| B - \int_a^{b_n} f(x)\,dx \right|,$$

it is easy to see that $|A - B| < \epsilon$ holds for all $\epsilon > 0$. Thus, $A = B$, and so the limit is independent of the chosen sequence. This shows that $\int_a^\infty f(x)\,dx$ exists, and the proof is finished. ∎

From the preceding theorem, and the inequality $|\int_s^t f(x)\,dx| \leq \int_s^t |f(x)|\,dx$ for $s < t$, we have the following:

Lemma 24.2. *If a function $f : [a, \infty) \to \mathbb{R}$ is Riemann integrable on every closed subinterval of $[a, \infty)$ and $\int_a^\infty |f(x)|\,dx$ exists then $\int_a^\infty f(x)\,dx$ also exists and $|\int_a^\infty f(x)\,dx| \leq \int_a^\infty |f(x)|\,dx$.*

The converse of this lemma is false. That is, there are functions f whose improper Riemann integrals $\int_a^\infty f(x)\,dx$ exist but for which $\int_a^\infty |f(x)|\,dx$ fails to exist. A well-known example is provided by the function $f(x) = \frac{\sin x}{x}$ (with $f(0) = 1$) over $[0, \infty)$. We shall see later that $\int_0^\infty \frac{\sin x}{x}\,dx = \frac{\pi}{2}$ but $\int_0^\infty \frac{|\sin x|}{x}\,dx$ does not exist. To see the latter, note that

$$\int_{(k-1)\pi}^{k\pi} \frac{|\sin x|}{x}\,dx \geq \frac{1}{k\pi} \int_{(k-1)\pi}^{k\pi} |\sin x|\,dx = \frac{2}{k\pi}$$

holds for each k. Therefore,

$$\int_0^{n\pi} \frac{|\sin x|}{x}\,dx = \sum_{k=1}^n \int_{(k-1)\pi}^{k\pi} \frac{|\sin x|}{x}\,dx \geq \frac{2}{\pi} \sum_{k=1}^n \frac{1}{k},$$

which shows that $\int_0^\infty \frac{|\sin x|}{x}\,dx$ does not exist in $\mathbb{R}$.

If the improper Riemann integral of a function exists, then it is natural to ask whether the function is, in fact, Lebesgue integrable. In general, this is not the case. However, if the improper Riemann integral of the absolute value of the function exists, then the function is Lebesgue integrable. The details follow.

Theorem 24.3. *Let $f : [a, \infty) \to \mathbb{R}$ be Riemann integrable on every closed subinterval of $[a, \infty)$. Then f is Lebesgue integrable if and only if the improper*

Riemann integral $\int_a^\infty |f(x)|\, dx$ exists. Moreover, in this case

$$\int f\, d\lambda = \int_a^\infty f(x)\, dx.$$

Proof. Assume that f is Lebesgue integrable over $[a, \infty)$. Then f^+ is also Lebesgue integrable over $[a, \infty)$. Let $\{r_n\}$ be a sequence of $[a, \infty)$ such that $\lim r_n = \infty$. For each n, let $f_n(x) = f^+(x)$ if $x \in [a, r_n]$ and $f_n(x) = 0$ if $x > r_n$. Then $\lim f_n(x) = f^+(x)$ and $0 \le f_n(x) \le f^+(x)$ hold for all $x \in [a, \infty)$. Moreover, by Theorem 23.6, $\{f_n\}$ is a sequence of Lebesgue integrable functions such that $\int f_n\, d\lambda = \int_a^{r_n} f^+(x)\, dx$. Thus, by the Lebesgue dominated convergence theorem

$$\int f^+\, d\lambda = \lim_{n\to\infty} \int f_n\, d\lambda = \lim_{n\to\infty} \int_a^{r_n} f^+(x)\, dx.$$

This shows that $\int_a^\infty f^+(x)\, dx$ exists and $\int_a^\infty f^+(x)\, dx = \int f^+\, d\lambda$. Similarly, $\int_a^\infty f^-(x)\, dx$ exists and $\int_a^\infty f^-(x)\, dx = \int f^-\, d\lambda$. But then, it follows from $f = f^+ - f^-$ and $|f| = f^+ + f^-$ that both improper Riemann integrals $\int_a^\infty f(x)\, dx$ and $\int_a^\infty |f(x)|\, dx$ exist. Moreover,

$$\int_a^\infty f(x)\, dx = \int f\, d\lambda \quad \text{and} \quad \int_a^\infty |f(x)|\, dx = \int |f|\, d\lambda.$$

For the converse, assume that the improper Riemann integral $\int_a^\infty |f(x)|\, dx$ exists. Clearly, $\lim \int_a^{a+n} |f(x)|\, dx = \int_a^\infty |f(x)|\, dx$. Define $f_n(x) = |f(x)|$ if $x \in [a, a+n]$ and $f_n(x) = 0$ if $x > a+n$ for each n. Note that $0 \le f_n(x) \uparrow |f(x)|$ holds for all $x \ge a$. Since f_n is Riemann integrable on $[a, a+n]$, f_n is an upper function on $[a, \infty)$ satisfying $\int f_n\, d\lambda = \int_a^{a+n} f_n(x)\, dx = \int_a^{a+n} |f(x)|\, dx$. Thus, by Theorem 21.6, $|f|$ is an upper function (and hence, Lebesgue integrable) such that $\int |f|\, d\lambda = \int_a^\infty |f(x)|\, dx$. The Lebesgue integrability of f now follows from Theorem 22.6 by observing that f is a measurable function, since it is a measurable function on every closed subinterval of $[a, \infty)$. The proof of the theorem is now complete. ∎

The next result deals with interchanging the processes of limit and integration.

Theorem 24.4. *Let $(X, \mathcal{S}, \mu)$ be a measure space, let J be a subinterval of $\mathbb{R}$, and let $f : X \times J \to \mathbb{R}$ be a function such that $f(\cdot, t)$ is a measurable function for each $t \in J$. Assume also that there exists an integrable function g*

such that for each $t \in J$ we have $|f(x, t)| \le g(x)$ for almost all x. If for some accumulation point t_0 (including possibly $\pm\infty$) of J there exists a function h such that $\lim_{t \to t_0} f(x, t) = h(x)$ exists in $\mathbb{R}$ for almost all x, then h defines an integrable function, and

$$\lim_{t \to t_0} \int f(x, t) \, d\mu(x) = \int \lim_{t \to t_0} f(x, t) \, d\mu(x) = \int h \, d\mu.$$

Proof. Assume that the function $f : X \times J \to \mathbb{R}$ satisfies the hypotheses of the theorem. Let $\{t_n\}$ be a sequence of J such that $\lim t_n = t_0$. Put $h_n(x) = f(x, t_n)$ for $x \in X$, and note that $|h_n| \le g$ a.e. holds for each n and $h_n \to h$ a.e. By Theorem 22.6, each h_n is integrable. Moreover, by the Lebesgue dominated convergence theorem, h defines an integrable function and

$$\lim_{n \to \infty} \int f(x, t_n) \, d\mu(x) = \lim_{n \to \infty} \int h_n \, d\mu = \int h \, d\mu$$

holds, from which our conclusion follows. ∎

If $f : X \times (a, b) \to \mathbb{R}$ is a function and $t_0 \in (a, b)$, then its difference quotient function at t_0 is defined by

$$D_{t_0}(x, t) = \frac{f(x, t) - f(x, t_0)}{t - t_0}$$

for all $x \in X$ and all $t \in (a, b)$ with $t \ne t_0$. To make the function D_{t_0} defined everywhere, we let $D_{t_0}(x, t_0) = 0$ for all $x \in X$.

As usual, $\lim_{t \to t_0} D_{t_0}(x, t)$, whenever the limit exists, is called the **partial derivative** of f with respect to t at the point (x, t_0) and is denoted by $\frac{\partial f}{\partial t}(x, t_0)$. That is,

$$\frac{\partial f}{\partial t}(x, t_0) = \lim_{t \to t_0} D_{t_0}(x, t) = \lim_{t \to t_0} \frac{f(x, t) - f(x, t_0)}{t - t_0}.$$

The next result deals with differentiation of a function defined by an integral.

Theorem 24.5. *Let (X, S, μ) be a measure space and let $f : X \times (a, b) \to \mathbb{R}$ be a function such that $f(\cdot, t)$ is Lebesgue integrable for every $t \in (a, b)$. Assume that for some $t_0 \in (a, b)$ the partial derivative $\frac{\partial f}{\partial t}(x, t_0)$ exists for almost all x. Suppose also that there exists an integrable function g and a neighborhood V of t_0 such that for each $t \in V$ we have $|D_{t_0}(x, t)| \le g(x)$ for almost all $x \in X$. Then*

1. *$\frac{\partial f}{\partial t}(\cdot, t_0)$ defines an integrable function, and*

2. *the function* $F : (a, b) \to \mathbb{R}$ *defined by* $F(t) = \int f(x, t) \, d\mu(x)$ *is differentiable at* t_0 *and*

$$F'(t_0) = \int \frac{\partial f}{\partial t}(x, t_0) \, d\mu(x).$$

Proof. Put $\frac{\partial f}{\partial t}(x, t_0) = 0$ at each point x where the partial derivative does not exist. Then $\lim_{t \to t_0} D_{t_0}(x, t) = \frac{\partial f}{\partial t}(x, t_0)$ holds for almost all x. Thus, by Theorem 24.4, $\frac{\partial f}{\partial t}(\cdot, t_0)$ defines an integrable function and

$$\frac{F(t) - F(t_0)}{t - t_0} = \int \frac{f(x, t) - f(x, t_0)}{t - t_0} \, d\mu(x)$$

$$= \int D_{t_0}(x, t) \, d\mu(x) \longrightarrow \int \frac{\partial f}{\partial t}(x, t_0) \, d\mu(x)$$

as $t \to t_0$. This shows that F is differentiable at t_0 and that $F'(t_0) = \int \frac{\partial f}{\partial t}(x, t_0) \, d\mu$ (x) holds. ∎

There is a criterion for testing the boundedness of the difference quotient function $D_{t_0}(x, t)$ by an integrable function that is very useful for applications. It requires the existence of a neighborhood V of t_0 satisfying the following two properties:

1. the partial derivative $\frac{\partial f}{\partial t}(x, t)$ exists for all x and all $t \in V$, and
2. there exists a non-negative integrable function g such that for each $t \in V$, we have $|\frac{\partial f}{\partial t}(x, t)| \leq g(x)$ for almost all x.

Indeed, if the previous two conditions hold, then by the mean value theorem, it easily follows that for each $t \in V$ we have $|D_{t_0}(x, t)| \leq g(x)$ for almost all x.

Next, as applications of the last two theorems we shall compute a number of classical improper Riemann integrals.

Theorem 24.6 (Euler).[10]   *We have* $\int_0^\infty e^{-x^2} \, dx = \frac{\sqrt{\pi}}{2}$.

Proof. The existence of the improper Riemann integral follows from the inequalities $0 \leq e^{-x^2} \leq e^{-x}$ for $x \geq 1$ and $\int_1^\infty e^{-x} \, dx = e^{-1}$. Also, since $e^{-x^2} \geq 0$ for all x, the improper Riemann integral is a Lebesgue integral over $[0, \infty)$.

---

[10]Leonhard Euler (1707–1783), a great Swiss mathematician. He was one of the most prolific writers throughout the history of science. He is considered (together with Gauss and Riemann) as one of the three greatest mathematicians of modern times.

For the computation of the integral, consider the real-valued functions on $\mathbb{R}$ defined by

$$f(t) = \left(\int_0^t e^{-x^2} \, dx \right)^2 \quad \text{and} \quad g(t) = \int_0^1 \frac{e^{-t^2(1+x^2)}}{1+x^2} \, dx.$$

The derivatives of the previous functions will be determined separately. For f we use the fundamental theorem of calculus and the chain rule. Thus,

$$f'(t) = 2e^{-t^2} \int_0^t e^{-x^2} \, dx.$$

For g observe that

$$\left| \frac{\partial}{\partial t} \left(\frac{e^{-t^2(1+x^2)}}{1+x^2} \right) \right| = \left| -2te^{-t^2(1+x^2)} \right| \le 2|t| \le M$$

holds for every $x \in [0, 1]$ and every $t \in \mathbb{R}$ in any bounded neighborhood of some fixed point t_0. The constant M depends, of course, upon the choice of the neighborhood of t_0. By Theorem 24.5, and the fact that the Lebesgue integrals are Riemann integrals, we get

$$g'(t) = \int_0^1 \frac{\partial}{\partial t} \left(\frac{e^{-t^2(1+x^2)}}{1+x^2} \right) dx = -2e^{-t^2} \int_0^1 te^{-x^2t^2} \, dx$$

for all $t \in \mathbb{R}$. Substituting $u = xt$ (for $t \ne 0$), we obtain

$$g'(t) = -2e^{-t^2} \int_0^t e^{-u^2} \, du = -2e^{-t^2} \int_0^t e^{-x^2} \, dx$$

for each $t \in \mathbb{R}$.

Thus, $f'(t) + g'(t) = 0$ holds for each $t \in \mathbb{R}$, and so, $f(t) + g(t) = c$ (a constant) holds for all t. In particular,

$$c = f(0) + g(0) = \int_0^1 \frac{dx}{1+x^2} = \frac{\pi}{4},$$

and so $f(t) + g(t) = \frac{\pi}{4}$ for each t. Now, observe that for each x and t we have

$$\left| \frac{e^{-t^2(1+x^2)}}{1+x^2} \right| \le \frac{1}{1+x^2},$$

and clearly

$$\lim_{t \to \infty} \frac{e^{-t^2(1+x^2)}}{1+x^2} = 0.$$

Thus, by Theorem 24.4, $\lim_{t \to \infty} g(t) = 0$. Consequently,

$$\frac{\pi}{4} = \lim_{t \to \infty} f(t) + \lim_{t \to \infty} g(t) = \left(\int_0^\infty e^{-x^2} dx \right)^2,$$

from which it follows that $\int_0^\infty e^{-x^2} dx = \frac{\sqrt{\pi}}{2}$.   ∎

Theorem 24.7.   *For each* $t \in \mathbb{R}$ *we have*

$$\int_0^\infty e^{-x^2} \cos(2xt) \, dx = \frac{\sqrt{\pi}}{2} e^{-t^2}.$$

Proof.   Let $F(t) = \int_0^\infty e^{-x^2} \cos(2xt) \, dx$ for all $t \in \mathbb{R}$. Since $|e^{-x^2} \cos(2xt)|$ $\leq e^{-x^2}$ holds for all x and t, it follows that the improper Riemann integral $F(t)$ exists and, moreover, is a Lebesgue integral over $[0, \infty)$. Now,

$$\left| \frac{\partial}{\partial t} \left[e^{-x^2} \cos(2xt) \right] \right| = \left| -2xe^{-x^2} \sin(2xt) \right| \leq 2xe^{-x^2} = g(x)$$

holds for all $x \geq 0$ and t. Hence, since the function g is positive on $[0, \infty)$ and the improper Riemann integral $\int_0^\infty g(x) \, dx$ exists (its value is 1), g is Lebesgue integrable over $[0, \infty)$. Therefore, by Theorem 24.5 (and the remarks after it), it follows that

$$F'(t) = \int_0^\infty \frac{\partial}{\partial t} \left[e^{-x^2} \cos(2xt) \right] dx = -2 \int_0^\infty xe^{-x^2} \sin(2xt) \, dx.$$

Integrating by parts, we obtain

$$-2 \int_0^\infty xe^{-x^2} \sin(2xt) \, dx = e^{-x^2} \sin(2xt) \Big|_0^\infty - 2t \int_0^\infty e^{-x^2} \cos(2xt) \, dx$$

$$= -2t \int_0^\infty e^{-x^2} \cos(2xt) \, dx.$$

Thus, $F'(t) = -2t F(t)$ holds for all t. Solving the differential equation, we get $F(t) = F(0)e^{-t^2}$. By Theorem 24.6, $F(0) = \frac{\sqrt{\pi}}{2}$, and so, $F(t) = \frac{\sqrt{\pi}}{2} e^{-t^2}$, as claimed.   ∎

For the next result, the value of $\frac{\sin x}{x}$ at zero will be assumed to be 1.

Theorem 24.8. *If $t \geq 0$, then*

$$\int_0^\infty \frac{\sin x}{x} e^{-xt} \, dx = \frac{\pi}{2} - \arctan t$$

holds.

Proof. Fix some $t_0 \geq 0$. The cases $t_0 > 0$ and $t_0 = 0$ will be treated separately.

Case I. Assume $t_0 > 0$.
For each fixed $t > 0$, note that $|e^{-xt} \frac{\sin x}{x}| \leq e^{-xt}$ holds for all $x \geq 0$. Thus, the improper Riemann integral exists as a Lebesgue integral over $[0, \infty)$. Let $F(t) = \int_0^\infty e^{-xt} \frac{\sin x}{x} \, dx$ for $t > 0$. Then F satisfies the following properties:

$$\lim_{t \to \infty} F(t) = 0, \tag{$\star$}$$

and

$$F'(t) = -\frac{1}{1 + t^2} \quad \text{for each } t > 0. \tag{$\star\star$}$$

To see $(\star)$, note that if $g(x) = 1$ for $x \in [0, 1]$ and $g(x) = e^{-x}$ for $x > 1$, then g is Lebesgue integrable over $[0, \infty)$ for each $t > 0$ and $|e^{-xt} \frac{\sin x}{x}| \leq g(x)$ holds for all $x \geq 0$ and all $t \geq 1$. On the other hand, $\lim_{t \to \infty} e^{-xt} \frac{\sin x}{x} = 0$ holds for all $x > 0$, and so, by Theorem 24.4, $\lim_{t \to \infty} F(t) = 0$.
To establish $(\star\star)$, note first that $\frac{\partial}{\partial t}[e^{-xt} \frac{\sin x}{x}] = -e^{-xt} \sin x$ holds for all $x \geq 0$, and that for each fixed $a > 0$, the inequality $|e^{-xt} \frac{\sin x}{x}| \leq e^{-ax}$ holds for each $t \geq a$ and all $x \geq 0$. By Theorem 24.5, $F'(t) = -\int_0^\infty e^{-xt} \sin x \, dx$ for all $t > a$ (and all $a > 0$), where the last equality holds since the improper Riemann integral is a Lebesgue integral. Thus, $F'(t) = -\int_0^\infty e^{-xt} \sin x \, dx$ holds for all $t > 0$. Since from elementary calculus we have

$$\int_0^r e^{-xt} \sin x \, dx = -\frac{e^{-rt}(t \sin r + \cos r)}{1 + t^2} + \frac{1}{1 + t^2},$$

by letting $r \to \infty$, we get $F'(t) = -\frac{1}{1+t^2}$ for $t > 0$.
Integrating $(\star\star)$ from t to t_0 yields

$$F(t_0) - F(t) = -\int_t^{t_0} \frac{dx}{1 + x^2} = \arctan t - \arctan t_0,$$

and by letting $t \to \infty$ it follows that $F(t_0) = \frac{\pi}{2} - \arctan t_0$.

Case II.   Assume $t_0 = 0$.

In this case, $\frac{\sin x}{x}$ is not Lebesgue integrable over $[0, \infty)$. However, the improper Riemann integral $\int_0^\infty \frac{\sin x}{x}\,dx$ exists. Indeed, if $0 < s < t$, then an integration by parts yields

$$\int_s^t \frac{\sin x}{x}\,dx = -\frac{\cos x}{x}\bigg|_s^t - \int_s^t \frac{\cos x}{x^2}\,dx = \frac{\cos s}{s} - \frac{\cos t}{t} - \int_s^t \frac{\cos x}{x^2}\,dx,$$

and thus,

$$\left|\int_s^t \frac{\sin x}{x}\,dx\right| \le \frac{1}{s} + \frac{1}{t} + \int_s^t \frac{dx}{x^2} = \frac{2}{s}.$$

By Theorem 24.1, $\int_0^\infty \frac{\sin x}{x}\,dx$ exists. In particular, note that $\lim_{n\to\infty} \int_n^{n+1} \frac{\sin x}{x}\,dx = 0$.

Now, for each n define $f_n(t) = \int_0^n e^{-xt} \frac{\sin x}{x}\,dx$ for $t \ge 0$, and note that the relation $|f_n(n)| \le \int_0^n e^{-xn}\,dx = \frac{1-e^{-n^2}}{n} \le \frac{1}{n}$ implies $\lim f_n(n) = 0$. By Theorem 24.5, we have

$$f_n'(t) = -\int_0^n e^{-xt} \sin x\,dx = \frac{e^{-nt}(t\sin n + \cos n) - 1}{1 + t^2},$$

and so, $\lim_{n\to\infty} f_n'(t) = -\frac{1}{1+t^2}$ holds for all $t > 0$. Also,

$$|f_n'(t)| \le \frac{1 + (1+t)e^{-t}}{1 + t^2}$$

holds for each $t > 0$, and the dominating function for the sequence $\{f_n'\}$ is Lebesgue integrable over $[0, \infty)$.

Let $g_n = f_n'\chi_{[0,n]}$, and note that $\{g_n\}$ is a sequence of Lebesgue integrable functions over $[0, \infty)$. Also, since $|g_n| \le |f_n'|$ and $\lim_{n\to\infty} g_n(t) = -\frac{1}{1+t^2}$ for $t > 0$, the Lebesgue dominated convergence theorem yields

$$\lim_{n\to\infty} \int_0^n f_n'(t)\,dt = \lim_{n\to\infty} \int_0^\infty g_n(t)\,dt = -\int_0^\infty \frac{dt}{1+t^2} = -\frac{\pi}{2}.$$

Since $\int_0^n f_n'(t)\,dt = f_n(n) - f_n(0)$ and $\lim f_n(n) = 0$, we get $\lim f_n(0) = \frac{\pi}{2}$. Finally, letting $n \to \infty$ in the identity

$$\int_0^{n+1} \frac{\sin x}{x}\,dx = \int_0^n \frac{\sin x}{x}\,dx + \int_n^{n+1} \frac{\sin x}{x}\,dx = f_n(0) + \int_n^{n+1} \frac{\sin x}{x}\,dx,$$

we easily get $\int_0^\infty \frac{\sin x}{x}\,dx = \frac{\pi}{2}$.   ∎

EXERCISES

1. Show that

$$\int_0^\infty x^{2n} e^{-x^2} \, dx = \frac{(2n)!}{2^{2n} n!} \cdot \frac{\sqrt{\pi}}{2}$$

holds for $n = 0, 1, 2, \ldots$.

2. Show that $\int_0^\infty e^{-tx^2} \, dx = \frac{1}{2} \sqrt{\frac{\pi}{t}}$ for each $t > 0$.

3. Show that $f(x) = \frac{\ln x}{x^2}$ is Lebesgue integrable over $[1, \infty)$ and that $\int f \, d\lambda = 1$.

4. Show that

$$\lim_{n \to \infty} \int_0^n \left(1 + \frac{x}{n} \right)^n e^{-2x} \, dx = 1.$$

5. Let $f : [0, \infty) \to (0, \infty)$ be a continuous, decreasing, and Lebesgue integrable function. Show that

$$\lim_{x \to \infty} \frac{1}{f(x)} \int_x^\infty f(s) \, ds = 0 \quad \text{if and only if} \quad \lim_{x \to \infty} \frac{f(x + t)}{f(x)} = 0 \quad \text{for each } t > 0.$$

6. Show that the improper Riemann integrals, $\int_0^\infty \cos(x^2) \, dx$ and $\int_0^\infty \sin(x^2) \, dx$ (which are known as the **Fresnel**[11] **integrals**) both exist. Also, show that $\cos(x^2)$ and $\sin(x^2)$ are not Lebesgue integrable over $[0, \infty)$.

7. Show that $\int_0^\infty \frac{\sin^2 x}{x^2} \, dx = \frac{\pi}{2}$.

8. Let (X, S, μ) be a measure space, T a metric space, and $f : X \times T \to \mathbb{R}$ a function. Assume that $f(\cdot, t)$ is a measurable function for each $t \in T$ and $f(x, \cdot)$ is a continuous function for each $x \in X$. Assume also that there exists an integrable function g such that for each $t \in T$ we have $|f(x, t)| \le g(x)$ for almost all $x \in X$. Show that the function $F : T \to \mathbb{R}$, defined by

$$F(t) = \int_X f(x, t) \, d\mu(x),$$

is a continuous function.

9. Show that

$$\int_0^\infty \frac{e^{-x} - e^{-xt}}{x} \, dx = \ln t$$

holds for each $t > 0$.

10. For each $t > 0$, let $F(t) = \int_0^\infty \frac{e^{-xt}}{1 + x^2} \, dx$.

   a. Show that the integral exists as an improper Riemann integral and as a Lebesgue integral.

   b. Show that F has a second-order derivative and that $F''(t) + F(t) = \frac{1}{t}$ holds for each $t > 0$.

---

[11] Augustin Jean Fresnel (1788–1827) was a French physicist who worked extensively in the field of optics. Using the integrals that bear his name, he was the first to demonstrate the wave conception of light.

11. Show that the improper Riemann integral, $\int_0^{\frac{\pi}{2}} \ln(t \cos x)\,dx$, exists for each $t > 0$ and that it is also a Lebesgue integral. Also, show that

$$\int_0^{\frac{\pi}{2}} \ln(t \cos x)\,dx = \frac{\pi}{2} \ln\left(\frac{t}{2}\right)$$

holds for all $t > 0$.

12. Show that for each $t \geq 0$, the improper Riemann integral $\int_0^\infty \frac{\sin xt}{x(1+x^2)}\,dx$ exists as a Lebesgue integral and that

$$\int_0^\infty \frac{\sin xt}{x(1+x^2)}\,dx = \frac{\pi}{2}(1 - e^{-t}).$$

13. The **Gamma function** for $t > 0$ is defined by an integral as follows:

$$\Gamma(t) = \int_0^\infty x^{t-1} e^{-x}\,dx.$$

a. Show that the integral

$$\int_0^\infty x^{t-1} e^{-x}\,dx = \lim_{\substack{r \to \infty \\ \epsilon \to 0^+}} \int_\epsilon^r x^{t-1} e^{-x}\,dx$$

exists as an improper Riemann integral (and hence, as a Lebesgue integral).

b. Show that $\Gamma(\frac{1}{2}) = \sqrt{\pi}$.

c. Show that $\Gamma(t + 1) = t\Gamma(t)$ holds for all $t > 0$, and use this conclusion to establish $\Gamma(n + 1) = n!$ for $n = 1, 2, \ldots$.

d. Show that Γ is differentiable at every $t > 0$ and that

$$\Gamma'(t) = \int_0^\infty x^{t-1} e^{-x} \ln x\,dx$$

holds.

e. Show that Γ has derivatives of all order and that

$$\Gamma^{(n)}(t) = \int_0^\infty x^{t-1} e^{-x} (\ln x)^n\,dx$$

holds for $n = 1, 2, \ldots$ and all $t > 0$.

14. Let $f : [0, 1] \to \mathbb{R}$ be a Lebesgue integrable function and define the real-valued function $F : [0, 1] \to \mathbb{R}$ by $F(t) = \int_0^1 f(x)\sin(xt)\,d\lambda(x)$.

a. Show that the integral defining F exists and that F is a uniformly continuous function.

b. Show that F has derivatives of all orders and that

$$F^{(2n)}(t) = (-1)^n \int_0^1 x^{2n} f(x)\sin(xt)\,d\lambda(x)$$

and

$$F^{(2n-1)}(t) = (-1)^n \int_0^1 x^{2n-1} f(x) \cos(xt) \, d\lambda(x)$$

for $n = 1, 2, \ldots$ and each $t \in [0, 1]$.

c.   Show that $F = 0$ (i.e., $F(t) = 0$ for all $t \in [0, 1]$) if and only if $f = 0$ a.e.

25.   APPROXIMATING INTEGRABLE FUNCTIONS

The following type of approximation problem is commonly encountered in analysis.

- Given an integrable function f, a family of $\mathcal{F}$ of integrable functions, and $\epsilon > 0$, determine whether there exists a function g in the collection $\mathcal{F}$ such that $\int |f - g| \, d\mu < \epsilon$.

By the definition of an integrable function, the following result is immediate.

Theorem 25.1.   *Let f be an integrable function and $\epsilon > 0$. Then there exists a step function ϕ such that $\int |f - \phi| \, d\mu < \epsilon$.*

Let us denote by L the collection of all step functions ϕ for which there exist sets $A_1, \ldots, A_n \in S$ all of finite measure, and real numbers $a_1, \ldots, a_n$ (all depending on ϕ) such that $\phi = \sum_{i=1}^n a_i \chi_{A_i}$ holds. Then L is a function space; see Exercise 1 of Section 21.

Theorem 25.2.   *Let f be an integrable function, and let $\epsilon > 0$. Then there exists a function $\phi \in L$ such that $\int |f - \phi| \, d\mu < \epsilon$.*

Proof.   Let $\mathcal{F}$ denote the collection of all integrable functions f such that for each $\epsilon > 0$, there exists some function $\phi \in L$ such that $\int |f - \phi| \, d\mu < \epsilon$. It should be clear that $\mathcal{F}$ is a vector space such that $\chi_A \in \mathcal{F}$ holds true for each $A \in S$. Now, a glance at Theorem 22.12 guarantees that $\mathcal{F}$ consists of all integrable functions. ∎

The next result deals with the approximation of integrable functions by continuous functions. Recall that if X is a topological space and $f : X \to \mathbb{R}$ is a function, then the closure of the set $\{x \in X : f(x) \neq 0\}$ is called the **support** of f (denoted by Supp f). If the support of f happens to be a compact set, then f is said to have **compact support**.

Theorem 25.3.   *Let X be a Hausdorff locally compact topological space, and let μ be a regular Borel measure on X. Assume that f is an integrable function*

with respect to the measure space $(X, \mathcal{B}, \mu)$. Then given $\epsilon > 0$, there exists a
continuous function $g : X \to \mathbb{R}$ with compact support such that $\int |f - g| \, d\mu < \epsilon$.

Proof. By Theorem 25.2, we can assume without loss of generality that f is
the characteristic function of some Borel set of finite measure. So, assume $f = \chi_A$
for some Borel set A with $\mu(A) < \infty$.

Since μ is a regular Borel measure, there exist a compact set K such that $K \subseteq A$
and $\mu(A) - \mu(K) < \epsilon$ (see Lemma 18.5) and an open set V such that $A \subseteq V$
and $\mu(V) - \mu(A) < \epsilon$. By Theorem 10.8, there exists a continuous function
$g : X \to [0, 1]$ (and hence, g is Borel measurable) with compact support such
that $g(x) = 1$ for each $x \in K$ and $\operatorname{Supp} g \subseteq V$. Clearly, g is integrable and
$|\chi_A - g| \le \chi_V - \chi_K$ holds. Therefore,

$$\int |\chi_A - g| \, d\mu \le \int (\chi_V - \chi_K) \, d\mu = \mu(V) - \mu(K) < 2\epsilon,$$

and the proof of the theorem is finished.   ∎

The following theorem describes an important property of the Lebesgue integrable functions on $\mathbb{R}$. It is usually referred to as the **Riemann–Lebesgue
Lemma**.

Theorem 25.4 (Riemann–Lebesgue).   *If $f : \mathbb{R} \to \mathbb{R}$ is Lebesgue integrable,*
then

$$\lim_{n \to \infty} \int f(x) \cos(nx) \, d\lambda(x) = \lim_{n \to \infty} \int f(x) \sin(nx) \, d\lambda(x) = 0.$$

Proof.   Note first that the inequality $|f(x) \cos(nx)| \le |f(x)|$ for all x, com-
bined with Theorem 22.6, shows that the function $f(x) \cos(nx)$ is a Lebesgue
integrable function for each n. Also, by Theorem 25.2, in order to establish
the theorem, it suffices to consider functions of the form $f = \chi_{[a,b)}$. Thus, let
$f = \chi_{[a,b)}$. In this case, the Lebesgue integrals are Riemann integrals, and

$$\left| \int f(x) \cos(nx) \, d\lambda(x) \right| = \left| \int_a^b \cos(nx) \, dx \right| = \frac{1}{n} |\sin(nb) - \sin(na)| \le \frac{2}{n} \to 0$$

as $n \to \infty$. Similarly, $\lim \int f(x) \sin(nx) \, d\lambda(x) = 0$.   ∎

A sequence of integrable functions $\{f_n\}$ is said to **converge in the mean** to some
function f if $\lim \int |f_n - f| \, d\mu = 0$ holds.

Theorem 25.5. *Let $\{f_n\}$ be a sequence of integrable functions. If f is an integrable function such that $\lim \int |f_n - f| \, d\mu = 0$, then there exists a subsequence $\{f_{k_n}\}$ of $\{f_n\}$ such that $f_{k_n} \to f$ a.e. holds.*

Proof. Let $\epsilon > 0$. For each n let $E_n = \{x \in X : |f_n(x) - f(x)| \geq \epsilon\}$, and note that each E_n is a measurable set of finite measure (see Theorem 22.5). Now, since $\epsilon \chi_{E_n} \leq |f_n - f|$ holds for each n, it follows that $\epsilon \mu^*(E_n) \leq \int |f_n - f| \, d\mu$ also holds for each n. Thus, $\lim \mu^*(E_n) = 0$, and so, $f_n \xrightarrow{\mu} f$. The conclusion now follows immediately from Theorem 19.4. ∎

A sequence of integrable functions that converges in the mean to some function need not converge pointwise to that function. An example of this situation is provided by the sequence $\{f_n\}$ of Example 19.6.

EXERCISES

1. Let $f : \mathbb{R} \to \mathbb{R}$ be a Lebesgue integrable function. Show that

$$\lim_{t \to \infty} \int f(x) \cos(xt) \, d\lambda(x) = \lim_{t \to \infty} \int f(x) \sin(xt) \, d\lambda(x) = 0.$$

2. A function $f : \mathcal{O} \to \mathbb{R}$ (where $\mathcal{O}$ is a nonempty open subset of $\mathbb{R}^n$) is said to be a C^∞-**function** if f has continuous partial derivatives of all orders.

   a. Consider the function $\rho : \mathbb{R} \to \mathbb{R}$ defined by $\rho(x) = \exp[\frac{1}{x^2-1}]$ if $|x| < 1$ and $\rho(x) = 0$ if $|x| \geq 1$. Show that ρ is a C^∞-function such that $\text{Supp}\, \rho = [-1, 1]$. (Induction and L'Hôpital's[12] rule are needed here.)

   b. For $\epsilon > 0$ and $a \in \mathbb{R}$ show that the function $f(x) = \rho(\frac{x-a}{\epsilon})$ is also a C^∞-function with $\text{Supp}\, f = [a - \epsilon, a + \epsilon]$.

3. Let $[a, b]$ be an interval, $\epsilon > 0$ such that $a + \epsilon < b - \epsilon$, and ρ as in the previous exercise. Define $h : \mathbb{R} \to \mathbb{R}$ by $h(x) = \int_a^b \rho(\frac{t-x}{\epsilon}) \, dt$ for all $x \in \mathbb{R}$. Then show that

   a. $\text{Supp}\, h \subseteq [a - \epsilon, b + \epsilon]$,

   b. $h(x) = c$ (a constant function) for all $x \in [a + \epsilon, b - \epsilon]$,

   c. h is a C^∞-function and $h^{(n)}(x) = \int_a^b \frac{\partial^n}{\partial x^n} \rho(\frac{t-x}{\epsilon}) \, dt$ holds for all $x \in \mathbb{R}$, and

   d. the C^∞-function $f = \frac{1}{c} h$ satisfies $0 \leq f(x) \leq 1$ for all $x \in \mathbb{R}$, $f(x) = 1$ for all $x \in [a + \epsilon, b - \epsilon]$, and $\int |\chi_{[a,b)} - f| \, d\lambda < 4\epsilon$.

4. Let $f : \mathbb{R} \to \mathbb{R}$ be an integrable function with respect to the Lebesgue measure, and let $\epsilon > 0$. Show that there exists a C^∞-function g such that $\int |f - g| \, d\lambda < \epsilon$. [HINT: Use Theorem 25.2 and the preceding exercise.]

---

[12]Guillaume-François-Antoine de L'Hôpital (1661–1704), a French mathematician. He is mainly remembered for the familiar rule of computing the limit of a fraction whose numerator and denominator tend simultaneously to zero (or to $\pm\infty$).

5. The purpose of this exercise is to establish the following general result: If $f : \mathbb{R}^n \to \mathbb{R}$
    is an integrable function (with respect to the Lebesgue measure) and $\epsilon > 0$, then there
    exists a C^∞-function g such that $\int |f - g| \, d\lambda < \epsilon$.

    a. Let $a_i < b_i$ for $i = 1, \ldots, n$, and put $I = \prod_{i=1}^{n} (a_i, b_i)$. Choose $\epsilon > 0$ such that
        $a_i + \epsilon < b_i - \epsilon$ for each i. Use Exercise 3 to select for each i a C^∞-function
        $f_i : \mathbb{R} \to \mathbb{R}$ such that $0 \leq f_i(x) \leq 1$ for all x, $f_i(x) = 1$ if $x \in [a + \epsilon, b_i - \epsilon]$,
        and Supp $f_i \subseteq [a_i - \epsilon, b_i + \epsilon]$. Now define $h : \mathbb{R}^n \to \mathbb{R}$ by $h(x_1, \ldots, x_n) = \prod_{i=1}^{n} f_i(x_i)$. Then show that h is a C^∞-function on $\mathbb{R}^n$ and that

        $$\int |\chi_I - h| \, d\lambda \leq 2 \left[\prod_{i=1}^{n} (b_i - a_i + 2\epsilon) - \prod_{i=1}^{n} (b_i - a_i) \right].$$

    b. Let $f : \mathbb{R}^n \to \mathbb{R}$ be Lebesgue integrable, and let $\epsilon > 0$. Then use part (a) to show
        that there exists a C^∞-function g with compact support such that

        $$\int |f - g| \, d\lambda < \epsilon.$$

6. Let μ be a regular Borel measure on $\mathbb{R}^n$, $f : \mathbb{R}^n \to \mathbb{R}$ a μ-integrable function, and
    $\epsilon > 0$. Show that there exists a C^∞-function $g : \mathbb{R}^n \to \mathbb{R}$ such that $\int |f - g| \, d\mu < \epsilon$.

7. Let $f : [a, b] \to \mathbb{R}$ be a Lebesgue integrable function, and let $\epsilon > 0$. Show that there
    exists a polynomial p such that $\int |f - p| \, d\lambda < \epsilon$, where the integral is considered, of
    course, over $[a, b]$.
    [HINT: Use the Stone–Weierstrass approximation theorem.]

26. PRODUCT MEASURES AND ITERATED INTEGRALS

Throughout this section, (X, S, μ) and (Y, Σ, ν) will be two fixed measure spaces.
The **product semiring** $S \otimes \Sigma$ of subsets of $X \times Y$ is defined by

$$S \otimes \Sigma = \{ A \times B : A \in S \text{ and } B \in \Sigma \}.$$

The members of $S \otimes \Sigma$ are called **rectangles**.
   The above collection $S \otimes \Sigma$ is indeed a semiring of subsets of $X \times Y$. This
follows immediately from the identities

1. $(A \times B) \cap (A_1 \times B_1) = (A \cap A_1) \times (B \cap B_1)$, and
2. $(A \times B) \setminus (A_1 \times B_1) = [(A \setminus A_1) \times B] \cup [(A \cap A_1) \times (B \setminus B_1)]$,

and the fact that $A \setminus A_1$ and $B \setminus B_1$ can be written as finite disjoint unions of
members of S and Σ, respectively. The product semiring was discussed in some
detail in Section 20.
   Now, define the set function $\mu \times \nu : S \otimes \Sigma \to [0, \infty]$ by $\mu \times \nu(A \times B) = \mu(A) \cdot \nu(B)$ for each $A \times B \in S \otimes \Sigma$ (keep in mind that $0 \cdot \infty = 0$). This set
function is a measure on the product semiring $S \otimes \Sigma$, called the **product measure**
of μ and ν. The details follow.

Theorem 26.1. *The set function* $\mu \times \nu : S \otimes \Sigma \to [0, \infty]$ *defined by*

$$\mu \times \nu(A \times B) = \mu(A) \cdot \nu(B)$$

for each $A \times B \in S \otimes \Sigma$ *is a measure. (This measure can be thought of as the generalization of the familiar formula "Length $\times$ Width" for computing the area of a rectangle in elementary geometry!)*

Proof. Clearly, $\mu \times \nu(\emptyset) = 0$. For the σ-additivity of $\mu \times \nu$, let $A \times B \in S \otimes \Sigma$ and let $\{A_n \times B_n\}$ be a sequence of mutually disjoint sets of $S \otimes \Sigma$ such that $A \times B = \bigcup_{n=1}^{\infty} A_n \times B_n$. It must be established that

$$\mu(A) \cdot \nu(B) = \sum_{n=1}^{\infty} \mu(A_n) \cdot \nu(B_n). \qquad (\star)$$

Obviously, $(\star)$ holds true if either A or B has measure zero. Thus, we can assume that $\mu(A) > 0$ and $\nu(B) > 0$.

Since $\chi_{A \times B} = \sum_{n=1}^{\infty} \chi_{A_n \times B_n}$, we see that

$$\chi_A(x) \cdot \chi_B(y) = \sum_{n=1}^{\infty} \chi_{A_n}(x) \cdot \chi_{B_n}(y)$$

holds true for all x and y. For each fixed $y \in B$ let $K = \{i \in \mathbb{N} : y \in B_i\}$ and note that $\chi_A(x) = \sum_{i \in K} \chi_{A_i}(x)$ for each $x \in X$. Since the collection $\{A_i : i \in K\}$ is pairwise disjoint (why?), the latter implies $\mu(A) = \sum_{i \in K} \mu(A_i)$. Therefore,

$$\mu(A) \cdot \chi_B(y) = \sum_{i \in K} \mu(A_i) \chi_{B_i}(y) = \sum_{n=1}^{\infty} \mu(A_n) \cdot \chi_{B_n}(y) \qquad (\star\star)$$

holds for all $y \in Y$. Since a term with $\mu(A_n) = 0$ does not alter the sum in $(\star)$ or $(\star\star)$, we can assume that $\mu(A_n) > 0$ for all n.

Now, if both A and B have finite measures, then by using Theorem 22.9 and integrating $(\star\star)$ term by term, we see that $(\star)$ is valid. On the other hand, if either A or B has infinite measure, then $\sum_{n=1}^{\infty} \mu(A_n) \cdot \nu(B_n) = \infty$ must hold. Indeed, if the last sum is finite, then, by Theorem 22.9, $\mu(A)\chi_B(y)$ defines an integrable function, which is impossible. Thus, in this case $(\star)$ holds with both sides infinite, and the proof is finished. ∎

The next few results will unravel the basic properties of the product measure $\mu \times \nu$. As usual $(\mu \times \nu)^*$ denotes the outer measure generated by the measure space $(X \times Y, S \otimes \Sigma, \mu \times \nu)$ on $X \times Y$.

Theorem 26.2.   *If $A \subseteq X$ and $B \subseteq Y$ are measurable sets of finite measure, then*

$$(\mu \times \nu)^*(A \times B) = \mu^* \times \nu^*(A \times B) = \mu^*(A) \cdot \nu^*(B).$$

Proof.   Clearly, $S \otimes \Sigma \subseteq \Lambda_\mu \otimes \Lambda_\nu$ holds. Now, let $\{A_n \times B_n\}$ be a sequence of $S \otimes \Sigma$ such that $A \times B \subseteq \bigcup_{n=1}^\infty A_n \times B_n$. Since, by Theorem 26.1, $\mu^* \times \nu^*$ is a measure on the semiring $\Lambda_\mu \otimes \Lambda_\nu$ and $A \times B \in \Lambda_\mu \otimes \Lambda_\nu$, it follows from Theorem 13.8 that

$$\mu^* \times \nu^*(A \times B) \le \sum_{n=1}^\infty \mu^* \times \nu^*(A_n \times B_n) = \sum_{n=1}^\infty \mu \times \nu(A_n \times B_n),$$

and so,

$$\mu^* \times \nu^*(A \times B) \le (\mu \times \nu)^*(A \times B).$$

On the other hand, if $\epsilon > 0$ is given, choose two sequences $\{A_n\} \subseteq S$ and $\{B_n\} \subseteq \Sigma$, with $A \subseteq \bigcup_{n=1}^\infty A_n$, $B \subseteq \bigcup_{n=1}^\infty B_n$, such that $\sum_{n=1}^\infty \mu(A_n) < \mu^*(A) + \epsilon$ and $\sum_{n=1}^\infty \nu(B_n) < \nu^*(B) + \epsilon$. Then, $A \times B \subseteq \bigcup_{n=1}^\infty \bigcup_{m=1}^\infty A_n \times B_m$ holds, and so

$$
\begin{aligned}
(\mu \times \nu)^*(A \times B) &\le \sum_{n=1}^\infty \sum_{m=1}^\infty \mu \times \nu(A_n \times B_m) = \sum_{n=1}^\infty \sum_{m=1}^\infty \mu(A_n) \cdot \nu(B_m) \\
&= \left[\sum_{n=1}^\infty \mu(A_n) \right] \cdot \left[\sum_{m=1}^\infty \nu(B_m) \right] < [\mu^*(A) + \epsilon] \cdot [\nu^*(B) + \epsilon]
\end{aligned}
$$

for all $\epsilon > 0$. That is,

$$(\mu \times \nu)^*(A \times B) \le \mu^*(A) \cdot \nu^*(B) = \mu^* \times \nu^*(A \times B).$$

Therefore, $(\mu \times \nu)^*(A \times B) = \mu^* \times \nu^*(A \times B)$ holds, as required.   ∎

It is expected that the members of $\Lambda_\mu \otimes \Lambda_\nu$ are $\mu \times \nu$-measurable subsets of $X \times Y$; that is, $\Lambda_\mu \otimes \Lambda_\nu \subseteq \Lambda_{\mu \times \nu}$ holds. The next theorem shows that this is actually the case.

Theorem 26.3.   *If A is a μ-measurable subset of X and B a ν-measurable subset of Y, then $A \times B$ is a $\mu \times \nu$-measurable subset of $X \times Y$.*

Proof.   Assume $A \in \Lambda_\mu$, $B \in \Lambda_\nu$, and fix $C \times D \in S \otimes \Sigma$ with $\mu \times \nu(C \times D) = \mu(C) \cdot \nu(D) < \infty$. In order to establish the $\mu \times \nu$-measurability of $A \times B$, by

Theorem 15.2, it suffices to show that

$$(\mu \times \nu)^*((C \times D) \cap (A \times B)) + (\mu \times \nu)^*((C \times D) \cap (A \times B)^c) \leq \mu \times \nu(C \times D).$$

If $\mu \times \nu(C \times D) = 0$, then the previous inequality is obvious (both sides are zero). So, we can assume $\mu(C) < \infty$ and $\nu(D) < \infty$. Clearly,

$$(C \times D) \cap (A \times B) = (C \cap A) \times (D \cap B), \text{ and}$$
$$(C \times D) \cap (A \times B)^c = [(C \cap A^c) \times (D \cap B)] \cup [(C \cap A) \times (D \cap B^c)]$$
$$\cup [(C \cap A^c) \times (D \cap B^c)]$$

hold with every member of the previous union having finite measure.

Now, the subadditivity of $(\mu \times \nu)^*$, combined with Theorem 26.2, yields

$$(\mu \times \nu)^*((C \times D) \cap (A \times B)) + (\mu \times \nu)^*((C \times D) \cap (A \times B)^c)$$
$$\leq \mu^*(C \cap A) \cdot \nu^*(D \cap B) + \mu^*(C \cap A^c) \cdot \nu^*(D \cap B)$$
$$+ \mu^*(C \cap A) \cdot \nu^*(D \cap B^c) + \mu^*(C \cap A^c) \cdot \nu^*(D \cap B^c)$$
$$= [\mu^*(C \cap A) + \mu^*(C \cap A^c)] \cdot [\nu^*(D \cap B) + \nu^*(D \cap B^c)]$$
$$= \mu(C) \cdot \nu(D) = \mu \times \nu(C \times D),$$

as required. ∎

In general, it is not true that the measure $\mu^* \times \nu^*$ is the only extension of $\mu \times \nu$ from $\mathcal{S} \otimes \Sigma$ to a measure on $\Lambda_\mu \otimes \Lambda_\nu$. However, if both $(X, \mathcal{S}, \mu)$ and (Y, Σ, ν) are σ-finite measure spaces, then $(X \times Y, \mathcal{S} \otimes \Sigma, \mu \times \nu)$ is likewise a σ-finite measure space, and therefore, by Theorem 15.10, $\mu^* \times \nu^*$ is the only extension of $\mu \times \nu$ to a measure on the semiring $\Lambda_\mu \otimes \Lambda_\nu$. Moreover, in view of $\Lambda_\mu \otimes \Lambda_\nu \subseteq \Lambda_{\mu \times \nu}$ and the fact that $(\mu \times \nu)^*$ is a measure on $\Lambda_{\mu \times \nu}$, it follows in this case that $(\mu \times \nu)^* = \mu^* \times \nu^*$ holds on $\Lambda_\mu \otimes \Lambda_\nu$.

We now turn our attention to measurability properties of arbitrary subsets of $X \times Y$. Let us recall a few definitions from Section 20. If A is a subset of $X \times Y$ and $x \in X$, then the x-**section** A_x of A is the subset of Y defined by

$$A_x = \{y \in Y : (x, y) \in A\}.$$

Similarly, if $y \in Y$, then the y-**section** A^y of A is the subset of X defined by

$$A^y = \{x \in X : (x, y) \in A\}.$$

The geometrical meanings of the x- and y-sections are shown in Figure 4.1. Regarding sections of sets, we have the following identities.

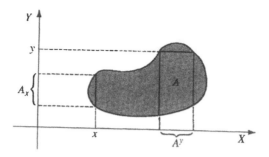

FIGURE 4.1.

a. $(\bigcup_{i \in I} A_i)_x = \bigcup_{i \in I}(A_i)_x$ and $(\bigcup_{i \in I} A_i)^y = \bigcup_{i \in I}(A_i)^y$;

b. $(\bigcap_{i \in I} A_i)_x = \bigcap_{i \in I}(A_i)_x$ and $(\bigcap_{i \in I} A_i)^y = \bigcap_{i \in I}(A_i)^y$;

c. $(A \setminus B)_x = A_x \setminus B_x$ and $(A \setminus B)^y = A^y \setminus B^y$.

The proofs of the previous identities are straightforward, and they are left as an exercise for the reader.

The next theorem demonstrates the relationship between the $\mu \times \nu$-measurable subsets of $X \times Y$ and the measurable subsets of X and Y, and it is a key result for this section.

Recall that an extended real-valued function f that is undefined on a set of measure zero is said to **define an integrable function** if there exists an integrable function g such that $f = g$ almost everywhere. That is, if arbitrary values are assigned to f on the points where it is undefined or attains an infinite value, then f becomes an integrable function. (The value of the integral does not depend, of course, upon the choices of these values.)

Theorem 26.4. *Let E be a $\mu \times \nu$-measurable subset of $X \times Y$ satisfying $(\mu \times \nu)^*(E) < \infty$. Then for μ-almost all x the set E_x is a ν-measurable subset of Y, and the function $x \mapsto \nu^*(E_x)$ defines an integrable function on X such that*

$$(\mu \times \nu)^*(E) = \int_X \nu^*(E_x) \, d\mu(x).$$

Similarly, for ν-almost all y, the set E^y is a μ-measurable subset of X, and the function $y \mapsto \mu^*(E^y)$ defines an integrable function on Y such that

$$(\mu \times \nu)^*(E) = \int_Y \mu^*(E^y) \, d\nu(y).$$

Proof. By the symmetry of the situation, it suffices to establish the first formula. The proof goes by steps.

Step I. Assume $E = A \times B \in S \otimes \Sigma$.

Clearly, $E_x = B$ if $x \in A$ and $E_x = \emptyset$ if $x \notin A$. Thus, E_x is a v-measurable subset of Y for each $x \in X$, and

$$v(E_x) = v(B)\chi_A(x) \qquad (\star)$$

holds for all $x \in X$. Since $(\mu \times v)^*(E) = (\mu \times v)(A \times B) = \mu(A) \cdot v(B) < \infty$, two possibilities arise:

a. *Both A and B have finite measure.*

In this case, $(\star)$ shows that $x \mapsto v^*(E_x)$ is an integrable function (actually, it is a step function) satisfying

$$\int_X v^*(E_x)\, d\mu(x) = \int v(B)\chi_A\, d\mu = \mu(A) \cdot v(B) = (\mu \times v)^*(E).$$

b. *Either A or B has infinite measure.*

In this case, the other set must have measure zero, and so, $(\star)$ guarantees $v(E_x) = 0$ for μ-almost all x. Thus, $x \mapsto v^*(E_x)$ defines the zero function, and hence

$$\int_X v^*(E_x)\, d\mu(x) = 0 = (\mu \times v)^*(E).$$

Step II. Assume that E is a σ-set of $S \otimes \Sigma$.

Choose a disjoint sequence $\{E_n\}$ of $S \otimes \Sigma$ such that $E = \bigcup_{n=1}^{\infty} E_n$. In view of $E_x = \bigcup_{n=1}^{\infty}(E_n)_x$ and the preceding step, it follows that E_x is a measurable subset of Y for each $x \in X$.

Now, define $f(x) = v^*(E_x)$ and $f_n(x) = \sum_{i=1}^{n} v((E_i)_x)$ for each $x \in X$ and all n. By Step I, each f_n defines an integrable function and

$$\int f_n\, d\mu = \sum_{i=1}^{n} \int_X v((E_i)_x)\, d\mu(x) = \sum_{i=1}^{n} \mu \times v(E_i) \uparrow (\mu \times v)^*(E) < \infty.$$

Since $\{(E_n)_x\}$ is a disjoint sequence of Σ, we have $v^*(E_x) = \sum_{n=1}^{\infty} v((E_n)_x)$, and so, $f_n(x) \uparrow f(x)$ holds for each $x \in X$. Thus, by Levi's theorem (Theorem 22.8), f defines an integrable function and

$$\int_X v^*(E_x)\, d\mu(x) = \int f\, d\mu = \lim_{n \to \infty} \int f_n\, d\mu = \sum_{i=1}^{\infty} \mu \times v(E_i) = (\mu \times v)^*(E).$$

Step III. Assume that E is a countable intersection of σ-sets of finite measure.

Choose a sequence $\{E_n\}$ of σ-sets such that $E = \bigcap_{n=1}^{\infty} E_n$, $(\mu \times \nu)^*(E_1) < \infty$, and $E_{n+1} \subseteq E_n$ for all n.

For each n, let $g_n(x) = 0$ if $\nu^*((E_n)_x) = \infty$ and $g_n(x) = \nu^*((E_n)_x$ if $\nu^*((E_n)_x < \infty$. By Step II, each g_n is an integrable function over X such that $\int g_n \, d\mu = (\mu \times \nu)^*(E_n)$ holds. In view of $E_x = \bigcap_{n=1}^{\infty}(E_n)_x$, it follows that E_x is a ν-measurable set for each $x \in X$. Also, since $\nu^*((E_1)_x) < \infty$ holds for μ-almost all x, it follows from Theorem 15.4 that $g_n(x) = \nu^*((E_n)_x) \downarrow \nu^*(E_x)$ holds for μ-almost all x. Thus, $x \mapsto \nu^*(E_x)$ defines an integrable function and

$$\int_X \nu^*(E_x) \, d\mu(x) = \lim_{n \to \infty} \int g_n \, d\mu = \lim_{n \to \infty} (\mu \times \nu)^*(E_n) = (\mu \times \nu)^*(E),$$

where the last equality holds again by virtue of Theorem 15.4.

Step IV: Assume that E is a null set, i.e., $(\mu \times \nu)^*(E) = 0$.

Arguing as in the proof of Theorem 15.11, we see that there exists a measurable set G, which is a countable intersection of σ-sets of finite measure, such that $E \subseteq G$ and $(\mu \times \nu)^*(G) = 0$.

By Step III, $\int_X \nu^*(G_x) \, d\mu(x) = (\mu \times \nu)^*(G) = 0$, and so, by Theorem 22.7(1), $\nu^*(G_x) = 0$ holds for μ-almost all x. In view of $E_x \subseteq G_x$ for all x, we must have $\nu^*(E_x) = 0$ for μ-almost all x. Therefore, E_x is ν-measurable for μ-almost all x and $x \mapsto \nu^*(E_x)$ defines the zero function. Thus,

$$\int_X \nu^*(E_x) \, d\mu(x) = 0 = (\mu \times \nu)^*(E).$$

Step V. The general case.

Choose a $\mu \times \nu$-measurable set F that is a countable intersection of σ-sets all of finite measure such that $E \subseteq F$ and $(\mu \times \nu)^*(F) = (\mu \times \nu)^*(E)$. Let $G = F \setminus E$. Then G is a null set, and thus, by Step IV, $\nu^*(G_x) = 0$ holds for μ-almost all x. Therefore, E_x is ν-measurable and $\nu^*(E_x) = \nu^*(F_x)$ holds for μ-almost all x. By Step III, $x \mapsto \nu^*(F_x)$ defines an integrable function, and so, $x \mapsto \nu^*(E_x)$ defines an integrable function and

$$(\mu \times \nu)^*(E) = (\mu \times \nu)^*(F) = \int_X \nu^*(F_x) \, d\mu(x) = \int_X \nu^*(E_x) \, d\mu(x)$$

holds. The proof of the theorem is now complete.   ∎

Now, let $f : X \times Y \to \mathbb{R}$ be a function. Then for each fixed $x \in X$, the symbol f_x will denote the function $f_x : Y \to \mathbb{R}$ defined by $f_x(y) = f(x, y)$ for all $y \in Y$. Similarly, for each $y \in Y$ the notation f^y denotes the function $f^y : X \to \mathbb{R}$ defined by $f^y(x) = f(x, y)$ for all $x \in X$.

Definition 26.5.   *Let* $f : X \times Y \rightarrow \mathbb{R}$ *be a function. Then the* **iterated integral**
$\int\int f \, d\mu d\nu$ *is said to exist if*
1. f^y *is an integrable function over X for* ν*-almost all y, and*
2.   *the function*

$$g(y) = \int f^y \, d\mu = \int_X f(x, y) \, d\mu(x)$$

   defines an integrable function over Y .

The value of the iterated integral $\int\int f \, d\mu d\nu$ *is computed by starting with the*
innermost integration and by continuing with the second as follows:

$$\int\int f \, d\mu d\nu = \int_Y \left[\int_X f(x, y) \, d\mu(x) \right] d\nu(y),$$

holding y fixed while the integration over X is performed.

The meaning of the iterated integral $\int\int f \, d\nu d\mu$ is analogous. That is,

$$\int\int f \, d\nu d\mu = \int_X \left[\int_Y f(x, y) \, d\nu(y) \right] d\mu(x).$$

If E is a $\mu \times \nu$-measurable subset of $X \times Y$ with $(\mu \times \nu)^*(E) < \infty$, then it
is readily seen from Theorem 26.4 that both iterated integrals $\int\int \chi_E \, d\mu d\nu$ and
$\int\int \chi_E \, d\nu d\mu$ exist, and that

$$\int\int \chi_E \, d\mu d\nu = \int\int \chi_E \, d\nu d\mu = \int \chi_E \, d(\mu \times \nu) = (\mu \times \nu)^*(E).$$

Since every $\mu \times \nu$-step function is a linear combination of characteristic functions
of $\mu \times \nu$-measurable sets of finite measure, it follows from the prior observation
that if ϕ is a $\mu \times \nu$-step function, then both iterated integrals $\int\int \phi \, d\mu d\nu$ and
$\int\int \phi \, d\nu d\mu$ exist and, moreover,

$$\int\int \phi \, d\mu d\nu = \int\int \phi \, d\nu d\mu = \int \phi \, d(\mu \times \nu).$$

The previous identities regarding iterated integrals are special cases of a more
general result known as Fubini's[13] theorem.

---

[13] Guido Fubini (1879–1943), an Italian mathematician. He made important contributions in analysis,
geometry, and mathematical physics.

Theorem 26.6 (Fubini). *Let $f : X \times Y \to \mathbb{R}$ be a $\mu \times \nu$-integrable function. Then both iterated integrals exist and*

$$\int f\, d(\mu \times \nu) = \iint f\, d\mu d\nu = \iint f\, d\nu d\mu.$$

Proof. Without loss of generality we can assume that $f(x, y) \geq 0$ holds for all x and y. Choose a sequence $\{\phi_n\}$ of step functions such that $0 \leq \phi_n(x, y) \uparrow f(x, y)$ holds for all x and y. Thus,

$$\int_X \left[\int_Y \phi_n(x, y)\, d\nu(y) \right] d\mu(x) = \int \phi_n\, d(\mu \times \nu) \uparrow \int f\, d(\mu \times \nu) < \infty. \quad (\star)$$

By Theorem 26.4, for each n the function

$$g_n(x) = \int (\phi_n)_x\, d\nu = \int_Y \phi_n(x, y)\, d\nu(y)$$

defines an integrable function over X; and clearly, $g_n(x) \uparrow$ holds for μ-almost all x. But then, by Levi's Theorem 22.8, there exists a μ-integrable function $g : X \to \mathbb{R}$ such that $g_n(x) \uparrow g(x)$ μ-a.e. holds. That is, there exists a μ-null subset A of X such that $\int (\phi_n)_x\, d\nu \uparrow g(x) < \infty$ holds for all $x \notin A$. Since $(\phi_n)_x \uparrow f_x$ holds for each x, it follows that f_x is ν-integrable for all $x \notin A$ and

$$g_n(x) = \int (\phi_n)_x\, d\nu = \int_Y \phi_n(x, y)\, d\nu(y) \uparrow \int_Y f_x\, d\nu$$

holds for all $x \notin A$.

Now, $(\star)$ combined with Theorem 21.6 implies that the function $x \mapsto \int_Y f_x\, d\nu$ defines an integrable function such that

$$\int f\, d(\mu \times \nu) = \int_X \left(\int_Y f_x\, d\nu \right) d\mu = \iint f\, d\nu d\mu.$$

Similarly, $\int f\, d(\mu \times \nu) = \iint f\, d\mu d\nu$, and the proof is complete. ∎

The existence of the iterated integrals is by no means enough to ensure that the function is integrable over the product space. For instance, let $X = Y = [0, 1]$, $\mu = \nu = \lambda$ (the Lebesgue measure), and $f(x, y) = (x^2 - y^2)/(x^2 + y^2)^2$ if

$(x, y) \neq (0, 0)$ and $f(0, 0) = 0$. Then, it is easy to see that

$$\iint f \, d\mu dv = -\frac{\pi}{4} \quad \text{and} \quad \iint f \, dv d\mu = \frac{\pi}{4}.$$

Fubini's theorem shows, of course, that f is not integrable over $[0, 1] \times [0, 1]$.

However, there is a converse to Fubini's theorem according to which the existence of one of the iterated integrals is sufficient for the integrability of the function over the product space. The theorem is known as Tonelli's[14] theorem, and this result is frequently used in applications.

Theorem 26.7 (Tonelli). *Let $(X, \mathcal{S}, \mu)$ and (Y, Σ, v) be two σ-finite measure spaces, and let $f : X \times Y \to \mathbb{R}$ be a $\mu \times v$-measurable function. If one of the iterated integrals $\iint |f| \, d\mu dv$ or $\iint |f| \, dv d\mu$ exists, then the function f is $\mu \times v$-integrable—and hence, the other iterated integral exists and*

$$\int f \, d(\mu \times v) = \iint f \, d\mu dv = \iint f \, dv d\mu.$$

Proof. We can assume without loss of generality that $f(x, y) \geq 0$ holds for all x and y. Since $(X, \mathcal{S}, \mu)$ and (Y, Σ, v) are σ-finite measure spaces, it is easy to see that the product measure space is also a σ-finite measure space. Choose a sequence $\{A_n\}$ of $\mu \times v$-measurable sets such that $(\mu \times v)^*(A_n) < \infty$ for each n and $A_n \uparrow X \times Y$. By Theorem 17.7, there exists a sequence $\{\psi_n\}$ of $\mu \times v$-simple functions such that $0 \leq \psi_n(x, y) \uparrow f(x, y)$ holds for all x and y. Let $\phi_n = \psi_n \cdot \chi_{A_n}$ for each n. Then $\{\phi_n\}$ is a sequence of $\mu \times v$-step functions such that $0 \leq \phi_n(x, y) \uparrow f(x, y)$ holds for all x and y.

Now, assume that $\iiint f \, d\mu dv$ exists. This means that for v-almost all y, the integral $\int f(x, y) d\mu(x)$ exists and defines a v-integrable function. From $\phi_n(x, y) \uparrow f(x, y)$ it follows that $\int \phi_n(x, y) d\mu(x) \uparrow \int f(x, y) d\mu(x)$ holds for v-almost all y. But then, by applying the Lebesgue dominated convergence theorem, we get

$$\int \phi_n d(\mu \times v) = \int_Y \left[\int_X \phi_n(x, y) d\mu(x) \right] dv(y) \uparrow \iint f \, d\mu dv < \infty.$$

This shows that f is a $\mu \times v$-upper function and that $\int f \, d(\mu \times v) = \iint f \, d\mu dv$ holds. The rest of the proof now follows immediately from Fubini's theorem. ∎

The Fubini and Tonelli theorems are usually referred to as "the method of computing a double integral by changing the order of integration."

---

[14]Leonida Tonelli (1885–1946), an Italian mathematician. He contributed to measure theory, the theory of integration, and to calculus of variations.

In general, it is a difficult problem to determine whether or not a given function $f : X \times Y \to \mathbb{R}$ is $\mu \times \nu$-measurable. However, in a number of applications the $\mu \times \nu$-measurability of f can be established from topological considerations. For instance, if $X = Y = \mathbb{R}$ and $\mu = \nu =$ the Lebesgue measure, then it should be clear that the product measure $\mu \times \nu$ on $\mathbb{R}^2$ is precisely the Lebesgue measure on $\mathbb{R}^2$. Therefore, every continuous real-valued function on $\mathbb{R}^2$ is necessarily $\mu \times \nu$-measurable. For more about the joint measurability of functions, see also Section 20.

EXERCISES

1. Let (X, S, μ) and (Y, Σ, ν) be two measure spaces, and let $A \times B \in \Lambda_\mu \otimes \Lambda_\nu$.

   a. Show that $\mu^*(A) \cdot \nu^*(B) \leq (\mu \times \nu)^*(A \times B)$.
   b. Show that if $\mu^*(A) \cdot \nu^*(B) \neq 0$, then $(\mu \times \nu)^*(A \times B) = \mu^*(A) \cdot \nu^*(B)$.
   c. Give an example for which $(\mu \times \nu)^*(A \times B) \neq \mu^*(A) \cdot \nu^*(B)$.

2. Let (X, S, μ) and (Y, Σ, ν) be two σ-finite measure spaces. Then show that

   $$(\mu \times \nu)^*(A \times B) = \mu^*(A) \cdot \nu^*(B)$$

   holds for each $A \times B \in \Lambda_\mu \otimes \Lambda_\nu$.

3. Let (X, S, μ) and (Y, Σ, ν) be two measure spaces. Assume that A and B are subsets of X and Y, respectively, such that $0 < \mu^*(A) < \infty$, and $0 < \nu^*(B) < \infty$. Then show that $A \times B$ is $\mu \times \nu$-measurable if and only if both A and B are measurable in their corresponding spaces. Is the above conclusion true if either A or B has measure zero?

4. Let (X, S, μ) and (Y, Σ, ν) be two σ-finite measure spaces, and let $f : X \times Y \to \mathbb{R}$ be a $\mu \times \nu$-measurable function. Show that for μ-almost all x, the function f_x is a ν-measurable function. Similarly, show that for ν-almost all y, the function f^y is μ-measurable.

5. Show that if $f(x, y) = (x^2 - y^2)/(x^2 + y^2)^2$, with $f(0, 0) = 0$, then

   $$\int_0^1 \left[\int_0^1 f(x, y)\, dx \right] dy = -\frac{\pi}{4} \quad \text{and} \quad \int_0^1 \left[\int_0^1 f(x, y)\, dy \right] dx = \frac{\pi}{4}.$$

6. Let $X = Y = \mathbb{N}$, $S = \Sigma =$ the collection of all subsets of $\mathbb{N}$, and $\mu = \nu =$ the counting measure. Give an interpretation of Fubini's theorem in this case.

7. Establish the following result, known as **Cavalieri's**[15] **principle**. Let (X, S, μ) and (Y, Σ, ν) be two measure spaces, and let E and F be two $\mu \times \nu$-measurable subsets of $X \times Y$ of finite measure. If $\nu^*(E_x) = \nu^*(F_x)$ holds for μ-almost all x, then

   $$(\mu \times \nu)^*(E) = (\mu \times \nu)^*(F).$$

8. For this exercise, λ denotes the Lebesgue measure on $\mathbb{R}$. Let (X, S, μ) be a σ-finite measure space, and let $f : X \to \mathbb{R}$ be a measurable function such that $f(x) \geq 0$

---

[15]Bonaventura Cavalieri (1589–1647), an Italian mathematician. He was a geometer who worked on problems regarding volumes of solids and wrote several monographs on this subject.

holds for all $x \in X$. Then show that

a. The set $A = \{(x, y) \in X \times \mathbb{R} : 0 \le y \le f(x)\}$, called the **ordinate set** of f, is a $\mu \times \lambda$-measurable subset of $X \times \mathbb{R}$.

b. The set $B = \{(x, y) \in X \times \mathbb{R} : 0 \le y < f(x)\}$ is a $\mu \times \lambda$-measurable subset of $X \times \mathbb{R}$ and $(\mu \times \lambda)^*(A) = (\mu \times \lambda)^*(B)$ holds.

c. The graph of f, i.e., the set $G = \{(x, f(x)) : x \in X\}$, is a $\mu \times \lambda$-measurable subset of $X \times \mathbb{R}$.

d. If f is μ-integrable, then $(\mu \times \lambda)^*(A) = \int f \, d\mu$ holds.

e. If f is μ-integrable, then $(\mu \times \lambda)^*(G) = 0$ holds.

9. Let $g : X \to \mathbb{R}$ be a μ-integrable function, and let $h : Y \to \mathbb{R}$ be a ν-integrable function. Define $f : X \times Y \to \mathbb{R}$ by $f(x, y) = g(x)h(y)$ for each x and y. Show that f is $\mu \times \nu$-integrable and that

$$\int f \, d(\mu \times \nu) = \left(\int_X g \, d\mu \right) \cdot \left(\int_Y h \, d\nu \right).$$

10. Use Tonelli's theorem to verify that

$$\int_\epsilon^r \frac{\sin x}{x} \, dx = \int_0^\infty \left(\int_\epsilon^r e^{-xy} \sin x \, dx \right) dy$$

holds for each $0 < \epsilon < r$. By letting $\epsilon \to 0^+$ and $r \to \infty$ (and justifying your steps) give another proof of the formula

$$\int_0^\infty \frac{\sin x}{x} \, dx = \frac{\pi}{2}.$$

11. Show that if $f(x, y) = ye^{-(1+x^2)y^2}$ for each x and y, then

$$\int_0^\infty \left[\int_0^\infty f(x, y) \, dx \right] dy = \int_0^\infty \left[\int_0^\infty f(x, y) \, dy \right] dx.$$

Use the previous equality to give an alternate proof of the formula

$$\int_0^\infty e^{-x^2} \, dx = \frac{\sqrt{\pi}}{2}.$$

12. Show that

$$\int_0^\infty \left(\int_0^r e^{-xy^2} \sin x \, dx \right) dy = \int_0^r \left(\int_0^\infty e^{-xy^2} \sin x \, dy \right) dx$$

holds for all $r > 0$. By letting $r \to \infty$ show that

$$\int_0^\infty \frac{\sin x}{\sqrt{x}} \, dx = \frac{\sqrt{2\pi}}{2}.$$

In a similar manner show that $\int_0^\infty \frac{\cos x}{\sqrt{x}} \, dx = \sqrt{2\pi}/2$.

13. Using the conclusions of the preceding exercise (and an appropriate change of variable), show that the values of the **Fresnel integrals** (see Exercise 6 of Section 24) are

$$\int_0^\infty \sin(x^2) \, dx = \int_0^\infty \cos(x^2) \, dx = \frac{\sqrt{2\pi}}{4}.$$

14. Let $X = Y = [0, 1]$, μ = the Lebesgue measure on $[0, 1]$, and ν = the counting measure on $[0, 1]$. Consider the "diagonal" $\Delta = \{(x, x) : x \in X\}$ of $X \times Y$. Show that:

    a. Δ is a $\mu \times \nu$-measurable subset of $X \times Y$, and hence, χ_Δ is a non-negative $\mu \times \nu$-measurable function.

    b.   Both iterated integrals $\iint \chi_\Delta \, d\mu d\nu$ and $\iint \chi_\Delta \, d\nu d\mu$ exist.

    c.   The function χ_Δ is not $\mu \times \nu$-integrable. Why doesn't this contradict Tonelli's theorem?

15. Let $f : \mathbb{R} \to \mathbb{R}$ be Borel measurable. Then show that the functions $f(x + y)$ and $f(x - y)$ are both $\lambda \times \lambda$-measurable.

    [HINT: Consider first $f = \chi_V$ for some open set V.]

CHAPTER 5

NORMED SPACES AND L_p-SPACES

The algebraic theory of vector spaces has been an integral part of modern mathematics for some time. In analysis, one studies vector spaces from the topological point of view, taking into consideration the already existing algebraic structure. The most fruitful study comes when one attaches to every vector a real number, called the *norm* of the vector. The norm can be thought of as a generalization of the concept of length. A normed space that is complete (in the metric induced by its norm) is called a Banach[1] space.

A variety of diverse problems from different branches of mathematics (and science in general) can be translated to the framework of Banach space theory and solved by applying its powerful techniques. For this reason the theory of Banach spaces is in the frontier of modern mathematical research.

This chapter presents a brief introduction to the theory of normed and Banach spaces. After developing the basic properties of normed spaces, the three cornerstones of functional analysis are proved—the principle of uniform boundedness, the open mapping theorem, and the Hahn–Banach theorem.[2] Then a section is devoted to the study of Banach lattices; that is, Banach spaces whose norms are compatible with the lattice structure of the spaces. As we shall see, many Banach lattices are actually old friends. Finally, the classical, L_p-spaces are investigated, and the theory of integration is placed in its appropriate perspective.

27. NORMED SPACES AND BANACH SPACES

A real-valued function $\|\cdot\|$ defined on a vector space X is called a **norm** if it satisfies the following three properties:

---

[1] Stefan Banach (1892–1945), a prominent Polish mathematician. He is the founder of the contemporary field of functional analysis.

[2] Hans Hahn (1879–1934), an Austrian mathematician and philosopher. He contributed decisively to functional analysis, general topology, and the foundations of mathematics.

1. $\|x\| \geq 0$ for each $x \in X$, and $\|x\| = 0$ if and only if $x = 0$;
2. $\|\alpha x\| = |\alpha| \cdot \|x\|$ for all $x \in X$ and $\alpha \in \mathbb{R}$;
3. $\|x + y\| \leq \|x\| + \|y\|$ for all $x, y \in X$.

Property (3) is referred to as the **triangle inequality**, and it is equivalent to the inequality

$$\|x - y\| \leq \|x - z\| + \|z - y\|$$

for all x, y, and z in X.

A vector space X equipped with a norm $\|\cdot\|$ is called a **normed vector space**, or simply a **normed space**. To avoid trivialities, the vector spaces will be tacitly assumed to be different from $\{0\}$. Also, they will be assumed to be real vector spaces.

On a normed space X, a metric is defined in terms of the norm $\|\cdot\|$ via the function $d(x, y) = \|x - y\|$. From the properties of the norm, it is a routine matter to verify that $d(\cdot, \cdot)$ is indeed a metric on X. We shall call this metric on X the **metric induced by the norm**. A sequence $\{x_n\}$ in X is said to converge in norm to x if $\lim \|x - x_n\| = 0$; that is, if $\{x_n\}$ converges to x with respect to the metric induced by the norm.

In view of the triangle inequality,

$$| \|x\| - \|y\| | \leq \|x - y\|$$

holds for all x and y in X. This readily implies that the norm considered as a function $x \mapsto \|x\|$ for X into $\mathbb{R}$ is uniformly continuous. Also, by the triangle inequality it is easy to prove, but important to observe, that if $x_n \to x$ and $y_n \to y$ in X and $\alpha_n \to \alpha$ in $\mathbb{R}$, then $x_n + y_n \to x + y$ and $\alpha_n x_n \to \alpha x$ hold.

A subset A of a normed space is said to be **norm bounded** (or simply **bounded**) if there exists some $M > 0$ such that $\|x\| \leq M$ holds for each $x \in A$. Every Cauchy sequence $\{x_n\}$ of a normed space is bounded. Indeed, to see this choose k such that $\|x_n - x_m\| < 1$ for all $n, m \geq k$, and let $M = \max\{1 + \|x_i\|: 1 \leq i \leq k\}$. Then the inequality $\|x_n\| \leq 1 + \|x_k\|$ for $n \geq k$ implies $\|x_n\| \leq M$ for all n.

A normed space X that is complete with respect to the metric induced by its norm is called a **Banach space**. In other words, X is a Banach space if for every Cauchy sequence $\{x_n\}$ of X there exists an element $x \in X$ such that $\lim \|x_n - x\| = 0$. Thus, Banach spaces are special examples of complete metric spaces.

Some examples of Banach spaces are presented next.

Example 27.1. The vector space $\mathbb{R}^n$ with the norm $\|x\| = (\sum_{i=1}^n x_i^2)^{\frac{1}{2}}$ for each $x = (x_1, \ldots, x_n) \in \mathbb{R}^n$ is a Banach space. This norm is called the **Euclidean norm**, and it gives the Euclidean metric.   ∎

Example 27.2. Let X be a nonempty set, and let $B(X)$ be the vector space of all bounded real-valued functions defined on X. Then $\|f\|_\infty = \sup\{|f(x)|: x \in X\}$ for each $f \in B(X)$

defines a norm on $B(X)$, called the **sup norm**. A glance at Example 6.12 guarantees that the vector space $B(X)$ with the sup norm is a Banach space.   ∎

Example 27.3. Let ℓ_1 denote the collection of all real sequences $x = (x_1, x_2, \ldots)$ such that $\sum_{n=1}^{\infty} |x_n| < \infty$. It is easy to see that ℓ_1 is a vector space under the algebraic operations

$$x + y = (x_1 + y_1, x_2 + y_2, \ldots) \quad \text{and} \quad \alpha x = (\alpha x_1, \alpha x_2, \ldots).$$

Moreover, if for every $x \in \ell_1$ we define $\|x\|_1 = \sum_{n=1}^{\infty} |x_n|$, then $\|\cdot\|_1$ is a norm on ℓ_1. The verification of the norm properties are left for the reader. We show next that ℓ_1 is actually a Banach space.

To this end, let $\{x_n\}$ be a Cauchy sequence of ℓ_1. That is, for every $\epsilon > 0$ there exists some k such that $\|x_n - x_m\|_1 < \epsilon$ holds for all $n, m > k$. It follows that there exists some $M > 0$ such that $\|x_n\|_1 \leq M$ for each n. Let $x_n = (x_1^n, x_2^n, \ldots)$ for each n. The relation

$$\left| x_i^n - x_i^m \right| \leq \sum_{i=1}^{\infty} \left| x_i^n - x_i^m \right| = \|x_n - x_m\|_1 < \epsilon$$

for $n, m > k$ implies that for each fixed i, the sequence of real numbers $\{x_i^n\}$ is a Cauchy sequence. Let $x_i = \lim_{n \to \infty} x_i^n$ for each i. Since for all $n > k$ and each p we have

$$\sum_{i=1}^{p} |x_i| \leq \sum_{i=1}^{p} |x_i - x_i^n| + \sum_{i=1}^{p} |x_i^n| \leq \sum_{i=1}^{p} |x_i - x_i^n| + \|x_n\|_1$$

$$= \lim_{m \to \infty} \left(\sum_{i=1}^{p} |x_i^m - x_i^n| \right) + \|x_n\|_1 \leq \epsilon + M < \infty,$$

it follows that $x = (x_1, x_2, \ldots) \in \ell_1$. Also, since $\sum_{i=1}^{p} |x_i^n - x_i^m| \leq \|x_n - x_m\|_1 < \epsilon$ for all $n, m > k$, we have $\sum_{i=1}^{p} |x_i - x_i^n| = \lim_{m \to \infty} (\sum_{i=1}^{p} |x_i^m - x_i^n|) \leq \epsilon$ for all p and all $n > k$. Hence, $\|x - x_n\|_1 \leq \epsilon$ holds for all $n > k$, so that $\{x_n\}$ converges to x in ℓ_1. That is, ℓ_1 is a Banach space.   ∎

Example 27.4. Consider an interval $[a, b]$ and a natural number k. Let $C^k[a, b]$ denote the collection of all real-valued functions defined on $[a, b]$ which have a continuous kth order derivative on $[a, b]$ (with right and left derivatives at the endpoints).

Clearly, $C^k[a, b]$ with the pointwise algebraic operations is a vector space. Moreover, if for each $f \in C^k[a, b]$ we let

$$\|f\| = \|f\|_\infty + \|f'\|_\infty + \cdots + \|f^{(k)}\|_\infty,$$

then $\|\cdot\|$ defines a norm under which $C^k[a, b]$ is a Banach space.

It is straightforward to verify that $\|\cdot\|$ is indeed a norm. To see that $C^k[a, b]$ is complete under $\|\cdot\|$, let $\{f_n\}$ be a Cauchy sequence of $C^k[a, b]$. Then, it is easy to see that there exist continuous functions $g_0, g_1, \ldots, g_k$ such that for each $i = 0, 1, \ldots, k$ the sequence of functions $\{f_n^{(i)}\}$ converges uniformly to g_i. Now, if $1 \leq i \leq k$, then

$$f_n^{(i-1)}(x) = f_n^{(i-1)}(a) + \int_a^x f_n^{(i)}(t)\, dt \qquad (\star)$$

holds for each $x \in [a, b]$. Thus, by the uniform convergence of the sequence $\{f_n^{(i)}\}$ it follows from $(\star)$ that

$$g_{i-1}(x) = g_{i-1}(a) + \int_a^x g_i(t)\, dt$$

holds for each $x \in [a, b]$. By the fundamental theorem of calculus, g_{i-1} is differentiable (for $1 \leq i \leq k$), and $g'_{i-1}(x) = g_i(x)$ for each $x \in [a, b]$. In particular, note that $g_i = g_0^{(i)}$ for $i = 1, \ldots, k$. Therefore, $g_0 \in C^k[a, b]$ and $\lim \| f_n - g_0 \| = 0$ holds, so that $C^k[a, b]$ is a Banach space. ∎

Two norms $\|\cdot\|_1$ and $\|\cdot\|_2$ on a vector space X are said to be **equivalent** if there exist two constants $K > 0$ and $M > 0$ such that

$$K \|x\|_1 \leq \|x\|_2 \leq M \|x\|_1$$

holds for each $x \in X$. The reader should stop and verify that two norms are equivalent if and only if they generate the same open sets.

In particular, note that if two norms on a vector space X are equivalent, then X is a Banach space with respect to one of them if and only if X is a Banach space with respect to the other. Also, observe that if $\|\cdot\|_1$ is equivalent to $\|\cdot\|_2$ and $\|\cdot\|_2$ is equivalent to $\|\cdot\|_3$, then $\|\cdot\|_1$ is equivalent to $\|\cdot\|_3$.

In a normed vector space, all open balls "look alike." More precisely, any two open balls are homeomorphic. Indeed, if $B(a, r)$ is an arbitrary open ball, then it is easy to see that the function $x \mapsto a + rx$ is a homeomorphism from $B(0, 1)$ onto $B(a, r)$. For this reason, the open ball $B(0, 1) = \{x \colon \|x\| < 1\}$ plays an important role in the study of normed spaces and is called the **open unit ball** of the space (and likewise $\{x \colon \|x\| \leq 1\}$ is called the **closed unit ball** of the space).

Example 27.5. It is instructive to consider various norms on $\mathbf{R}^2$. For $v = (x, y) \in \mathbf{R}^2$ define

$$\|v\|_1 = |x| + |y|,$$
$$\|v\|_2 = (x^2 + y^2)^{\frac{1}{2}},$$
$$\|v\|_\infty = \max\{|x|, |y|\},$$
$$\|v\| = \left(\frac{x^2}{a^2} + \frac{y^2}{b^2} \right)^{\frac{1}{2}},$$

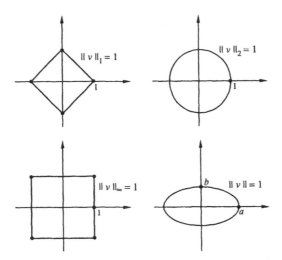

FIGURE 5.1.   The Unit Balls of Various Norms

where a and b are two, fixed, positive real numbers. The reader should stop and verify that the previous functions are norms and that they are, in fact, all equivalent. The geometric shape of the closed unit ball for each norm is illustrated in Figure 5.1.   ∎

The fact that all norms in the preceding example were equivalent is not accidental.

Theorem 27.6.   *In a finite dimensional vector space all norms are equivalent.*

Proof.   Without loss of generality, we can assume that the finite dimensional space is $\mathbb{R}^n$. Let $\|\cdot\|_2$ denote its Euclidean norm; that is, $\|x\|_2 = (x_1^2 + \cdots + x_n^2)^{\frac{1}{2}}$. Also, let $\|\cdot\|$ be another norm on $\mathbb{R}^2$. It suffices to establish that $\|\cdot\|$ is equivalent to $\|\cdot\|_2$.

If $x = (x_1, \ldots, x_n) = x_1 e_1 + \cdots + x_n e_n$ (where the e_n are the standard basic unit vectors), then by the triangle inequality we have

$$\|x\| = \left\| \sum_{i=1}^{n} x_i e_i \right\| \leq \sum_{i=1}^{n} |x_i| \cdot \|e_i\| \leq \left(\sum_{i=1}^{n} \|e_i\| \right) \cdot \|x\|_2.$$

Thus, if $M = \sum_{i=1}^{n} \|e_i\|$, then

$$\|x\| \leq M \|x\|_2$$

holds for all $x \in \mathbb{R}^n$. This establishes one-half of the desired inequality. In addition, the inequalities

$$\mid \|x\| - \|y\| \mid \le \|x - y\| \le M \|x - y\|_2$$

show that the function $x \mapsto \|x\|$ from $\mathbb{R}^n$ with the Euclidean norm into $\mathbb{R}$ is (uniformly) continuous.

Let $S = \{x \in \mathbb{R}^n : \|x\|_2 = 1\}$ be the "unit sphere" for the Euclidean norm. Then S is closed and bounded, and hence, by the Heine–Borel Theorem 7.4, S is compact for the Euclidean norm. In particular, the continuous function $x \mapsto \|x\|$ attains a minimum value on S, say at x_0. Thus, $\|x\| \ge \|x_0\|$ holds for all $x \in S$. Let $K = \|x_0\|$. Since $\|x_0\|_2 = 1$, it follows that $x_0 \ne 0$, and so $K = \|x_0\| > 0$. Now, if $x \in \mathbb{R}^n$ is nonzero, then $\|x\|/\|x\|_2 = \|x/\|x\|_2\| \ge K$, and so

$$K \|x\|_2 \le \|x\|$$

holds for all $x \in \mathbb{R}^n$. This establishes the other half of the desired inequality and completes the proof of the theorem. ∎

As an application of the last theorem, let us establish the following useful result:

Theorem 27.7. *Every finite dimensional vector subspace of a normed space is closed.*

Proof. Let Y be a finite dimensional vector subspace of a normed space X. Then Y can be identified (linearly isomorphically) with some $\mathbb{R}^n$. Hence, by Theorem 27.6, the norm of X restricted to Y must be equivalent to the Euclidean norm. In particular, Y is a complete metric space, and so (by Theorem 6.13) Y is closed. ∎

The structure of the open (or closed) unit ball reflects the topological and geometrical properties of all balls. For example, a normed space is locally compact if and only if its closed unit ball is compact. (Recall that a topological space is locally compact if each point has a neighborhood whose closure is compact.)

The next theorem tells us that the finite dimensional normed spaces are the only locally compact normed spaces.

Theorem 27.8. *A normed space is locally compact if and only if it is finite dimensional.*

Proof. If the normed space is finite dimensional, then by Theorem 27.6, its norm is equivalent to the Euclidean norm. By Theorem 7.4, the closed unit ball is compact with respect to the Euclidean metric, and from this it follows (how?) that the space is locally compact.

For the converse, assume that X is a locally compact normed space. Since all closed balls are homeomorphic, the closed unit ball $V = \{x \in X: \|x\| \le 1\}$ must be compact. Choose $x_1, \ldots, x_n \in V$ such that $V \subseteq \bigcup_{i=1}^{n} B(x_i, \frac{1}{2})$, and let Y be the linear subspace generated by $\{x_1, \ldots, x_n\}$. We shall show that $X = Y$, and this will establish that X is finite dimensional.

Assume by way of contradiction that $X \ne Y$. Thus, there exists some $x_0 \in X$ with $x_0 \notin Y$. Since (by Theorem 27.7) the vector subspace Y is closed, it follows that $d(x_0, Y) = \inf\{\|x_0 - y\|: y \in Y\} > 0$ (also, see Exercise 1 of Section 10). Pick some $y \in Y$ with $\|x_0 - y\| < 2d(x_0, Y)$. Then for some $1 \le i \le n$ we have $(x_0 - y)/\|x_0 - y\| \in B(x_i, \frac{1}{2})$. Since $y + \|x_0 - y\|x_i$ belongs to Y, it follows that

$$\frac{1}{2} > \left\| \frac{x_0 - y}{\|x_0 - y\|} - x_i \right\| = \frac{\|x_0 - (y + \|x_0 - y\|x_i)\|}{\|x_0 - y\|} \ge \frac{d(x_0, Y)}{\|x_0 - y\|} > \frac{1}{2}$$

which is impossible. This contradiction completes the proof of the theorem. ∎

EXERCISES

1. Let X be a normed space. Then show that X is a Banach space if and only if its unit sphere $\{x \in X: \|x\| = 1\}$ is a complete metric space (under the induced metric $d(x, y) = \|x - y\|$).

2. Let X be a normed vector space. Fix $a \in X$ and a nonzero scalar α.

   a. Show that the mappings $x \mapsto a + x$ and $x \mapsto \alpha x$ are both homeomorphisms.
   b. If A and B are two sets with either A or B open and α and β are nonzero scalars, then show that $\alpha A + \beta B$ is an open set.

3. Let X be a normed vector space, and let $B = \{x \in X: \|x\| < 1\}$ be its open unit ball. Show that $\overline{B} = \{x \in X: \|x\| \le 1\}$.

4. Let X be a normed space, and let $\{x_n\}$ be a sequence of X such that $\lim x_n = x$ holds. If $y_n = n^{-1}(x_1 + \cdots + x_n)$ for each n, then show that $\lim y_n = x$. (See also Exercise 11 of Section 4.)

5. Assume that two vectors x and y in a normed space satisfy $\|x + y\| = \|x\| + \|y\|$. Then show that

$$\|\alpha x + \beta y\| = \alpha \|x\| + \beta \|y\|$$

   holds for all scalars $\alpha \ge 0$ and $\beta \ge 0$.

6. Let X be the vector space of all real-valued functions defined on $[0, 1]$ having continuous first-order derivatives. Show that $\|f\| = |f(0)| + \|f'\|_\infty$ is a norm on X that is equivalent to the norm $\|f\|_\infty + \|f'\|_\infty$.

7. A series $\sum_{n=1}^{\infty} x_n$ in a normed space is said to **converge** to x if $\lim \|x - \sum_{i=1}^{n} x_i\| = 0$. As usual, we write $x = \sum_{n=1}^{\infty} x_n$. A series $\sum_{n=1}^{\infty} x_n$ is said to be **absolutely summable** if $\sum_{n=1}^{\infty} \|x_n\| < \infty$ holds.

   Show that a normed space X is a Banach space if and only if every absolutely summable series is convergent.

8. Show that a closed proper vector subspace of a normed vector space is nowhere dense.

9. Assume that $f: [0, 1] \to \mathbb{R}$ is a continuous function which is not a polynomial. By Corollary 11.6, we know that there exists a sequence of polynomials $\{p_n\}$ that converges uniformly to f. Show that the set of natural numbers

$$\{k \in \mathbb{N}: \ k = \text{degree of } p_n \text{ for some } n\}$$

is countable.

[HINT: Show that the vector subspace of $C[0, 1]$ consisting of all polynomials of degree less than or equal to some m is closed.]

10. This exercise describes some classes of important subsets of a vector space. A nonempty subset A of a vector space X is said to be:

a. **symmetric**, if $x \in A$ implies $-x \in A$, i.e., if $A = -A$;
b. **convex**, if $x, y \in A$ implies $\lambda x + (1 - \lambda)y \in A$ for all $0 \le \lambda \le 1$, i.e., for every two vectors $x, y \in A$ the line segment joining x and y lies in A;
c. **circled** (or **balanced**) if $x \in A$ implies $\lambda x \in A$ for each $|\lambda| \le 1$.

Establish the following:

i.   A circled set is symmetric.
ii. A convex and symmetric set containing zero is circled.
iii. A nonempty subset B of a vector space is convex if and only if $aB + bB = (a+b)B$ holds for all scalars $a \ge 0$ and $b \ge 0$.
iv. If A is a convex subset of a normed space, then the closure $\overline{A}$ and the interior A° of A are also convex sets.

11. This exercise describes all norms on a vector space X that are equivalent to a given norm. So, let $(X, \|\cdot\|)$ be a normed vector space. Let A be a bounded convex symmetric subset of X having zero as an interior point (relative to the topology generated by the norm $\|\cdot\|$). Define the function $p_A: X \to \mathbb{R}$ by

$$p_A(x) = \inf\{\lambda > 0: \ x \in \lambda A\}.$$

Establish the following:

a. The function p_A is a well-defined norm on X.
b. The norm p_A is equivalent to $\|\cdot\|$, i.e., there exist two constants $C > 0$ and $K > 0$ such that $C\|x\| \le p_A(x) \le K\|x\|$ holds true for each $x \in X$.
c. The closed unit ball of p_A is the closure of A, i.e., $\{x \in X: \ p_A(x) \le 1\} = \overline{A}$.
d. Let $\|\|\cdot\|\|$ be a norm on X which is equivalent to $\|\cdot\|$, and consider the norm bounded nonempty symmetric convex set $B = \{x \in X: \ \|\|x\|\| \le 1\}$. Then zero is an interior point of B and $\|\|x\|\| = p_B(x)$ holds for each $x \in X$.

28. OPERATORS BETWEEN BANACH SPACES

In this section, we shall study linear operators (or transformations). Recall that a function $T: X \to Y$ between two vector spaces is called a **linear operator** (or simply an **operator**) if $T(\alpha x + \beta y) = \alpha T(x) + \beta T(y)$ holds for all $x, y \in X$, and all $\alpha, \beta \in \mathbb{R}$. Observe that every linear operator T satisfies $T(0) = 0$.

If X and Y are two normed spaces, then the symbol $\|\cdot\|$ will be used (as usual) to denote the norm on both spaces.

Definition 28.1. *Let $T: X \to Y$ be a linear operator between two normed spaces. Then the **operator norm** of T is defined by*

$$\|T\| = \sup\{\|T(x)\|: \|x\| = 1\}.$$

*If $\|T\|$ is finite, then T is called a **bounded operator** (and, of course, if $\|T\| = \infty$, then T is called an **unbounded operator**).*

Observe that

$$\|T(x)\| \leq \|T\| \cdot \|x\|$$

holds for all $x \in X$. To see this, note that if $x \neq 0$, then the vector $y = x/\|x\|$ satisfies $\|y\| = 1$, and so

$$\frac{\|T(x)\|}{\|x\|} = \left\| T\left(\frac{x}{\|x\|}\right) \right\| = \|T(y)\| < \|T\|$$

holds. In particular, it follows from the previous inequality that

$$\|T\| = \sup\{\|T(x)\|: \|x\| \leq 1\}.$$

Next, some concrete examples of operators are presented.

Example 28.2.   Let $X = C[0, 1]$ with the sup norm. Define $T: X \to X$ by $T(f)(x) = xf(x)$ for each $f \in X$ and $x \in [0, 1]$. Clearly, T is a linear operator such that $\|T(f)\|_\infty \leq \|f\|_\infty$ for each $f \in X$. This shows that $\|T\| \leq 1$ (in fact, $\|T\| = 1$), and so T is a bounded operator.   ■

The next example presents an unbounded operator.

Example 28.3.   Let X be the vector space of all real-valued functions on $[0, 1]$ that have continuous derivatives with the sup norm. Also, let $Y = C[0, 1]$ with the sup norm. Define $D: X \to Y$ by $D(f) = f'$ (the **differential operator**). It is easy to see that D is a linear operator. However, $\|D\| = \infty$. Indeed, if $f_n(x) = x^n$, then $\|f_n\|_\infty = 1$ and $\|D(f_n)\|_\infty = \sup\{nx^{n-1}: x \in [0, 1]\} = n$ hold for each n, implying $\|D\| = \infty$. Hence, D is an unbounded operator.   ■

The next example is drawn from the important class of operators known as "integral operators."

Example 28.4.   Let $[a, b]$ be a (finite) closed interval, and let $K: [a, b] \times [a, b] \to \mathbb{R}$ be a continuous function. Consider the vector space $C[a, b]$ with the sup norm, and then define

$T: C[a, b] \to C[a, b]$ by

$$T(f)(x) = \int_a^b K(x, y) f(y) \, dy$$

for each $f \in C[a, b]$. (The uniform continuity of K on $[a, b] \times [a, b]$ can be invoked here to verify that indeed $T(f) \in C[a, b]$ for each $f \in C[a, b]$.) For obvious reasons, the operator T is called an **integral operator** and the function K is referred to as the **kernel** of T.

Clearly, T is a linear operator. On the other hand, if

$$M = \sup\{|K(x, y)|: (x, y) \in [a, b] \times [a, b]\} < \infty,$$

then the estimate $|T(f)(x)| \leq M(b - a) \|f\|_\infty$ shows that $\|T(f)\|_\infty \leq M(b - a) \|f\|_\infty$. Thus, $\|T\| \leq M(b - a) < \infty$, and so T is a bounded operator.

It is worth verifying that T also satisfies the following important property:

- If $B = \{f \in C[a, b]: \|f\|_\infty < 1\}$ is the open unit ball of $C[a, b]$, then $\overline{T(B)}$ is a compact subset of $C[a, b]$.

To see this, observe first that $\overline{T(B)}$ is closed and bounded. Thus, by the Ascoli–Arzelà Theorem 9.10, we need only to show that $T(B)$ is an equicontinuous subset of $C[a, b]$. To see this, fix $x_0 \in [a, b]$, and let $\epsilon > 0$. Since (by Theorem 7.7) K is uniformly continuous, there exists some $\delta > 0$ such that $|K(x_1, y) - K(x_2, y)| < \epsilon$ holds whenever $|x_1 - x_2| < \delta$. Therefore, if $x \in [a, b]$ satisfies $|x - x_0| < \delta$ and $f \in B$, then

$$|T(f)(x) - T(f)(x_0)| = \left| \int_a^b [K(x, y) - K(x_0, y)] f(y) \, dy \right| \leq (b - a)\epsilon.$$

This shows that $T(B)$ is equicontinuous at x_0. Since x_0 is arbitrary, $T(B)$ is equicontinuous everywhere. Thus, $\overline{T(B)}$ is compact.

The preceding property is expressed simply by saying that T is a compact operator. In general, an operator $T: X \to Y$ between two Banach spaces is said to be a **compact operator** if $\overline{T(B)}$ is a compact subset of Y (where B is the open unit ball of X). ∎

Our next example is borrowed from the theory of differential equations.

Example 28.5. Let $C^k[a, b]$ be the Banach space of Example 27. Consider $C[a, b]$ with the sup norm $\|\cdot\|_\infty$, and fix k functions $p_0, p_1, \ldots, p_{k-1}$ in $C[a, b]$. Now, define $L: C^k[a, b] \to C[a, b]$ by

$$L(y) = y^{(k)} + p_{k-1} y^{(k-1)} + \cdots + p_1 y' + p_0 y$$

for each $y \in C^k[a, b]$.

Clearly, L is a linear (differential) operator. On the other hand, if

$$M = 1 + \|p_0\|_\infty + \|p_1\|_\infty + \cdots + \|p_{k-1}\|_\infty,$$

then it is easy to see that

$$\|L(y)\|_\infty \le M[\|y^{(k)}\|_\infty + \|y^{(k-1)}\|_\infty + \cdots + \|y'\|_\infty + \|y\|_\infty] = M\|y\|$$

holds. This shows that L is a bounded operator.

It is also interesting to observe that L is onto. This follows from the standard existence theorem of solutions to an ordinary linear differential equation. ∎

The bounded operators are precisely the continuous linear operators. The next theorem clarifies the situation.

Theorem 28.6. *For a linear operator $T: X \to Y$ between two normed spaces, the following statements are equivalent:*

1. *T is a bounded operator.*
2. *There exists a real number $M \ge 0$ such that $\|T(x)\| \le M\|x\|$ holds for all $x \in X$.*
3. *T is continuous at zero.*
4. *T is continuous.*

Proof. (1) $\Longrightarrow$ (2) We have seen before that $\|T(x)\| \le \|T\| \cdot \|x\|$ holds for all $x \in X$. Thus, if T is a bounded operator, then (2) holds for any choice of the real number M with $M \ge \|T\|$.

(2) $\Longrightarrow$ (3) Clearly, $\lim\|x_n\| = 0$ combined with the inequality $\|T(x_n)\| \le M\|x_n\|$ implies $\lim\|T(x_n)\| = 0$. That is, T is continuous at zero.

(3) $\Longrightarrow$ (4) The (uniform) continuity of T follows immediately from the identity $\|T(x) - T(y)\| = \|T(x - y)\|$ and the simple fact that $\lim x_n = x$ holds in a normed space if and only if $\lim(x_n - x) = 0$.

(4) $\Longrightarrow$ (1) Assume by way of contradiction that $\|T\| = \infty$. Then there exists a sequence $\{x_n\}$ of X with $\|x_n\| = 1$ and $\|T(x_n)\| \ge n$ for each n. Let $y_n = \frac{x_n}{n}$ for each n, and note that $\|y_n\| = \frac{1}{n}$ implies $\lim y_n = 0$. But then, by the continuity of T, we must have $\lim T(y_n) = 0$, contrary to $\|T(y_n)\| = n^{-1}\|T(x_n)\| \ge 1$ for each n. Thus, $\|T\| < \infty$ holds, and the proof of the theorem is complete. ∎

Let X and Y be two normed spaces. Then the collection of all bounded linear operators from X into Y will be denoted by $L(X, Y)$. By Theorem 28.6, it follows that $L(X, Y)$ under the algebraic operations

$$(S + T)(x) = S(x) + T(x) \quad \text{and} \quad (\alpha T)(x) = \alpha T(x)$$

is a vector space. In fact, $L(X, Y)$ is a normed vector space under the operator norm $\|T\| = \sup\{\|T(x)\|: \|x\| = 1\}$. The details follow.

Theorem 28.7. *Let X and Y be two normed spaces. Then $L(X, Y)$ is a normed vector space. Moreover, if Y is a Banach space, then $L(X, Y)$ is likewise a Banach space.*

Proof. We verify first that $T \mapsto \|T\| = \sup\{\|T(x)\|: \|x\| = 1\}$ is a norm on $L(X, Y)$. Clearly, from its definition $\|T\| \geq 0$ holds for all $T \in L(X, Y)$. Also, the inequality $\|T(x)\| \leq \|T\| \cdot \|x\|$ shows that $\|T\| = 0$ if and only if $T = 0$. The proof of the identity $\|\alpha T\| = |\alpha| \cdot \|T\|$ is straightforward. For the triangle inequality, let $S, T \in L(X, Y)$, and let $x \in X$ with $\|x\| = 1$. Then

$$\|(S + T)(x)\| = \|S(x) + T(x)\| \leq \|S(x)\| + \|T(x)\| \leq \|S\| + \|T\|$$

holds, which shows that $\|S + T\| \leq \|S\| + \|T\|$. Thus, $L(X, Y)$ is a normed vector space.

Now, assume that Y is a Banach space. To complete the proof, we have to show that $L(X, Y)$ is a Banach space. To this end, let $\{T_n\}$ be a Cauchy sequence of $L(X, Y)$. From the inequality $\|T_n(x) - T_m(x)\| \leq \|T_n - T_m\| \cdot \|x\|$, it follows that for each $x \in X$, the sequence $\{T(x_n)\}$ of Y is Cauchy and thus convergent in Y. Let $T(x) = \lim T_n(x)$ for each $x \in X$, and note that T defines a linear operator from X into Y. Since $\{T_n\}$ is a Cauchy sequence, there exists some $M > 0$ such that $\|T_n\| \leq M$ for all n. But then, the inequality $\|T_n(x)\| \leq \|T_n\| \cdot \|x\| \leq M \|x\|$, coupled with the continuity of the norm, shows that $\|T(x)\| \leq M \|x\|$ holds for all $x \in X$. Therefore, by Theorem 28.6, $T \in L(X, Y)$.

Finally, we show that $\lim T_n = T$ holds in $L(X, Y)$. Let $\epsilon > 0$. Choose k such that $\|T_n - T_m\| < \epsilon$ for all $n, m \geq k$. Now, the relation

$$\|T_m(x) - T_n(x)\| \leq \|T_m - T_n\| \cdot \|x\| \leq \epsilon \|x\|$$

for all $n, m \geq k$ implies

$$\|T(x) - T_n(x)\| = \lim_{m \to \infty} \|T_m(x) - T_n(x)\| \leq \epsilon \|x\|$$

for all $n \geq k$ and $x \in X$. That is, we have $\|T - T_n\| \leq \epsilon$ for all $n \geq k$, and therefore, $\lim T_n = T$ holds in $L(X, Y)$.   ∎

A subset A of $L(X, Y)$ is said to be **pointwise bounded** if for every $x \in X$ the subset $\{T(x): T \in A\}$ of Y is norm bounded.

If a subset A of $L(X, Y)$ is norm bounded (i.e., if there exists some $M > 0$ such that $\|T\| \le M$ holds for each $T \in A$), then in view of the relation

$$\|T(x)\| \le \|T\| \cdot \|x\| \le M \|x\|$$

for each $T \in A$, it follows that A is pointwise bounded. That is, every norm bounded set of operators is pointwise bounded. The converse of this statement is also true, provided that X is a Banach space. The result is known as "the principle of uniform boundedness," or as "the Banach–Steinhaus theorem," and it is a very powerful theorem.

Theorem 28.8 (The Principle of Uniform Boundedness). *Let X be a Banach space, and let Y be a normed space. Then a subset of $L(X, Y)$ is norm bounded if and only if it is pointwise bounded.*

Proof. We already know that every norm bounded subset of $L(X, Y)$ is pointwise bounded. For the converse, let $A \subseteq L(X, Y)$ be pointwise bounded.

Start by observing that for each n the set

$$E_n = \{x \in X \colon \|T(x)\| \le n \text{ for all } T \in A\}$$

is norm closed. Also, since A is pointwise bounded, it follows that $X = \bigcup_{n=1}^{\infty} E_n$ holds. In view of the completeness of X and Theorem 6.18, there exists some k with $(E_k)^{\circ} \ne \emptyset$. Choose $y \in E_k$ and $r > 0$ such that $\|x - y\| \le r$ implies $x \in E_k$.

Now, let $T \in A$, and let $x \in X$ with $\|x\| = 1$. Since $\|(y + rx) - y\| = r$, it follows that $y + rx \in E_k$, and so

$$r\|T(x)\| = \|T(rx)\| = \|T(y + rx) - T(y)\| \le \|T(y + rx)\| + \|T(y)\| \le 2k.$$

Thus, $\|T(x)\| \le 2kr^{-1} = M$ holds for all $x \in X$ with $\|x\| = 1$, and therefore, $\|T\| \le M$ holds for all $T \in A$. That is, A is a norm bounded subset of $L(X, Y)$. ∎

In applications, the next result is an often-utilized special case of the "Principle of Uniform Boundedness."

Corollary 28.9. *Let X be a Banach space, Y a normed space, and $\{T_n\}$ a sequence of $L(X, Y)$. If $\lim T_n(x) = T(x)$ holds for each $x \in X$, then T is a bounded operator.*

The hypothesis that X is a Banach space in Theorem 28.8 is essential. The next example clarifies the situation.

Example 28.10.   Consider the vector space X consisting of all (real) sequences whose terms are eventually zero. Then X equipped with the sup norm is a normed space. On the other hand, the reader can verify easily that X is not a Banach space. Now, for each n define $f_n: X \to \mathbb{R}$ by

$$f_n(x) = \sum_{k=1}^{n} k x_k$$

for each $x = (x_1, x_2, \ldots)$ in X. From $|f_n(x)| \leq (\sum_{k=1}^{n} k) \|x\|_\infty$, it follows that $\{f_n\} \subseteq L(X, \mathbb{R})$. Moreover, $\lim f_n(x) = \sum_{k=1}^{\infty} k x_k$ holds for each $x \in X$. (Observe that the series has in fact a finite number of nonzero terms.) Thus, $\{f_n\}$ is pointwise bounded. However, it is easy to see that $\|f_n\| \geq n$ hold for each n, so that $\{f_n\}$ is not a norm bounded sequence. Hence, in general the Principle of Uniform Boundedness does not hold true if X is not assumed to be complete. ∎

Another useful variation of the Principle of Uniform Boundedness (which is also due to S. Banach and H. Steinhaus) is known as the "Principle of Condensation of Singularities," and is stated next.

Theorem 28.11 (The Principle of Condensation of Singularities).   *For each pair of natural numbers n and m, let $T_{n,m}: X \to Y$ be a bounded operator from a Banach space X to a normed space Y. If for each m we have $\lim \sup_n \|T_{n,m}\| = \infty$, then the set*

$$\left\{ x \in X: \ \limsup_{n \to \infty} \|T_{n,m}(x)\| = \infty \ \text{for each } m \right\}$$

is a dense subset of X.

Proof.   For every $n, m,$ and k, let

$$A_{nmk} = \{x \in X: \ \|T_{p,m}(x)\| > k \ \ \text{for some} \ \ p > n\}.$$

Clearly, A_{nmk} is an open set and we claim that it is also dense. We must show that if $x \in X$ and $r > 0$, then $B(x, r) \cap A_{nmk} \neq \emptyset$. So, let $x \in X$ and $r > 0$.

By the Uniform Boundedness Principle there exists a vector x_0 such that

$$\limsup_{p \to \infty} \|T_{p,m}(x_0)\| = \infty.$$

Pick some $\epsilon > 0$ such that $\epsilon \|x_0\| < r$. From

$$\epsilon \|T_{p,m}(x_0)\| = \|T_{p,m}(x + \epsilon x_0) - T_{p,m}(x)\| \leq \|T_{p,m}(x + \epsilon x_0)\| + \|T_{p,m}(x)\|,$$

it easily follows that either $\|T_{p,m}(x)\| > k$ or else $\|T_{p,m}(x + \epsilon x_0)\| > k$ for some $p > n$. That is, either $x \in B(x, r) \cap A_{nmk}$ or $x + \epsilon x_0 \in B(x, r) \cap A_{nmk}$. Hence, $B(x, r) \cap A_{nmk} \neq \emptyset$, and so A_{nmk} is an open dense set.

Now, notice that the Baire Category Theorem 6.17 guarantees that

$$\bigcap \{A_{nmk}: \ n, m, k = 1, 2, \ldots\}$$

is a dense set. Clearly, any point in this set satisfies $\limsup_n \|T_{n,m}(x)\| = \infty$ for all m. ∎

Our next objective is to establish another important result of analysis known as the "open mapping theorem." It asserts that a surjective continuous linear operator between two Banach spaces is an open mapping (in the sense that it carries open sets onto open sets). To establish this result we need a lemma.

Lemma 28.12. *Let X and Y be two Banach spaces, and let $T: X \to Y$ be an onto continuous linear operator. If zero is an interior point of a subset A of X, then zero is also an interior point of $T(A)$.*

Proof. Let $V = \{x \in X: \|x\| \leq 1\}$, and observe that $rV = \{rx: x \in V\}$ is the closed ball with center at zero and radius r. Since zero is assumed to be an interior point of A, there exists some $r > 0$ with $rV \subseteq A$. By the linearity of T, we must have $T(rV) = rT(V) \subseteq T(A)$. Hence, to establish the result it suffices to show that zero is an interior point of $T(V)$.

Clearly, $X = \bigcup_{n=1}^{\infty} nV$, and since T is an onto linear operator, $Y = \bigcup_{n=1}^{\infty} nT(V)$ also holds. By Theorem 6.18, there exists some k such that $\overline{kT(V)}$ has a nonempty interior. Since $\overline{kT(V)} = k\overline{T(V)}$, it follows that $\overline{T(V)}$ has an interior point. That is, there exist some $y_0 \in \overline{T(V)}$ and $r > 0$ such that $B(y_0, 2r) \subseteq \overline{T(V)}$. Now, if $y \in Y$ satisfies $\|y\| < 2r$, then $y - y_0 \in \overline{T(V)}$. Therefore,

$$y = (y - y_0) + y_0 \in \overline{T(V)} + \overline{T(V)} \subseteq 2\overline{T(V)},$$

where the last inclusion follows easily from the identity $V + V = 2V$. That is, $\{y \in Y: \|y\| < r\} \subseteq \overline{T(V)}$. By the linearity of T, it follows that

$$\{y \in Y: \|y\| < r2^{-n}\} \subseteq 2^{-n}\overline{T(V)} = \overline{T(2^{-n}V)}$$

holds for each n.

Now, let $y \in Y$ be fixed such that $\|y\| < r2^{-1}$. Since $y \in \overline{T(2^{-1}V)}$, there exists some $x_1 \in 2^{-1}V$ such that $\|y - T(x_1)\| < r2^{-2}$. Now proceed inductively. Assume that x_n has been selected such that $x_n \in 2^{-n}V$ and $\|y - \sum_{i=1}^{n} T(x_i)\| < r2^{-n-1}$.

Clearly, $y - \sum_{i=1}^{n} T(x_i) \in \overline{T(2^{-n-1}V)}$, and so there exists some $x_{n+1} \in 2^{-n-1}V$ with $\|y - \sum_{i=1}^{n+1} T(x_i)\| < r2^{-n-2}$. Thus, a sequence $\{x_n\}$ is selected such that $\|x_n\| \le 2^{-n}$ and

$$\left\| y - \sum_{i=1}^{n} T(x_i) \right\| = \left\| y - T\left(\sum_{i=1}^{n} x_i \right) \right\| < r2^{-n-1}$$

hold for all n. Next, define $s_n = x_1 + \cdots + x_n$ for each n, and note that

$$\|s_{n+p} - s_n\| = \left\| \sum_{i=n+1}^{n+p} x_i \right\| \le \sum_{i=n+1}^{n+p} \|x_i\| \le 2^{-n}$$

for all n and p shows that $\{s_n\}$ is a Cauchy sequence. Let $x = \lim s_n$ in X. Then $\|x\| \le \sum_{n=1}^{\infty} \|x_n\| \le 1$ (i.e., $x \in V$), and by the continuity and linearity of T, we see that

$$T(x) = \lim_{n \to \infty} T(s_n) = \lim_{n \to \infty} \sum_{i=1}^{n} T(x_i) = y.$$

That is, $y \in T(V)$, and so $\{y \in Y: \|y\| < \frac{r}{2}\} \subseteq T(V)$. The proof of the lemma is now complete.   ∎

The open mapping theorem is stated next. The result is due to S. Banach.

Theorem 28.13 (The Open Mapping Theorem). *Let X and Y be two Banach spaces, and let $T: X \to Y$ be a bounded linear operator. If T is onto, then T is an open mapping (and hence, if in addition T is one-to-one, then it is a homeomorphism).*

Proof. Let $\mathcal{O}$ be an open subset of X, and let $y \in T(\mathcal{O})$. Pick a vector $x \in \mathcal{O}$ such that $y = T(x)$, and note that $y - T(\mathcal{O}) = T(x - \mathcal{O})$ holds. Now observe that zero is an interior point of $x - \mathcal{O}$, and hence, by Lemma 28.12, zero is also an interior point of $T(x - \mathcal{O}) = y - T(\mathcal{O})$. This implies that y is an interior point of $T(\mathcal{O})$. Since y is arbitrary, $T(\mathcal{O})$ is an open set, and the proof of the theorem is complete.   ∎

Consider two normed spaces X and Y. Then a norm can be defined on $X \times Y$ by $\|(x, y)\| = \|x\| + \|y\|$. This norm is called the **product norm**. You should stop and check that indeed this function satisfies the properties of a norm. Some other (frequently used) norms equivalent to the product norm are

$$\|(x, y)\| = (\|x\|^2 + \|y\|^2)^{\frac{1}{2}}$$

and

$$\|(x, y)\| = \max\{\|x\|, \|y\|\}.$$

It should be noted that $\lim(x_n, y_n) = (x, y)$ holds in $X \times Y$ with respect to the product norm if and only if $\lim x_n = x$ in X and $\lim y_n = y$ in Y both hold. Moreover, if both X and Y are Banach spaces, then $X \times Y$ with the product norm is likewise a Banach space. Unless otherwise specified, the Cartesian product of two normed spaces will be considered as a normed space under its product norm.

The next example illustrates an application of the open mapping theorem to differential equations.

Example 28.14. Consider the differential equation

$$y^{(k)} + p_{k-1}y^{(k-1)} + \cdots + p_1 y' + p_0 y = q, \tag{1}$$

where $p_0, p_1, \ldots, p_{k-1}$ and q are fixed continuous functions on a (finite) closed interval $[a, b]$. If y is a solution of (1) and $c \in [a, b]$, then the k real numbers $y(c)$, $y'(c)$, $\ldots$, $y^{(k-1)}(c)$ are called the **initial values** of y at c. By the standard theorems of ordinary differential equations, it is well known that given $c \in [a, b]$ and k real numbers $\alpha_1, \ldots, \alpha_k$, there exists precisely one solution y of (1) having $\alpha_1, \ldots, \alpha_k$ as its initial values at c, i.e., y satisfies $y(c) = \alpha_1$, $y'(c) = \alpha_2, \ldots, y^{(k-1)}(c) = \alpha_k$.

What we would like to demonstrate here (by using the open mapping theorem) is that the solutions of (1) depend continuously on their initial values. That is, we would like to show that "small perturbations" in the initial values cause "small perturbations" to the solutions. To do this, we need to translate the problem to the framework of Banach spaces.

Consider the Banach space $C^k[a, b]$ of Example 27. Clearly, its norm

$$\|y\| = \|y\|_\infty + \|y'\|_\infty + \cdots + \|y^{(k)}\|_\infty$$

controls the "sizes" of the functions and their derivatives. Also, consider the bounded linear operator $L: C^k[a, b] \to C[a, b]$ of Example 28; that is,

$$L(y) = y^{(k)} + p_{k-1}y^{(k-1)} + \cdots + p_1 y' + p_0 y.$$

Now, fix some $c \in [a, b]$ and define $T: C^k[a, b] \to C[a, b] \times \mathbb{R}^k$ by

$$T(y) = (L(y), y(c), y'(c), \ldots, y^{(k-1)}(c))$$

for each $y \in C^k[a, b]$. Then it is easy to check that T is linear and continuous. On the other hand, the existence and uniqueness of the solutions of (1) with prescribed initial values guarantees that T is one-to-one and onto. Thus, by the open mapping theorem, T^{-1} (which exists and is linear) must be continuous.

The continuity of T^{-1} means that given $\epsilon > 0$ there exists some $\delta > 0$ such that $\|T^{-1}(f, x) - T^{-1}(g, z)\| < \epsilon$ if (f, x) and (g, z) in $C[a, b] \times \mathbb{R}^k$ satisfy $\|f - g\|_\infty + \|x - z\|_\infty < \delta$. Translating this statement back to the differential equation (1), we get the following: Suppose that y_1 and y_2 are two solutions of (1) satisfying the initial conditions

$$y_1(c) = \alpha_1, \quad y_1'(c) = \alpha_2, \quad \ldots, \quad y_1^{(k-1)}(c) = \alpha_k$$

and

$$y_2(c) = \beta_1, \quad y_2'(c) = \beta_2, \quad \ldots, \quad y_2^{(k-1)}(c) = \beta_k.$$

If $|\alpha_i - \beta_i| < \delta$ holds for each $i = 1, \ldots, k$, then

$$\|y_1 - y_2\|_\infty + \|y_1' - y_2'\|_\infty + \cdots + \|y_1^{(k-1)} - y_2^{(k-1)}\|_\infty < \epsilon.$$

That is, the solutions of the differential equation (1) depend continuously on their initial values. ∎

Recall that if $T: X \to Y$ is a function, then the subset $G = \{(x, T(x)): x \in X\}$ of $X \times Y$ is called the **graph** of T. Now, if X and Y are vector spaces and T is a linear operator, then G is a vector subspace of $X \times Y$. Moreover, if T is a bounded operator, then it is a routine matter to verify that G is a closed subspace of $X \times Y$. The converse of this last statement is true for Banach spaces. The result is known as "the closed graph theorem," and it has numerous applications in analysis.

Theorem 28.15 (The Closed Graph Theorem). *Let X and Y be two Banach spaces, and let $T: X \to Y$ be a linear operator. If the graph of T is a closed subspace of $X \times Y$, then T is a bounded operator.*

Proof. Since the graph G of T is a closed subspace of the Banach space $X \times Y$, it is a Banach space in its own right. The function $(x, T(x)) \mapsto x$ is a linear operator from G onto X that is clearly one-to-one and bounded. (It is bounded because $\|x\| \le \|x\| + \|T(x)\| = \|(x, T(x))\|$.) By the open mapping theorem, $(x, T(x)) \mapsto x$ is a homeomorphism. That is, $x \mapsto (x, T(x))$ is a continuous function from X onto G. Now, observe that $(x, T(x)) \to T(x)$ is continuous. Therefore, $x \mapsto T(x)$ from X into Y is continuous since it is a composition of two continuous functions, and the proof of the theorem is complete. ∎

EXERCISES

1. Let X and Y be two Banach spaces and let $T: X \to Y$ be a bounded linear operator. Show that either T is onto or else $T(X)$ is a meager set.

2. Let X be a Banach space, $T: X \to X$ a bounded operator, and I the identity operator on X. If $\|T\| < 1$, then show that $I - T$ is invertible.
   [HINT: Show that $(I - T)^{-1} = \sum_{n=0}^{\infty} T^n$.]

3. On $C[0, 1]$ consider the two norms

$$\|f\|_\infty = \sup\{|f(x)|: x \in [0, 1]\} \quad \text{and} \quad \|f\|_1 = \int_0^1 |f(x)| \, dx.$$

Then show that the identity operator $I: (C[0, 1], \|\cdot\|_\infty) \to (C[0, 1], \|\cdot\|_1)$ is continuous, onto, but not open. Why doesn't this contradict the open mapping theorem?

4. Let X be the vector space of all real-valued functions on $[0, 1]$ that have continuous derivatives with the sup norm. Also, let $Y = C[0, 1]$ with the sup norm. Define the mapping $D: X \to Y$ by $D(f) = f'$.

   a. Show that D is an unbounded linear operator.
   b. Show that D has a closed graph.
   c. Why doesn't the conclusion in (b) contradict the closed graph theorem?

5. Consider the mapping $T: C[0, 1] \to C[0, 1]$ defined by $Tf(x) = x^2 f(x)$ for all $f \in C[0, 1]$ and each $x \in [0, 1]$.

   a. Show that T is a bounded linear operator.
   b. If $I: C[0, 1] \to C[0, 1]$ denotes the identity operator (i.e., $I(f) = f$ for each $f \in C[0, 1]$), then show that $\|I + T\| = 1 + \|T\|$.

6. Let X be a vector space which is complete in each of the two norms $\|\cdot\|_1$ and $\|\cdot\|_2$. If there exists a real number $M > 0$ such that $\|x\|_1 \leq M\|x\|_2$ holds for all $x \in X$, then show that the two norms must be equivalent.

7. Let X, Y, and Z be three Banach spaces. Assume that $T: X \to Y$ is a linear operator and $S: Y \to Z$ is a bounded, one-to-one linear operator. Show that T is a bounded operator if and only if the composite linear operator $S \circ T$ (from X into Z) is bounded.
   [HINT: Use the closed graph theorem.]

8. An operator $P: V \to V$ on a vector space is said to be a **projection** if $P^2 = P$ holds. Also, a closed vector subspace Y of a Banach space is said to be **complemented** if there exists another closed subspace Z of X such that $Y \oplus Z = X$.
   Show that a closed subspace of a Banach space is complemented if and only if it is the range of a continuous projection.
   [HINT: Use the closed graph theorem.]

29. LINEAR FUNCTIONALS

In this section, we shall discuss the basic properties of continuous linear functionals. Recall that a linear operator $f: X \to \mathbb{R}$, where X is a vector space, is called a **linear functional** on X.

One of our objectives is to show that for every normed vector space X there are enough bounded linear functionals to separate the points of X. That is, for every pair of distinct vectors x and y of X, there exists a bounded linear functional on X such that $f(x) \neq f(y)$. This result (as well as many others) rests upon a classical result known as the "Hahn–Banach theorem," which is one of the cornerstones of modern analysis. We shall discuss this theorem next.

A mapping $p: X \to \mathbb{R}$, where X is a vector space, is called a **sublinear mapping** if it satisfies the following two properties:

a. $p(x + y) \le p(x) + p(y)$ for all $x, y \in X$; and
b. $p(\alpha x) = \alpha p(x)$ for all $x \in X$ and $\alpha \ge 0$.

The crux of the proof of the Hahn–Banach theorem lies in the following lemma:

Lemma 29.1.   *Let p be a sublinear mapping on a vector space X, Y a vector subspace of X, and $x_0 \notin Y$. If f is a linear functional on Y such that $f(x) \le p(x)$ holds for all $x \in Y$, then f can be extended to a linear functional g on the vector subspace Z generated by Y and x_0 satisfying $g(x) \le p(x)$ for all $x \in Z$.*

Proof.   Clearly, $Z = \{x + \alpha x_0 : x \in Y$ and $\alpha \in \mathbb{R}\}$. Assume that g is a linear functional on Z that agrees with f on Y. Then $g(x + \alpha x_0) = f(x) + \alpha g(x_0)$ holds for all $x \in Y$ and $\alpha \in \mathbb{R}$, and so g is determined uniquely by the value $g(x_0)$. Let $c = g(x_0)$. Thus, every real number c gives rise to a linear functional on Z that agrees with f on Y. Our objective is to show that there exists some value of c such that

$$f(x) + \alpha c \le p(x + \alpha x_0) \tag{$\star$}$$

holds for all $x \in Y$ and $\alpha \in \mathbb{R}$. For $\alpha > 0$ and $x \in Y$, the inequality $(\star)$ is equivalent to $c \le p(\alpha^{-1}x + x_0) - f(\alpha^{-1}x)$ and to $-f(\alpha^{-1}x) - p(-\alpha^{-1}x - x_0) \le c$ for $x \in Y$ and $\alpha < 0$. Certainly, these inequalities [and hence, $(\star)$] will be satisfied by a choice of c for which

$$-f(x) - p(-x - x_0) \le c \le p(x + x_0) - f(x) \tag{$\star\star$}$$

holds for all $x \in Y$.

If $x, y \in Y$, then the relations

$$f(y) - f(x) = f(y - x) \le p(y - x) = p(y + x_0 + (-x - x_0))$$
$$\le p(y + x_0) + p(-x - x_0)$$

show that $-f(x) - p(-x - x_0) \le p(y + x_0) - f(y)$ holds for all $x, y \in Y$. Consequently, if

$$s = \sup\{-f(x) - p(-x - x_0): x \in Y\} \quad \text{and} \quad t = \inf\{p(y + x_0) - f(y): y \in Y\},$$

then s and t are both real numbers and $s \le t$. But then any real number c such that $s \le c \le t$ satisfies $(\star\star)$, and hence, $(\star)$. This completes the proof of the lemma.   ∎

The classical Hahn–Banach theorem is stated next. This theorem is the heart of modern analysis, and it has far-reaching applications.

Theorem 29.2 (Hahn–Banach). *Let p be a sublinear mapping on a vector space X, and let Y be a vector subspace of X. If f is a linear functional on Y such that $f(x) \le p(x)$ holds for all $x \in Y$, then f can be extended to a linear functional g on all of X satisfying $g(x) \le p(x)$ for every $x \in X$.*

Proof. Let C be the collection of all pairs (g, Z) such that Z is a vector subspace of X containing Y and g is a linear functional on Z satisfying $g(x) = f(x)$ for all $x \in Y$ and $g(x) \le p(x)$ for all $x \in Z$. The collection C is nonempty since $(f, Y) \in C$. Define an order relation on C as follows: $(g_2, Z_2) \ge (g_1, Z_1)$ whenever $Z_1 \subseteq Z_2$ and $g_2(x) = g_1(x)$ for all $x \in Z_1$. It is easy to verify that $\ge$ is indeed an order relation on C.

Now, consider a chain $\{(g_i, Z_i): i \in I\}$ of C (i.e., for every pair $i, j \in I$ either $(g_i, Z_i) \ge (g_j, Z_j)$ or $(g_j, Z_j) \ge (g_i, Z_i)$ holds). Let $Z = \bigcup_{i \in I} Z_i$, and note that Z is a vector subspace of X. Now, define $g: Z \to \mathbb{R}$ by letting $g(x) = g_i(x)$ if $x \in Z_i$. Since $\{(g_i, Z_i): i \in I\}$ is a chain, the value of $g(x)$ is independent of the chosen index i. Clearly, g is a linear functional. In addition, $g(x) = f(x)$ holds for all $x \in Y$, and $g(x) \le p(x)$ for each $x \in Z$. Thus, $(g, Z) \in C$, and clearly, $(g, Z) \ge (g_i, Z_i)$ holds for all $i \in I$. Therefore, every chain of C has an upper bound in C. By Zorn's lemma, C has a maximal element, say (g, Z).

To complete the proof, it suffices to show that $Z = X$. Indeed, if $Z \ne X$, then there exists some $x_0 \in X$ with $x_0 \notin Z$. Let M be the vector subspace generated by Z and x_0. By Lemma 29.1, there exists a linear functional h on M such that $h(x) = g(x)$ for all $x \in Z$ and $h(x) \le p(x)$ for all $x \in M$. But then $(h, M) \in C$, and $(h, M) \ge (g, Z)$ holds with $(h, M) \ne (g, Z)$, contrary to the maximality property of (g, Z). Hence, $Z = X$ holds, and the proof of the theorem is complete. ∎

The next three theorems are applications of the Hahn–Banach theorem. The first result tells us that a continuous linear functional defined on a subspace can be extended to a continuous linear functional on the whole space with preservation of its original norm.

Theorem 29.3. *Let Y be a vector subspace of a normed space X, and let f be a continuous linear functional on Y. Then f can be extended to a continuous linear functional g on X such that $\|g\| = \|f\|$.*

Proof. Let $\|f\| = \sup\{|f(y)|: y \in Y$ and $\|y\| \le 1\} < \infty$. Define $p: X \to \mathbb{R}$ by $p(x) = \|f\| \cdot \|x\|$ for each $x \in X$. Note that p is a sublinear mapping on X such that $f(x) \le p(x)$ holds for all $x \in Y$. By the Hahn–Banach theorem there exists a linear extension g of f to all of X such that $g(x) \le p(x)$ holds for all $x \in X$. This implies $|g(x)| \le \|f\| \cdot \|x\|$ for all $x \in X$, and so $\|g\| \le \|f\|$. On the other hand,

$$\|f\| = \sup\{|f(y)|: \ y \in Y \ \text{and} \ \|y\| \le 1\}$$
$$\le \sup\{|g(x)|: \ x \in X \ \text{and} \ \|x\| \le 1\}$$
$$= \|g\|$$

also holds, so that $\|g\| = \|f\|$. Thus, g is a desired extension of f to all of X. ∎

For a normed space X, the Banach space $L(X, \mathbb{R})$ is called the **norm dual** of X and is denoted by X^*. That is, X^* consists of all functions $f: X \to \mathbb{R}$ that are both linear and continuous. The members of X^* are called the continuous linear functionals on X. The norm dual X^* plays an important role in the theory of Banach spaces, and part of this theory deals with the properties and structure of X^*.

The vectors of a normed vector space are always separated by its continuous linear functionals.

Theorem 29.4. *If x is a vector in some normed space X, then there exists a continuous linear functional f on X such that $\|f\| = 1$ and $f(x) = \|x\|$. In particular, X^* separates the points of X.*

Proof. If $x = 0$, then any $f \in X^*$ with $\|f\| = 1$ satisfies $f(x) = \|x\|$. (When $X \ne \{0\}$, Theorem 29.3 guarantees $X^* \ne \{0\}$.) Thus, assume $x \ne 0$.

Let $Y = \{\alpha x: \alpha \in \mathbb{R}\}$, the vector subspace generated by x. Then, the formula $g(\alpha x) = \alpha \|x\|$ defines a continuous linear functional on Y such that $\|g\| = 1$ and $g(x) = \|x\|$. By Theorem 29.3, there exists a continuous linear extension f of g to all of X with $\|f\| = \|g\| = 1$. Clearly, $f(x) = \|x\|$, as desired.

To see that X^* separates the points of X, let $x, y \in X$ with $x \ne y$. By the preceding, there exists some $f \in X^*$ such that $f(x) - f(y) = f(x - y) = \|x - y\| \ne 0$. Hence, $f(x) \ne f(y)$ holds, and the proof is complete. ∎

A vector always can be separated from a closed subspace that does not contain it by a continuous linear functional. The details are included in the next result.

Theorem 29.5. *Let Y be a vector subspace of a normed vector space X, and let $x_0 \notin \overline{Y}$. Then there exists some $f \in X^*$ such that $f(x) = 0$ for all $x \in Y$ and $f(x_0) = 1$.*

Proof. Since $x_0 \notin \overline{Y}$, there exists some $r > 0$ such that $\|x - x_0\| > r$ holds for all $x \in Y$. Let $Z = \{x + \alpha x_0: x \in Y \ \text{and} \ \alpha \in \mathbb{R}\}$, the vector subspace generated by Y and x_0. Now, define $f: Z \to \mathbb{R}$ by $f(x + \alpha x_0) = \alpha$. Then f is a linear functional and

$$r|f(x + \alpha x_0)| = r|\alpha| \leq |\alpha| \cdot \|\alpha^{-1}x + x_0\| = \|x + \alpha x_0\|$$

holds for all $x \in Y$ and $\alpha \in \mathbb{R}$. It follows that f is a continuous linear functional on Z whose norm does not exceed r^{-1}. Also, $f(x) = 0$ holds for each $x \in Y$ and $f(x_0) = 1$. Now, apply Theorem 29.3 to extend f to a continuous linear functional on X.   ∎

Let X be a normed space. As we have seen in Theorem 28.7, the norm dual X^* $[= L(X, \mathbb{R})]$ of X is always a Banach space. In particular, the norm dual $(X^*)^*$ of X^* is likewise a Banach space. This Banach space is called the **second dual** of X and is denoted by X^{**}. That is, $X^{**} = (X^*)^*$.

Every vector $x \in X$ gives rise to a continuous linear functional $\hat{x}$ on X^* via the formula

$$\hat{x}(f) = f(x)$$

for all $f \in X^*$. Indeed, clearly, $\hat{x}$ is linear on X^*, and the inequality

$$|\hat{x}(f)| = |f(x)| \leq \|f\| \cdot \|x\| = \|x\| \cdot \|f\|$$

shows that $\hat{x} \in X^{**}$ and $\|\hat{x}\| \leq \|x\|$. On the other hand, by Theorem 29.4 there exists some $f \in X^*$ with $\|f\| = 1$ and $f(x) = \|x\|$. Thus, $\|x\| = f(x) = |\hat{x}(f)| \leq \|\hat{x}\|$, and so $\|\hat{x}\| = \|x\|$ holds for all $x \in X$.

The mapping $x \mapsto \hat{x}$ (from X into X^{**}) is called the **natural embedding** of X into its second dual X^{**}. Summarizing the previous, we have the following theorem:

Theorem 29.6. *The natural embedding $x \mapsto \hat{x}$ of a normed space X into its second dual X^{**} is a norm preserving linear operator (and hence, X can be considered as a subspace of X^{**}).*

A linear operator $T: X \to Y$ between normed spaces that satisfies $\|T(x)\| = \|x\|$ for all $x \in X$ is called a **linear isometry**. In this terminology the preceding theorem can be phrased as follows: *The natural embedding $x \mapsto \hat{x}$ is a linear isometry.*

In general, $x \mapsto \hat{x}$ is not surjective, and hence, X (when embedded in X^{**}) is, in general, a proper subspace of X^{**}. If X is not a Banach space, then $x \mapsto \hat{x}$ cannot be onto, simply because X^{**} is a Banach space. If the natural embedding of a Banach space X into its second dual X^{**} is onto, then X is called a **reflexive Banach space**, and this is denoted by $X^{**} = X$. The properties of the natural embedding of X into its second dual X^{**} are utilized in many applications, some of which are illustrated below.

If Y is a vector subspace of a normed space X, then it is easy to see that its closure $\overline{Y}$ is also a vector subspace (why?). In particular, if X is a Banach space, then $\overline{Y}$ is the completion of Y. These observations will be used in the proof of the next theorem.

Theorem 29.7. *The completion of a normed space is a Banach space.*

Proof. Let X be a normed space. Consider X embedded in X^{**} by its natural embedding. Then X^{**} is a Banach space and induces on X its original norm. Thus, $\overline{X}$ is the completion of X, which is likewise a Banach space.   ■

The next result can be viewed as a "dual" of the Uniform Boundedness Principle.

Theorem 29.8. *Let A be a subset of a normed space X such that $\{f(x)\colon x \in A\}$ is a bounded subset of $\mathbb{R}$ for each $f \in X^*$. Then A is a norm bounded subset of X.*

Proof. Consider X embedded in X^{**}. Then A as a subset of X^{**} is pointwise bounded on the Banach space X^*. Thus, by the Principle of Uniform Boundedness (Theorem 28.8), A is norm bounded in X^{**}. It follows that A is also norm bounded in X, and the proof is finished.   ■

EXERCISES

1. Let $f\colon X \to \mathbb{R}$ be a linear functional defined on a vector space X. The **kernel** of f is the vector subspace

    $$\text{Ker } f = f^{-1}(\{0\}) = \{x \in X\colon f(x) = 0\}.$$

    If X is a normed space and $f\colon X \to \mathbb{R}$ is nonzero linear functional, establish the following:

    a. f is continuous if and only if its kernel is a closed subspace of X.
    b. f is discontinuous if and only if its kernel is dense in X.

2. Show that a linear functional f on a normed space X is discontinuous if and only if for each $a \in X$ and each $r > 0$ we have

    $$f(B(a,r)) = \{f(x)\colon \|a - x\| < r\} = \mathbb{R}.$$

3. Let $f, f_1, f_2, \ldots, f_n$ be linear functionals defined on a common vector space X. Show that there exist constants $\lambda_1, \ldots, \lambda_n$ satisfying $f = \sum_{i=1}^{n} \lambda_i f_i$ (i.e., f lies in the linear span of $f_1 \ldots, f_n$) if and only if $\bigcap_{i=1}^{n} \text{Ker } f_i \subseteq \text{Ker } f$.

4. Prove the converse of Theorem 28.7. That is, show that if X and Y are (nontrivial) normed spaces and $L(X, Y)$ is a Banach space, then Y is a Banach space.

[HINT: Let $\{y_n\}$ be a Cauchy sequence of Y. Pick $f \in X^*$ with $f \neq 0$, and then consider the sequence $\{T_n\}$ of $L(X, Y)$ defined by $T_n(x) = f(x)y_n$.]

5. The Banach space $B(\mathbb{N})$ is denoted by ℓ_∞. That is, ℓ_∞ is the Banach space consisting of all bounded sequences with the sup norm. Consider the collections of vectors

$$c_0 = \{x = (x_1, x_2, x_3, \ldots) \in \ell_\infty \colon x_n \to 0\}, \text{ and}$$

$$c = \{x = (x_1, x_2, x_3, \ldots) \in \ell_\infty \colon \lim x_n \text{ exists in } \mathbb{R}\}.$$

Show that c_0 and c are both closed vector subspaces of ℓ_∞.

6. Let c denote the vector subspace of ℓ_∞ consisting of all convergent sequences (see Exercise 5 above). Define the **limit functional** $L \colon c \to \mathbb{R}$ by

$$L(x) = L(x_1, x_2, \ldots) = \lim_{n \to \infty} x_n,$$

and $p \colon \ell_\infty \to \mathbb{R}$ by $p(x) = p(x_1, x_2, \ldots) = \limsup x_n$.

a. Show that L is a continuous linear functional, where c is assumed equipped with the sup norm.
b. Show that p is sublinear and that $L(x) = p(x)$ holds for each $x \in c$.
c. By the Hahn–Banach Theorem 29.2 there exists a linear extension of L to all of ℓ_∞ (which we shall denote by L again) satisfying $L(x) \leq p(x)$ for all $x \in \ell_\infty$. Establish the following properties of the extension L:

    i. For each $x \in \ell_\infty$, we have

$$\liminf_{n \to \infty} x_n \leq L(x) \leq \limsup_{n \to \infty} x_n.$$

    ii. L is a positive linear functional, i.e., $x \geq 0$ implies $L(x) \geq 0$.

    iii. L is a continuous linear functional (and in fact $\|L\| = 1$).

7. Generalize Exercise 6 above as follows. Show that there exists a linear functional $\mathcal{L}im \colon \ell_\infty \to \mathbb{R}$ satisfying the following properties:

a. $\mathcal{L}im$ is a positive linear functional of norm one.
b. For each $x = (x_1, x_2, \ldots) \in \ell_\infty$, we have

$$\liminf_{n \to \infty} \frac{x_1 + x_2 + \cdots + x_n}{n} \leq \mathcal{L}im(x) \leq \limsup_{n \to \infty} \frac{x_1 + x_2 + \cdots + x_n}{n}.$$

In particular, $\mathcal{L}im$ is an extension of the limit functional L.
c. For each $x = (x_1, x_2, \ldots) \in \ell_\infty$, we have

$$\mathcal{L}im(x_1, x_2, x_3, \ldots) = \mathcal{L}im(x_2, x_3, x_4, \ldots).$$

Any such linear functional $\mathcal{L}im$ is called a **Banach–Mazur**[3] **limit**.
[HINT: Define $p \colon \ell_\infty \to \mathbb{R}$ by $p(x) = \limsup \frac{x_1 + x_2 + \cdots + x_n}{n}$ and note that p is sublinear satisfying $L(x) = p(x)$ for all $x \in c$.]

8. Let X be a normed vector space. Show that if X^* is separable (in the sense that it contains a countable dense subset), then X is also separable.

---

[3] Stanislaw Mazur (1905–1981), a Polish mathematician and a close collaborator of Stefan Banach. He made important contributions to functional analysis and probability theory.

[HINT: Let $\{f_1, f_2, \ldots\}$ be a countable dense subset of X^*. For each n choose $x_n \in X$ with $\|x_n\| = 1$ and $|f_n(x_n)| \geq \frac{1}{2}\|f_n\|$, and let Y be the closed subspace generated by $\{x_1, x_2, \ldots\}$. Use Theorem 29.5 to show that $Y = X$.]

9. Show that a Banach space X is reflexive if and only if X^* is reflexive.
    [HINT: If $X \neq X^{**}$, then by Theorem 29.5 there exists a nonzero $F \in X^{***}$ such that $F(x) = 0$ for all $x \in X$.]

10. This problem describes the adjoint of a bounded operator. If $T: X \to Y$ is a bounded operator between two normed spaces, then its **adjoint** is the operator $T^*: Y^* \to X^*$ defined by $(T^*f)(x) = f(Tx)$ for all $f \in Y^*$ and all $x \in X$. (Writing $h(x) = \langle x, h \rangle$, the definition of the adjoint operator is written in "duality" notation as

$$\langle Tx, f \rangle = \langle x, T^*f \rangle$$

for all $f \in Y^*$ and all $x \in X$.)

   a. Show that $T^*: Y^* \to X^*$ is a well-defined bounded linear operator whose norm coincides with that of T, i.e., $\|T^*\| = \|T\|$.

   b. Fix some $g \in X^*$ and some $u \in Y$ and define $S: X \to Y$ by $S(x) = g(x)u$. Show that S is a bounded linear operator satisfying $\|S\| = \|g\| \cdot \|u\|$. (Any such operator S is called a **rank-one** operator.)

   c. Describe the adjoint of the operator S defined in the preceding part (b).

   d. Let $A = [a_{ij}]$ be an $m \times n$ matrix with real entries. As usual, we consider the adjoint operator A^* as a (bounded) linear operator from $\mathbb{R}^n$ to $\mathbb{R}^m$. Describe A^*.

30. BANACH LATTICES

As we have seen, both the continuous functions and the measurable functions have a natural ordering under which they are lattices. For this reason, it is important to consider Banach spaces that are also lattices. Let us begin by reviewing a few of the basic facts that will be needed about vector lattices from Section 9.

A **partially ordered vector space** is a real vector space X equipped with an order relation $\geq$ that is compatible with the algebraic structure as follows:

1. If $x \geq y$, then $x + z \geq y + z$ holds for all $z \in X$.
2. If $x \geq y$, then $\alpha x \geq \alpha y$ holds for all $\alpha \geq 0$.

The set $X^+ = \{x \in X: x \geq 0\}$ is called the **positive cone** of X, and its members are called the **positive vectors** of X. Clearly, the sum of two positive vectors is again a positive vector.

A partially ordered vector space X is called a **vector lattice** (or a **Riesz**[4] **space**) if for every pair of vectors $x, y \in X$ both $\sup\{x, y\}$ and $\inf\{x, y\}$ exist. As usual, $\sup\{x, y\}$ is denoted by $x \vee y$ and $\inf\{x, y\}$ by $x \wedge y$. That is, $x \vee y = \sup\{x, y\}$ and $x \wedge y = \inf\{x, y\}$.

---

[4]Frigyes (Frédéric) Riesz (1880–1956), a distinguished Hungarian mathematician. His name is closely associated with the development of functional analysis in the first half of the twentieth century.

In a vector lattice, the **positive part**, the **negative part**, and the **absolute value** of a vector x are defined by

$$x^+ = x \vee 0, \quad x^- = (-x) \vee 0, \quad \text{and} \quad |x| = x \vee (-x),$$

respectively. By Theorem 9.1 the following identities hold:

$$x = x^+ - x^- \quad \text{and} \quad |x| = x^+ + x^-.$$

A number of useful inequalities are stated in the next theorem.

Theorem 30.1. *If x, y, and z are vectors in a vector lattice, then the following inequalities hold*:

1. $|x + y| \leq |x| + |y|$;
2. $||x| - |y|| \leq |x - y|$;
3. $|x^+ - y^+| \leq |x - y|$;
4. $|x \vee z - y \vee z| \leq |x - y|$;
5. $|x \wedge z - y \wedge z| \leq |x - y|$.

Proof. (1) From $x \leq |x|$ and $y \leq |y|$, it follows that $x + y \leq |x| + |y|$. Similarly, $-(x + y) \leq |x| + |y|$ holds, and thus

$$|x + y| = (x + y) \vee [-(x + y)] \leq |x| + |y|.$$

(2) By (1) we see that $|x| = |x - y + y| \leq |x - y| + |y|$, and so $|x| - |y| \leq |x - y|$. Similarly, $|y| - |x| \leq |x - y|$, and hence, $||x| - |y|| \leq |x - y|$.

(3) Note that $x^+ = \frac{1}{2}(x + |x|)$. Then to establish the required inequality, use (1) and (2) as follows:

$$|x^+ - y^+| = \left| \frac{1}{2}(x + |x|) - \frac{1}{2}(y + |y|) \right| = \frac{1}{2}|(x - y) + (|x| - |y|)|$$

$$\leq \frac{1}{2}|x - y| + \frac{1}{2}||x| - |y|| \leq |x - y|.$$

(4) Note that

$$x \vee z - y \vee z = [(x - z) \vee 0 + z] - [(y - z) \vee 0 + z] = (x - z)^+ - (y - z)^+$$

holds. (See the identities in Section 9 before Theorem 9.1.) Thus, by (3) we have

$$|x \vee z - y \vee z| = |(x - z)^+ - (y - z)^+| \leq |(x - z) - (y - z)| = |x - y|.$$

(5) Since $x \wedge z = -[(-x) \vee (-z)]$ and $y \wedge z = -[(-y) \vee (-z)]$, it follows from (4) that

$$|x \wedge z - y \wedge z| = |(-y) \vee (-z) - (-x) \vee (-z)| \leq |-y-(-x)| = |x - y|,$$

and the proof of the theorem is complete. ∎

Let X be a vector lattice. A subset A of X is called **order bounded** if there exists an element $y \in X$ such that $|x| \leq y$ holds for all $x \in A$. A linear functional on X is said to be **order bounded** if it carries order bounded subsets of X onto order bounded subsets of $\mathbb{R}$. That is, a linear functional $f: X \to \mathbb{R}$ is order bounded if for every $y \in X^+$ there exists some $M > 0$ such that $|f(x)| \leq M$ holds for all $x \in X$ with $|x| \leq y$.

A linear functional f on X is called **positive** if $f(x) \geq 0$ holds for each $x \in X^+$. Clearly, every positive linear functional is order bounded. As we shall see in the next theorem, every order bounded linear functional can be written as a difference of two positive linear functionals.

The collection of all order bounded linear functionals on a vector lattice X is denoted by $X^\sim$ and is called the **order dual** of X. Obviously, $X^\sim$ (under the usual algebraic operations) is a vector space. Moreover, if we define $f \geq g$ whenever $f(x) \geq g(x)$ holds for all $x \in X^+$, then it is easy to see that $X^\sim$ equipped with $\geq$ is a partially ordered vector space. In actuality, $X^\sim$ is a vector lattice, as the next result of F. Riesz shows.

Theorem 30.2 (F. Riesz). *If X is a vector lattice, then its order dual $X^\sim$ is likewise a vector lattice. Moreover,*

$$f^+(x) = \sup\{f(y): 0 \leq y \leq x\},$$
$$f^-(x) = \sup\{-f(y): 0 \leq y \leq x\}, \quad and$$
$$|f|(x) = \sup\{f(y): |y| \leq x\}$$

hold for each $f \in X^\sim$ and all $x \in X^+$.

Proof. In view of the identities

$$f \vee g = (f - g)^+ + g \quad \text{and} \quad f \wedge g = -[(-f) \vee (-g)],$$

we establish that $X^\sim$ is a vector lattice by proving that f^+ exists for each $f \in X^\sim$. To this end, let $f \in X^\sim$. Define $g: X^+ \to \mathbb{R}$ by

$$g(x) = \sup\{f(y): 0 \leq y \leq x\}$$

for each $x \in X^+$. The supremum is finite since f is order bounded. Clearly, $g(x) \geq 0$, $g(x) \geq f(x)$, and $g(\alpha x) = \alpha g(x)$ hold for all $x \in X^+$ and $\alpha \geq 0$.

We claim that $g(x + y) = g(x) + g(y)$ holds for all $x, y \in X^+$. To see this, let $x, y \in X^+$. If u and v satisfy $0 \leq u \leq x$ and $0 \leq v \leq y$, then $0 \leq u + v \leq x + y$ holds, and so $f(u) + f(v) = f(u + v) \leq g(x + y)$; consequently, $g(x) + g(y) \leq g(x + y)$. On the other hand, if $0 \leq z \leq x + y$, then let $u = x \wedge z$, $v = z - u$, and note (by using the identities in Section 9 before Theorem 9.1) that

$$0 \leq v = z - x \wedge z = z + (-x) \vee (-z) = (z - x) \vee 0 \leq y \vee 0 = y.$$

Therefore, $0 \leq u \leq x$ and $0 \leq v \leq y$. This implies

$$f(z) = f(u + v) = f(u) + f(v) \leq g(x) + g(y),$$

from which it follows that $g(x + y) \leq g(x) + g(y)$. That is, $g(x + y) = g(x) + g(y)$.

Now, for arbitrary $x \in X$ define $g(x) = g(x^+) - g(x^-)$. Note that if $x = u - v$ with u, v positive, then $x^+ + v = u + x^-$, and so $g(x^+) + g(v) = g(u) + g(x^-)$ holds by the additivity of g on X^+. Therefore, $g(x) = g(x^+) - g(x^-) = g(u) - g(v)$, and so the value of $g(x)$ does not depend upon the particular representation of x as a difference of two positive vectors. The preceding observation, combined with $g(x + y) = g(x) + g(y)$, and $g(\alpha x) = \alpha g(x)$ for all $x, y \in X^+$ and $\alpha \geq 0$, implies that g is a linear functional on X.

Finally, we show that $g = f^+$ holds in $X^\sim$. Indeed, if h is another positive linear functional satisfying $f \leq h$, then since $f(y) \leq h(y) \leq h(x)$ holds for all $0 \leq y \leq x$, it follows that $g(x) \leq h(x)$ for each $x \in X^+$. Therefore, g is the least upper bound of f and zero. That is, $g = f^+$ holds in $X^\sim$. This shows that $X^\sim$ is a vector lattice and that $f^+(x) = \sup\{f(y): 0 \leq y \leq x\}$ holds.

The formula for f^- follows from $f^- = (-f)^+$, and the formula for the absolute value follows from $|f| = f^+ + f^-$. ∎

The formula $|f|(|x|) = \sup\{f(y): |y| \leq |x|\}$ implies the following useful inequality

$$|f(x)| \leq |f|(|x|),$$

for all $f \in X^\sim$ and all $x \in X$.

In view of the identity $f \vee g = (f - g)^+ + g$ and Theorem 30.2, the following identities hold:

$$f \vee g(x) = \sup\{f(y) + g(x - y): 0 \leq y \leq x\}$$

and

$$f \wedge g(x) = \inf\{f(y) + g(x - y): 0 \le y \le x\}$$

for all $f, g \in X^\sim$ and $x \in X^+$.

Sometimes it is important to know that the "dual" formulas of Theorem 30.2 are also true. More precisely we have the following:

Theorem 30.3. *Let X be a vector lattice, and let f be a positive linear functional. Then for every $x \in X$ the following identities hold:*

$$f(x^+) = \sup\{g(x): g \in X^\sim \text{ and } 0 \le g \le f\},$$
$$f(x^-) = \sup\{-g(x): g \in X^\sim \text{ and } 0 \le g \le f\}, \quad \text{and}$$
$$f(|x|) = \sup\{g(x): g \in X^\sim \text{ and } |g| \le f\}.$$

Proof. We establish the formula for $f(x^+)$. Note first that if $0 \le g \le f$ holds, then $g(x) \le g(x^+) \le f(x^+)$, and so $\sup\{g(x): g \in X^\sim \text{ and } 0 \le g \le f\} \le f(x^+)$.

For the reverse inequality consider the function $p: X \to \mathbb{R}$ defined by $p(u) = f(u^+)$. It is easy to see that p is a sublinear mapping on X such that $p(u) \ge 0$ holds for each $u \in X$. Now, let $Y = \{\alpha x: \alpha \in \mathbb{R}\}$, and define $h: Y \to \mathbb{R}$ by $h(\alpha x) = \alpha f(x^+)$. Obviously, $h(u) \le p(u)$ holds for all $u \in Y$, and so, by the Hahn–Banach Theorem 29.2, h can be extended linearly to all of X so as to preserve the inequality $h(u) \le p(u)$ for all $u \in X$. Next, observe that if $u \ge 0$, then $h(u) \le p(u) = f(u^+) = f(u)$ holds, and moreover, $-h(u) = h(-u) \le p(-u) = f((-u)^+) = f(0) = 0$. That is, $0 \le h \le f$ holds. Therefore, $f(x^+) = h(x) \le \sup\{g(x): g \in X^\sim \text{ and } 0 \le g \le f\}$, so that

$$f(x^+) = \sup\{g(x): g \in X^\sim \text{ and } 0 \le g \le f\}.$$

The other two formulas now follow easily from the relations $f(x^-) = f((-x)^+)$ and $f(|x|) = f(x^+) + f(x^-)$. ∎

A norm $\|\cdot\|$ on a vector lattice X is said to be a **lattice norm** whenever $|x| \le |y|$ in X implies $\|x\| \le \|y\|$. A **normed vector lattice** is a vector lattice equipped with a lattice norm. If a normed vector lattice X is complete, then X is referred to as a **Banach lattice**.

Let X be a normed vector lattice. Then it should be clear that $\|x\| = \||x|\|$ holds for all $x \in X$. Also, in view of Theorem 30.1, the following inequalities are valid for all $x, y \in X$:

$$\|x^+ - y^+\| \le \|x - y\| \quad \text{and} \quad \| |x| - |y| \| \le \|x - y\|.$$

In particular, they imply that the mappings $x \mapsto x^+$ and $x \mapsto |x|$ (both from X into X) are uniformly continuous.

It is interesting to observe that most of the normed spaces one encounters in analysis are normed lattices or Banach lattices. Some examples follow.

Example 30.4. Let X be a nonempty set. The collection of all real-valued bounded functions defined on X is denoted by $B(X)$. Then $B(X)$ is a vector lattice (in fact a function space) under the ordering $f \geq g$ whenever $f(x) \geq g(x)$ holds for all $x \in X$. Also, under the sup norm $\|f\|_\infty = \sup\{|f(x)|: x \in X\}$, the vector lattice $B(X)$ is a Banach lattice. See Example 6.12. ∎

Example 30.5. Let X be a topological space. Denote (as usual) by $C_c(X)$ the vector space of all continuous real-valued functions on X that have compact support. In other words, $f \in C(X)$ belongs to $C_c(X)$ if and only if the set $\{x \in X: f(x) \neq 0\}$ has compact closure. (When X is compact, $C_c(X) = C(X)$.) Then $C_c(X)$ under the pointwise ordering (i.e., $f \geq g$ if $f(x) \geq g(x)$ for all $x \in X$), and the sup norm is a normed vector lattice. ∎

Example 30.6. Consider $C[0, 1]$ with the pointwise ordering, and define a norm by $\|f\| = \int_0^1 |f(x)| \, dx$. Then $C[0, 1]$ under this norm is a normed vector lattice, but not a Banach lattice. ∎

Example 30.7. Let ℓ_1 denote the vector space of all (real) sequences $x = (x_1, x_2, \ldots)$ such that $\sum_{n=1}^{\infty} |x_n| < \infty$. With the pointwise algebraic and lattice operations ℓ_1 is a function space. Under the norm $\|x\| = \sum_{n=1}^{\infty} |x_n|$, the vector lattice ℓ_1 becomes a Banach lattice (see Example 27.3). ∎

A vector subspace Y of a vector lattice X is called a **vector sublattice** of X if for every pair x and y of Y, the elements $x \vee y$ and $x \wedge y$ belong to Y. That is, Y is a vector sublattice of X if it is closed under the lattice operations of X. A vector subspace V of a vector lattice X is said to be an **ideal** whenever $|x| \leq |y|$ and $y \in V$ imply $x \in V$. In view of the identity $x \vee y = \frac{1}{2}(x + y + |x - y|)$, it follows that every ideal is a vector sublattice.

Let X be a normed vector lattice and let A be an order-bounded subset of X. Pick some $y \in X$ such that $|x| \leq y$ for each $x \in A$, and consequently, $\|x\| \leq \|y\|$ holds for all $x \in A$. That is, every order-bounded set is norm bounded. It follows from this that every continuous linear functional on X carries order-bounded subsets of X onto bounded subsets of $\mathbb{R}$. In other words, the norm dual X^* is a vector subspace of the order dual $X^\sim$. The next result shows that X^* is, in actuality, an ideal of $X^\sim$.

Theorem 30.8. *The norm dual X^* of a normed vector lattice X is an ideal of the order dual $X^\sim$ (and hence, X^* is a vector lattice in its own right).*

Proof. Assume $|f| \leq |g|$ holds in $X^\sim$ with $g \in X^*$ and $f \in X^\sim$. We must show that $f \in X^*$.

Let $x \in X$ satisfy $\|x\| = 1$. Then for every $y \in X$ with $|y| \leq |x|$ we have $\|y\| \leq \|x\| = 1$. Therefore,

$$|f(x)| \leq |f|(|x|) \leq |g|(|x|) = \sup\{g(y): |y| \leq |x|\} \leq \|g\|,$$

where the equality in the middle holds by virtue of Theorem 30.2. This shows that $f \in X^*$ and $\|f\| \leq \|g\|$. ∎

The preceding arguments also show that for a normed vector lattice X the relation $|f| \leq |g|$ in X^* implies $\|f\| \leq \|g\|$. That is, X^* is a normed vector lattice. Since X^* is in addition a Banach space (Theorem 28.7), the following result should be immediate.

Theorem 30.9. *The norm dual of a normed vector lattice is a Banach lattice.*

If X is a normed vector lattice, then Theorem 30.8 shows that $X^* \subseteq X^{\sim}$ holds. However, the norm dual of a Banach lattice always coincides with its order dual.

Theorem 30.10. *If X is a Banach lattice, then $X^* = X^{\sim}$ holds.*

Proof. Let $f \in X^{\sim}$. Assume by way of contradiction that f is not continuous. That is, assume that $\|f\| = \sup\{|f(x)|: \|x\| = 1\} = \infty$. Then there exists a sequence of vectors $\{x_n\}$ such that $\|x_n\| = 1$ and $|f(x_n)| \geq n^3$ for all n. Let $y_n = \sum_{k=1}^{n} k^{-2}|x_k|$. Then

$$\|y_{n+p} - y_n\| = \left\| \sum_{k=n+1}^{n+p} k^{-2}|x_k| \right\| \leq \sum_{k=n+1}^{n+p} k^{-2}$$

holds for all n and p, and hence, $\{y_n\}$ is a Cauchy sequence of X. Since X is complete, there exists some $y \in X$ with $\lim y_n = y$.

Clearly, $0 \leq y_n \leq y_{n+1}$, and we claim that $0 \leq y_n \leq y$ holds for each n. Indeed, note first that the inequalities

$$0 \leq (y_n - y)^+ \leq (y_{n+p} - y)^+ \leq |y_{n+p} - y| \quad \text{imply} \quad \|(y_n - y)^+\| \leq \|y_{n+p} - y\|$$

for all n and p. Therefore, $0 \leq \|(y_n - y)^+\| \leq \lim_{p\to\infty}\|y_{n+p} - y\| = 0$, and so $(y_n - y)^+ = 0$ holds for each n. But then, $y_n - y \leq (y_n - y) \vee 0 = (y_n - y)^+ = 0$ implies $0 \leq y_n \leq y$ for all n.

Now, since $n^{-2}|x_n| \leq y_n \leq y$, it follows that

$$n \leq n^{-2}|f(x_n)| \leq n^{-2}|f|(|x_n|) \leq |f|(y_n) \leq |f|(y) < \infty$$

for each n, which is impossible. Thus, f is continuous, and so $X^* = X^{\sim}$. ∎

Let X be a normed vector lattice, and let Y be a vector sublattice of X. We already know that $\overline{Y}$ is a vector subspace of X, and in view of the continuity of $x \mapsto x^+$, it is easy to see that $\overline{Y}$ is a vector sublattice of X.

We also know that the natural embedding $x \mapsto \hat{x}$ of a normed space into its second dual is linear and norm preserving. If X is a normed vector lattice, then $x \mapsto \hat{x}$ is in addition lattice preserving. Indeed, for $x \in X$ and $0 \leq f \in X^*$, Theorems 30.2 and 30.3 applied consecutively in connection with the fact that X^* is an ideal of $X^\sim$ give

$$\begin{aligned}
(\hat{x})^+(f) &= \sup\{\hat{x}(g): g \in X^* \text{ and } 0 \leq g \leq f\} \\
&= \sup\{g(x): g \in X^* \text{ and } 0 \leq g \leq f\} \\
&= \sup\{g(x): g \in X^\sim \text{ and } 0 \leq g \leq f\} \\
&= f(x^+) = \widehat{x^+}(f).
\end{aligned}$$

That is, $(\hat{x})^+ = \widehat{x^+}$ holds for each $x \in X$, and this shows that $x \mapsto \hat{x}$ preserves the lattice operations. Thus, X "sits" in X^{**} with its norm, algebraic, and lattice structures preserved. Since the closure of X in the Banach lattice X^{**} is the completion of X, it follows from the preceding that the completion of X is a Banach lattice. Summarizing, we have the following result:

Theorem 30.11. *The completion of a normed vector lattice is a Banach lattice.*

A linear isometry $T: X \to Y$ between two normed vector lattices that satisfies $T(x \vee y) = T(x) \vee T(y)$ for all $x, y \in X$ is called a **lattice isometry**. Two normed vector lattices X and Y are said to be **isomorphic** if there exists a lattice isometry from X onto Y. In other words, X and Y are isomorphic if there exists a mapping from X onto Y that preserves all structures: the norm, the algebraic, and the lattice.

An operator $T: X \to Y$ between two partially ordered vector spaces is called **positive** if $T(x) \geq 0$ holds for all $x \geq 0$. When X is a Banach lattice and Y a normed vector lattice, then every positive operator from X into Y is necessarily continuous. (To see this, repeat the arguments of the proof of Theorem 30.10.)

Here is an example of a classical positive operator:

Example 30.12 (The Laplace Transform). In this example, we shall denote by dt the Lebesgue measure on $[0, \infty)$. Consider the operator $\mathcal{L}: L_1([0, \infty)) \to C_b([0, \infty))$ defined by

$$\mathcal{L}f(s) = \int_0^\infty e^{-st} f(t)\, dt, \quad s \geq 0,$$

where $C_b([0, \infty))$ denotes the Banach lattice of all (uniformly) bounded continuous functions on $[0, \infty)$. Since $|e^{-st} f(t)| \leq |f(t)|$ holds for all $s, t \geq 0$ and for each $s \geq 0$ the function $e^{-st} f(t)$ is measurable, it follows from Theorem 22.6 that the function $e^{-st} f(t)$

is Lebesgue integrable over $[0, \infty)$ for each $s \geq 0$. Moreover, it should be clear that $|\mathcal{L}f(s)| \leq \|f\|_1$ for each $s \geq 0$ so that $\mathcal{L}f$ is a uniformly bounded function. In addition, since $\lim_{s \to s_0} e^{-st} f(t) = e^{-s_0 t} f(t)$ holds for each $s_0 \geq 0$, it follows from Theorem 24.4 that

$$\lim_{s \to s_0} \mathcal{L}f(s) = \lim_{s \to s_0} \int_0^\infty e^{-st} f(t)\, d = \int_0^\infty \lim_{s \to s_0} e^{-st} f(t)\, dt$$

$$= \int_0^\infty e^{-s_0 t} f(t)\, dt = \mathcal{L}f(s_0),$$

which shows that $\mathcal{L}f$ is indeed a bounded continuous function. (The preceding argument also shows that $\lim_{s \to \infty} \mathcal{L}f(s) = 0$.) It is now a routine matter to verify that $\mathcal{L}$ is a linear positive operator—and hence, a continuous operator. The operator $\mathcal{L}$ is known as the **Laplace[5] transform** and plays an important role in applications.

It is useful to notice that the Laplace transform is also one-to-one. To see this, assume that some function $f \in L_1([0, \infty))$ satisfies $\mathcal{L}f(s) = \int_0^\infty e^{-st} f(t)\, dt = 0$ for each $s \geq 0$. Making the change of variable $x = e^{-t}$, we get $\int_0^1 x^{s-1} f(-\ln x)\, dx = 0$ for all $s \geq 0$. In particular, we have $\int_0^1 x^k f(-\ln x)\, dx = 0$ for all $k = 0, 1, 2, \ldots$. This implies $f(-\ln x) = 0$ for almost all $x \in [0, 1]$ (see Exercise 21 of Section 22). Using the fact that the function $\ln x$ carries null sets to null sets (see Exercise 9 of Section 18), we infer that $f = 0$ a.e. Thus, $\mathcal{L}$ is one-to-one. ∎

We shall close this section with a remarkable convergence property (due to P. P. Korovkin[6]) about sequences of positive operators on $C[0, 1]$. Korovkin's result demonstrates the usefulness of the order structures.

If $T: C[0, 1] \to C[0, 1]$ is an operator, then (for simplicity) we shall write Tf instead of $T(f)$. For our discussion below, $C[0, 1]$ will be considered equipped with the sup norm $\|\cdot\|_\infty$. For instance, $\lim f_n = f$ in $C[0, 1]$ will mean $\lim \|f_n - f\|_\infty = 0$, i.e., that $\{f_n\}$ converges uniformly to f. Also, $\mathbf{1}$, x, and x^2 will denote the three functions of $C[0, 1]$ defined by $\mathbf{1}(t) = 1$, $x(t) = t$, and $x^2(t) = t^2$ for each $t \in [0, 1]$.

Theorem 30.13 (Korovkin). *Let $\{T_n\}$ be a sequence of positive operators from $C[0, 1]$ into $C[0, 1]$. If $\lim T_n f = f$ holds when f equals $\mathbf{1}$, x, and x^2, then $\lim T_n f = f$ holds for all $f \in C[0, 1]$.*

Proof. Let $f \in C[0, 1]$. Our objective is to show that given $\epsilon > 0$ there exist constants C_1, C_2, and C_3 (depending on ϵ) such that

$$\|T_n f - f\|_\infty \leq \epsilon + C_1 \|T_n x^2 - x^2\|_\infty + C_2 \|T_n x - x\|_\infty + C_3 \|T_n \mathbf{1} - \mathbf{1}\|_\infty.$$

---

[5]Pierre Simon Marquis de Laplace (1749–1827), a distinguished French mathematician, physicist, and astronomer. He made fundamental contributions to many fields including electricity, magnetism, planetary motion, and the theory of probability.

[6]Pavel Petrovich Korovkin (1913–1985), a Russian mathematician. He worked in approximation theory.

If this is done, then our hypothesis implies that $\|T_n f - f\|_\infty < 2\epsilon$ must hold for all sufficiently large n. Therefore, this will establish that $\lim T_n f = f$ holds for all $f \in C[0, 1]$.

To this end, start by observing that for each t in the interval $[0, 1]$ the function $0 \le g_t \in C[0, 1]$ defined by $g_t(s) = (s - t)^2$ satisfies $g_t = x^2 - 2tx + t^2 \mathbf{1}$. Since each T_n is positive, $T_n g_t$ is a positive function. In particular, we have

$$
\begin{aligned}
0 \le T_n g_t(t) &= (T_n x^2 - 2t T_n x + t^2 T_n \mathbf{1})(t) \\
&= (T_n x^2 - x^2)(t) - 2t(T_n x - x)(t) + t^2(T_n \mathbf{1} - \mathbf{1})(t) \\
&\le \|T_n x^2 - x^2\|_\infty + 2\|T_n x - x\|_\infty + \|T_n \mathbf{1} - \mathbf{1}\|_\infty
\end{aligned}
\tag{1}
$$

for each $t \in [0, 1]$.

Now, let $M = \|f\|_\infty$, and let $\epsilon > 0$. By the uniform continuity of f on $[0, 1]$, there exists some $\delta > 0$ such that $-\epsilon < f(s) - f(t) < \epsilon$ holds whenever $s, t \in [0, 1]$ satisfy $|s - t| < \delta$. Next, observe that

$$
-\epsilon - \frac{2M}{\delta^2}(s - t)^2 \le f(s) - f(t) \le \epsilon + \frac{2M}{\delta^2}(s - t)^2
\tag{2}
$$

holds for all $s, t \in [0, 1]$. Indeed, if $|s - t| < \delta$, then (2) follows from

$$
-\epsilon < f(s) - f(t) < \epsilon.
$$

On the other hand, if $|s - t| \ge \delta$, then (2) follows from the inequalities

$$
-\frac{2M}{\delta^2}(s - t)^2 \le -2M \le f(s) - f(t) \le 2M \le \frac{2M}{\delta^2}(s - t)^2.
$$

Since each T_n is positive and linear, it follows from (2) that

$$
-\epsilon T_n \mathbf{1} - \frac{2M}{\delta^2} T_n g_t \le T_n f - f(t) T_n \mathbf{1} \le \epsilon T_n \mathbf{1} + \frac{2M}{\delta^2} T_n g_t.
\tag{3}
$$

Next, let $K = 2M/\delta^2$ and evaluate (3) at t to get

$$
\begin{aligned}
|[T_n f - f(t) T_n \mathbf{1}](t)| &\le \epsilon T_n \mathbf{1}(t) + K T_n g_t(t) \\
&= \epsilon + \epsilon[(T_n \mathbf{1} - \mathbf{1})(t)] + K T_n g_t(t) \\
&\le \epsilon + \epsilon \|T_n \mathbf{1} - \mathbf{1}\|_\infty + K T_n g_t(t).
\end{aligned}
$$

In particular, note that

$$
\begin{aligned}
|(T_n f - f)(t)| &\le |[T_n f - f(t) T_n \mathbf{1}](t)| + |f(t)| \cdot |(T_n \mathbf{1} - \mathbf{1})(t)| \\
&\le \epsilon + K T_n g_t(t) + (M + \epsilon)\|T_n \mathbf{1} - \mathbf{1}\|_\infty
\end{aligned}
\tag{4}
$$

holds for each $t \in [0, 1]$. Thus, by taking into account (1), it follows from (4) that

$$\|T_n f - f\|_\infty \leq \epsilon + K \|T_n x^2 - x^2\|_\infty + 2K \|T_n x - x\|_\infty + (K + M + \epsilon)\|T_n 1 - 1\|_\infty.$$

This completes the proof of the theorem.   ∎

EXERCISES

1. Let X be a vector lattice, and let $f: X^+ \to [0, \infty)$ be an additive function (that is, $f(x + y) = f(x) + f(y)$ holds for all $x, y \in X^+$). Then show that there exists a unique linear functional g on X such that $g(x) = f(x)$ holds for all $x \in X^+$.
   [HINT: Use the arguments of the proof of Lemma 18.7 to show first that $f(rx) = rf(x)$ holds for each $x \in X^+$ and each rational number $r \geq 0$. Then define $g(x) = f(x^+) - f(x^-)$ for each $x \in X$.]

2. A vector lattice is called **order complete** if every nonempty subset that is bounded from above has a least upper bound (also called the supremum of the set).
   Show that if X is a vector lattice, then its order dual $X^\sim$ is an order complete vector lattice.
   [HINT: If A is a nonempty set of positive linear functionals such that $f \leq g$ holds in $X^\sim$ for each $f \in A$, put $h(x) = \sup\{(\vee_{i=1}^n f_i)(x): f_i \in A\}$ for each $x \in X^+$. Use the preceding exercise to show that h extends to a positive linear functional and that $h = \sup A$.]

3. Show that the collection of all bounded functions on $[0, 1]$ is an ideal of $\mathbf{R}^{[0,1]}$. Also, show that $C[0, 1]$ is a vector sublattice of $\mathbf{R}^{[0,1]}$ but not an ideal.

4. Let X be a vector lattice. Show that a norm $\|\cdot\|$ on X is a lattice norm if and only if it satisfies the following two properties:

   a. If $0 \leq x \leq y$, then $\|x\| \leq \|y\|$, and
   b. $\|x\| = \| |x| \|$ holds for all $x \in X$.

5. Show that in a normed vector lattice X, its positive cone X^+ is a closed set.
   [HINT: $X^+ = \{x \in X: x^- = 0\}$.]

6. Let X be a normed vector lattice. Assume that $\{x_n\}$ is a sequence of X such that $x_n \leq x_{n+1}$ holds for all n. Show that if $\lim x_n = x$ holds in X, then the vector x is the least upper bound of the sequence $\{x_n\}$ in X. In symbols, $x_n \uparrow x$ holds.
   [HINT: Observe that $x_{n+p} - x_n \geq 0$ for all n and p and use the conclusion of the preceding exercise.]

7. Assume that $x_n \to x$ holds in a Banach lattice and let $\{\epsilon_n\}$ be a sequence of strictly positive real numbers, i.e., $\epsilon_n > 0$ for each n. Show that there exists a subsequence $\{x_{k_n}\}$ of $\{x_n\}$ and some positive vector u such that $|x_{k_n} - x| \leq \epsilon_n u$ holds for each n.
   [HINT: Pick a subsequence $\{y_n\}$ of $\{x_n\}$ satisfying $\|y_n - x\| < \epsilon_n 2^{-n}$ for each n and let $u = \sum_{n=1}^\infty (\epsilon_n)^{-1}|y_n - x|$. Now, use the preceding exercise to conclude that $(\epsilon_n)^{-1}|y_n - x| \leq u$ holds for each n.]

8. Let $T: X \to Y$ be a positive operator between two normed vector lattices. If X is a Banach lattice, then show that T is continuous.
   [HINT: If T is not continuous, then there exist a sequence $\{x_n\}$ of X and some $\epsilon > 0$ satisfying $x_n \to 0$ and $\|T x_n\| \geq \epsilon$ for each n. By the preceding exercise, there exists

a subsequence $\{y_n\}$ of $\{x_n\}$ and some $u \in X^+$ satisfying $|y_n| \leq \frac{1}{n}u$ for each n. Now, note that $\|T y_n\| \leq \frac{1}{n}\|T u\|$ holds for each n.]

9. Show that any two complete lattice norms on a vector lattice must be equivalent. [HINT: Apply the previous exercise.]

10. The **averaging operator** $A: \ell_\infty \to \ell_\infty$ is defined by

$$A(x) = \left(x_1, \frac{x_1 + x_2}{2}, \frac{x_1 + x_2 + x_3}{3}, \ldots, \frac{x_1 + x_2 + \cdots + x_n}{n}, \ldots\right)$$

for each $x = (x_1, x_2, \ldots) \in \ell_\infty$. Establish the following:

a. A is a positive operator.
b. A is a continuous operator.
c. The vector space

$$V = \left\{x = (x_1, x_2, \ldots) \in \ell_\infty: \left\{\tfrac{x_1 + x_2 + \cdots + x_n}{n}\right\} \text{ converges in } \mathbb{R}\right\}$$

is a closed subspace of ℓ_∞. Is $V = \ell_\infty$?

11. This exercise shows that for a normed vector lattice X, its norm dual X^* may be a proper ideal of its order dual $X^\sim$. Let X be the collection of all sequences $\{x_n\}$ such that $x_n = 0$ for all but a finite number of terms (depending on the sequence). Show that:

a. X is a function space.
b. X equipped with the sup norm is a normed vector lattice, but not a Banach lattice.
c. If $f: X \to \mathbb{R}$ is defined by $f(x) = \sum_{n=1}^{\infty} n x_n$ for each $x = \{x_n\} \in X$, then f is a positive linear functional on X that is not continuous.

12. Determine the norm completion of the normed vector lattice of the preceding exercise.

13. Determine the norm completion of the normed vector lattice of Example 30.5 when X is a Hausdorff locally compact topological space.

14. Let X and Y be two vector lattices, and let $T: X \to Y$ be a linear operator. Show that the following statements are equivalent:

a. $T(x \vee y) = T(x) \vee T(y)$ holds for all $x, y \in X$.
b. $T(x \wedge y) = T(x) \wedge T(y)$ holds for all $x, y \in X$.
c. $T(x) \wedge T(y) = 0$ holds in Y whenever $x \wedge y = 0$ holds in X.
d. $|T(x)| = T(|x|)$ holds for all $x \in X$.

(A linear operator T that satisfies the preceding equivalent statements is referred to as a **lattice homomorphism**.)

15. Let ℓ_∞ be the Banach lattice of all bounded real sequences, that is, $\ell_\infty = B(\mathbb{N})$, and let $\{r_1, r_2, \ldots\}$ be an enumeration of the rational numbers of $[0, 1]$. Show that the mapping $T: C[0, 1] \to \ell_\infty$ defined by $T(f) = (f(r_1), f(r_2), \ldots)$ is a lattice isometry that is not onto.

16. Let X be a normed vector lattice. Then show that an element $x \in X$ satisfies $x \geq 0$ if and only if $f(x) \geq 0$ holds for each continuous positive linear functional f on X. [HINT: For the "if" part use the second formula of Theorem 30.3 to obtain $f(x^-) = 0$ for each continuous positive linear functional f.]

17. Let X be a Banach lattice. If $0 \leq x \in X$, then show that

$$\|x\| = \sup\{f(x): 0 \leq f \in X^* \text{ and } \|f\| = 1\}.$$

18. Assume that $\varphi: [0, 1] \to \mathbb{R}$ is a strictly monotone continuous function and that $T: C[0, 1] \to C[0, 1]$ is a continuous linear operator. If $T(\varphi f) = \varphi T(f)$ holds for each $f \in C[0, 1]$ (where φf denotes the pointwise product of φ and f). Show that there exists a unique function $h \in C[0, 1]$ satisfying $T(f) = hf$ for all $f \in C[0, 1]$.

19. If $f \in C[0, 1]$, then the polynomials

$$B_n(x) = \sum_{k=0}^{n} \binom{n}{k} f\left(\frac{k}{n}\right) x^k (1-x)^{n-k},$$

where $\binom{n}{k}$ is the binomial coefficient defined by $\binom{n}{k} = \frac{n!}{k!(n-k)!}$, are known as the **Bernstein**[7] **polynomials** of f.

Show that if $f \in C[0, 1]$, then the sequence $\{B_n\}$ of Bernstein polynomials of f converges uniformly to f.

[HINT: Consider the sequence $\{T_n\}$ of positive operators defined by

$$T_n f(x) = \sum_{k=0}^{n} \binom{n}{k} f\left(\frac{k}{n}\right) x^k (1-x)^{n-k}$$

and apply Korovkin's theorem.]

20. Let $T: C[0, 1] \to C[0, 1]$ be a positive operator. Show that if $Tf = f$ holds true when f equals $\mathbf{1}$, x, and x^2, then T is the identity operator (that is, $Tf = f$ holds for each $f \in C[0, 1]$).

21. **(Korovkin)** Let $\{T_n\}$ be a sequence of positive operators from $C[0, 1]$ into $C[0, 1]$ satisfying $T_n \mathbf{1} = \mathbf{1}$. If there exists some $c \in [0, 1]$ such that $\lim T_n g = 0$ holds for the function $g(t) = (t-c)^2$, then show that $\lim T_n f = f(c) \cdot \mathbf{1}$ holds for all $f \in C[0, 1]$.

31. L_p-SPACES

Our attention is now turned from the study of general normed spaces to function spaces. Many of the classical spaces in analysis consist of measurable functions, and most of the important norms on such spaces are defined by integrals. The theory of integration enables us to study the remarkable properties of these spaces. Here the classical L_p-spaces will be considered. As we shall see, they are special examples of Banach lattices.

Throughout this section (X, S, μ) will be a fixed measure space, and unless otherwise specified, all properties of functions will refer to this measure space. It

---

[7] Sergei Natanovich Bernstein (1880–1968), a Russian mathematician. He contributed to the approximation of functions and probability theory.

is important to keep in mind that if f is a measurable function, the $|f|^p$ is also measurable for each $p > 0$.

Definition 31.1. *Let $0 < p < \infty$. Then the collection of all measurable functions f for which $|f|^p$ is integrable will be denoted by $L_p(\mu)$.*

If clarity requires the measure space X to be indicated, then $L_p(\mu)$ will be denoted by $L_p(X)$ or even by $L_p(X, \mathcal{S}, \mu)$.

It is easy to see that $L_p(\mu)$ is a vector space. Indeed, if $f \in L_p(\mu)$, then clearly, $\alpha f \in L_p(\mu)$ holds for all $\alpha \in \mathbb{R}$. On the other hand, the elementary inequality among the real numbers

$$|a + b|^p \le 2^p(|a|^p + |b|^p)$$

shows that $L_p(\mu)$ is closed under addition.[8] Moreover, if $f \in L_p(\mu)$, then the inequalities $0 \le f^+ \le |f|$ and $0 \le f^- \le |f|$ imply that f^+, f^-, and $|f|$ belong to $L_p(\mu)$. In other words, $L_p(\mu)$ is a vector lattice.

For each $f \in L_p(\mu)$ let

$$\|f\|_p = \left(\int |f|^p \, d\mu \right)^{\frac{1}{p}}.$$

The number $\|f\|_p$ is called the L_p-**norm** of f. Obviously, $\|f\|_p \ge 0$ and $\|\alpha f\|_p = |\alpha| \cdot \|f\|_p$ hold for all $f \in L_p(\mu)$ and $\alpha \in \mathbb{R}$.

To obtain additional properties of the L_p-norms, we need an inequality.

Lemma 31.2. *If $0 < \lambda < 1$, then*

$$a^\lambda b^{1-\lambda} \le \lambda a + (1 - \lambda)b$$

holds for every pair of nonnegative real numbers a and b.

Proof. The inequality is trivial if either a or b equals zero. Hence, assume $a > 0$ and $b > 0$. Consider the function $f: [0, \infty) \to \mathbb{R}$ defined by $f(x) = 1 - \lambda + \lambda x - x^\lambda$ for $x > 0$. Then $f'(x) = \lambda(1 - x^{\lambda-1})$, and so $x = 1$ is the only critical point of f. It follows that f attains its minimum at $x = 1$. Thus, $f(1) = 0 \le 1 - \lambda + \lambda x - x^\lambda$ holds for all $x > 0$. Now, let $x = a/b$ to obtain the desired inequality. ∎

---

[8]To verify this inequality, note that $|a| = (|a|^p)^{\frac{1}{p}} \le (|a|^p + |b|^p)^{\frac{1}{p}}$, and so

$$|a + b| \le |a| + |b| \le 2(|a|^p + |b|^p)^{\frac{1}{p}}.$$

An important inequality between L_p-norms, known as **Hölder's[9] inequality**, is stated next.

Theorem 31.3 (Hölder's Inequality). *Let* $1 < p < \infty$ *and* $1 < q < \infty$ *be such that* $\frac{1}{p} + \frac{1}{q} = 1$. *If* $f \in L_p(\mu)$ *and* $g \in L_q(\mu)$, *then* $fg \in L_1(\mu)$ *and*

$$\int |fg|\,d\mu \le \left(\int |f|^p\,d\mu \right)^{\frac{1}{p}} \cdot \left(\int |g|^q\,d\mu \right)^{\frac{1}{q}} = \|f\|_p \cdot \|g\|_q.$$

Proof. If $f = 0$ a.e. or $g = 0$ a.e. holds, then the inequality is trivial. So, assume $f \ne 0$ a.e. and $g \ne 0$ a.e. Then $\|f\|_p > 0$ and $\|g\|_q > 0$. Now, apply Lemma 31.2 with

$$\lambda = \frac{1}{p}, \quad a = (|f(x)|/\|f\|_p)^p, \quad \text{and} \quad b = (|g(x)|/\|g\|_q)^q$$

to obtain

$$\frac{|f(x)g(x)|}{\|f\|_p \cdot \|g\|_q} \le \frac{1}{p} \frac{|f(x)|^p}{(\|f\|_p)^p} + \frac{1}{q} \frac{|g(x)|^q}{(\|g\|_q)^q}.$$

By Theorem 22.6, $fg \in L_1(\mu)$, and by integrating, we get

$$\frac{\int |fg|\,d\mu}{\|f\|_p \cdot \|g\|_q} \le \frac{1}{p} + \frac{1}{q} = 1.$$

That is, $\int |fg|\,d\mu \le \|f\|_p \cdot \|g\|_q$, as claimed. ∎

For the special case $p = q = 2$, Hölder's inequality is known as the Cauchy–Schwarz[10] inequality; see also Theorem 32.2. The triangle inequality of the function $\|\cdot\|_p$ is referred to as the **Minkowski[11] inequality**. The details follow.

Theorem 31.4 (Minkowski's Inequality). *Let* $1 \le p < \infty$. *Then for every pair* $f, g \in L_p(\mu)$ *the following inequality holds:*

$$\|f + g\|_p \le \|f\|_p + \|g\|_p.$$

---

[9]Otto Ludwig Hölder (1859–1937), a German mathematician. He worked in group theory and geometry. He also contributed to philosophical matters concerning the foundation of mathematics.

[10]Hermann Amandus Schwarz (1843–1921), a German mathematician. He worked in complex analysis and made several contributions to the theory of minimal surfaces.

[11]Hermann Minkowski (1864–1909), a German mathematician. He studied extensively the geometric properties of convex sets. His ideas in mathematical physics contributed greatly to the creation of the theory of relativity.

Proof. For $p = 1$ the inequality is clearly true. Thus, we can assume $1 < p < \infty$. Let $1 < q < \infty$ be such that $\frac{1}{p} + \frac{1}{q} = 1$.

We already know that if f and g belong to $L_p(\mu)$, then $f + g$ likewise belongs to $L_p(\mu)$. Next, observe that since $(p-1)q = p$, it follows that $|f + g|^{p-1} \in L_q(\mu)$. Thus, by Theorem 31.3 both functions

$$|f| \cdot |f + g|^{p-1} \quad \text{and} \quad |g| \cdot |f + g|^{p-1}$$

belong to $L_1(\mu)$ and we have the inequalities

$$\int |f| \cdot |f + g|^{p-1} \, d\mu \leq \|f\|_p \cdot \left(\int |f + g|^{(p-1)q} \, d\mu \right)^{\frac{1}{q}}$$

$$= \|f\|_p \cdot (\|f + g\|_p)^{\frac{p}{q}},$$

$$\int |g| \cdot |f + g|^{p-1} \, d\mu \leq \|g\|_p \cdot (\|f + g\|_p)^{\frac{p}{q}}.$$

So, from $|f + g|^p = |f + g||f + g|^{p-1} \leq (|f| + |g|)|f + g|^{p-1}$, we get

$$(\|f + g\|_p)^p = \int |f + g|^p \, d\mu \leq \int |f| \cdot |f + g|^{p-1} \, d\mu$$

$$+ \int |g| \cdot |f + g|^{p-1} \, d\mu$$

$$\leq \|f\|_p \cdot (\|f + g\|_p)^{\frac{p}{q}} + \|g\|_p \cdot (\|f + g\|_p)^{\frac{p}{q}}$$

$$= (\|f\|_p + \|g\|_p) \cdot (\|f + g\|_p)^{\frac{p}{q}}.$$

This easily implies

$$\|f + g\|_p = (\|f + g\|_p)^{p - \frac{p}{q}} \leq \|f\|_p + \|g\|_p.$$

The proof of the theorem is now complete. ∎

Summarizing the preceding discussion: If $1 \leq p < \infty$, then

a. $\|f\|_p \geq 0$.
b. $\|\alpha f\|_p = |\alpha| \cdot \|f\|_p$, and
c. $\|f + g\|_p \leq \|f\|_p + \|g\|_p$

hold for all $f, g \in L_p(\mu)$ and $\alpha \in \mathbb{R}$.

Obviously, by Theorem 22.7, $\|f\|_p = 0$ if and only if $f = 0$ a.e. holds. Thus, unfortunately, the function $\|\cdot\|_p$ on $L_p(\mu)$ fails to satisfy the norm requirement that $\|f\|_p = 0$ imply $f = 0$. To avoid this difficulty, it is customary to call two functions of $L_p(\mu)$ equivalent if they are equal almost everywhere. Clearly, this introduces an equivalence relation on $L_p(\mu)$, and $\|\cdot\|_p$ becomes a norm on the equivalence classes. In other words, $L_p(\mu)$, for $1 \le p < \infty$, is a normed space if we do not distinguish between functions that are equal almost everywhere. This means that $L_p(\mu)$ in reality consists of equivalence classes of functions, but this should not pose a problem. In actual practice, the equivalence classes are relegated to the background, and the elements of $L_p(\mu)$ are thought of as functions (where two functions are considered identical if they are equal almost everywhere). An important advantage of the identification of functions that are equal almost everywhere is the following: A function of $L_p(\mu)$ can assume infinite values or even be left undefined on a null set—since by assigning finite values to these points, the function becomes equivalent to a real-valued function of $L_p(\mu)$.

Also, it should be clear that if g is a measurable function and $f \in L_p(\mu)$ satisfies $|g| \le |f|$ a.e., then $g \in L_p(\mu)$ and $\|g\|_p \le \|f\|_p$ holds. In other words, $\|\cdot\|_p$ is a lattice norm. Therefore, for $1 \le p < \infty$ each $L_p(\mu)$ is a normed vector lattice and, in fact, a Banach lattice, as the next result of F. Riesz and E. Fischer[12] shows.

Theorem 31.5 (Riesz–Fischer). *If* $1 \le p < \infty$, *then* $L_p(\mu)$ *is a Banach lattice.*

Proof. Let $\{f_n\}$ be a Cauchy sequence. By passing to a subsequence if necessary, we can assume without loss of generality that $\|f_{n+1} - f_n\|_p < 2^{-n}$ holds for each n. We must establish the existence of some $f \in L_p(\mu)$ such that $\lim \|f - f_n\|_p = 0$.

Let $g_1 = 0$ and $g_n = |f_1| + |f_2 - f_1| + \cdots + |f_n - f_{n-1}|$ for $n \ge 2$. Then $0 \le g_n \uparrow$ and

$$\int (g_n)^p \, d\mu = (\|g_n\|_p)^p \le \left[\|f_1\|_p + \sum_{i=2}^{\infty} \|f_i - f_{i-1}\|_p \right]^p \le [\|f_1\|_p + 1]^p$$

holds for all n. By Levi's Theorem 22.8, there exists some $g \in L_p(\mu)$ such that $0 \le g_n \uparrow g$ a.e.

From

$$|f_{n+k} - f_n| = \left| \sum_{i=n+1}^{n+k} (f_i - f_{i-1}) \right| \le \sum_{i=n+1}^{n+k} |f_i - f_{i-1}| = g_{n+k} - g_n,$$

---

[12]Ernst Sigismund Fischer (1875–1954), an Austrian mathematician. He spent his scientific career in Germany and studied orthonormal sequences of functions.

it follows that $\{f_n\}$ converges pointwise a.e. to some function f. Since

$$|f_n| = \left| f_1 + \sum_{i=2}^{n}(f_i - f_{i-1}) \right| \le g_n \le g \text{ a.e.,}$$

it follows that $|f| \le g$ a.e. holds, and hence, $f \in L_p(\mu)$. Now, in view of $|f - f_n| \le 2g$ a.e. and $\lim|f_n - f|^p = 0$, the Lebesgue dominated convergence theorem implies $\lim\|f - f_n\|_p = 0$, and the proof is finished. $\blacksquare$

A glance at the preceding proof reveals also the following interesting property of convergent sequences in L_p-spaces.

Lemma 31.6. *If a sequence $\{f_n\} \subseteq L_p(\mu)$, where $1 \le p < \infty$, satisfies $\lim\|f - f_n\|_p = 0$, then there exist a subsequence $\{f_{k_n}\}$ of $\{f_n\}$ and some $g \in L_p(\mu)$ such that $f_{k_n} \to f$ a.e. and $|f_{k_n}| \le g$ a.e. for each n.*

In general, it is not true that $\lim\|f - f_n\|_p = 0$ implies $f_n \to f$ a.e. For instance, the sequence $\{f_n\}$ of Example 19.6 satisfies $\lim\|f_n\|_p = 0$ (for each $1 \le p < \infty$), but $\{f_n(x)\}$ does not converge for any $x \in [0, 1]$.

Also, it is easy to construct an example of a sequence in an L_p-space that converges pointwise to some function of the space, but fails to converge in the L_p-norm. For instance, consider $\mathbb{R}$ with the Lebesgue measure and $f_n = \chi_{(n,n+1)}$ for each n. Then $f_n(x) \to 0$ holds for each $x \in \mathbb{R}$ and $f_n \in L_p(\mathbb{R})$ for all n and all $1 \le p < \infty$. On the other hand, $\|f_n\|_p = 1$ holds for each n and $1 \le p < \infty$, and so $\{f_n\}$ does not converge to zero with respect to any L_p-norm.

The next useful result gives a condition for pointwise convergence to imply norm convergence in L_p-spaces.

Theorem 31.7. *Assume $1 \le p < \infty$. Let $f \in L_p(\mu)$ and let $\{f_n\}$ be a sequence of $L_p(\mu)$ such that $f_n \to f$ a.e. If $\lim\|f_n\|_p = \|f\|_p$, then $\lim\|f - f_n\|_p = 0$.*

Proof. Start by observing that $(a + b)^p \le 2^{p-1}(a^p + b^p)$ holds for each pair of nonnegative real numbers a and b. Indeed, for $p = 1$ the inequality is trivial. On the other hand, if $1 < p < \infty$, then the convexity of the function $g(x) = x^p$ $(x \ge 0)$ implies

$$\left(\frac{a+b}{2} \right)^p \le \frac{1}{2}(a^p + b^p),$$

and hence, $(a + b)^p \le 2^{p-1}(a^p + b^p)$ holds. In particular, for each pair of real numbers a and b we have

$$|a - b|^p \le 2^{p-1}(|a|^p + |b|^p).$$

Thus, $0 \le 2^{p-1}(|f_n|^p + |f|^p) - |f_n - f|^p$ a.e., and by applying Fatou's lemma (Theorem 22.10) and using the assumption $\lim \int |f_n|^p \, d\mu = \int |f|^p \, d\mu$, we get

$$
\begin{aligned}
2^p \int |f|^p \, d\mu &= \int \lim_{n \to \infty} [2^{p-1}(|f_n|^p + |f|^p) - |f_n - f|^p] \, d\mu \\
&\le \liminf_{n \to \infty} \int [2^{p-1}(|f_n|^p + |f|^p) - |f_n - f|^p] \, d\mu \\
&= 2^{p-1} \int |f|^p \, d\mu + 2^{p-1} \lim_{n \to \infty} \int |f_n|^p \, d\mu \\
&\quad + \liminf_{n \to \infty} \left[-\int |f_n - f|^p \, d\mu \right] \\
&= 2^p \int |f|^p \, d\mu - \limsup_{n \to \infty} \int |f_n - f|^p \, d\mu.
\end{aligned}
$$

Now, since $\int |f|^p \, d\mu < \infty$, the last inequality yields $\limsup \int |f_n - f|^p \, d\mu \le 0$. Hence, $\limsup \int |f_n - f|^p \, d\mu = \liminf \int |f_n - f|^p \, d\mu = 0$, so that

$$\lim_{n \to \infty} \int |f_n - f| \, d\mu = 0.$$

Therefore, $\lim \|f_n - f\|_p = 0$ holds, as required. ∎

A real number M is said to be an **essential bound** for a function f whenever $|f(x)| \le M$ holds for almost all x. A function is called **essentially bounded** if it has an essential bound. Therefore, a function is essentially bounded if it is bounded except possibly on a set of measure zero. The **essential supremum** of a function f is defined by

$$\|f\|_\infty = \inf\{M \ge 0 : |f(x)| \le M \text{ holds for almost all } x\}.$$

If f does not have any essential bound, then it is understood that $\|f\|_\infty = \infty$. Observe that $|f(x)| \le \|f\|_\infty$ holds for almost all x.

The following properties are easily verified, and they are left as exercises for the reader.

1. If $f = g$ a.e., then $\|f\|_\infty = \|g\|_\infty$.
2. $\|f\|_\infty \ge 0$ for each function f, and $\|f\|_\infty = 0$ if and only if $f = 0$ a.e.

3. $\|\alpha f\|_\infty = |\alpha| \cdot \|f\|_\infty$ for all $\alpha \in \mathbb{R}$.
4. $\|f + g\|_\infty \le \|f\|_\infty + \|g\|_\infty$.
5. If $|f| \le |g|$, then $\|f\|_\infty \le \|g\|_\infty$.

Definition 31.8. *The collection of all essentially bounded measurable functions is denoted by* $L_\infty(\mu)$.

Here again, two functions are considered identical if they are equal almost everywhere. It should be obvious that with the usual algebraic and lattice operations $L_\infty(\mu)$ is a vector lattice. Moreover, according to the above listed properties, $L_\infty(\mu)$ equipped with $\|\cdot\|_\infty$ is a normed vector lattice that is actually a Banach lattice.

Theorem 31.9. $L_\infty(\mu)$ *is a Banach lattice.*

Proof. Let $\{f_n\}$ be a Cauchy sequence of $L_\infty(\mu)$. We have to show that there exists some $f \in L_\infty(\mu)$ such that $\lim \|f - f_n\|_\infty = 0$.

Since for each pair m and n we have $|f_n(x) - f_m(x)| \le \|f_n - f_m\|_\infty$ for almost all x, it follows that there exists a null set A such that $|f_n(x) - f_m(x)| \le \|f_n - f_m\|_\infty$ holds for all m and n and all $x \notin A$. But then $\lim f_n(x) = f(x)$ exists in $\mathbb{R}$ for all $x \notin A$, and moreover, f is measurable and essentially bounded. That is, $f \in L_\infty(\mu)$.

Now, let $\epsilon > 0$. Choose k such that $\|f_n - f_m\|_\infty < \epsilon$ for all $n, m > k$. Since $|f(x) - f_n(x)| = \lim_{m \to \infty} |f_m(x) - f_n(x)| \le \epsilon$ holds for all $x \notin A$ and $n > k$, it follows that $\|f - f_n\|_\infty \le \epsilon$ for each $n > k$. This shows that $\lim \|f - f_n\|_\infty = 0$, and the proof of the theorem is complete. ∎

It is easy to verify that each step function belongs to every L_p-space. Moreover, the collection of all step functions forms a vector sublattice of every L_p-space. In addition, by Theorem 25.1 this vector sublattice is norm dense in $L_1(\mu)$. The next result tells us that actually the vector lattice of step functions is norm dense in every $L_p(\mu)$ with $1 \le p < \infty$.

Theorem 31.10. *For every* $1 \le p < \infty$, *the collection of all step functions is norm dense in* $L_p(\mu)$.

Proof. Let $0 \le f \in L_p(\mu)$. By Theorem 17.7, there exists a sequence $\{\phi_n\}$ of simple functions such that $0 \le \phi_n \uparrow f$ a.e. Clearly, each ϕ_n is a step function and $(f - \phi_n)^p \downarrow 0$ a.e. holds. By the Lebesgue dominated convergence theorem we get $\|f - \phi_n\|_p = (\int |f - \phi_n|^p \, d\mu)^{\frac{1}{p}} \downarrow 0$.

Since every function of $L_p(\mu)$ can be written as a difference of two positive functions of $L_p(\mu)$, it easily follows that the step functions are norm dense in $L_p(\mu)$. ∎

In case the measure is a regular Borel measure, the continuous functions with compact support are also norm dense in each $L_p(\mu)$ for $1 \leq p < \infty$. The details are included in the next theorem.

Theorem 31.11. *Let μ be a regular Borel measure on a Hausdorff locally compact topological space X. Then the collection of all continuous functions with compact support is norm dense in $L_p(\mu)$ for every $1 \leq p < \infty$.*

Proof. Clearly, every continuous function with compact support belongs to each $L_p(\mu)$. Now, let $1 \leq p < \infty$, $f \in L_p(\mu)$, and $\epsilon > 0$. We must show that there exists some continuous function g with compact support such that $\|f - g\|_p < \epsilon$. By Theorem 31.10 it suffices to assume that $f = \chi_A$, where A is a measurable set such that $\mu^*(A) < \infty$.

By Theorem 25.3, there exists a continuous function $g \colon X \to [0, 1]$ with compact support such that $\int |\chi_A - g| \, d\mu < 2^{-p}\epsilon^p$. (Note that $|\chi_A - g| \leq 2$ holds.) But then

$$\|\chi_A - g\|_p = \left(\int |\chi_A - g|^p \, d\mu \right)^{\frac{1}{p}} = \left(\int |\chi_A - g| \cdot |\chi_A - g|^{p-1} \, d\mu \right)^{\frac{1}{p}}$$

$$\leq 2 \left(\int |\chi_A - g| \, d\mu \right)^{\frac{1}{p}} < 2 \cdot 2^{-1} \cdot \epsilon = \epsilon,$$

and the proof is finished. ∎

Consider $\mathbb{R}$ equipped with the measure μ that assigns to every subset of $\mathbb{R}$ the value zero, that is, $\mu = 0$. Then μ is a regular Borel measure, and obviously, any two functions on $\mathbb{R}$ are equal μ-almost everywhere. Thus, in this case, all functions on $\mathbb{R}$ can be identified with the zero function, a situation that is not very useful.

Therefore, it is desirable to deal with regular Borel measures for which distinct continuous functions are not equivalent. To do this, we need to know where the measure is "concentrated" in the space.

Theorem 31.12. *Let μ be a regular Borel measure on a Hausdorff locally compact topological space X. Then there exists a unique closed subset E of X with the following two properties:*

1. *$\mu(E^c) = 0$, and*
2. *if V is an open set such that $E \cap V \neq \emptyset$, then $\mu(E \cap V) > 0$.*

Proof. Let $\mathcal{O} = \bigcup \{V \colon V \text{ is open and } \mu(V) = 0\}$. Clearly, $\mathcal{O}$ is an open set, and we claim that $\mu(\mathcal{O}) = 0$.

To see this, let K be a compact subset of $\mathcal{O}$. From the definition of $\mathcal{O}$ it follows that there exist open sets $V_1, \ldots, V_n$, all of measure zero, such that $K \subseteq \bigcup_{i=1}^{n} V_i$.

Hence, $\mu(K) = 0$. Our claim now follows from $\mu(\mathcal{O}) = \sup\{\mu(K): K$ compact and $K \subseteq \mathcal{O}\}$; see Definition 18.4.

Now, let $E = \mathcal{O}^c$. Then E is a closed set and $\mu(E^c) = \mu(\mathcal{O}) = 0$. On the other hand, if V is an open set such that $E \cap V \neq \emptyset$, then $\mu(E \cap V) > 0$ must hold true. Otherwise, if $\mu(E \cap V) = 0$ holds, then $\mu(V) = \mu(E \cap V) + \mu(E^c \cap V) = 0$ also holds, implying $V \subseteq \mathcal{O} = E^c$, contrary to $E \cap V \neq \emptyset$.

For the uniqueness of E assume that another closed set F satisfies (1) and (2). From (1) it follows at once that $F^c \subseteq \mathcal{O}$, and so $E = \mathcal{O}^c \subseteq F$ holds. On the other hand, since $\mu(\mathcal{O} \cap F) = 0$, it follows from (2) that $\mathcal{O} \cap F = \emptyset$. Hence, $F \subseteq \mathcal{O}^c = E$, so that $F = E$, and the proof is finished. ∎

The unique set E determined by Theorem 31.12 is called the **support** of μ and is denoted by Supp μ. That is, Supp $\mu = E$. If we think of the measure space as a set over which some material has been distributed, then Supp μ represents the parts of the set at which the material has been placed.

How "large" can the support of a regular Borel measure be? If $X = \mathbb{R}^n$, then for example, the support of the zero measure is the empty set, while the support of the Lebesgue measure λ satisfies Supp $\lambda = \mathbb{R}^n$.

Let μ be a regular Borel measure on a Hausdorff locally compact topological space X with Supp $\mu = X$. Then two continuous real-valued functions f and g on X satisfy $f = g$ a.e. if and only if $f(x) = g(x)$ holds for all $x \in X$. This follows immediately by observing that if $f(a) \neq g(a)$ holds for some $a \in X$, then $f(x) \neq g(x)$ holds for all x in some nonempty open set V (a neighborhood of a). Since $\mu(V) > 0$, it is impossible for $f = g$ a.e. to be true. In particular, it follows that $\|f\|_p = (\int |f|^p d\mu)^{\frac{1}{p}}$ defines a lattice norm on $C_c(X)$, the function space of all continuous real-valued functions on X with compact support. In general, $C_c(X)$ equipped with an L_p-norm is not a Banach lattice. However, by Theorem 31.11 the following result should be immediate.

Theorem 31.13. *Let μ be a regular Borel measure on a Hausdorff locally compact topological space X with* Supp $\mu = X$. *Then for each $1 \le p < \infty$, the completion of $C_c(X)$ with the L_p-norm is the Banach lattice $L_p(\mu)$.*

In general, the L_p-spaces are not "comparable." As an example, let $X = (0, \infty)$ with the Lebesgue measure. Then the function $f(x) = x^{-\frac{1}{2}}$ if $0 < x \le 1$ and $f(x) = 0$ if $x > 1$ belongs to $L_1(\mu)$, but it does not belong to $L_2(\mu)$. On the other hand, the function $g(x) = 0$ if $0 < x < 1$ and $g(x) = x^{-1}$ if $x \ge 1$ belongs to $L_2(\mu)$, but not to $L_1(\mu)$.

Two comparison results of the L_p-spaces are presented next. The first one is for the case that $(X, \mathcal{S}, \mu)$ is a finite measure space.

Theorem 31.14.   *Let (X, S, μ) be a finite measure space, and assume that $1 \le p < q \le \infty$. Then $L_q(\mu) \subseteq L_p(\mu)$ holds.*

Proof.   Clearly, in this case $L_\infty(\mu) \subseteq L_p(\mu)$ holds for each $1 \le p < \infty$. Thus, assume $1 \le p < q < \infty$.

Let $r = q/p > 1$, and then choose $s > 1$ such that $\frac{1}{r} + \frac{1}{s} = 1$. If $f \in L_q(\mu)$, then clearly, $|f|^p \in L_r(\mu)$. Since the constant function **1** belongs to $L_s(\mu)$, it follows from Theorem 31.3 that $|f|^p = |f|^p \cdot \mathbf{1} \in L_1(\mu)$. That is, $f \in L_p(\mu)$, and the proof of the theorem is complete.   ∎

It should be observed that if $L_q(\mu) \subseteq L_p(\mu)$ holds, then $L_q(\mu)$ is an ideal of the vector lattice $L_p(\mu)$.

Some important examples of L_p-spaces are provided by considering the counting measure on $\mathbb{N}$. In this case, the functions on $\mathbb{N}$ are denoted as sequences, and integration is replaced by summation. These L_p-spaces are called the **little L_p-spaces**, and they are denoted by ℓ_p. In other words, if $0 < p < \infty$, then ℓ_p consists of all sequences $x = (x_1, x_2, \ldots)$ such that $\sum_{n=1}^{\infty} |x_n|^p < \infty$, and in this case $\|x\|_p = (\sum_{n=1}^{\infty} |x_n|^p)^{\frac{1}{p}}$. Similarly, ℓ_∞ is the vector space of all bounded sequences with the sup norm.

The ℓ_p-spaces, unlike the general L_p-spaces, are always comparable. Note the contrast between the next theorem and the preceding one.

Theorem 31.15.   *If $1 \le p < q \le \infty$, then $\ell_p \subseteq \ell_q$ holds. Moreover, the inclusion is proper.*

Proof.   Observe that if $x = (x_1, x_2, \ldots)$ belongs to some ℓ_p-space with $1 \le p < \infty$, then $\{x_n\}$ must be a bounded sequence (actually, convergent to zero), and hence, $x \in \ell_\infty$. That is, $\ell_p \subseteq \ell_\infty$ holds for all $1 \le p < \infty$.

Thus, assume $1 < q < \infty$. Let $x = (x_1, x_2, \ldots) \in \ell_p$. Since $\sum_{n=1}^{\infty} |x_n|^p < \infty$, there exists some k such that $|x_n| < 1$ for all $n \ge k$. This implies $|x_n|^q \le |x_n|^p$ for all $n \ge k$, and this shows that $\sum_{n=1}^{\infty} |x_n|^q < \infty$. Therefore, $x \in \ell_q$, and hence, $\ell_p \subseteq \ell_q$.

For the last part note that if we let $x_n = n^{-\frac{1}{p}}$ for all n, then $x = (x_1, x_2, \ldots) \in \ell_q$ but $x \notin \ell_p$.   ∎

Two numbers p and q in $[1, \infty]$ are called **conjugate exponents** if $\frac{1}{p} + \frac{1}{q} = 1$. We adhere to the convention $1/\infty = 0$, so that 1 and ∞ are conjugate exponents.

Let p and q be two conjugate exponents. If $g \in L_q(\mu)$, then it follows from Theorem 31.3 that $fg \in L_1(\mu)$ for each $f \in L_p(\mu)$. Therefore, for each fixed $g \in L_q(\mu)$ a real-valued function F_g can be defined on $L_p(\mu)$ by

$$F_g(f) = \int fg \, d\mu$$

for all $f \in L_p(\mu)$. Clearly, F_g is a linear functional and, in fact, as the next result shows, a bounded linear functional.

Theorem 31.16. *Let $1 < p \leq \infty$, let q be its conjugate exponent, and let $g \in L_q(\mu)$. Then the linear functional defined by*

$$F_g(f) = \int fg \, d\mu$$

for $f \in L_p(\mu)$ is a bounded linear functional on $L_p(\mu)$ satisfying $\|F_g\| = \|g\|_q$.

Proof. First, we consider the case $p = \infty$ and $q = 1$. From $|F_g(f)| \leq \|g\|_1 \cdot \|f\|_\infty$ for each $f \in L_\infty(\mu)$, it follows that F_g is a bounded linear functional and that $\|F_g\| \leq \|g\|_1$ holds. On the other hand, let $f = \text{Sgn}\, g$, where $\text{Sgn}\, g(x) = 1$ if $g(x) \geq 0$ and $\text{Sgn}\, g(x) = -1$ if $g(x) < 0$. Then f belongs to $L_\infty(\mu)$ and satisfies $\|f\|_\infty = 1$ and $F_g(f) = \int |g| \, d\mu = \|g\|_1$. Therefore, $\|F_g\| = \|g\|_1$.

Now, we consider $1 < p < \infty$. By Hölder's inequality

$$|F_g(f)| = \left| \int fg \, d\mu \right| \leq \|g\|_q \cdot \|f\|_p$$

holds for all $f \in L_p(\mu)$. Hence, F_g is a bounded linear functional, and $\|F_g\| \leq \|g\|_q$ holds. Now, let $f = |g|^{q-1} \text{Sgn}\, g$. Clearly, f is a measurable function, and $|f|^p = |g|^{p(q-1)} = |g|^q$ holds, so that $f \in L_p(\mu)$. Since $fg = |g|^q$, it follows that

$$F_g(f) = \int fg \, d\mu = \int |g|^q \, d\mu = \left(\int |g|^q \, d\mu \right)^{\frac{1}{p}} \cdot \left(\int |g|^q \, d\mu \right)^{\frac{1}{q}}$$

$$= \left(\int |f|^p \, d\mu \right)^{\frac{1}{p}} \cdot \left(\int |g|^q \, d\mu \right)^{\frac{1}{q}} = \|f\|_p \cdot \|g\|_q.$$

That is, $\|F_g\| \geq \|g\|_q$. Thus, $\|F_g\| = \|g\|_q$ holds, and the proof is complete. ∎

The preceding theorem shows that for each $1 < p \leq \infty$ a linear isometry $g \mapsto F_g$ can be defined from $L_q(\mu)$ into $L_p^*(\mu)$, the norm dual of $L_p(\mu)$. Observe that this isometry is also lattice preserving. Indeed, if $0 \leq f \in L_p(\mu)$, then by Theorem 30.2

$$(F_g)^+(f) = \sup\{F_g(h) \colon 0 \leq h \leq f\} = \sup \left\{ \int hg \, d\mu \colon 0 \leq h \leq f \right\}$$

$$= \int fg^+ \, d\mu = F_{g^+}(f).$$

Therefore, $(F_g)^+ = F_{g^+}$ holds, which implies that $g \mapsto F_g$ is also a lattice isometry.

Is every bounded linear functional on $L_p(\mu)$ representable, as in Theorem 31.16, by a function of $L_q(\mu)$? The answer is yes if $1 < p < \infty$. This is a classical result of F. Riesz. A proof of this theorem, as well as some of its applications, is deferred until Section 37. Therefore, $L_p^*(\mu)$ and $L_q(\mu)$ can be considered (under the above isomorphism) as identical Banach lattices. This is usually expressed by saying that for $1 < p < \infty$ the norm dual of $L_p(\mu)$ is $L_q(\mu)$; in symbols, $L_p^*(\mu) = L_q(\mu)$.

When $p = \infty$ the lattice isometry $g \mapsto F_g$ from $L_1(\mu)$ to $L_\infty^*(\mu)$ is rarely onto. The following example will clarify the situation.

Example 31.17. Let $(X, \mathcal{S}, \mu)$ be a measure space such that there exists a disjoint sequence of measurable sets $\{E_n\}$ with $\mu^*(E_n) > 0$ for each n and $X = \bigcup_{n=1}^\infty E_n$. Let L be the collection of all real-valued functions f defined on X that are constant on each E_n, assuming on each E_n the value $f(E_n)$, and for which $\lim f(E_n)$ exists in $\mathbf{R}$. Clearly, L is a vector sublattice of $L_\infty(\mu)$.

Now, define a linear functional F on L by $F(f) = \lim f(E_n)$ for each $f \in L$. It is clear that $|F(f)| \leq \|f\|_\infty$ holds for all $f \in L$, and so F is a continuous linear functional. By Theorem 29.3, F can be extended to $L_\infty(\mu)$ with preservation of its original norm. Denote this extension by F again.

We claim that F cannot be represented by a function of $L_1(\mu)$. To see this, assume by way of contradiction that there exists some $g \in L_1(\mu)$ satisfying $F(f) = \int f g \, d\mu$ for all $f \in L_\infty(\mu)$. Let $G_n = (\bigcup_{i=1}^n E_i)^c$ and $f_n = \chi_{G_n}$. Then $\{f_n\}$ is a sequence of L, and $F(f_n) = 1$ holds for each n. On the other hand, because $|f_n g| \leq |g|$ and $f_n g \to 0$, it follows from the Lebesgue dominated convergence theorem that $F(f_n) = \int f_n g \, d\mu \to 0$, which is impossible.

Therefore, the lattice isometry $g \mapsto F_g$ from $L_1(\mu)$ to $L_\infty^*(\mu)$ is not onto. ∎

Later, we shall see (Theorem 37.10) that if the measure is σ-finite, then the norm dual of $L_1(\mu)$ coincides with $L_\infty(\mu)$.

The representation theorem for the bounded linear functionals on the ℓ_p-spaces can be proved directly.

Theorem 31.18. *Let $1 \leq p < \infty$, and let f be a continuous linear functional on ℓ_p. Then there exists a unique $y = (y_1, y_2, \ldots) \in \ell_q$ (where $\frac{1}{p} + \frac{1}{q} = 1$) such that*

$$f(x) = \sum_{n=1}^\infty x_n y_n$$

holds for every $x = (x_1, x_2, \ldots) \in \ell_p$.

Proof. For each n, let e_n be the sequence having the value one at the nth coordinate and zero at every other. Clearly, if $x = (x_1, x_2, \ldots) \in \ell_p$, then

$\lim \|x - \sum_{i=1}^{n} x_i e_i\|_p = 0$. Thus, $f(x) = \sum_{n=1}^{\infty} x_n f(e_n)$ holds. Let $y_n = f(e_n)$ for each n. To complete the proof, we have to show that $y = (y_1, y_2, \ldots) \in \ell_q$. If $p = 1$, then $|y_n| = |f(e_n)| \le \|f\|$, so that $y \in \ell_\infty$.

Now, for $1 < p < \infty$, let $a_n = y_n \cdot |y_n|^{q-2}$ if $y_n \ne 0$ and $a_n = 0$ if $y_n = 0$. Then $|a_n|^p = |y_n|^q = a_n y_n$ holds for all n. Moreover,

$$\sum_{i=1}^{n} |y_i|^q = \sum_{i=1}^{n} a_i y_i = \sum_{i=1}^{n} a_i f(e_i) = f\left(\sum_{i=1}^{n} a_i e_i\right)$$

$$\le \|f\| \cdot \left\|\sum_{i=1}^{n} a_i e_i\right\|_p = \|f\| \cdot \left(\sum_{i=1}^{n} |a_i|^p\right)^{\frac{1}{p}}$$

$$= \|f\| \cdot \left(\sum_{i=1}^{n} |y_i|^q\right)^{\frac{1}{p}}.$$

Thus, $\left(\sum_{i=1}^{n} |y_i|^q\right)^{1-\frac{1}{p}} = \left(\sum_{i=1}^{n} |y_i|^q\right)^{\frac{1}{q}} \le \|f\| < \infty$ holds for each n. This implies that $y = (y_1, y_2, \ldots)$ belongs to ℓ_q and $f(x) = \sum_{n=1}^{\infty} x_n y_n$. ∎

For a given normed space, it is often useful to have a characterization of its compact subsets. The Ascoli–Arzelà theorem provided such a criterion for the compact subsets of $C(X)$-spaces. Next, we shall characterize the compact subsets of the Banach spaces $L_p([0, 1])$. To do this, we need some preliminary discussion.

Every function $f \in L_p([0, 1])$ will be considered defined on all of $\mathbb{R}$ by $f(t) = 0$ if $t \notin [0, 1]$. Also, for simplicity, we shall write $\int_a^b f(x) \, dx$ instead of $\int_{[a,b]} f \, d\lambda$. If $1 \le p \le \infty$, then for $f \in L_p([0, 1])$ and $h > 0$ we define

$$f_h(t) = \frac{1}{2h} \int_{t-h}^{t+h} f(x) \, dx$$

for each $t \in [0, 1]$. Note that the integral exists, since by Theorem 31.14 we have $L_p([0, 1]) \subseteq L_1([0, 1])$.

Each f_h is a continuous function. Indeed, if $t_n \to t$, then $g_n = f \chi_{(t_n-h, t_n+h)} \to f \chi_{(t-h, t+h)}$ holds. Hence, in view of $|g_n| \le |f|$, the Lebesgue dominated convergence theorem implies that

$$\lim_{n \to \infty} f_h(t_n) = \frac{1}{2h} \lim_{n \to \infty} \int_{t_n-h}^{t_n+h} f(x) \, dx = \frac{1}{2h} \int_{t-h}^{t+h} f(x) \, dx = f_h(t).$$

In particular, note that since $C[0, 1] \subseteq L_p([0, 1])$, it follows that $f_h \in L_p([0, 1])$ for each $h > 0$.

Lemma 31.19.   *Let* $1 \leq p < \infty$, *and let* $f \in L_p([0, 1])$. *Then for each* $h > 0$ *the continuous function* f_h *satisfies*

a. $|f_h(t)| \leq (2h)^{-\frac{1}{p}} \|f\|_p$ *for all* $t \in [0, 1]$, *and*
b. $\|f_h\|_p \leq \|f\|_p$.

Proof.   If $p > 1$, then choose $1 < q < \infty$ with $\frac{1}{p} + \frac{1}{q} = 1$ and apply Hölder's inequality to get

$$\|f_h(t)\|^p = \frac{1}{(2h)^p} \left| \int_{t-h}^{t+h} \mathbf{1} \cdot f(x) \, dx \right|^p$$

$$\leq \frac{1}{(2h)^p} \left(\int_{t-h}^{t+h} \mathbf{1} \, dx \right)^{\frac{p}{q}} \cdot \int_{t-h}^{t+h} |f(x)|^p \, dx$$

$$= \frac{1}{2h} \int_{t-h}^{t+h} |f(x)|^p \, dx.$$

Therefore,

$$|f_h(t)|^p \leq \frac{1}{2h} \int_{t-h}^{t+h} |f(x)|^p \, dx \tag{1}$$

holds for each $1 < p < \infty$ and all $t \in [0, 1]$. Also, (1) is obviously true for $p = 1$, and thus, statement (a) follows immediately.

On the other hand, it follows from (1) that

$$\int_0^1 |f_h(t)|^p \, dt \leq \frac{1}{2h} \int_0^1 \left[\int_{t-h}^{t+h} |f(x)|^p \, dx \right] dt$$

$$= \frac{1}{2h} \int_0^1 \left[\int_{-h}^h |f(t+y)|^p \, dy \right] dt. \tag{2}$$

Here we have used the substitution $x = t + y$; see Exercise 16 of Section 22. Since the function $f(t + y)$ is a Lebesgue measurable function on $\mathbb{R}^2$ (see Exercise 15 of Section 26), it follows from Tonelli's Theorem 26.7 that

$$\int_0^1 \left[\int_{-h}^h |f(t+y)|^p \, dy \right] dt = \int_{-h}^h \left[\int_0^1 |f(t+y)|^p \, dt \right] dy$$

$$\leq 2h \int_0^1 |f(x)|^p \, dx.$$

Thus, (2) implies

$$\int_0^1 |f_h(t)|^p \, dt \leq \int_0^1 |f(x)|^p \, dx,$$

so that $\| f_h \|_p \leq \| f \|_p$ holds. ∎

A. N. Kolmogorov[13] characterized the compact subsets of $L_p([0, 1])$-spaces as follows:

Theorem 31.20 (Kolmogorov). *Let $1 \leq p < \infty$, and let A be a closed and bounded subset of $L_p([0, 1])$. Then the following statements are equivalent:*

1. *The set A is compact (for the L_p-norm).*
2. *For each $\epsilon > 0$ there exists some $\delta > 0$ such that $\| f - f_h \|_p < \epsilon$ holds for all $f \in A$ and all $0 < h < \delta$.*

Proof. (1) $\Longrightarrow$ (2) Let $\epsilon > 0$. Since (by Theorem 31.11) $C[0, 1]$ is dense (for the L_p-norm) in $L_p([0, 1])$ and A is compact, it is easy to see that there exist continuous functions $f_1, \ldots, f_n$ such that $A \subseteq \bigcup_{i=1}^n B(f_i, \epsilon)$.

By the uniform continuity of each f_i, there exists some $\delta > 0$ such that $|f_i(t) - f_i(x)| < \epsilon$ holds for each $1 \leq i \leq n$ whenever $t, x \in [0, 1]$ satisfy $|x - t| < \delta$. In particular, if $0 < h < \delta$, then

$$|f_i(t) - (f_i)_h(t)| = \frac{1}{2h} \left| \int_{t-h}^{t+h} [f_i(t) - f_i(x)] \, dx \right| \leq \epsilon$$

holds. Thus, $\| f_i - (f_i)_h \|_p \leq \epsilon$ for each $1 \leq i \leq n$ and all $0 < h < \delta$.

Now, if $f \in A$, then choose $1 \leq i \leq n$ with $f \in B(f_i, \epsilon)$. By Lemma 31.19 we have $\| f_h - (f_i)_h \|_p \leq \| f - f_i \|_p < \epsilon$. Therefore,

$$\| f - f_h \|_p \leq \| f - f_i \|_p + \| f_i - (f_i)_h \|_p + \| (f_i)_h - f_h \|_p < 3\epsilon$$

holds for all $f \in A$ and all $0 < h < \delta$.

(2) $\Longrightarrow$ (1) According to Theorem 7.8, it is enough to show that A is totally bounded (for the L_p-norm).

To this end, let $\epsilon > 0$. Fix some $h > 0$ such that $\| f - f_h \|_p < \epsilon$ holds for all $f \in A$. Next, choose $M > 0$ with $\| f \|_p \leq M$ for all $f \in A$. Then by Lemma 31.19 it follows that

$$|f_h(t)| \leq M(2h)^{-\frac{1}{p}} = K$$

---

[13] Andrey Nikolayevich Kolmogorov (1903–1987), a prominent Russian mathematician. He is the founder of modern probability theory. In addition, he studied the theory of turbulent flows and worked on dynamical systems in relation to planetary motion.

holds for all $t \in [0, 1]$ and all $f \in A$. Let $A_h = \{f_{hh}: f \in A\}$, where

$$f_{hh}(t) = \frac{1}{2h} \int_{t-h}^{t+h} f_h(x)\, dx.$$

Clearly, $|f_{hh}(t)| \leq K$ holds for all $t \in [0, 1]$ and $f \in A$, and hence, A_h is a uniformly bounded set. Next, we claim that the set of continuous functions A_h is equicontinuous.

To see this, note that if $f \in A$ and $t < s$, then

$$
\begin{aligned}
|f_{hh}(s) - f_{hh}(t)| &= \frac{1}{2h} \left| \int_{s-h}^{s+h} f_h(x)\, dx - \int_{t-h}^{t+h} f_h(x)\, dx \right| \\
&= \frac{1}{2h} \left| \int_{t+h}^{s+h} f_h(x)\, dx - \int_{t-h}^{s-h} f_h(x)\, dx \right| \\
&\leq \frac{1}{2h} \left[\int_{t+h}^{s+h} |f_h(x)|\, dx + \int_{t-h}^{s-h} |f_h(x)|\, dx \right] \\
&\leq \frac{1}{2h} [2K(s-t)] = \frac{K}{h}(s-t)
\end{aligned}
$$

holds, and this shows that A_h is an equicontinuous set.

Now, by the Ascoli–Arzelà theorem, A_h is a totally bounded subset of $C[0, 1]$ (for the sup norm). Choose functions $f_1, \ldots, f_n \in A$ such that for each $f \in A$ there exists some $1 \leq i \leq n$ with $\|f_{hh} - (f_i)_{hh}\|_\infty < \epsilon$. In particular, note that

$$
\begin{aligned}
\|f - f_i\|_p &\leq \|f - f_h\|_p + \|f_h - f_{hh}\|_p + \|f_{hh} - f_i\|_p < 2\epsilon + \|f_{hh} - f_i\|_p \\
&< 2\epsilon + \|f_{hh} - (f_i)_{hh}\|_p + \|(f_i)_{hh} - (f_i)_h\|_p + \|(f_i)_h - f_i\|_p \\
&< 5\epsilon.
\end{aligned}
$$

Thus, A is totally bounded (for the L_p-norm), and the proof is finished. ∎

EXERCISES

1. Let $f \in L_p(\mu)$, and let $\epsilon > 0$. Show that

    $$\mu^*(\{x \in X: |f(x)| \geq \epsilon\}) \leq \epsilon^{-p} \int |f|^p\, d\mu.$$

2. Let $\{f_n\}$ be a sequence of some $L_p(\mu)$-space with $1 \leq p < \infty$. Show that if $\lim \|f_n - f\|_p = 0$ holds in $L_p(\mu)$, then $\{f_n\}$ converges in measure to f.
    [HINT: Use the preceding exercise.]

3. Let $(X, \mathcal{S}, \mu)$ be a measure space and consider the set

    $$E = \{\chi_A: A \in \Lambda_\mu \text{ with } \mu^*(A) < \infty\}.$$

Show that E is a closed subset of $L_1(\mu)$ (and hence, a complete metric space in its own right). Use this conclusion and the identity $\mu(A\Delta B) = \int |\chi_A - \chi_B|\,d\mu$ to provide an alternate solution to Exercise 12(c) of Section 14.

4. Show that equality holds in the inequality of Lemma 31.2 if and only if $a = b$. Use this to show that if $f \in L_p(\mu)$ and $g \in L_q(\mu)$, where $1 < p < \infty$ and $\frac{1}{p} + \frac{1}{q} = 1$, then $\int |fg|\,d\mu = \|f\|_p \cdot \|g\|_q$ holds if and only if there exist two constants C_1 and C_2 (not both zero) such that $C_1|f|^p = C_2|g|^q$ holds.

5. Assume that $\mu^*(X) = 1$ and $0 < p < q \le \infty$. If f is in $L_q(\mu)$, then show that $\|f\|_p \le \|f\|_q$ holds.
   [HINT: Use Hölder's inequality.]

6. Let $f \in L_1(\mu) \cap L_\infty(\mu)$. Then show that:

   a. $f \in L_p(\mu)$ for each $1 < p < \infty$,
   b. If $\mu^*(X) < \infty$, then $\lim_{p\to\infty} \|f\|_p = \|f\|_\infty$ holds.

   [HINT: For (b), let $\epsilon > 0$. Then $E = \{x \in X: |f(x)| > \|f\|_\infty - \epsilon\}$ has positive measure, and $(\|f\|_\infty - \epsilon) \cdot \chi_E \le |f|$ holds.]

7. Let $f \in L_2[0, 1]$ satisfy $\|f\|_2 = 1$ and $\int_0^1 f(x)\,d\lambda(x) \ge \alpha > 0$. Also, for each $\beta \in \mathbb{R}$, let $E_\beta = \{x \in [0, 1]: f(x) \ge \beta\}$. If $0 < \beta < \alpha$, show that $\lambda(E_\beta) \ge (\beta - \alpha)^2$.

8. Show that for $1 \le p < \infty$ each ℓ_p is a separable Banach lattice.

9. Show that ℓ_∞ is not separable.
   [HINT: Consider the collection of all sequences having zeros and ones as their entries.]

10. Show that $L_\infty([0, 1])$ (with the Lebesgue measure) is not separable.
    [HINT: Consider the set $\{\chi_{[0,x]}: 0 < x < 1\}$.]

11. Let X be a Hausdorff locally compact topological space, and fix a point $a \in X$. Let μ be the measure on X defined on all subsets of X by $\mu(A) = 1$ if $a \in A$ and $\mu(A) = 0$ if $a \notin A$. In other words, μ is the Dirac measure (see Example 13.4). Show that μ is a regular Borel measure and that $\text{Supp}\,\mu = \{a\}$.

12. If $g \in C^1[a, b]$ and $f \in L_1[a, b]$, then

    a. show that the function $F: [a, b] \to \mathbb{R}$ defined by $F(x) = \int_a^x f(t)\,d\lambda(t)$ is continuous, and
    b. establish the following "Integration by Parts" formula:

    $$\int_a^b g(x)f(x)\,d\lambda(x) = g(x)F(x)\Big|_a^b - \int_a^b g'(x)F(x)\,dx.$$

    [HINT: For (a) use Exercise 6 of Section 22. For (b) use Theorem 25.3 in connection with Lemma 31.6 and the Lebesgue dominated convergence theorem.]

13. Let μ be a regular Borel measure on $\mathbb{R}^n$. Then show that the collection of all real-valued functions on $\mathbb{R}^n$ that are infinitely many times differentiable is norm dense in $L_p(\mu)$ for each $1 \le p < \infty$.
    [HINT: See Exercise 5 of Section 25.]

14. Let $(X, \mathcal{S}, \mu)$ be a measure space with $\mu^*(X) = 1$. Assume that a function $f \in L_1(\mu)$ satisfies $f(x) \ge M > 0$ for almost all x. Then show that $\ln(f) \in L_1(\mu)$ and that $\int \ln(f)\,d\mu \le \ln(\int f\,d\mu)$ holds.
    [HINT: Let $t = f(x)/\|f\|_1$ in the inequality $1 - \frac{1}{t} \le \ln t \le t - 1$, and integrate.]

15. Show with an example that Theorem 31.7 is false when $p = \infty$.

16. This exercise shows that Theorem 31.16 is false when $p = 1$ and presents a necessary and sufficient condition for the mapping $g \mapsto F_g$ from $L_\infty(\mu)$ into $L_1^*(\mu)$ to be an isometry.

    a.   Show that for each $g \in L_\infty(\mu)$, the linear functional $F_g(f) = \int fg \, d\mu$ for $f \in L_1(\mu)$ is a bounded linear functional on $L_1(\mu)$ such that $\|F_g\| \le \|g\|_\infty$ holds.

    b.   Consider a nonempty set X and μ the measure defined on every subset of X by $\mu(\emptyset) = 0$ and $\mu(A) = \infty$ if $A \ne \emptyset$. Then show that $L_1(\mu) = \{0\}$ and $L_\infty(\mu) = B(X)$ [the bounded functions on X] and conclude from this that $g \in L_\infty(\mu)$ satisfies $\|F_g\| = \|g\|_\infty$ if and only if $g = 0$.

    c.   Let us say that a measure space (X, S, μ) has the **finite subset property** whenever every measurable set of infinite measure has a measurable subset of finite positive measure. (A measure with the finite subset property is also called a **locally finite measure**.)

            Show that the linear mapping $g \mapsto F_g$ from $L_\infty(\mu)$ into $L_1^*(\mu)$ is a lattice isometry if and only if (X, S, μ) has the finite subset property.

17. Let (X, S, μ) be a measure space. Assume that there exist measurable sets $E_1, \ldots, E_n$ such that $0 < \mu(E_i) < \infty$ for $1 \le i \le n$, $X = \bigcup_{i=1}^n E_i$, and each E_i does not contain any proper nonempty measurable set. Then show that $L_\infty^*(\mu) = L_1(\mu)$; that is, show that $g \mapsto F_g$ from $L_1(\mu)$ to $L_\infty^*(\mu)$ is onto.
[HINT: If $F \in L_\infty^*(\mu)$, let $c_i = F(\chi_{E_i})$, $g = \sum_{i=1}^n [c_i/\mu^*(E_i)] \cdot \chi_{E_i}$, and then show that $F = F_g$ holds.]

18. Let (X, S, μ) be a measure space, and let $0 < p < 1$.

    a.   Show by a counterexample that $\|\cdot\|_p$ is no longer a norm on $L_p(\mu)$.

    b.   For each $f, g \in L_p(\mu)$ let $d(f, g) = \int |f - g|^p \, d\mu = (\|f - g\|_p)^p$. Show that d is a metric on $L_p(\mu)$ and that $L_p(\mu)$ equipped with d is a complete metric space.

[HINT: For the triangle inequality, observe that $(a + b)^p \le a^p + b^p$ holds for every pair of non-negative real numbers a and b.]

19. Let (X, S, μ) be a finite measure space. Then show that Theorem 31.10 is true for $p = \infty$. That is, show that the step functions are norm dense in $L_\infty(\mu)$.

20. If K is a compact subset of a metric space X, then show that there exists a regular Borel measure μ on X such that $\mathrm{Supp}\,\mu = K$.

21. If $\{f_n\}$ is a norm bounded sequence of $L_2(\mu)$, then show that $f_n/n \to 0$ a.e.

22. Let (X, S, μ) be a measure space such that $\mu^*(X) = 1$. If $f, g \in L_1(\mu)$ are two positive functions satisfying $f(x)g(x) \ge 1$ for almost all x, then show that

$$\left(\int f \, d\mu \right) \cdot \left(\int g \, d\mu \right) \ge 1.$$

23. Consider a measure space (X, S, μ) with $\mu^*(X) = 1$, and let $f, g \in L_2(\mu)$. If $\int f \, d\mu = 0$, then show that

$$\left(\int fg \, d\mu \right)^2 \le \left[\int g^2 \, d\mu - \left(\int g \, d\mu \right)^2 \right] \int f^2 \, d\mu.$$

24. If two functions $f, g \in L_3(\mu)$ satisfy $\|f\|_3 = \|g\|_3 = \int f^2 g\, d\mu = 1$, then show that $g = |f|$ a.e.

25. For a function $f \in L_1(\mu) \cap L_2(\mu)$ establish the following properties:

   a. $f \in L_p(\mu)$ for each $1 \le p \le 2$, and
   b. $\lim_{p \to 1+} \|f\|_p = \|f\|_1$.

26. Assume that the positive real numbers $\alpha_1, \ldots, \alpha_n$ satisfy $0 < \alpha_i < 1$ for each i and $\sum_{i=1}^{n} \alpha_i = 1$. If $f_1, \ldots, f_n$ are positive integrable functions on some measure space, then show that

   a. $f_1^{\alpha_1} f_2^{\alpha_2} \cdots f_n^{\alpha_n} \in L_1(\mu)$, and
   b. $\int f_1^{\alpha_1} f_2^{\alpha_2} \cdots f_n^{\alpha_n}\, d\mu \le (\|f_1\|_1)^{\alpha_1} (\|f_2\|_1)^{\alpha_2} \cdots (\|f_n\|_1)^{\alpha_n}$.

27. Let (X, S, μ) be a measure space and let $\{A_n\}$ be a sequence of measurable sets satisfying $0 < \mu^*(A_n) < \infty$ for each n and $\lim \mu^*(A_n) = 0$. Fix $1 < p < \infty$ and let $g_n = [\mu^*(A_n)]^{-\frac{1}{q}} \chi_{A_n}$ for each n, where $\frac{1}{p} + \frac{1}{q} = 1$. Prove that $\lim \int f g_n\, d\mu = 0$ for each $f \in L_p(\mu)$.

28. Let (X, S, μ) be a measure space such that $\mu^*(X) = 1$. For each $1 < p < \infty$ define the set

$$\mathcal{E}_p = \left\{ f \in L_1(\mu) : \int |f|\, d\mu = 1 \text{ and } \int |f|^p\, d\mu = 2 \right\}.$$

Show that for each $0 < \epsilon < 1$ there exists some $\delta_p > 0$ such that

$$\mu^*(\{x \in X : |f(x)| > \epsilon\}) \ge \delta_p$$

for each $f \in \mathcal{E}_p$.

29. Let (X, S, μ) be a measure space and let $1 \le p < \infty$ and $0 < \eta < p$.

   a. Show that the nonlinear function $\psi : L_p(\mu) \to L_{\frac{p}{\eta}}(\mu)$, where $\psi(f) = |f|^\eta$, is norm continuous.
   b. If $f_n \to f$ and $g_n \to g$ hold in $L_p(\mu)$, then show that

$$\lim_{n \to \infty} \int |f_n|^{p-\eta} |g_n|^\eta\, d\mu = \int |f|^{p-\eta} |g|^\eta\, d\mu.$$

30. Let $T : L_p(\mu) \to L_p(\mu)$ be a continuous operator, where $1 < p < \infty$, and let $0 \le \eta \le p$. Show that:

   a. If $f \in L_p(\mu)$, then $|f|^{p-\eta} |Tf|^\eta \in L_1(\mu)$ and

$$\int |f|^{p-\eta} |Tf|^\eta\, d\mu \le \|T\|^\eta (\|f\|_p)^p.$$

   b. If for some $f \in L_p(\mu)$ with $\|f\|_p \le 1$ we have $\int |f|^{p-\eta} |Tf|^\eta\, d\mu = \|T\|^\eta$, then $|Tf| = \|T\| |f|$.

31. Let (X, S, μ) be a measure space and let $f \in L_p(\mu)$ for some $1 \le p < \infty$. Show that the function $g : [0, \infty) \to [0, \infty)$ defined by

$$g(t) = p t^{p-1} \mu^*(\{x \in X : |f(x)| \ge t\})$$

is Lebesgue integrable over $[0, \infty)$ and that

$$\int |f|^p \, d\mu = \int_{[0,\infty)} g(t) \, d\lambda(t) = p \int_0^\infty t^{p-1} \mu^*(\{x \in X: |f(x)| \geq t\}) \, dt.$$

32. Let $(X, \mathcal{S}, \mu)$ be a measure space and let $f: X \to \mathbb{R}$ be a measurable function. If $\mu^*(\{x \in X: |f(x)| \geq t\}) \leq e^{-t}$ for all $t \geq 0$, then show that $f \in L_p(\mu)$ holds for each $1 \leq p < \infty$.

33. Consider the vector space of functions

$$E = \{f: \mathbb{R}^n \to \mathbb{R}| \ f \text{ is a } C^\infty\text{-function with compact support and } \int_{\mathbb{R}^n} f \, d\lambda = 0\}.$$

Show that for each $1 < p < \infty$ the vector space E is dense in $L_p(\mathbb{R}^n)$. Is E dense in $L_1(\mathbb{R}^n)$?

34. Let $(0, \infty)$ be equipped with the Lebesgue measure, and let $1 < p < \infty$. For each $f \in L_p(\lambda)$ let $T(f)(x) = x^{-1} \int f \chi_{(0,x)} \, d\lambda$ for $x > 0$. Then show that T defines a one-to-one bounded linear operator from $L_p((0, \infty))$ into itself such that $\|T\| = \frac{p}{p-1}$. [HINT: Assume that $0 \leq f \in C_c((0, \infty))$. Then integrate by parts, and use Hölder's inequality to get

$$(\|T(f)\|_p)^p = \int_0^\infty \left(\frac{1}{x} \int_0^x f(t) \, dt \right)^p dx = \frac{1}{1-p} \int_0^\infty \left(\int_0^x f(t) \, dt \right)^p d(x^{1-p})$$

$$= \frac{p}{p-1} \int_0^\infty f(x)[\, T(f)(x)\,]^{p-1} \, dx$$

$$\leq \frac{p}{p-1} \|f\|_p \cdot \left(\int_0^\infty [\, T(f)(x)\,]^{q(p-1)} \, dx\right)^{\frac{1}{q}}$$

$$= \frac{p}{p-1} \|f\|_p \cdot (\|T(f)\|_p)^{\frac{p}{q}}.$$

Now, use Theorem 31.11. For $\|T\|$ use the sequence of functions $\{f_n\}$ defined by $f_n(x) = x^{(n^{-1}-1)p^{-1}}$ if $0 < x < 1$ and $f_n(x) = 0$ if $x \geq 1$.]

CHAPTER 6

HILBERT SPACES

The notion of norm can be thought of as a generalization of Euclid's concept of length. So, normed or Banach spaces can be viewed as finite or infinite dimensional analogues of the classical two- or three-dimensional spaces of Euclidean geometry. In the two- or three-dimensional spaces of Euclidean geometry there is, however, one more important ingredient that describes the length of a vector: it can be obtained by means of the familiar inner product. That is, the length of an arbitrary vector x is given by

$$\|x\| = \sqrt{x \cdot x},$$

where $x \cdot y$ denotes the inner product of the vectors x and y. In general, the inner product $x \cdot y = \sum_{i=1}^{n} x_i y_i$ of two vectors x, $y \in \mathbb{R}^n$ can be considered as a function of two variables that satisfies certain linearity and homogeneity properties.

Besides using the inner product to evaluate the norm of a vector, the inner product also allows us to compute angles between vectors and express projections of vectors along lines and planes. For this reason, although all norms on $\mathbb{R}^n$ are equivalent (Theorem 28.8), the Euclidean norm is the one that is more suitable for physical applications.

Normed spaces whose norms are obtained (as above) by means of an inner product are called *inner product spaces*. If they are also complete, they are referred to as *Hilbert*[1] *spaces*. Naturally, these spaces inherit the most important properties of Euclidean spaces, and for physical applications they are the most suitable infinite dimensional analogues of Euclidean spaces.

This chapter presents a brief introduction to Hilbert spaces. In the first section we study the basic properties of an inner product, and in the second section we introduce Hilbert spaces. We discuss orthonomal bases in the third section. The final (fourth) section deals with some natural applications of Hilbert spaces to Fourier analysis. In particular, it discusses a few classical convergence properties

---

[1] David Hilbert (1862–1943), a German mathematician. He was a towering influential mathematical figure of modern times. In his 1900 address at the International Congress of Mathematicians, he proposed a list of 23 mathematical problems that have since stimulated and shaped the direction of present mathematical research.

of Fourier series. Fourier analysis is today one of the basic tools of every engineer and applied scientist working in linear systems, antennas, mechanical vibrations, optics, bioengineering, various random processes, boundary value problems, and the design of filters used in the field of image and signal processing.

Since the natural setting of Hilbert spaces is within the framework of complex vector spaces, in this chapter we shall deal mainly with complex vector spaces. The complex conjugate of a complex number $\alpha = a + b\iota$ will be denoted by $\bar{\alpha}$, i.e., $\bar{\alpha} = a - b\iota$.

32. INNER PRODUCT SPACES

An **inner product** on a real vector space X is a real-valued function of two variables $(\cdot, \cdot): X \times X \to \mathbb{R}$ such that:

1. $(\cdot, \cdot)$ *is linear in the first variable, i.e.,* $(\alpha x + \beta y, z) = \alpha(x, z) + \beta(y, z)$ *for all* $x, y, z \in X$ *and all real numbers* α *and* β;
2. $(\cdot, \cdot)$ *is symmetric, i.e.,* $(x, y) = (y, x)$ *for all* $x, y \in X$; *and*
3. $(x, x) \geq 0$ *for each* $x \in X$ *and* $(x, x) = 0$ *if and only if* $x = 0$.

From (2) and (1), we see that every inner product on a real vector space is also linear in the second variable. A **real inner product space** is a real vector space equipped with an inner product.

In case X is a complex vector space, a complex-valued function of two variables $(\cdot, \cdot): X \times X \to \mathbb{C}$ is an **inner product** on X if

a. $(\cdot, \cdot)$ *is linear in the first variable, i.e.,* $(\alpha x + \beta y, z) = \alpha(x, z) + \beta(y, z)$ *for all* $x, y, z \in X$ *and all complex numbers* α *and* β;
b. $(x, y) = \overline{(y, x)}$ *for all* $x, y \in X$; *and*
c. $(x, x) \geq 0$ *for each* $x \in X$ *and* $(x, x) = 0$ *if and only if* $x = 0$.

A **complex inner product space** is a complex vector space equipped with an inner product. It is easy to see that the inner product of a complex inner product space is additive in the second variable and "conjugate homogeneous" in the second variable. That is,

- $(x, y + z) = (x, y) + (x, z)$ *for all* $x, y, z \in X$, *and*
- $(x, \alpha y) = \bar{\alpha}(x, y)$ *for all* $x, y \in X$ *and all* $\alpha \in \mathbb{C}$.

Every real vector space X can be "extended" to a complex vector space by means of its "complexification." If we define formally

$$X_{\mathbf{c}} = X \oplus \iota X = \{x + \iota y \colon x, y \in X\},$$

and equip it with the algebraic operations

$$(x + \iota y) + (x_1 + \iota y_1) = x + x_1 + \iota(y + y_1), \text{ and}$$
$$(\alpha + \iota\beta)(x + \iota y) = \alpha x - \beta y + \iota(\beta x + \alpha y),$$

then X_c is a complex vector space—called the **complexification** of X. The reader should stop and verify that X_c with these operations is indeed a complex vector space. Also, it should be clear that the vector space X can be identified with the vector subspace $\{x + \iota 0: x \in X\}$ of X_c.

Now, if X is a real inner product space with complexification X_c, then we can define a function $\langle \cdot, \cdot \rangle: X_c \times X_c \to C$ via the formula

$$\langle x + \iota y, x_1 + \iota y_1 \rangle = (x, x_1) + (y, y_1) + \iota[(y, x_1) - (x, y_1)].$$

It is a routine matter to verify that $\langle \cdot, \cdot \rangle$ is indeed an inner product on X_c, and that

$$\langle x + \iota 0, x_1 + \iota 0 \rangle = (x, x_1).$$

That is, the inner product $\langle \cdot, \cdot \rangle$ is an extension of the inner product $(\cdot, \cdot)$ from X to its complexification X_c.

Since most interesting results regarding inner product spaces are associated with complex inner product spaces and since every real inner product space can be extended (as above) to a complex inner product space, from now on, unless otherwise stated, all inner product spaces will be assumed to be complex inner product spaces. In particular, if X is a real inner product space, then we shall denote for brevity its complexification X_c by X again. In case we consider a real inner product space, we shall refer to specifically as a "real inner product space."

Here are two examples of inner product spaces:

Example 32.1. The classical Euclidean spaces $\mathbb{R}^n$ provide the simplest examples of real inner product spaces. The inner product in $\mathbb{R}^n$ is defined by $(x, y) = \sum_{i=1}^{n} x_i y_i$ for all vectors $x = (x_1, \ldots, x_n)$ and $y = (y_1, \ldots, y_n)$. Its complexification is simply C^n whose inner product is defined by

$$(x, y) = \sum_{i=1}^{n} x_i \overline{y_i}$$

for all $x, y \in C^n$.

The second example of an inner product space is the vector space $C[a, b]$ of all continuous real-valued functions on a closed interval $[a, b]$. The inner product is given by the formula $(f, g) = \int_a^b f(x)g(x)\,dx$. Its complexification is the vector space of all continuous complex-valued functions defined on $[a, b]$ and its inner product is given by

$$(f, g) = \int_a^b f(x) \overline{g(x)} \, dx$$

for all continuous functions $f, g: [a, b] \to \mathbf{C}$.   ∎

In an inner product space X, the **norm** of a vector $x \in X$ is defined by

$$\|x\| = \sqrt{(x, x)}.$$

Our immediate objective is to show that the preceding formula indeed defines a norm on X. To do this, we need to establish the classical Cauchy–Schwarz inequality.

Theorem 32.2 (The Cauchy–Schwarz Inequality). *If x and y are two arbitrary vectors in an inner product space, then*

$$|(x, y)| \leq \|x\| \, \|y\|.$$

Proof. Let x and y be two vectors in an inner product space. If $x = 0$ or $y = 0$, then the inequality is obvious. So, we can assume that $x \neq 0$ and $y \neq 0$ and the desired inequality can be written as $|(\frac{x}{\|x\|}, \frac{y}{\|y\|})| \leq 1$. This implies that we can suppose without loss of generality that $\|x\| = \|y\| = 1$. Now notice that

$$
\begin{aligned}
0 &\leq \|x - (x, y)y\|^2 \\
&= (x - (x, y)y, x - (x, y)y) \\
&= (x, x) - (x, y)\overline{(x, y)} - (x, y)\overline{(x, y)} + (x, y)\overline{(x, y)}(y, y) \\
&= 1 - |(x, y)|^2 - |(x, y)|^2 + |(x, y)|^2 \\
&= 1 - |(x, y)|^2.
\end{aligned}
$$

Therefore $|(x, y)|^2 \leq 1$ or $|(x, y)| \leq 1$, as desired.   ∎

We are now ready to establish that the formula $\|x\| = \sqrt{(x, x)}$ defines indeed a norm.

Theorem 32.3. *If X is an inner product space, then the formula*

$$\|x\| = \sqrt{(x, x)}$$

*defines a norm on X (called the **norm induced by the inner product**).*

Proof. Start by observing that $\|x\| \geq 0$ for each $x \in X$ and that $\|x\| = 0$ if and only if $x = 0$. Now, if $\alpha \in \mathbb{C}$ and $x \in X$, then

$$|\alpha|^2 \|x\|^2 = \alpha\overline{\alpha}(x, x) = (\alpha x, \alpha x) = \|\alpha x\|^2,$$

and so $\|\alpha x\| = |\alpha| \|x\|$. Finally, for the triangle inequality use the Cauchy–Schwarz inequality to get

$$
\begin{aligned}
\|x + y\|^2 &= (x + y, x + y) \\
&= (x, x) + (x, y) + (y, x) + (y, y) \\
&= (x, x) + 2\operatorname{Re}(x, y) + (y, y) \\
&\leq \|x\|^2 + 2|(x, y)| + \|y\|^2 \\
&\leq \|x\|^2 + 2\|x\| \|y\| + \|y\|^2 \\
&= (\|x\| + \|y\|)^2.
\end{aligned}
$$

This implies $\|x + y\| \leq \|x\| + \|y\|$, and the proof is complete. ∎

The inner product is a jointly continuous function.

Lemma 32.4. *The inner product is a jointly continuous function with respect to the induced norm. That is,*

$$\|x_n - x\| \to 0 \ \text{and} \ \|y_n - y\| \to 0 \implies (x_n, y_n) \to (x, y).$$

Proof. The conclusion follows from the inequality

$$|(x_n, y_n) - (x, y)| = |(x_n - x, y_n) + (x, y_n - y)| \leq \|y_n\| \|x_n - x\| + \|x\| \|y_n - y\|,$$

and the fact that if $\|y_n - y\| \to 0$, then $\{\|y_n\|\}$ is a bounded sequence. ∎

A well known theorem of Euclidean geometry states that the sum of the squares of the sides of a parallelogram equals the sum of the squares of its diagonals. This fact is also true in any inner product space and is known as the *parallelogram law*.

Theorem 32.5 (The Parallelogram Law). *If x and y are two vectors in an inner product space, then*

$$\|x + y\|^2 + \|x - y\|^2 = 2(\|x\|^2 + \|y\|^2).$$

Proof. If x and y are two vectors in an inner product space, then note that

$$\begin{aligned}
\|x + y\|^2 + \|x - y\|^2 &= (x + y, x + y) + (x - y, x - y) \\
&= [(x, x) + (x, y) + (y, x) + (y, y)] \\
&\quad + [(x, x) - (x, y) - (y, x) + (y, y)] \\
&= 2(x, x) + 2(y, y) \\
&= 2(\|x\|^2 + \|y\|^2),
\end{aligned}$$

as claimed.   ∎

For a given norm $\|\cdot\|$ on a vector space X there exists at most one inner product on X that induces the norm $\|\cdot\|$. To see this, assume that two inner products $(\cdot, \cdot)$ and $\langle \cdot, \cdot \rangle$ on X induce $\|\cdot\|$. That is, assume that $(x, x) = \langle x, x \rangle = \|x\|^2$ for all $x \in X$. In particular, we have

$$\begin{aligned}
(x + y, x + y) &= (x, x) + (x, y) + (y, x) + (y, y) \\
&= \langle x + y, x + y \rangle \\
&= \langle x, x \rangle + \langle x, y \rangle + \langle y, x \rangle + \langle y, y \rangle,
\end{aligned}$$

and therefore

$$(x, y) + (y, x) = \langle x, y \rangle + \langle y, x \rangle \qquad\qquad (\star)$$

for all x and y. If the vector space X is real, then the latter easily implies

$$(x, y) = \langle x, y \rangle$$

for all x and y. If X is a complex vector space, then replacing x by $\imath x$ in $(\star)$, we get $\imath(x, y) - \imath(y, x) = \imath\langle x, y \rangle - \imath\langle y, x \rangle$ or

$$(x, y) - (y, x) = \langle x, y \rangle - \langle y, x \rangle$$

for all x and y. Adding this to $(\star)$ yields $2(x, y) = 2\langle x, y \rangle$, or $(x, y) = \langle x, y \rangle$ for all x and y. This establishes that a given norm is the induced norm of at-most one inner product.

Which norms are induced by inner products? Remarkably, as we shall see next, the ones that satisfy the parallelogram law.

Theorem 32.6. *A norm $\|\cdot\|$ on a vector space is induced by an inner product if and only if it satisfies the parallelogram law, i.e., if and only if*

$$\|x + y\|^2 + \|x - y\|^2 = 2(\|x\|^2 + \|y\|^2)$$

holds for all vectors x and y. Moreover, the inner product $(\cdot, \cdot)$ *that induces* $\|\cdot\|$
in the case of a real vector space is given by

$$(x, y) = \frac{1}{4}(\|x + y\|^2 - \|x - y\|^2),$$

and in the case of a complex vector space by

$$(x, y) = \frac{1}{4}(\|x + y\|^2 - \|x - y\|^2 + \iota\|x + \iota y\|^2 - \iota\|x - \iota y\|^2).$$

Proof. Assume that a real normed space X satisfies the parallelogram law. We
shall show that the formula

$$(x, y) = \frac{1}{4}(\|x + y\|^2 - \|x - y\|^2),$$

is an inner product which induces the norm $\|\cdot\|$. Clearly, $(x, x) = \|x\|^2 \geq 0$ for each
x and $(x, x) = 0$ if and only if $x = 0$. Also, it should be clear that $(x, y) = (y, x)$
for all x and y. Next, we shall show that the function $(\cdot, \cdot)$ is additive in the first
variable. To do this, note first that

$$\begin{aligned}
4(u + v, w) + 4(u - v, w) &= (\|u + v + w\|^2 - \|u + v - w\|^2) \\
&\quad + (\|u - v + w\|^2 - \|u - v - w\|^2) \\
&= (\|u + w + v\|^2 + \|u + w - v\|^2) \\
&\quad - (\|u - w + v\|^2 + \|u - w - v\|^2) \\
&= (2\|u + w\|^2 + 2\|v\|^2) - (2\|u - w\|^2 + 2\|v\|^2) \\
&= 2(\|u + w\|^2 - \|u - w\|^2) \\
&= 8(u, w).
\end{aligned}$$

Thus, for all $u, v, w \in X$, we have

$$(u + v, w) + (u - v, w) = 2(u, w). \tag{$\star\star$}$$

When $v = u$, $(\star\star)$ yields $(2u, w) = 2(u, w)$. Now, letting $u = \frac{1}{2}(x+y)$, $v = \frac{1}{2}(x-y)$
and $w = z$ in $(\star\star)$, we get

$$(x, z) + (y, z) = (u + v, z) + (u - v, z) = 2(u, z) = (2u, z) = (x + y, z),$$

which is the additivity of $(\cdot, \cdot)$ in the first variable.

Now, as in the proof of Lemma 18.7, we can establish that $(rx, y) = r(x, y)$
holds for each rational number r and all $x, y \in X$. Since $(\cdot, \cdot)$, as defined above,

is a jointly continuous function (relative to the norm $\|\cdot\|$), it easily follows that $(\alpha x, y) = \alpha(x, y)$ for all $\alpha \in \mathbb{R}$ and all $x, y \in X$. This completes the proof of the real case. We leave the complex case as an exercise for the reader. ∎

In an inner product space the notion of orthogonality is introduced by means of the inner product. Two vectors x and y are said to be **orthogonal**, in symbols $x \perp y$, if $(x, y) = 0$. A set of vectors S in an inner product space is said to be an **orthogonal set** if $0 \notin S$ and any two distinct vectors of S are orthogonal. An orthogonal set consisting of unit vectors is called an **orthonormal set**. Notice that if S is an orthogonal set, then the set of vectors $\{x/\|x\|: x \in S\}$ is automatically an orthonormal set.

In the case of a real inner product space, we can also define the angle between two vectors. Indeed, if x and y are two non-zero vectors in a real inner product space, then (in view of the Cauchy–Schwarz inequality) there exists a unique angle θ satisfying $0 \leq \theta \leq 180°$ and

$$\cos \theta = \frac{(x, y)}{\|x\| \, \|y\|}.$$

The angle θ is called the **angle between the vectors** x and y. So, in the real case, two vectors are orthogonal if and only if the angle between the vectors is $90°$ (we also define the angle between the zero vector and any other vector to be $90°$).

One of the oldest and most famous theorems in mathematics is the Pythagorean[2] theorem. It states that the square of the hypotenuse of a right triangle equals the sum of the squares of its two legs. This result also carries over to inner product spaces.

Theorem 32.7 (The Pythagorean Theorem). *If in an inner product space the vectors $x_1, x_2, \ldots, x_n$ are pairwise orthogonal, then*

$$\|x_1 + x_2 + \cdots + x_n\|^2 = \|x_1\|^2 + \|x_2\|^2 + \cdots + \|x_n\|^2.$$

---

[2]Pythagoras of Samos (ca 580–500 BC), a great Greek philosopher and mathematician. Born on the Greek island of Samos, he is credited with many discoveries in mathematics and science and was the first to associate number theory with musical sounds. Pythagoras moved to the Greek city of Crotona in southern Italy (a well known city in ancient times for its medical school and for many great athletes who won in the ancient olympic games) and created the famous Pythagorean school, which was in essence a secret brotherhood. Following Pythagoras' maxim "everything was not to be told to everybody," what was done and taught among the members of the brotherhood was kept in profound secrecy from the outside world. Due to this secrecy, the extend of knowledge and discoveries of the Pythagorean school is fragmented and highly speculative.

Proof. If the vectors $x_1, x_2, \ldots, x_n$ are pairwise orthogonal in an inner product space, i.e., $(x_i, x_j) = 0$ for $i \neq j$, then note that

$$\left\| \sum_{i=1}^{n} x_i \right\|^2 = \left(\sum_{i=1}^{n} x_i, \sum_{j=1}^{n} x_j \right) = \sum_{i=1}^{n} \sum_{j=1}^{n} (x_i, x_j) = \sum_{i=1}^{n} (x_i, x_i) = \sum_{i=1}^{n} \|x_i\|^2,$$

as claimed.　∎

An immediate consequence of the Pythagorean theorem is that non-zero orthogonal vectors are linearly independent.

Corollary 32.8. *In an inner product space any set of non-zero pairwise orthogonal vectors is linearly independent.*

Proof. Assume that $x_1, x_2, \ldots, x_n$ are non-zero pairwise orthogonal vectors and let $\sum_{i=1}^{n} \lambda_i x_i = 0$. Since the vectors $\lambda_1 x_1, \lambda_2 x_2, \ldots, \lambda_2 x_n$ are also pairwise orthogonal, it follows from the Pythagorean theorem that $\sum_{i=1}^{n} |\lambda_i|^2 \|x_i\|^2 = \|\sum_{i=1}^{n} \lambda_i x_i\|^2 = 0$. This implies $\lambda_i = 0$ for each i, so that the vectors $x_1, x_2 \ldots, x_n$ are linearly independent.　∎

Recall that if $\{\lambda_i\}_{i \in I}$ is a family of non-negative real numbers, then the symbol $\sum_{i \in I} \lambda_i$ denotes the supremum of the collection of all possible finite sums of the λ_i. That is, if $\mathcal{F}$ denotes the collection of all finite subsets of I, then

$$\sum_{i \in I} \lambda_i = \sup_{\Phi \in \mathcal{F}} \sum_{i \in \Phi} \lambda_i,$$

where the supremum is taken in the extended real numbers.

Lemma 32.9. *If $\{\lambda_i\}_{i \in I}$ is a family of non-negative real numbers and $\sum_{i \in I} \lambda_i < \infty$, then $\lambda_i \neq 0$ holds for at most countably many indices i (in which case $\sum_{i \in I} \lambda_i$ is either a finite sum or a series).*

Proof. If $F_n = \{i \in I : \lambda_i > \frac{1}{n}\}$, then F_n is a finite set (why?) and moreover, $\{i \in I : \lambda_i > 0\} = \bigcup_{n=1}^{\infty} F_n$. Hence, $\{i \in I : \lambda_i > 0\}$ is at-most countable.　∎

One of the most basic properties regarding orthonormal sets is Bessel's[3] inequality. As we shall see later on, almost every aspect regarding the structure of orthonormal sets depends upon the next result.

---

[3]Friedrich Wilhelm Bessel (1784–1846), a notable German astronomer and mathematician. By introducing new methods, he determined the positions of several stars and planets quite accurately. He is also well known for a special class of functions which is today the indispensable tool of every scientist working in applied mathematics, physics, or engineering.

Theorem 32.10 (Bessel's Inequality). *If $\{x_i\}_{i \in I}$ is an orthonormal family of vectors in an inner product space, then*

$$\sum_{i \in I} |(x, x_i)|^2 \leq \|x\|^2$$

holds for each vector x. In particular, for each x all but an at most countable number of the (x, x_i) vanish.

Proof. Fix some vector x in an inner product space and let $x_1, x_2, \ldots, x_n$ be a finite orthonormal set. Then we have

$$0 \leq \left(x - \sum_{i=1}^{n} (x, x_i)x_i, x - \sum_{j=1}^{n} (x, x_j)x_j \right)$$

$$= \|x\|^2 - \sum_{i=1}^{n} (x, x_i)(x_i, x) - \sum_{j=1}^{n} (x, (x, x_j)x_j) + \sum_{i=1}^{n} \sum_{j=1}^{n} ((x, x_i)x_i, (x, x_j)x_j)$$

$$= \|x\|^2 - \sum_{i=1}^{n} |(x, x_i)|^2 - \sum_{j=1}^{n} |(x, x_j)|^2 + \sum_{i=1}^{n} |(x, x_i)|^2$$

$$= \|x\|^2 - \sum_{i=1}^{n} |(x, x_i)|^2.$$

Hence, $\sum_{i=1}^{n} |(x, x_i)|^2 \leq \|x\|^2$. Now, if $\{x_i\}_{i \in I}$ is an arbitrary orthonormal set, then the above conclusion easily implies $\sum_{i \in I} |(x, x_i)|^2 \leq \|x\|^2$. In particular, $(x, x_i) \neq 0$ must hold true for at most countably many indices i.   ∎

There is a standard procedure called the **Gram[4]–Schmidt[5] orthogonalization process** for converting a countable (or a finite) set of linearly independent vectors into an orthogonal set of vectors. As usual, let us use the symbol $Span\{z_1, \ldots, z_n\}$ to denote the vector subspace generated by the vectors $z_1, \ldots, z_n$.

Theorem 32.11 (The Gram–Schmidt Orthogonalization Process). *Let $x_1, x_2, \ldots$ be a sequence of linearly independent vectors in an inner product space. If we define the sequence of vectors $y_1, y_2, \ldots$ inductively by*

$$y_1 = x_1 \quad \text{and} \quad y_{n+1} = x_{n+1} - \sum_{i=1}^{n} \frac{(x_{n+1}, y_i)}{\|y_i\|^2} y_i \quad \text{for } n = 1, 2, \ldots,$$

---

[4]Jorgen Pedersen Gram (1850–1916), a Danish mathematician. He taught at the University of Copenhagen and was the chairman of the Danish Insurance Council.

[5]Erhard Schmidt (1876–1959), a German mathematician. He was a student of David Hilbert and worked on integral equations and the geometry of Hilbert spaces.

then we have the following:

1. *The set* $\{y_1, y_2, \ldots\}$ *is orthogonal, i.e.,* $y_n \neq 0$ *for each n and* $(y_i, y_j) = 0$ *for* $i \neq j$, *and*
2. $Span\{x_1, x_2, \ldots, x_n\} = Span\{y_1, y_2, \ldots, y_n\}$ *for* $n = 1, 2, \ldots$.

Moreover, the vectors $y_1, y_2, \ldots$ *are (aside of scalar factors) uniquely determined in the sense that if* $z_1, z_2, \ldots$ *is another sequence of orthogonal vectors satisfying*

$$Span\{x_1, x_2, \ldots, x_n\} = Span\{z_1, z_2, \ldots, z_n\}$$

for each n, then for each n there exists a non-zero scalar λ_n *such that* $z_n = \lambda_n y_n$.

Proof. First, we shall prove (1) and (2) by induction. For $n = 1$, the claims (1) and (2) are obvious. So, for the induction step, assume that the vectors $y_1, \ldots, y_n$ (as constructed by the above method) are all non-zero, pairwise orthogonal and satisfy

$$Span\{x_1, x_2, \ldots, x_n\} = Span\{y_1, y_2, \ldots, y_n\}. \tag{$\dagger$}$$

Now, define the vector $y_{n+1} = x_{n+1} - \sum_{i=1}^{n} \frac{(x_{n+1}, y_i)}{\|y_i\|^2} y_i$ and note that for each $1 \leq j \leq n$ we have

$$(y_{n+1}, y_j) = (x_{n+1}, y_j) - \sum_{i=1}^{n} \frac{(x_{n+1}, y_i)}{\|y_i\|^2}(y_i, y_j) = (x_{n+1}, y_j) - (x_{n+1}, y_j) = 0,$$

so that the vectors $y_1, \ldots, y_n, y_{n+1}$ are pairwise orthogonal. Now, using ($\dagger$), the identity $x_{n+1} = y_{n+1} + \sum_{i=1}^{n} \frac{(x_{n+1}, y_i)}{\|y_i\|^2} y_i$ and that $x_1, \ldots, x_{n+1}$ are linearly independent, we easily infer that $y_{n+1} \neq 0$ and $Span\{x_1, x_2, \ldots, x_n, x_{n+1}\} = Span\{y_1, y_2, \ldots, y_n, y_{n+1}\}$. This completes the induction and the proof of (1) and (2).

To establish the uniqueness of the sequence $\{y_n\}$, assume that another sequence $\{z_n\}$ of pairwise orthogonal vectors satisfies

$$Span\{x_1, x_2, \ldots, x_n\} = Span\{y_1, y_2, \ldots, y_n\} = Span\{z_1, z_2, \ldots, z_n\}$$

for each n. Clearly, $z_n \neq 0$ for each n. We shall prove by induction that for each n there exists a non-zero scalar λ_n such that $z_n = \lambda_n y_n$. For $n = 1$, the conclusion is obvious. So, assume that there exist non-zero scalars $\lambda_1, \ldots, \lambda_n$ such that $z_i = \lambda_i y_i$ for each $i = 1, \ldots, n$. From

$$Span\{y_1, y_2, \ldots, y_n, y_{n+1}\} = Span\{z_1, z_2, \ldots, z_n, z_{n+1}\},$$

we can write $z_{n+1} = \sum_{i=1}^{n+1} c_i y_i$ with $c_{n+1} \neq 0$. Now from $z_{n+1} - c_{n+1} y_{n+1} = \sum_{i=1}^{n} c_i y_i$ and the fact that the vectors z_{n+1} and y_{n+1} are both orthogonal to $Span\{y_1, y_2, \ldots, y_n\}$, it easily follows that $z_{n+1} - c_{n+1} y_{n+1} \perp z_{n+1} - c_{n+1} y_{n+1}$. Thus, $z_{n+1} - c_{n+1} y_{n+1} = 0$ or $z_{n+1} = c_{n+1} y_{n+1}$. This completes the induction and the proof of the theorem. ∎

EXERCISES

1. Let $c_1, c_2, \ldots, c_n$ be n (strictly) positive real numbers. Show that the function of two variables $(\cdot, \cdot): \mathbb{R}^n \times \mathbb{R}^n \to \mathbb{R}$, defined by $(x, y) = \sum_{i=1}^{n} c_i x_i y_i$, is an inner product on $\mathbb{R}^n$.

2. Let $(X, (\cdot, \cdot))$ be a real inner product vector space with complexification X_c. Show that the function $\langle \cdot, \cdot \rangle: X_c \times X_c \to \mathbb{C}$ defined via the formula

$$\langle x + \imath y, x_1 + \imath y_1 \rangle = (x, x_1) + (y, y_1) + \imath [(y, x_1) - (x, y_1)].$$

   is an inner product on X_c. Also, show that the norm induced by the inner product $\langle \cdot, \cdot \rangle$ on X_c is given by

$$\|x + \imath y\| = \sqrt{(x, x) + (y, y)} = (\|x\|^2 + \|y\|^2)^{\frac{1}{2}}.$$

3. Let Ω be a Hausdorff compact topological space and let μ be a regular Borel measure on Ω such that $Supp\, \mu = \Omega$. Show that the function $(\cdot, \cdot): C(\Omega) \times C(\Omega) \to \mathbb{R}$, defined by

$$(f, g) = \int_\Omega fg \, d\mu,$$

   is an inner product. Also, describe the complexification of $C(\Omega)$ and the extension of the inner product to the complexification of $C(\Omega)$.

4. Show that equality holds in the Cauchy–Schwarz inequality (i.e., $|(x, y)| = \|x\| \, \|y\|$) if and only if x and y are linearly dependent vectors.

5. If x is a vector in an inner product space, then show that $\|x\| = \sup\{|(x, y)|: \|y\| = 1\}$.

6. Show that in a real inner product space $x \perp y$ holds if and only if $\|x + y\|^2 = \|x\|^2 + \|y\|^2$. Does $\|x + y\|^2 = \|x\|^2 + \|y\|^2$ in a complex inner product space imply $x \perp y$?

7. Assume that a sequence $\{x_n\}$ in an inner product space satisfy $(x_n, x) \to \|x\|^2$ and $\|x_n\| \to \|x\|$. Show that $x_n \to x$.

8. Let S be an orthogonal subset of an inner product space. Show that there exists a complete orthogonal subset C such that $S \subseteq C$.

9. Show that the norms of the following Banach spaces cannot be induced by inner products:

   a. The norm $\|x\| = \max\{|x_1|, |x_2|, \ldots, |x_n|\}$ on $\mathbb{R}^n$.
   b. The sup norm on $C[a, b]$.
   c. The L_p-norm on any $L_p(\mu)$-space for each $1 \leq p \leq \infty$ with $p \neq 2$.

10. Prove Theorem 32.6 for complex normed spaces.

11. Let X be a complex inner product space and let $T: X \to X$ be a linear operator. Show that $T = 0$ if and only if $(Tx, x) = 0$ for each $x \in X$. Is this result true for real inner product spaces?

12. If $\{x_n\}$ is an orthonormal sequence in an inner product space, then show that $\lim(x_n, y) = 0$ for each vector y.

13. The **orthogonal complement** of a nonempty subset A of an inner product space X is defined by

$$A^\perp = \{x \in X: x \perp y \text{ for all } y \in A\}.$$

We shall denote $(A^\perp)^\perp$ by $A^{\perp\perp}$. Establish the following properties regarding orthogonal complements:

a. $A^\perp$ is a closed subspace of X, $A \subseteq A^{\perp\perp}$ and $A \cap A^\perp = \{0\}$.

b. If $A \subseteq B$, then $B^\perp \subseteq A^\perp$.

c. $A^\perp = \overline{A}^\perp = [\mathcal{L}(A)]^\perp = [\overline{\mathcal{L}(A)}]^\perp$, where $\mathcal{L}(A)$ denotes the vector subspace generated by A in X.

d. If M and N are two vector subspaces of X, then $M^{\perp\perp} + N^{\perp\perp} \subseteq (M + N)^{\perp\perp}$.

e. If M is a finite dimensional subspace, then $X = M \oplus M^\perp$.

14. Let V be a vector subspace of a real inner product space X. A linear operator $L: V \to X$ is said to be **symmetric** if $(Lx, y) = (x, Ly)$ holds for all $x, y \in V$.

a. Consider the real inner product space $C[a, b]$ and let

$$V = \{f \in C^2[a, b]: f(a) = f(b) = 0\}.$$

Also, let $p \in C^1[a, b]$ and $q \in C[a, b]$ be two fixed functions. Show that the linear operator $L: V \to C[a, b]$, defined by

$$L(f) = (pf')' + qf,$$

is a symmetric operator.

b. Consider $\mathbb{R}^n$ equipped with its standard inner product and let $A: \mathbb{R}^n \to \mathbb{R}^n$ be a linear operator. As usual, we identify the operator with the matrix $A = [a_{ij}]$ representing it, where the jth column of the matrix A is the column vector Ae_j. Show that A is a symmetric operator if and only if A is a symmetric matrix. (Recall that an $n \times n$ matrix $B = [b_{ij}]$ is said to be **symmetric** if $b_{ij} = b_{ji}$ holds for all i and j.)

c. Let $L: V \to X$ be a symmetric operator. Then L extends naturally to a linear operator $L: V_{\mathbf{c}} = \{x + \iota y: x, y \in V\} \to X_{\mathbf{c}}$ via the formula $L(x + \iota y) = Lx + \iota Ly$. Show that L also satisfies $(Lu, v) = (u, Lv)$ for all $u, v \in V_{\mathbf{c}}$ and that the eigenvalues of L are all real numbers.

d. Show that eigenvectors of a symmetric operator corresponding to distinct eigenvalues are orthogonal.

15. Let $(\cdot, \cdot)$ denote the standard inner product on $\mathbb{R}^n$, i.e., $(x, y) = \sum_{i=1}^{n} x_i y_i$ for all $x, y \in \mathbb{R}^n$. Recall that an $n \times n$ matrix A is said to be **positive definite** if $(x, Ax) > 0$ holds for all non-zero vectors $x \in \mathbb{R}^n$.

Show that a function of two variables $\langle \cdot, \cdot \rangle: \mathbb{R}^n \times \mathbb{R}^n \to \mathbb{R}$ is an inner product on $\mathbb{R}^n$ if and only if there exists a unique real symmetric positive definite matrix A such that

$$\langle x, y \rangle = (x, Ay)$$

holds for all $x, y \in \mathbb{R}^n$. (It is known that a symmetric matrix is positive definite if and only if its eigenvalues are all positive.)

33. HILBERT SPACES

We start by introducing the notion of a Hilbert space.

Definition 33.1.   *A **Hilbert space** is an inner product space which is complete under the norm induced by its inner product.*

Again, we distinguish Hilbert spaces into real and complex. If H is a real Hilbert space, then its complexification $H_c = H \oplus \imath H$ is a complex Hilbert space whose induced norm is given by

$$\|x + \imath y\| = \sqrt{\|x\|^2 + \|y\|^2}.$$

Here are two classical examples of Hilbert spaces:

Example 33.2.   The real Banach space ℓ_2 is a real Hilbert space under the inner product defined by

$$(x, y) = \sum_{n=1}^{\infty} x_n y_n,$$

for all $x = (x_1, x_2, \ldots)$ and $y = (y_1, y_2, \ldots)$ in ℓ_2.

Its complexification, which we shall denote by ℓ_2 again, consists of all square summable sequences of complex numbers. In this case, the inner product satisfies

$$(x, y) = \sum_{n=1}^{\infty} x_n \overline{y_n}.$$

for all $x = (x_1, x_2, \ldots)$ and $y = (y_1, y_2, \ldots)$ in ℓ_2.   ■

Example 33.3.   The real Banach space $L_2(\mu)$ is a real Hilbert space. Its complexification, which we shall denote again by $L_2(\mu)$, consists of all complex valued measurable functions such that $\int |f|^2 \, d\mu < \infty$ holds (where, as usual, two functions are identical if they are equal almost everywhere). Then $L_2(\mu)$ is a complex Hilbert space with the inner product defined by

$$(f, g) = \int f(x)\overline{g(x)} \, d\mu.$$

for all $f, g \in L_2(\mu)$.   ■

Example 33.4. Let $(X, \mathcal{S}, \mu)$ be a measure space and let $\rho: X \to (0, \infty)$ be a measurable function—called a **weight function**. Then the collection of measurable functions

$$L_2(\rho) = \left\{ f \in \mathcal{M}: \int \rho |f|^2 \, d\mu < \infty \right\}$$

is clearly a real vector space. Now, if we define $(\cdot, \cdot): L_2(\rho) \times L_2(\rho) \to \mathbb{R}$ by

$$(f, g) = \int \rho f g \, d\mu,$$

then $(\cdot, \cdot)$ is a well-defined inner product. Indeed, since $f \in L_2(\rho)$ is equivalent to having $\sqrt{\rho} f \in L_2(\mu)$, it follows from Hölder's inequality that

$$\left| \int \rho f g \, d\mu \right| \leq \left(\int \rho |f|^2 \, d\mu \right)^{\frac{1}{2}} \left(\int \rho |g|^2 \, d\mu \right)^{\frac{1}{2}} < \infty,$$

and so $(\cdot, \cdot)$ is well-defined. We leave it as an exercise for the reader to verify that $L_2(\rho)$ is a Hilbert space. As usual, we shall denote the complexification of $L_2(\rho)$ by $L_2(\rho)$ again.

The reader should also notice that $L_2(\rho)$ is exactly the Hilbert space $L_2(\nu)$ for the measure $\nu: \Lambda_\mu \to [0, \infty]$ defined by

$$\nu(A) = \int_A \rho(x) \, d\mu(x)$$

for each $A \in \Lambda_\mu$. ∎

The norm completion of an inner product space is a Hilbert space.

Theorem 33.5. *The norm completion $\hat{X}$ of an inner product space X is a Hilbert space. Moreover, if $x, y \in \hat{X}$ and two sequences $\{x_n\}$ and $\{y_n\}$ of X satisfy $x_n \to x$ and $y_n \to y$ in $\hat{X}$, then*

$$(x, y) = \lim_{n \to \infty} (x_n, y_n).$$

Proof. By Theorem 29.7 we know that $\hat{X}$ is a Banach space. We leave it as an exercise for the reader to verify that the formula $(\cdot, \cdot)$ as defined above is a well-defined inner product that induces $\|\cdot\|$ on $\hat{X}$. ∎

Recall that in a normed space the distance from a vector x to a set A is defined by

$$d(x, A) = \inf\{\|x - y\|: y \in A\}.$$

Our next important result asserts that a closed convex subset of a Hilbert space contains a unique vector of smallest distance from a given vector.

Theorem 33.6. *If A is a non-empty closed convex subset of a Hilbert space H, then for each $x \in H$ there exists a unique $y \in A$ such that*

$$d(x, A) = \|x - y\|.$$

Proof. Let A be a non-empty closed convex subset of a Hilbert space H, let $x \in H$, and put $d = d(x, A)$. Choose a sequence $\{y_n\} \subseteq A$ such that

$$d = \lim_{n \to \infty} \|x - y_n\|.$$

We claim that the sequence $\{y_n\}$ is a Cauchy sequence. To see this, note first that by the Parallelogram law, we have

$$
\begin{aligned}
\|y_n - y_m\|^2 &= \|(y_n - x) - (y_m - x)\|^2 \\
&= 2\|y_n - x\|^2 + 2\|y_m - x\|^2 - \|(y_n - x) + (y_m - x)\|^2 \\
&= 2\|y_n - x\|^2 + 2\|y_m - x\|^2 - 4\left\|\frac{y_n + y_m}{2} - x\right\|^2.
\end{aligned}
$$

Since A is convex, it follows that $\frac{y_n + y_m}{2} \in A$, and so $d \le \|\frac{y_n + y_m}{2} - x\|$. Therefore,

$$\|y_n - y_m\|^2 \le 2\|y_n - x\|^2 + 2\|y_m - x\|^2 - 4d^2 \xrightarrow[n,m \to \infty]{} 0.$$

This shows that $\{y_n\}$ is a Cauchy sequence. The completeness of H guarantees that $\{y_n\}$ converges to some vector y in H. Since A is closed, $y \in A$ and, clearly, $d = d(x, A) = \|x - y\|$.

For the uniqueness of y, assume that another vector $z \in H$ satisfies $\|x - z\| = d(x, A)$. Since A is a convex set, $\frac{1}{2}(y + z) \in A$, and by the Parallelogram Law

$$
\begin{aligned}
d^2 &\le \left\|x - \frac{1}{2}(y + z)\right\|^2 = \frac{1}{4}\|(x - y) + (x - z)\|^2 \\
&= \frac{1}{4}[2\|x - y\|^2 + 2\|x - z\|^2 - \|(x - y) - (x - z)\|^2] \\
&= \frac{1}{4}(2d^2 + 2d^2 - \|z - y\|^2) \\
&= d^2 - \frac{1}{4}\|z - y\|^2 \le d^2.
\end{aligned}
$$

This implies $\frac{1}{4}\|y - z\|^2 = 0$ or $y = z$, and the proof is finished. ∎

Now, let X be an inner product space. If A is a non-empty subset of X, then the **orthogonal complement** $A^\perp$ of A is the set of all vectors that are orthogonal to every vector of A. That is,

$$A^\perp = \{x \in X : x \perp y \text{ for all } y \in A\}.$$

From the linearity and continuity of the inner product it should be clear that $A^\perp$ is always a closed subspace of X satisfying $A^\perp = (\overline{A})^\perp$ and $A \cap A^\perp = \{0\}$; see also Exercise 13 of Section 32.

Remarkably, when X is a Hilbert space and A is a closed subspace, then A together with $A^\perp$ span the whole Hilbert space.

Theorem 33.7. *If M is a closed subspace of a Hilbert space H, then we have $H = M \oplus M^\perp$.*

Proof. Since $M \cap M^\perp = \{0\}$, it suffices to show that every vector $x \in H$ can be written in the form $x = y + z$ with $y \in M$ and $z \in M^\perp$. So, let $x \in X$.

Since M is a closed and (as a vector subspace) convex set, there exists (by Theorem 33.6) a unique vector $y \in M$ such that $d(x, M) = \|x - y\|$. Let $z = x - y$ and note that $x = y + z$. We shall finish the proof by establishing that $z \perp M$.

To this end, let $w \in M$ and write $(z, w) = re^{i\theta}$ with $r \geq 0$. Then, for each real λ we have

$$\|z\|^2 \leq \|z + \lambda e^{i\theta} w\|^2 = \|z\|^2 + 2r\lambda + \lambda^2 \|w\|^2.$$

This implies $2r\lambda + \lambda^2 \|w\|^2 \geq 0$ for each real λ, and so $2r \leq -\lambda \|w\|^2$ for each $\lambda < 0$. Hence, $r \leq 0$, and therefore $r = 0$. Consequently, $(z, w) = 0$. ∎

In a Hilbert space the dense subspaces have a nice characterization in terms of an orthogonality condition.

Corollary 33.8. *A vector subspace M of a Hilbert space is dense if and only if the zero vector is the one and only vector orthogonal to M.*

Proof. Let M be a vector subspace of a Hilbert space H and let $\overline{M}$ be its norm closure. Assume first that $\overline{M} = H$ and let $x \perp M$. Pick a sequence $\{x_n\} \subseteq M$ such that $x_n \to x$ and note that $0 = (x_n, x) \to (x, x)$ implies $(x, x) = 0$, or $x = 0$.

For the converse, assume that $M^\perp = \{0\}$. Since $(\overline{M})^\perp = M^\perp$, it follows from Theorem 33.7 that $\overline{M} = \overline{M} \oplus \{0\} = \overline{M} \oplus (\overline{M})^\perp = H$, and so M is dense in H. ∎

Since the inner product is a jointly continuous function, it follows that every vector y in an inner product space X defines a continuous linear functional $f_y : X \to \mathbf{C}$

via the formula

$$f_y(x) = (x, y).$$

If X is a Hilbert space, then (as we shall see next) all continuous linear functionals on X are of this form.

Theorem 33.9 (F. Riesz). *If H is a Hilbert space and $f: H \to C$ is a continuous linear functional, then there exists a unique vector $y \in H$ such that*

$$f(x) = (x, y)$$

holds for all $x \in H$. Moreover, we have $\|f\| = \|y\|$.

Proof. Let $f: H \to C$ be a continuous linear functional on a Hilbert space and let M be its kernel, i.e.,

$$M = \text{Ker } f = f^{-1}(\{0\}) = \{x \in H: f(x) = 0\}.$$

Since f is a continuous linear functional, it follows that M is a closed subspace. If $M = H$, then $y = 0$ satisfies $f(x) = (x, y) = 0$ for each $x \in H$. So, we can assume that M is a proper closed subspace of H. Then, there exist some $x_0 \in H$ with $f(x_0) = 1$ and (by Theorem 33.7) some non-zero vector $w \in M^\perp$. Now, notice that if $x \in H$, then $x - f(x)x_0 \in M$, and so $(x - f(x)x_0, w) = 0$ or $f(x)(x_0, w) = (x, w)$. This implies $f(w)(x_0, w) = (w, w) > 0$, and so $(x_0, w) \neq 0$. Now, notice that the vector $y = w/(w, x_0)$ satisfies $f(x) = (x, y)$ for all $x \in H$.

For the uniqueness note that if $(x, y) = (x, y_1)$ for each $x \in H$, then $(x, y - y_1) = 0$ for each $x \in H$, and so, by letting $x = y - y_1$, we get $(y - y_1, y - y_1) = 0$ or $y = y_1$. Finally, to establish the norm equality, note first that the Cauchy–Schwarz inequality

$$|f(x)| = |(x, y)| \leq \|x\| \|y\|$$

implies $\|f\| \leq \|y\|$. On the other hand, if $y \neq 0$, then letting $x = y/\|y\|$, we see that $\|x\| = 1$ and so $\|f\| \geq |f(x)| = |(y/\|y\|, y)| = \|y\|$. Thus, $\|f\| = \|y\|$. ∎

If H is a Hilbert space, then Theorem 33.9 shows that a function $y \mapsto f_y$ where $f_y(x) = (x, y)$ can be defined from H onto H^*. In view of the properties

$$f_y + f_z = f_{y+z}, \quad \alpha f_y = f_{\bar{\alpha}y} \quad \text{and} \quad \|f_y\| = \|y\|,$$

it easily follows that this mapping is a "conjugate" linear isometry from H onto

H^*. By means of this isometry, we can establish that Hilbert spaces are reflexive Banach spaces.

Corollary 33.10. *Every Hilbert space is reflexive.*

Proof. Let H be a Hilbert space and let $F: H^* \to \mathbf{C}$ be a continuous linear functional. Define $\phi: H \to \mathbf{C}$ via the formula $\phi(y) = \overline{F(f_y)}$. From

$$\phi(y + z) = \overline{F(f_{y+z})} = \overline{F(f_y + f_z)} = \overline{F(f_y)} + \overline{F(f_z)} = \phi(y) + \phi(z),$$

$$\phi(\alpha y) = \overline{F(f_{\alpha y})} = \overline{F(\overline{\alpha} f_y)} = \overline{\overline{\alpha} F(f_y)} = \alpha \overline{F(f_y)} = \alpha \phi(y),$$

and $|\phi(y)| = |\overline{F(f_y)}| = |F(f_y)| \le \|F\| \, \|f_y\| = \|F\| \, \|y\|$, we see that $\phi \in H^*$. So, by Theorem 33.9, there exists a unique $x \in X$ such that $(y, x) = \phi(y) = \overline{F(f_y)}$ for all $y \in H$. This implies $\hat{x}(f_y) = f_y(x) = (x, y) = F(f_y)$ for each $y \in H$, and so $F = \hat{x}$. This shows that the natural embedding $x \mapsto \hat{x}$ of H into its double dual H^{**} is onto, and consequently H is a reflexive Banach space. ∎

We shall close this section with a few more facts concerning orthogonal and orthonormal sets in Hilbert spaces. Let us say that an orthogonal set S of an inner product space is **complete** if $x \perp s$ for each $s \in S$ implies $x = 0$. In view of Corollary 33.8, it should be noted that an orthogonal set S in a Hilbert space is complete if and only if the vector space generated by S is dense.

Theorem 33.11. *Every inner product space has a complete orthogonal set (and hence, it also has a complete orthonormal set).*

Proof. The conclusion follows from Zorn's lemma by observing that the collection of all orthogonal sets, ordered by inclusion, has a maximal element. To finish the proof, notice that an orthogonal set is maximal if and only if it is complete. ∎

It is useful to note that if S is a complete orthogonal subset of a real Hilbert space H, then S is also a complete orthogonal set in the complexification $H_{\mathbf{c}} = H \oplus \imath H$ of H. Indeed, if some $x + \imath y$ satisfies $x + \imath y \perp S$, then $(x + \imath y, s + \imath 0) = (x, s) + \imath(y, s) = 0$ for all $s \in S$, and so $(x, s) = (y, s) = 0$ for each $s \in S$. Since S is complete, the latter implies $x = y = 0$, and so $x + \imath y = 0$.

Now, assume that the vector subspace M generated in a Hilbert space H by a countable collection of vectors $x_1, x_2, \ldots$ is dense in H. If $y_1, y_2, \ldots$ is the orthogonal sequence of vectors produced by applying the Gram–Schmidt orthogonalization process to the sequence $x_1, x_2, \ldots$, then the orthogonal set $\{y_1, y_2, \ldots\}$ is automatically complete. Indeed, if $y \perp y_n$ for each n, then $y \perp x_n$ for each n, and so $y \perp M$. Since M is dense in H, we get $y \perp \overline{M} = H$, and thus $y = 0$. This observation is used extensively in constructing complete orthogonal (or complete orthonormal) sequences of vectors. The next result illustrates this point.

Theorem 33.12. *Assume that $\rho : [a, b] \rightarrow (0, \infty)$ is a measurable weight function such that $\{1, x, x^2, x^3, \ldots\} \subseteq L_2(\rho)$. Also, let a sequence $P_0, P_1, P_2, \ldots$ of non-zero polynomials be such that*

a. *each P_n is of degree n, and*

b. *$(P_n, P_m) = \int_a^b \rho(x)P_n(x)\overline{P_m(x)}\,dx = 0$ for $n \neq m$.*

Then the orthogonal sequence $P_0, P_1, P_2, \ldots$ is complete and coincides (aside of scalar factors) with the sequence of orthogonal functions of $L_2(\rho)$ that is obtained by applying the Gram–Schmidt orthogonalization process to the sequence of linearly independent functions $\{1, x, x^2, x^3, \ldots\}$.

Proof. We shall prove first that

$$Span\{1, x, x^2, \ldots, x^n\} = Span\{P_0, P_1, P_2, \ldots, P_n\} \qquad (\star)$$

holds for each n. For $n = 0$, the claim is obvious. So, assume that $(\star)$ is true for some n and let P be a polynomial of degree $n + 1$. Since P_{n+1} is a non-zero polynomial of degree $n + 1$, there exists some non-zero scalar c_{n+1} such that $P - c_{n+1}P_{n+1}$ is a polynomial of degree less than or equal to n, i.e., $P - c_{n+1}P_{n+1} \in Span\{1, x, x^2, \ldots, x^n\}$. So, from $(\star)$, we can write $P - c_{n+1}P_{n+1} = \sum_{i=0}^n c_i P_i$, or $P = \sum_{i=0}^{n+1} c_i P_i$. This shows that

$$Span\{1, x, x^2, \ldots, x^n, x^{n+1}\} \subseteq Span\{P_0, P_1, P_2, \ldots, P_n, P_{n+1}\}.$$

Since the reverse inclusion is obvious, the validity of $(\star)$ for $n + 1$ follows. Thus, $(\star)$ is true for each n. Now, a glance at Theorem 32.11 guarantees that (aside of scalar factors) the sequence of polynomials $P_0, P_1, P_2, \ldots$ is the one obtained by applying the Gram–Schmidt orthogonalization process to the sequence $\{1, x, x^2, \ldots\}$.

Finally, in order establish that $P_0, P_1, P_2, \ldots$ is a complete sequence, it suffices to show that the linear subspace M generated by $\{1, x, x^2, \ldots\}$ is dense in $L_2(\rho)$. To see this, note first that $M \subseteq C[a, b] \subseteq L_2(\rho)$ and that M (by Corollary 11.6) is dense with respect to the sup norm in $C[a, b]$. This implies (how?) that $C[a, b]$ is dense in the norm induced by the inner product in $L_2(\rho)$. Next, notice that Theorem 31.11 guarantees (how?) that $C[a, b]$ is dense in $L_2(\rho)$, and so M is dense in $L_2(\rho)$. This shows that $P_0, P_1, P_2, \ldots$ is a complete orthogonal sequence in $L_2(\rho)$. ∎

The following example illustrates the preceding theorem:

Example 33.13. Let $L_2([-1, 1])$ be the Hilbert space of all square integrable functions on the closed interval $[-1, 1]$ equipped with the Lebesgue measure. Then the functions $1, x, x^2, \ldots$ are linearly independent. If the Gram–Schmidt orthogonalization process is applied to these polynomials, then the polynomials of the resulting orthogonal sequence are known as **Legendre[6] polynomials**. We claim that, aside of scalar factors, the Legendre polynomials are given by the formula $\frac{d^n}{dx^n}(x^2 - 1)^2$, the nth-order derivative of the polynomial $(x^2 - 1)^n$. It is a custom to normalize the polynomials at 1, in which case the Legendre polynomials are given by

$$P_n(x) = \frac{1}{2^n n!} \frac{d^n}{dx^n} (x^2 - 1)^n.$$

Clearly, each Legendre polynomial P_n is of degree n (since it is the nth-order derivative of a polynomial of degree $2n$) and satisfies the condition $P_n(1) = 1$.

By Theorem 33.12, in order to prove that the Legendre polynomials result from the Gram–Schmidt orthogonalization process, we need only to show they are mutually orthogonal. Pick n and m such that $n > m$. To verify that P_n is orthogonal to P_m, it suffices show that $\int_{-1}^{1} x^k P_n(x)\, dx = 0$ for all $k < n$.

The proof of this claim is based upon the observation that the function $(x^2 - 1)^n = (x - 1)^n (x + 1)^n$ and all of its derivatives of order less than or equal to $n - 1$ vanish at the points ± 1. So, integrating by parts k times (where $k < n$) and using the preceding observation, we get

$$\begin{aligned}
\int_{-1}^{1} x^k P_n(x)\, dx &= \frac{1}{2^n n!} \int_{-1}^{1} x^k \frac{d^n}{dx^n} (x^2 - 1)^n\, dx \\
&= \frac{(-1)^k k!}{2^n n!} \int_{-1}^{1} \frac{d^{n-k}}{dx^{n-k}} (x^2 - 1)^n\, dx \\
&= \frac{(-1)^k k!}{2^n n!} \frac{d^{n-k-1}}{dx^{n-k-1}} (x^2 - 1)^n \Big|_{-1}^{1} = 0.
\end{aligned}$$

Therefore $P_n \perp P_m$ holds for all $n \neq m$. By Theorem 33.12, $\{P_n\}$ is a complete orthogonal sequence in the Hilbert space $L_2([-1, 1])$.

Finally, it is not difficult to see that the norms of the Legendre polynomials are given by $\|P_n\| = \sqrt{\frac{2}{2n+1}}$. Therefore, the sequence of polynomials $Q_0, Q_1, Q_2, \ldots$, where

$$Q_n(x) = \frac{1}{2^n n!} \sqrt{\frac{2n + 1}{2}} \frac{d^n}{dx^n} (x^2 - 1)^n,$$

is a complete orthonormal sequence of the Hilbert space $L_2([-1, 1])$. ∎

---

[6]Adrien-Marie Legendre (1752–1833), a French mathematician. His major contribution was on elliptic integrals that provided the analytical tools for modern mathematical physics. He spent over 40 years of his life attempting to prove Euclid's parallel postulate.

EXERCISES

1. Verify that the inner product space $L_2(\rho)$ of Example 33 is a Hilbert space.
2. Show that the Hilbert space $L_2[0, \infty)$ is separable.
3. Let $\{\psi_n\}$ be an orthonormal sequence of functions in the Hilbert space $L_2[a, b]$ which is also uniformly bounded. If $\{\alpha_n\}$ is a sequence of scalars such that $\alpha_n \psi_n \to 0$ a.e., then show that $\lim \alpha_n = 0$.
4. Let $\{\phi_n\}$ be an orthonormal sequence of functions in the Hilbert space $L_2[-1, 1]$. Show that the sequence of functions $\{\psi_n\}$, where

$$\psi_n(x) = \left(\frac{2}{b-a}\right)^{\frac{1}{2}} \phi_n\left(\frac{2}{b-a}\left(x - \frac{b+a}{2}\right)\right),$$

   is an orthonormal sequence in the Hilbert space $L_2[a, b]$.
5. Show that the function $(\cdot, \cdot): \hat{X} \times \hat{X} \to \mathbf{C}$ defined in Theorem 33.5 is a well-defined inner product on $\hat{X}$ that induces the norm of $\hat{X}$.
6. Show that the closed unit ball of ℓ_2 is not a norm compact set.
7. Show that the **Hilbert cube** (the set of all $x = (x_1, x_2, \ldots) \in \ell_2$ such that $|x_n| \leq \frac{1}{n}$ holds for all n) is a compact subset of ℓ_2.
8. Show that every subspace M of a Hilbert space satisfies $\overline{M} = M^{\perp\perp}$.
9. For two arbitrary vector subspaces M and N of a Hilbert space, establish the following:
   a. $(M + N)^{\perp} = M^{\perp} \cap N^{\perp}$, and
   b. if M and N are both closed, then $(M \cap N)^{\perp} = \overline{M^{\perp} + N^{\perp}}$.
10. Let X be an inner product space such that $M = M^{\perp\perp}$ holds for every closed subspace M. Show that X is a Hilbert space.
11. Consider the linear operator $V: L_2[a, b] \to L_2[a, b]$ defined by

$$V f(x) = \int_a^x f(t)\, dt.$$

   Show that the norm of the operator satisfies $\|V\| \leq b - a$.
12. Let $\{x_n\}$ be a norm bounded sequence of vectors in the Hilbert space ℓ_2, where $x_n = (x_1^n, x_2^n, x_3^n, \ldots)$. If for each coordinate k we have $\lim_{n\to\infty} x_k^n = 0$, then show that

$$\lim_{n\to\infty} (x_n, y) = 0$$

   holds for each vector $y \in \ell_2$.
13. Let H be a Hilbert space and let $\{x_n\}$ be a sequence satisfying

$$\lim_{n\to\infty} (x_n, y) = (x, y)$$

   for each $y \in H$. Show that there exists a subsequence $\{x_{k_n}\}$ of $\{x_n\}$ such that

$$\lim_{n\to\infty} \left\| x - \frac{1}{n} \sum_{i=1}^n x_{k_i} \right\| = 0.$$

14. Let $\rho: [a, b] \to (0, \infty)$ be a measurable and essentially bounded function and for each $n = 0, 1, 2, \ldots$ let P_n be a non-zero polynomial of degree n. Assume that

$$\int_a^b \rho(x) P_n(x) \overline{P_m(x)} \, dx = 0 \text{ for } n \neq m.$$

Show that each P_n has n distinct real roots all lying in the open interval (a, b).

15. In Example 33 we defined the sequence $P_0, P_1, P_2, \ldots$ of Legendre polynomials by the formulas

$$P_n(x) = \frac{1}{2^n n!} \frac{d^n}{dx^n} (x^2 - 1)^n.$$

We also proved that these are (aside of scalar factors) the polynomials obtained by applying the Gram–Schmidt orthogonalization process to the sequence of linearly independent functions $\{1, x, x^2, \ldots\}$ in the Hilbert space $L_2([-1, 1])$. Show that for each n we have

$$P_n(1) = 1 \quad \text{and} \quad \|P_n\| = \sqrt{\frac{2}{2n + 1}}.$$

16. Let $\{T_\alpha\}_{\alpha \in A}$ be a family of linear continuous operators from a complex Hilbert space X into another complex Hilbert space Y. Assume that for each $x \in X$ and each $y \in Y$, the set of complex numbers $\{(T_\alpha(x), y): \alpha \in A\}$ is bounded. Show that the family of operators $\{T_\alpha\}_{\alpha \in A}$ is uniformly norm bounded, i.e., show that there exists some constant $M > 0$ satisfying $\|T_\alpha\| \leq M$ for all $\alpha \in A$.

17. Let $\{\phi_n\}$ be an orthonormal sequence in a Hilbert space H and consider the operator $T: H \to H$ defined by

$$T(x) = \sum_{n=1}^{\infty} \alpha_n (x, \phi_n) \phi_n,$$

where $\{\alpha_n\}$ is a sequence of scalars satisfying $\lim \alpha_n = 0$. Show that T is a compact operator.

18. Assume that $T, T^*: H \to H$ are two functions on a Hilbert space satisfying

$$(Tx, y) = (x, T^*y)$$

for all $x, y \in H$. Show that T and T^* are both bounded linear operators satisfying

$$\|T\| = \|T^*\| \quad \text{and} \quad \|TT^*\| = \|T\|^2.$$

19. Show that if $T: H \to H$ is a bounded linear operator on a Hilbert space, then there exists a unique bounded operator $T^*: H \to H$ (called the **adjoint operator** of T) satisfying

$$(Tx, y) = (x, T^*y)$$

for all $x, y \in H$. Moreover, show that $\|T\| = \|T^*\|$.

34. ORTHONORMAL BASES

Recall that an orthogonal set S in an inner product space is said to be **orthonormal** if $\|s\| = 1$ holds for each $s \in S$. A **complete orthonormal set** is a maximal orthonormal set. By Theorem 33.11, we know that every inner product space has a complete orthonormal set.

> **Definition 34.1.**   *A complete orthonormal set in a Hilbert space is known as an* **orthonormal basis**. *An orthonormal basis will be denoted by* $\{e_i\}_{i \in I}$.

That is, a family of vectors $\{e_i\}_{i \in I}$ of a Hilbert space is an orthonormal basis if and only if

1. $(e_i, e_j) = \delta_{ij}$, where δ_{ij} is Kronecker's[7] delta (defined by $\delta_{ij} = 1$ if $i = j$ and $\delta_{ij} = 0$ if $i \neq j$); and
2. $(x, e_i) = 0$ for all i implies $x = 0$.

The orthonormal sets that are bases are characterized as follows:

> **Theorem 34.2.**   *For an orthonormal family* $\{e_i\}_{i \in I}$ *of vectors in a Hilbert space the following statements are equivalent.*
>
> 1. *The family* $\{e_i\}_{i \in I}$ *is an orthonormal basis.*
> 2. *If* $x \perp e_i$ *for each* $i \in I$, *then* $x = 0$.
> 3. *For each vector* x *we have* $(x, e_i) \neq 0$ *for at-most countably many indices* i *and* $x = \sum_{i \in I}(x, e_i)e_i$, *where the series converges in the norm.*
> 4. *For each pair of vectors* x *and* y *we have* $(x, e_i) \neq 0$ *and* $(y, e_j) \neq 0$ *for at-most countably many indices and* $(x, y) = \sum_{i \in I}(x, e_i)\overline{(y, e_i)}$.
> 5. **(Parseval's[8] Identity)** *For each vector* x *we have* $\|x\|^2 = \sum_{i \in I} |(x, e_i)|^2$.

Proof.   (1) $\implies$ (2) This is an immediate consequence of the definition of the orthonormal basis.

(2) $\implies$ (3) Fix some vector x. By Bessel's inequality (Theorem 32.10), we know that $(x, e_i) \neq 0$ for at most countably many indices i. Let $i_1, i_2, \ldots$ be an enumeration of the indices for which $(x, e_i) \neq 0$. (Here we consider the case where the set of indices $\{i \in I : (x, e_i) \neq 0\}$ is countable; the finite case can be dealt in a similar manner.) From Bessel's inequality, it also follows that $\sum_{k=1}^{\infty} |(x, e_{i_k})|^2 \leq \|x\|^2 < \infty$.

---

[7]Leopold Kronecker (1823–1891), a German mathematician. He made important contributions to the theory of equations and to the theory of algebraic numbers. Due to his belief that mathematics should deal only with "finite numbers," he vigorously opposed Cantor's introduction of set theory. He is also famous for the quotation "God created the integers, all else is the work of man."

[8]Marc-Antoine Parseval des Chênes (1755–1836), a French mathematician. He published very little and he is only known for this identity. He was a royalist who was imprisoned in 1792 for publishing poetry against Napoleon's regime.

Now, the Pythagorean Theorem 32.7 guarantees

$$\left\| \sum_{k=n}^{m} (x, e_{i_k}) e_{i_k} \right\|^2 = \sum_{k=n}^{m} |(x, e_{i_k})|^2$$

for all n and m. This shows that the sequence $\{\sum_{k=1}^{n}(x, e_{i_k})e_{i_k}\}$ is a Cauchy sequence, and thus the series $\sum_{k=1}^{\infty}(x, e_{i_k})e_{i_k} = \sum_{i \in I}(x, e_i)e_i$ is norm convergent. Let $y = x - \sum_{k=1}^{\infty}(x, e_{i_k})e_{i_k}$. Since $(x - \sum_{k=1}^{n+m}(x, e_{i_k})e_{i_k}, e_{i_n}) = 0$ for all $m \geq 1$, it follows from the continuity of the inner product that $(y, e_{i_n}) = 0$ for all n. Consequently, $(y, e_i) = 0$ for each $i \in I$, and so, by our hypothesis, $y = 0$ or $x = \sum_{i \in I}(x, e_i)e_i$.

(3) $\implies$ (4) Let $x = \sum_{i \in I}(x, e_i)e_i$ and $y = \sum_{j \in I}(y, e_j)e_j$. Since there are only at most a countable number of non-zero terms in each sum and the two series converge in norm to x and y, respectively, it follows from the joint continuity of the inner product that

$$(x, y) = \left(\sum_{i \in I}(x, e_i)e_i, \sum_{j \in I}(y, e_j)e_j \right)$$

$$= \sum_{i \in I} \sum_{j \in J}(x, e_i)\overline{(y, e_j)}(e_i, e_j)$$

$$= \sum_{i \in I}(x, e_i)\overline{(y, e_j)}.$$

(4) $\implies$ (5) The desired identity follows immediately by letting $y = x$.

(5) $\implies$ (1) Suppose that $\{e_i\}_{i \in I}$ is not an orthonormal basis. This means that $\{e_i\}_{i \in I}$ is not a maximal orthonormal set. Thus, there exists a non-zero vector y that is orthogonal to each e_i. But then, our hypothesis implies $\|y\|^2 = \sum_{i \in I}|(y, e_i)|^2 = 0$ contrary to the fact that y is a non-zero vector. Therefore, $\{e_i\}_{i \in I}$ is an orthonormal basis, and the proof of the theorem is finished. ∎

Definition 34.3. *If $\{e_i\}_{i \in I}$ is an orthonormal basis in a Hilbert space and x is an arbitrary vector, then the family of scalars $\{(x, e_i)\}_{i \in I}$ is referred to as the family of* **Fourier**[9] **coefficients** *of x relative to the orthonormal basis $\{e_i\}_{i \in I}$.*

The Hilbert spaces with countable orthonormal bases are precisely the infinite-dimensional separable Hilbert spaces.

---

[9]Jean Baptiste Joseph Fourier (1768–1830), a French mathematician. He is the founder of Fourier analysis. In his 1807 memoir, *On the Propagation of Heat in Solid Bodies*, he introduced the idea of expanding functions as trigonometric series. Although this was a controversial revolutionary idea at that time, today it is the basis of modern "Fourier analysis."

Theorem 34.4. *An infinite-dimensional Hilbert space H is separable if and only if it has a countable orthonormal basis. Moreover, in this case, every orthonormal basis of H is countable.*

Proof. Let $\{e_1, e_2, \ldots\}$ be a countable orthonormal basis in a Hilbert space H. Then the set of all finite linear combinations of the e_n with rational coefficients (call a complex number rational if its real and imaginary parts are both rational) is a countable dense subset of H. Therefore, H is a separable Hilbert space.

For the converse, assume that H is a separable Hilbert space and let $\{x_1, x_2, \ldots\}$ be a countable dense subset of H. We claim that there exists a strictly increasing sequence of natural numbers $k_1 < k_2 < k_3 \cdots$ such that $\{x_{k_1}, x_{k_2}, \ldots\}$ is a linearly independent sequence satisfying

$$Span\{x_{k_1}, x_{k_2}, \ldots, x_{k_n}\} = Span\{x_1, x_2, \ldots, x_{k_n}\}.$$

for each n. We shall establish the existence of such a sequence by induction.

Start by observing that we can assume $x_n \neq 0$ for each n. Let $k_1 = 1$. For the induction step, assume that we selected natural numbers $k_1 < k_2 < \cdots < k_n$ such that the vectors $\{x_{k_1}, x_{k_2}, \ldots, x_{k_n}\}$ are linearly independent and

$$Span\{x_{k_1}, x_{k_2}, \ldots, x_{k_n}\} = Span\{x_1, x_2, x_3, \ldots, x_{k_n}\}.$$

If every vector x_i with $i > k_n$ lies in the linear span of $\{x_{k_1}, x_{k_2}, \ldots, x_{k_n}\}$, then the set $\{x_1, x_2, \ldots\}$ spans a finite dimensional vector subspace, and so it cannot be dense in the infinite dimensional Hilbert space H, a contradiction. Hence, there exists some vector x_i with $i > k_n$ which does not lie in the linear span of $\{x_{k_1}, x_{k_2}, \ldots, x_{k_n}\}$. Let

$$k_{n+1} = \min\{i \in \mathbb{N}: x_i \text{ is not in the linear span of } \{x_{k_1}, x_{k_2}, \ldots, x_{k_n}\}\}.$$

Clearly,

$$Span\{x_{k_1}, x_{k_2}, \ldots, x_{k_n}, x_{k_{n+1}}\} = Span\{x_1, x_2, x_3, \ldots, x_{k_{n+1}}\}.$$

This completes the induction and the proof of the existence of the sequence $\{x_{k_n}\}$.

Since $\{x_1, x_2, \ldots\}$ is dense in H, it follows that the linear span of $\{x_{k_n}\}$ is also dense in H. So, if we apply the Gram–Schmidt orthogonalization process to the sequence $\{x_{k_n}\}$, we get a complete orthogonal sequence. Normalizing this orthogonal sequence yields a countable orhonormal basis for H.

For last part, assume that $\{b_1, b_2, \ldots\}$ is a countable orthonormal basis of a Hilbert space H and that $\{e_i\}_{i \in I}$ is another orthonormal basis of H. For each n let

$$I_n = \left\{ i \in I \colon |(e_i, b_n)| > \frac{1}{n} \right\}.$$

From Parseval's identity $1 = \|b_n\|^2 = \sum_{i \in I} |(b_n, e_i)|^2$, it follows that I_n is a finite set. On the other hand, if $i \in I$, then using Parseval's identity once more, we see that $1 = \|e_i\|^2 = \sum_{n=1}^{\infty} |(e_i, b_n)|^2$, and so $(e_i, b_n) \neq 0$ for some n. Therefore, $I = \bigcup_{n=1}^{\infty} I_n$. This shows that I is at most countable—and hence, a countable set. ∎

In Example 33.13, we saw that the sequence of normalized Legendre polynomials $Q_0, Q_1, Q_2, \ldots$, where

$$Q_n(x) = \frac{1}{2^n n!} \sqrt{\frac{2n + 1}{2}} \frac{d^n}{dx^n} (x^2 - 1)^n,$$

is an orthonormal basis of the Hilbert space $L_2([-1, 1])$. We now present two more examples of orthonormal bases in some classical Hilbert spaces. The first example exhibits an orthonormal basis in the Hilbert space $L_2([0, \infty))$.

Example 34.5. Let us denote the differential operator by D, i.e., $Df = f'$ and $D^k f = f^{(k)}$, the kth derivative of f. The **Laguerre**[10] **polynomials** are defined by the formulas

$$L_n(x) = e^x D^n(x^n e^{-x}) = (-1)^n x^n + \sum_{k=0}^{n-1} (-1)^k \binom{n}{k} n(n-1) \cdots (k+1) x^k,$$

for $n = 0, 1, 2, \ldots$. Clearly, each Laguerre polynomial $L_n(x)$ is of degree n. Next, we define the **Laguerre functions** $\phi_0, \phi_1, \phi_2, \ldots$ by

$$\phi_n(x) = \frac{1}{n!} e^{-\frac{x}{2}} L_n(x), \quad n = 0, 1, 2, \ldots.$$

We claim that $\{\phi_0, \phi_2, \phi_3, \ldots\}$ is an orthonormal basis in the Hilbert space $L_2([0, \infty))$ of all square integrable functions on the interval $[0, \infty)$ equipped with the Lebesgue measure. This claim will be established by steps.

STEP I: *The sequence* $\{\phi_0, \phi_2, \phi_3, \ldots\}$ *is orthogonal.*

Let $m < n$. Then we need to show that $\int_0^{\infty} L_m(x) L_n(x) e^{-x}\, dx = 0$. Since $L_m(x)$ is a polynomial of degree m, it suffices to verify that for all $k < n$ we have

$$\int_0^{\infty} x^k L_n(x) e^{-x}\, dx = 0. \tag{$\star$}$$

---

[10] Edmond Nicolas Laguerre (1834–1886), a French mathematician. He worked in geometry and on approximation methods in analysis.

To establish $(\star)$, integrate by parts k times to get

$$\int_0^\infty x^k L_n(x) e^{-x} \, dx = \int_0^\infty x^k D^n(x^n e^{-x}) \, dx = (-1)^k k! \int_0^\infty D^{n-k}(x^n e^{-x}) \, dx = 0.$$

Therefore, the Laguerre functions are mutually orthogonal.

STEP II: *The sequence* $\{\phi_0, \phi_2, \phi_3, \ldots\}$ *is orthonormal.*

Taking into account $(\star)$ and integrating by parts n times, we get

$$
\begin{aligned}
\left\| \frac{1}{n!} e^{-\frac{x}{2}} L_n \right\|^2 &= \frac{1}{(n!)^2} \int_0^\infty L_n(x) L_n(x) e^{-x} \, dx = \frac{1}{(n!)^2} \int_0^\infty (-1)^n x^n L_n(x) e^{-x} \, dx \\
&= \frac{1}{(n!)^2} \int_0^\infty (-1)^n x^n D^n(x^n e^{-x}) \, dx = \frac{1}{(n!)^2} (n!) \int_0^\infty x^n e^{-x} \, dx \\
&= \frac{1}{(n!)^2} (n!)(n!) = 1.
\end{aligned}
$$

STEP III: *The sequence* $\{\phi_0, \phi_2, \phi_3, \ldots\}$ *is complete.*

Assume that a function $f \in L_2([0, \infty))$ satisfies $f \perp \phi_n$ for each n, i.e.,

$$\int_0^\infty f(x) e^{-\frac{x}{2}} L_n(x) \, dx = 0 \quad \text{for } n = 0, 1, 2, \ldots .$$

We must show $f = 0$ a.e. To this end, let $g(x) = f(x) e^{-\frac{x}{2}}$ and note $\int_0^\infty g(x) x L_n(x) \, dx = 0$ holds for all $n = 0, 1, \ldots$. This implies

$$\int_0^\infty x^n g(x) \, dx = 0, \quad \text{for } n = 0, 1, 2, \ldots . \tag{$\star\star$}$$

To show that $f = 0$ a.e., it suffices to prove that $g = 0$ a.e.

To see this, start by considering the complex valued function

$$F(s) = \int_0^\infty e^{-sx} g(x) \, dx$$

define for all complex numbers s with real part $\mathrm{Re}\, s > 0$. Next, fix a complex number s_0 with $\mathrm{Re}\, s_0 > 0$. Then there exist some neighborhood V of s_0 and some $\alpha > 0$ such that $\mathrm{Re}\, s > \alpha$ holds for all $s \in V$. This implies that there exists some $M > 0$ such that

$$\left| \frac{\partial e^{-sx} g(x)}{\partial s} \right| = \left| -x e^{-sx} g(x) \right| \le M |g(x)|$$

holds for all $s \in V$ and all $x \in [0, \infty)$. Now, a glance at Theorem 24.5 (and the discussion after its proof) guarantees that F is differentiable at s_0 and that $F'(s_0) = -\int_0^\infty x e^{-sx} g(x) \, dx$. In other words, F is an analytic function.

Next, observe that for each n, Hölder's inequality implies

$$\int_0^\infty |x^n g(x)|\, dx = \int_0^\infty |f(x)| x^n e^{-\frac{x}{2}}\, dx \le \|f\| \left(\int_0^\infty x^{2n} e^{-x}\, dx \right)^{\frac{1}{2}} = \|f\| \sqrt{(2n)!}.$$

In particular, for each $s \ge 0$ and each n, we have

$$\int_0^\infty \frac{s^n}{n!} |x^n g(x)|\, dx \le \|f\| \frac{s^n}{n!} \sqrt{(2n)!} = a_n.$$

Since $\lim_{n \to \infty} \frac{a_{n+1}}{a_n} = 2s$, it follows that the series $\sum_{n=0}^\infty a_n$ converges for all $0 < s < \frac{1}{2}$. But then, Theorem 22.9 (see also Exercise 10 of Section 22) implies that the series $\sum_{n=0}^\infty (-1)^n \frac{(sx)^n}{n!} g(x) = e^{-sx} g(x)$ defines an integrable function and that

$$F(s) = \int_0^\infty e^{-sx} g(x)\, dx = \sum_{n=0}^\infty (-1)^n \frac{s^n}{n!} \int_0^\infty x^n g(x)\, dx$$

holds for all $0 < s < \frac{1}{2}$. Now, a glance at (**) yields

$$F(s) = \int_0^\infty e^{-sx} g(x)\, dx = 0, \quad \text{for } 0 < s < \frac{1}{2}.$$

Since F is an analytic function, the identity theorem for analytic functions implies $F(s) = 0$ for all complex numbers s with $\operatorname{Re} s > 0$; see for instance [2, Theorem 16.25, p. 462]. That is, the Laplace transform of g equals zero and so $g = 0$ a.e. (see Example 30.12), and the proofs of our claims are finished. ∎

Example 34.6. Consider the weighted Hilbert space $L_2(\rho)$ with the weight $\rho(x) = e^{-\frac{1}{2}x^2}$ defined over the whole real line $(-\infty, \infty)$. The functions

$$1, x, x^2, x^3, \ldots$$

belong to $L_2(\rho)$ and they are clearly linearly independent. We claim that the linear span of this sequence is dense in $L_2(\rho)$. To see this, assume that a function $f \in L_2(\rho)$ satisfies $\int_{-\infty}^\infty f(x) x^n e^{-\frac{1}{2}x^2}\, dx = 0$ for each $n = 0, 1, 2, \ldots$. Now, consider the function $g(x) = f(x) e^{-\frac{1}{2}x^2}$. Since the functions $h(x) = f(x) e^{-\frac{1}{4}x^2}$ and $\theta(x) = e^{-\frac{1}{4}x^2}$ belong to $L_2(\mathbb{R})$, it follows that $g = h\theta \in L_1(\mathbb{R})$ and

$$\int_{-\infty}^\infty x^n g(x)\, dx = 0, \quad n = 0, 1, 2, \ldots.$$

We must show that $f = 0$ a.e. or, equivalently, that $g = 0$ a.e.

To see this, we introduce the complex-valued function $G \colon \mathbb{C} \to \mathbb{C}$ by

$$G(z) = \int_{-\infty}^\infty f(x) e^{-\frac{1}{2}x^2} e^{ixz}\, dx.$$

It should be clear that this (complex) integral is well defined and that (as in the previous example) G is differentiable at each point (and so G is an entire function). At $z = 0$, we have

$$G^{(n)}(0) = i^n \int_\infty^\infty f(x)e^{-\frac{1}{2}x^2}x^n\, dx = 0$$

for each $n = 0, 1, 2, \ldots$. Therefore, $G(z) = 0$ for each $z \in \mathbf{C}$, and consequently

$$\int_{-\infty}^\infty f(x)e^{-\frac{1}{2}x^2}e^{-isx}\, dx = \int_{-\infty}^\infty g(x)e^{-isx}\, dx = 0$$

for all $-\infty < s < \infty$. In other words, the "Fourier transform" of g is zero, and this guarantees that $g = 0$ a.e.; see, for instance [15, p. 408] or [26, p. 187].

We now claim that if we apply the Gram–Schmidt orthogonalization process to the sequence of functions $1, x, x^2, \ldots$, then the members of the resulting complete orthogonal sequence $\{H_n\}$ (aside of scalar factors) are the **Hermite**[11] **polynomials** which are defined by

$$H_n(x) = (-1)^n e^{\frac{1}{2}x^2} D^n\left(e^{-\frac{1}{2}x^2}\right),$$

where D denotes the differentiation operator. Clearly, each $H_n(x)$ is a polynomial of degree n. Now, if $k < n$, then—integrating by parts k times—yields

$$\int_{-\infty}^\infty x^k H_n(x)e^{-\frac{1}{2}x^2}\, dx = (-1)^n \int_{-\infty}^\infty x^k D^n\left(e^{-\frac{1}{2}x^2}\right) dx$$

$$= (-1)^{n+k} k! \int_{-\infty}^\infty D^{n-k}\left(e^{-\frac{x^2}{2}}\right) dx = 0.$$

This implies

$$(H_m, H_n) = \int_{-\infty}^\infty H_m(x)H_n(x)e^{-\frac{x^2}{2}}\, dx = 0$$

for $n \neq m$. Finally, a glance at Theorem 32.11 guarantees that (aside of scalar factors) the sequence $\{H_n\}$ is indeed the one obtained by applying the Gram–Schmidt orthogonalization process to the sequence of linearly independent functions $1, x, x^2, x^3, \ldots$. Normalizing the sequence of functions $\{H_n\}$ we get to a complete orthonormal basis for $L_2(\rho)$. ∎

Next, we shall use Theorem 34.2 to present a concrete realization of an arbitrary Hilbert space. To do this, we need some preliminary discussion.

Recall that a linear operator $T: X \to Y$ between two normed spaces is called **norm preserving** (or an **isometry**) if $\|Tx\| = \|x\|$ holds for all $x \in X$. Similarly, a linear operator $L: H_1 \to H_2$ between two Hilbert spaces is said to be **inner product preserving** if $(Lx, Ly) = (x, y)$ holds true for all $x, y \in H_1$.

---

[11]Charles Hermite (1822–1901), a French mathematician. He worked in the theory of functions. He was the first to prove (in 1873) that e is a transcendental number.

From Theorem 32.6 the following result should be immediate:

Lemma 34.7. *A linear operator $L: H_1 \to H_2$ between two Hilbert spaces is norm preserving if and only if it is inner product preserving.*

Now, let Q be an arbitrary nonempty set. If $x: Q \to \mathbf{C}$ is an arbitrary complex-valued function (which it will also be denoted as a family $\{x(q)\}_{q \in Q}$), then its ℓ_2-**norm** is defined by

$$\|x\|_2 = \left(\sum_{q \in Q} |x(q)|^2 \right)^{\frac{1}{2}}.$$

A complex-valued function $x: Q \to \mathbf{C}$ is said to be **square summable** if $\|x\|_2 < \infty$. This is, of course, equivalent to saying that $x(q) \neq 0$ for at most countably many q and if $q_1, q_2, \ldots$ is an enumeration of the set $\{q \in Q: x(q) \neq 0\}$, then $\sum_{n=1}^{\infty} |x(q_n)|^2 < \infty$. It should be clear that the collection $\ell_2(Q)$ of all square summable functions with the pointwise algebraic operations and the ℓ_2-norm is a normed vector space. In actuality, it can be easily shown that $\ell_2(Q)$ is a Hilbert space.

Lemma 34.8. *The vector space $\ell_2(Q)$ of all square summable complex-valued functions defined on a nonempty set Q under the inner product*

$$(x, y) = \sum_{q \in Q} x(q)\overline{y(q)}$$

is a Hilbert space.

We are now ready to establish that every Hilbert space is linearly isometric to a concrete $\ell_2(Q)$ Hilbert space.

Theorem 34.9. *Every Hilbert space H is linearly isometric to a Hilbert space of the form $\ell_2(Q)$. Specifically, if $\{e_i\}_{i \in I}$ is an orthonormal basis of H, then the linear operator $L: H \to \ell_2(I)$, defined by*

$$L(x) = \{(x, e_i)\}_{i \in I},$$

is a surjective linear isometry.

Proof. Clearly, L is linear and by Parseval's Identity (Theorem 34.2(5)) it is also an isometry. We leave it as an exercise for the reader to show that L is also a surjective linear operator. ∎

Corollary 34.10. *An infinite dimensional Hilbert space H is separable if and only if it is linearly isometric to ℓ_2.*

EXERCISES

1. Let $\{e_i\}_{i\in I}$ and $\{f_j\}_{j\in J}$ be two orthonormal bases of a Hilbert space. Show that I and J have the same cardinality.

2. Let $\{e_i\}_{i\in I}$ be an orthonormal basis in a Hilbert space H. If D is a dense subset of H, then show that the cardinality of D is at least as large as that of I. Use this conclusion to provide an alternate proof of Theorem 34.4 by proving that for an infinite-dimensional Hilbert space H the following statements are equivalent:

   a. H has a countable orthonormal basis.
   b. H is separable.
   c. H is linearly isometric to ℓ_2.

3. Let I be an arbitrary nonempty set, and for each $i \in I$ let $e_i = \chi_{\{i\}}$. Show that the family of functions $\{e_i\}_{i\in I}$ is an orthonormal basis for the Hilbert space $\ell_2(I)$.

4. Let $\{e_i\}_{i\in i}$ be an orthonormal basis in a Hilbert space and let x be a unit vector, i.e, $\|x\| = 1$. Show that for each $k \in \mathbb{N}$ the set $\{i \in I : |(x, e_i)| \geq \frac{1}{k}\}$ has at most k^2 elements.

5. Let M be a closed vector subspace of a Hilbert space H and let $\{e_i\}_{i\in I}$ be an orthonormal basis of M; where M is now considered as a Hilbert space in its own right under the induced operations. If $x \in H$, then show that the unique vector of M closest to x (which is guaranteed by Theorem 33.6) is the vector $y = \sum_{i\in I}(x, e_i)e_i$.

6. Let $\{e_n\}$ be an orthonormal basis of a separable Hilbert space. For each n, let $f_n = e_{n+1} - e_n$. Show that the vector subspace generated by the sequence $\{f_n\}$ is dense.

7. Prove Lemma 34.7.

8. Prove Lemma 34.8.

9. Complete the details of the proof of Theorem 34.9.

10. Let $\{e_n\}$ be an orthonormal sequence of vectors in the Hilbert space $L_2[0, 2\pi]$. Suppose that for each continuous function f in $L_2[0, 2\pi]$ we have $f = \sum_{n=1}^{\infty}(f, e_n)e_n$. Show that $\{e_n\}$ is an orthonormal basis.

11. Let $\{\phi_n\}$ be an orthonormal sequence of vectors in the Hilbert space $L_2[0, 2\pi]$. Suppose that for each continuous function f in $L_2[0, 2\pi]$ we have $\|f\|^2 = \sum_{n=1}^{\infty}|(f, \phi_n)|^2$. Show that $\{\phi_n\}$ is an orthonormal basis.

12. Let $\{\phi_1, \phi_2, \ldots\}$ be an orthonormal basis of the Hilbert space $L_2(\mu)$, where μ is a finite measure. Fix a function $f \in L_2(\mu)$ and let $\{\alpha_1, \alpha_2, \ldots\}$ be its sequence of Fourier coefficients relative to $\{\phi_n\}$, i.e., $\alpha_n = \int f\, \overline{\phi_n}\, d\mu$. Show that (although the series $\sum_{n=1}^{\infty} \alpha_n \phi_n$ need not converge pointwise almost everywhere to f) the Fourier series $\sum_{n=1}^{\infty} \alpha_n \phi_n$ can be integrated term-by-term in the sense that for every measurable set E we have

$$\int_E f\, d\mu = \sum_{n=1}^{\infty} \alpha_n \int_E \phi_n\, d\mu.$$

[HINT: If $s_n = \sum_{k=1}^{n} \alpha_n \phi_n$, then the Cauchy–Schwarz inequality implies

$$\left| \int_E f \, d\mu - \int_E s_n \, d\mu \right|^2 \leq \left(\int_E |f - s_n| \, d\mu \right)^2 \leq \|f - s_n\|^2 \mu^*(E). \,]$$

13. Establish the following "perturbation" property of orthonormal bases. If $\{e_i\}_{i \in I}$ is an orthonormal basis and $\{f_i\}_{i \in I}$ is an orthonormal family satisfying

$$\sum_{i \in I} \|e_i - f_i\|^2 < \infty,$$

then $\{f_i\}_{i \in I}$ is also an orthonormal basis.

35. FOURIER ANALYSIS

In this section, we shall study some basic properties of periodic functions. Recall that a function $f : \mathbb{R} \to \mathbb{C}$ is said to be **periodic** if there exists some $p > 0$ such that $f(x + p) = f(x)$ for each $x \in \mathbb{R}$. The positive number $p > 0$ is called a **period** of f. In essence, a periodic function is completely determined by its values on the closed interval $[0, p]$. Any function $f : [0, p] \to \mathbb{C}$ that satisfies $f(0) = f(p)$ gives rise to a periodic function via the formula $f(x + p) = f(x)$ for each $x \in \mathbb{R}$. Notice that if $f : \mathbb{R} \to \mathbb{C}$ is a periodic function, then the function $g : \mathbb{R} \to \mathbb{C}$, defined by $g(x) = f(\frac{p}{2\pi} x)$, is a periodic function with period 2π.

From now on, unless otherwise stated, all periodic functions encountered in this section will be assumed to have period 2π. As a matter of fact, they will be defined only on $[0, 2\pi]$ (and they will be tacitly assumed defined on all of $\mathbb{R}$ via the formula $f(x + 2\pi) = f(x)$). That is, a periodic function is any function $f : [0, 2\pi] \to \mathbb{C}$ satisfying $f(0) = f(2\pi)$.

The closed interval $[0, 2\pi]$ will always be considered equipped with the Lebesgue measure; we shall write dx instead of $d\lambda(x)$. The Hilbert space $L_2[0, 2\pi]$ is assumed equipped with the inner product

$$(f, g) = \int_0^{2\pi} f(x) \overline{g(x)} \, dx,$$

and so it induces the standard L_2-norm

$$\|f\|_2 = \sqrt{(f, f)} = \left(\int_0^{2\pi} |f(x)|^2 \, dx \right)^{\frac{1}{2}}.$$

A straightforward computation shows that

$$\int_0^{2\pi} e^{\imath n x} e^{-\imath m x} \, dx = \begin{cases} 0 & \text{if } n \neq m \\ 2\pi & \text{if } n = m. \end{cases}$$

This means that the countable collection of functions

$$e^{inx} = \cos nx + i \sin nx, \quad n = 0, \pm 1, \pm 2, \pm 3, \ldots,$$

is an orthogonal subset of $L_2[0, 2\pi]$. Therefore, the collection

$$\left\{ \frac{1}{\sqrt{2\pi}} e^{inx} : \ n = 0, \pm 1, \pm 2, \pm 3, \ldots \right\}$$

is an orthonormal set of functions in $L_2[0, 2\pi]$. However, for historical reasons, these functions are normally used in their non-normalized form. We shall establish here that the orthogonal set

$$\{ e^{inx} : \ n = 0, \pm 1, \pm 2, \ldots \}$$

is, in fact, a complete orthogonal set—and so $\{ \frac{1}{\sqrt{2\pi}} e^{inx} : n = 0, \pm 1, \pm 2, \ldots \}$ is an orthonormal basis for the Hilbert space $L_2[0, 2\pi]$.

A **trigonometric polynomial** is any periodic function P of the form

$$P(x) = \frac{1}{2}a_0 + \sum_{n=1}^{m}(a_n \cos nx + b_n \sin nx) \tag{1}$$

where $a_0, a_1, \ldots, a_n$ and $b_1, \ldots, b_n$ are complex numbers. From $e^{ix} = \cos x + i \sin x$ it follows that

$$\sin nx = \frac{e^{inx} - e^{-inx}}{2i} \quad \text{and} \quad \cos nx = \frac{e^{inx} + e^{-inx}}{2},$$

and so every trigonometric polynomial P can also be written in the form

$$P(x) = \sum_{n=-m}^{m} c_n e^{inx} \tag{2}$$

for appropriate complex coefficients c_n. Similarly, using the identities $e^{inx} = \cos nx + i \sin nx$, $\cos(-nx) = \cos nx$, and $\sin(-nx) = -\sin nx$, we see that every expression $P(x)$ as in (2) can be written as in (1). Thus, we have shown that every trigonometric polynomial P can be written in the following two ways:

$$P(x) = \frac{1}{2}a_0 + \sum_{n=1}^{m}(a_n \cos nx + b_n \sin nx) = \sum_{n=-m}^{m} c_n e^{inx},$$

where the coefficients are related by the identities

$$a_0 = 2c_0, \quad a_n = c_n - c_{-n}, \quad b_n = i(c_n - c_{-n})$$

and

$$c_0 = \frac{1}{2}a_0, \quad c_n = \frac{1}{2}(a_n - \imath b_n), \quad c_{-n} = \frac{1}{2}(a_n + \imath b_n).$$

Now, taking the inner product of a trigonometric polynomial P with the function $e^{\imath kn}$, it follows from the orthogonality properties that

$$c_n = \frac{1}{2\pi} \int_0^{2\pi} P(x)e^{-\imath nx} \, dx, \quad n = 0, \pm 1, \pm 2 \ldots, \pm m.$$

Next, assume that $f \in L_1[0, 2\pi]$ (nothing precludes f from being also complex-valued). Then the inequality $|f(x)e^{\imath nx}| \le |f(x)|$ coupled with Theorem 22.6 guarantees that $f(x)e^{\imath nx} \in L_1[0, 2\pi]$ for each integer n. This observation will be used in the next definition.

Definition 35.1. *The **Fourier coefficients** of a function $f \in L_1[0, 2\pi]$ are the terms of the double sequence of complex numbers $\ldots, c_{-2}, c_{-1}, c_0, c_1, c_2, \ldots$ defined by*

$$c_n = \frac{1}{2\pi} \int_0^{2\pi} f(x)e^{-\imath nx} \, dx$$

*for each $n = 0, \pm 1, \pm 2, \ldots$. The **Fourier series** of a function $f \in L_1[0, 2\pi]$ is the formal series*

$$\sum_{n=-\infty}^{\infty} c_n e^{\imath nx}.$$

An important consequence of the Riemann–Lebesgue lemma (Theorem 25.4) is that the Fourier coefficients of an integrable function converge to zero.

Theorem 35.2. *If $f \in L_1[0, 2\pi]$, then its Fourier coefficients satisfy*

$$\lim_{n \to \infty} c_n = \lim_{n \to \infty} c_{-n} = 0.$$

Proof. Let $f \in L_1[0, 2\pi]$. Then, by the Riemann–Lebesgue lemma (Theorem 25.4), we have

$$\lim_{n \to \infty} \int_0^{2\pi} f(x) \sin nx \, dx = \lim_{n \to \infty} \int_0^{2\pi} f(x) \cos nx \, dx = 0.$$

So, from the identity

$$c_n = \frac{1}{2\pi} \int_0^{2\pi} f(x) e^{-\imath nx}\, dx = \frac{1}{2\pi} \int_0^{2\pi} f(x)[\cos nx - \imath \sin nx]\, dx$$

$$= \frac{1}{2\pi} \int_0^{2\pi} f(x) \cos nx\, dx - \frac{\imath}{2\pi} \int_0^{2\pi} f(x) \sin nx\, dx,$$

it easily follows that $c_n \to 0$ and $c_{-n} \to 0$.   ∎

Using that the set of functions $\{e^{\imath nx}: n = 0, \pm1, \pm2, \ldots\}$ is orthogonal and the formulas

$$\cos nx = \frac{e^{\imath nx} + e^{-\imath nx}}{2} \quad \text{and} \quad \sin nx = \frac{e^{\imath nx} - e^{-\imath nx}}{2\imath},$$

we see that the functions

$$\tfrac{1}{2}, \cos x, \sin x, \cos 2x, \sin 2x, \ldots, \cos nx, \sin nx, \ldots$$

are mutually orthogonal in $L_2[0, 2\pi]$. With respect to this orthogonal set of functions, the Fourier series of any function $f \in L_1[0, 2\pi]$ can also be written as

$$\sum_{n=-\infty}^{\infty} c_n e^{\imath nx} = \frac{1}{2} a_0 + \sum_{n=1}^{\infty} (a_n \cos nx + b_n \sin nx),$$

where the coefficients $a_0, a_1, a_2, \ldots$ and $b_1, b_2, \ldots$ are given by

$$a_0 = 2c_0 = \frac{1}{\pi} \int_0^{2\pi} f(x)\, dx,$$

$$a_n = c_n + c_{-n} = \frac{1}{\pi} \int_0^{2\pi} f(x) \cos nx\, dx, \quad \text{and}$$

$$b_n = \imath(c_n - c_{-n}) = \frac{1}{\pi} \int_0^{2\pi} f(x) \sin nx\, dx.$$

If we express the Fourier series of a function in terms of the cosine and sine functions, then we shall also refer to the coefficients $a_0, a_1, a_2, \ldots$ and $b_1, b_2, \ldots$ as the **Fourier coefficients** of f.

We are now ready to establish a result that will guarantee that the set of orthogonal functions $\{e^{\imath nx}: n = 0, \pm1, \pm2, \ldots\}$ is complete in $L_2[0, 2\pi]$.

Theorem 35.3. *A Lebesgue integrable function over* $[0, 2\pi]$ *whose Fourier coefficients all vanish is almost everywhere zero.*

Proof. Let $f \in L_1[0, 2\pi]$ be a function satisfying

$$\int_0^{2\pi} f(x)e^{inx}\, dx = 0 \tag{†}$$

for all $n = 0, \pm 1, \pm 2, \ldots$. Assume at the beginning that f is also real valued. We shall establish that $f = 0$ a.e. holds true by considering two cases. First, we shall assume that f is also continuous and then we shall consider f to be an arbitrary integrable function.

CASE I: *Assume that f is a continuous function satisfying* (†) *for all integers n.*

Assume by way of contradiction that $f \neq 0$. This implies that there exists some $x_0 \in (0, 2\pi)$ satisfying $f(x_0) \neq 0$. Replacing f by $-f$ (if necessary), we can assume that $f(x_0) > 0$. Choose some constant $C > 0$ and some $0 < \epsilon < \min\{1, x_0, 2\pi - x_0\}$ such that $f(x) > C$ holds for all $x \in I = (x_0 - \epsilon, x_0 + \epsilon)$.

Now, we need a trigonometric polynomial that is greater than one on the interval $(x_0 - \epsilon, x_0 + \epsilon)$ and less than one on the complement $I^c = [0, 2\pi] \setminus I$. The reader should verify that the trigonometric polynomial

$$P(x) = \frac{1 + \cos(x - x_0)}{1 + \cos \epsilon}.$$

satisfies $P(x) > 1$ for all $x \in I$ and $0 \leq P(x) \leq 1$ on I^c. From (†), it follows that

$$\int_I f(x)[P(x)]^n\, dx + \int_{I^c} f(x)[P(x)]^n\, dx = \int_0^{2\pi} f(x)[P(x)]^n\, dx = 0$$

for $n = 0, 1, 2, 3, \ldots$. Therefore,

$$\int_I f(x)[P(x)]^n\, dx = -\int_{I^c} f(x)[P(x)]^n\, dx \tag{$\star$}$$

holds for all $n = 0, 1, 2, 3, \ldots$.

We claim that the right-hand side of equation ($\star$) is bounded while the left-hand side is unbounded. To see this, first observe that since $0 \leq P(x) \leq 1$ holds for each $x \in I^c$, we have

$$\left| \int_{I^c} f(x)[P(x)]^n\, dx \right| \leq \int_{I^c} |f(x)[P(x)]^n|\, dx \leq \int_0^{2\pi} |f(x)|\, dx.$$

Thus, the right-hand side of $(\star)$ is bounded. On the other hand, let J be a closed subinterval of I. Since P is strictly greater than one on the open interval I, it follows that there exists some $\alpha > 1$ satisfying $P(x) \geq \alpha > 1$ for each $x \in J$. Therefore, for each n we have

$$\int_I f(x)[P(x)]^n \, dx \geq \int_J f(x)[P(x)]^n \, dx \geq C\alpha^n \lambda(J).$$

This shows that the left-hand side of equation $(\star)$ is unbounded, a contradiction. This contradiction establishes that $f = 0$.

CASE II: *Assume that $f \in L_1[0, 2\pi]$ satisfies* (†) *for all integers n.*

Consider the function $F: [0, 2\pi] \to \mathbb{R}$ defined by

$$F(x) = \int_0^x f(t) \, dt.$$

Clearly, F is a continuous function and from $F(0) = 0$ and $F(2\pi) = \int_0^{2\pi} f(t) \, dt = 0$, it also follows that F is a periodic function. We claim that $F \perp e^{\imath n x}$ holds for all $n \neq 0$. To see this, apply the integration by parts formula (see Exercise 12 of Section 31) to get

$$0 = \int_0^{2\pi} e^{-\imath n x} f(x) \, dx = e^{-\imath n x} F(x) \big|_0^{2\pi} - \int_0^{2\pi} (-\imath n) e^{-\imath n x} F(x) \, dx$$

$$= F(0) - F(2\pi) - \imath n \int_0^{2\pi} F(x) e^{-\imath n x} \, dx$$

$$= -\imath n \int_0^{2\pi} F(x) e^{-\imath n x} \, dx.$$

Therefore,

$$\int_0^{2\pi} F(x) e^{-\imath n x} \, dx = 0 \text{ for } n = \pm 1, \pm 2, \ldots.$$

Next, let $C = \frac{1}{2\pi} \int_0^{2\pi} F(x) \, dx$, and consider the continuous function $G: [0, 2\pi] \to \mathbb{R}$ defined by

$$G(x) = F(x) - C.$$

Since $1 \perp e^{\imath n x}$, it follows that G is perpendicular to $e^{\imath n x}$ for all $n \neq 0$. Any easy inspection also shows that $G(x) \perp 1$ holds. That is, $G \perp e^{\imath n x}$ holds for all integers

$n = 0, \pm 1, \pm 2, \pm 3, \ldots$. But then, by CASE I, $G(x) = 0$ for all $x \in [0, 2\pi]$. This implies $F(x) = \int_0^x f(t)\,dt = C$ for all $x \in [0, 2\pi]$. Since $F(0) = 0$, it follows that $C = 0$, and so $\int_0^x f(t)\,dt = 0$ for all $x \in [0, 2\pi]$. This implies that $f = 0$ a.e. (see Exercise 19 of Section 22), and the proof of CASE II is complete.

Finally, consider the complex case. That is, assume that $f = g + \imath h$ is an integrable function whose Fourier coefficients all vanish. Then, it easily follows that both g and h have their Fourier coefficients all equal to zero. Therefore, by the preceding conclusion, $g = h = 0$ a.e. This implies $f = 0$ a.e., and the proof of the theorem is complete. ∎

Corollary 35.4. *The two sets of orthogonal functions*

1. $\{e^{\imath nx}: n = 0, \pm 1, \pm 2, \ldots\}$, *and*
2. $\{\frac{1}{2}, \cos x, \sin x, \cos 2x, \sin 2x, \ldots, \cos nx, \sin nx, \ldots\}$

are both complete in the Hilbert space $L_2[0, 2\pi]$.

Proof. By Theorem 34.2, it suffices to show that if $f \in L_2[0, 2\pi]$ is perpendicular to $e^{\imath nx}$ for each integer n, then $f = 0$. So, assume that $f \in L_2[0, 2\pi]$ satisfies $f \perp e^{\imath nx}$ for each integer n, i.e., assume that the Fourier coefficients of f are all zero. Since $L_2[0, 2\pi] \subseteq L_1[0, 2\pi]$ holds, it follows from Theorem 35.3 that $f = 0$. Therefore, the set of functions $\{e^{\imath nx}: n = 0, \pm 1, \pm 2, \ldots\}$ is a complete orthogonal subset of the Hilbert space $L_2[0, 2\pi]$, and the proof is complete. ∎

Parseval's identity for the complete sequence of orthogonal functions $\{e^{-\imath nx}\}$ takes the following form:

Corollary 35.5. *If $f \in L_2[0, 2\pi]$, then its Fourier series $\sum_{n=-\infty}^{\infty} c_n e^{-\imath nx}$ is norm convergent to f in $L_2[0, 2\pi]$ and*

$$\frac{1}{2\pi} \int_0^{2\pi} |f(x)|^2\,dx = \sum_{n=-\infty}^{\infty} |c_n|^2.$$

Proof. From Corollary 35.4, we know that the collection of functions

$$\left\{ \frac{1}{\sqrt{2\pi}} e^{-\imath nx}: n = 0, \pm 1, \pm 2, \ldots \right\}$$

is an orthonormal basis for the Hilbert space $L_2[0, 2\pi]$. The conclusions now follow from Theorem 34.2. ∎

Corollary 35.6. *If f belongs to $L_2[0, 2\pi]$, then its Fourier series*

$$\sum_{n=-\infty}^{\infty} c_n e^{-\imath nx} = \frac{a_0}{2} + \sum_{n=1}^{\infty} (a_n \cos nx + b_n \sin nx)$$

can be integrated term-by-term in any closed subinterval $[u, v]$ *of* $[0, 2\pi]$ *in the sense that*

$$\int_u^v f(x)\, dx = \frac{1}{2}a_0(v - u) + \sum_{n=1}^{\infty} \int_u^v (a_n \cos nx + b_n \sin nx)\, dx.$$

Proof. Let $s_n(x) = \frac{a_0}{2} + \sum_{k=1}^{n}(a_n \cos kx + b_n \sin kx)$. From Theorem 34.2, we also know that $\| f - s_n \| \to 0$. Now, if $[u, v]$ is a closed subinterval of $[0, 2\pi]$, then

$$\left| \int_u^v f(x)\, dx - \int_u^v s_n(x)\, dx \right| \leq \int_u^v |f(x) - s_n(x)|\, dx$$

$$\leq \left(\int_u^v |f(x) - s_n(x)|^2\, dx \right)^{\frac{1}{2}} \left(\int_u^v 1^2\, dx \right)^{\frac{1}{2}}$$

$$= \| f - s_n \| \sqrt{v - u}.$$

This implies

$$\frac{1}{2}a_0(v - u) + \sum_{k=1}^{n} \int_u^v (a_n \cos nx + b_n \sin nx)\, dx = \int_u^v s_n(x)\, dx \longrightarrow \int_u^v f(x)\, dx,$$

and the desired conclusion follows.   ∎

The fact that the Fourier series of a function $f \in L_2[0, 2\pi]$ is norm convergent in the Hilbert space $L_2[0, 2\pi]$ should not be interpreted to mean that the partial sums of the Fourier series converge pointwise to f almost everywhere. Pointwise convergence of the Fourier series is a much more difficult and delicate problem. The study of the convergence of the Fourier series is the subject of inquiry of the field of Fourier analysis with special emphasis on the following two basic problems:

1. **The Convergence Problem:** *When does the Fourier series converge at a given point x (or at each point of a given set)?*
2. **The Representation Problem:** *If the Fourier series converges at some point x, what is the relationship between the sum of the series and the value of the function $f(x)$ at the point x?*

We remark that it is possible for the Fourier series to diverge at every point and, moreover, there exist continuous functions whose Fourier series diverge on an uncountable set. It is even possible to construct examples of continuous periodic functions whose Fourier series diverge at a countable number of points in the interval $[0, 2\pi]$; see also Example 35.11 at the end of this section. The most striking result concerning the pointwise convergence of the Fourier series was established by the Swedish mathematician Lennart Carleson in 1966. He proved that the Fourier series of any function in $L_2[0, 2\pi]$ converges almost everywhere.[12]

The unpleasant situation regarding the pointwise convergence of the Fourier series can be corrected by considering the sequence of arithmetic means of the partial sums of the Fourier series. Our final objective in this section is to prove that the sequence of arithmetic means of the partial sums of the Fourier series of a periodic function f converges to the value of the function at every point of continuity. To do this, we need some preliminary discussion.

For the rest of our discussion, $f : [0, 2\pi] \to \mathbb{C}$ is a fixed periodic integrable function which will also be assumed defined on all of $\mathbb{R}$ via the formula $f(x + 2\pi) = f(x)$. Its Fourier series is

$$\sum_{n=-\infty}^{\infty} c_n e^{-inx} = \frac{a_0}{2} + \sum_{n=1}^{\infty} (a_n \cos nx + b_n \sin nx).$$

The nth partial sum of the Fourier series is

$$
\begin{aligned}
s_n(x) &= \frac{1}{2} a_0 + \sum_{k=1}^{n} (a_k \cos kx + b_k \sin kx) \\
&= \frac{1}{2} \left[\frac{1}{\pi} \int_0^{2\pi} f(t) \, dt \right] + \sum_{k=1}^{n} \left[\frac{1}{\pi} \int_0^{2\pi} f(t) \cos(kt) \, dt \right] \cos kx \\
&\quad + \sum_{k=1}^{n} \left[\frac{1}{\pi} \int_0^{2\pi} f(t) \sin(kt) \, dt \right] \sin kx \\
&= \frac{1}{\pi} \int_0^{2\pi} f(t) \left[\frac{1}{2} + \sum_{k=1}^{n} (\cos kx \cos kt + \sin kx \sin kt) \right] dt \\
&= \frac{1}{\pi} \int_0^{2\pi} f(t) \left[\frac{1}{2} + \sum_{k=1}^{n} \cos k(x - t) \right] dt \\
&= \frac{1}{\pi} \int_0^{2\pi} f(t) D_n(x - t) \, dt,
\end{aligned}
$$

where

$$D_n(t) = \frac{1}{2} + \sum_{k=1}^{n} \cos kt.$$

---

[12] On the convergence and growth of partial sums of Fourier series, *Acta Mathematica* **116** (1966), 135–157.

The trigonometric polynomial D_n is called the **Dirichlet kernel** of order n. The Dirichlet kernel can be expressed in a closed form.

Lemma 35.7. *The Dirichlet kernel satisfies*

$$D_n(t) = \frac{\sin\left(n + \frac{1}{2}\right)t}{2\sin\frac{t}{2}}.$$

Proof. Notice that

$$
\begin{aligned}
D_n(t) &= \frac{1}{2} + \sum_{k=1}^{n} \cos kt = \frac{1}{2}\sum_{k=-n}^{n} e^{ikt} = \frac{1}{2}e^{-int}\sum_{k=0}^{2n} e^{ikt} \\
&= \frac{1}{2}e^{-int}\frac{[e^{i(2n+1)t} - 1]}{e^{it} - 1} = \frac{e^{i(n+1)t} - e^{-int}}{2e^{i\frac{t}{2}}(e^{i\frac{t}{2}} - e^{-i\frac{t}{2}})} \\
&= \frac{e^{i(n+\frac{1}{2})t} - e^{-i(n+\frac{1}{2})t}}{2(e^{i\frac{t}{2}} - e^{-i\frac{t}{2}})} \\
&= \frac{\sin\left(n + \frac{1}{2}\right)t}{2\sin\frac{t}{2}},
\end{aligned}
$$

as desired. ∎

We now consider the sequence $\{\sigma_n(x)\}$ of the arithmetic means of the partial sums of the Fourier series of the function f. That is, $s_0(x) = \frac{1}{2}a_0$ and

$$\sigma_n(x) = \frac{s_0(x) + s_1(x) + s_2(x) + \cdots + s_n(x)}{n + 1}, \quad \text{for } n = 0, 1, 2\ldots.$$

Using the Dirichlet kernel and some algebraic simplifications, we see that

$$
\begin{aligned}
\sigma_n(x) &= \frac{1}{n+1}\left[\frac{1}{\pi}\int_0^{2\pi} f(t)D_0(x-t)\,dt + \frac{1}{\pi}\int_0^{2\pi} f(t)D_1(x-t)\,dt + \cdots \right. \\
&\qquad \left. \cdots + \frac{1}{\pi}\int_0^{2\pi} f(t)D_n(x-t)\,dt\right] \\
&= \frac{1}{\pi}\int_0^{2\pi} f(t)\left[\frac{1}{n+1}\sum_{k=0}^{n} D_k(x-t)\right]dt \\
&= \frac{1}{\pi}\int_0^{2\pi} f(t)K_n(x-t)\,dt,
\end{aligned}
$$

where the expression

$$K_n(t) = \frac{1}{n+1} \sum_{k=0}^{n} D_k(t) = \frac{1}{2(n+1)\sin\frac{t}{2}} \sum_{k=0}^{n} \sin\left(k + \frac{1}{2}\right)t$$

is now called the **Fejér**[13] **kernel** of order n. Thus, the sequence $\{\sigma_n\}$ of means of the partial sums of the Fourier series of f are given by the formula

$$\sigma_n(x) = \frac{1}{\pi} \int_0^{2\pi} f(t) K_n(x - t)\, dt.$$

It is not difficult to express the Fejér kernel in a close form. Multiplying the numerator and the denominator of $K_n(t)$ by $2\sin\frac{t}{2}$ and using the trigonometric identity $2\sin(k + \frac{1}{2}t)\sin\frac{1}{2}t = \cos kt - \cos(k+1)t$, we get

$$K_n(t) = \frac{1 - \cos(n+1)t}{4(n+1)\sin^2\frac{t}{2}} = \frac{1}{2(n+1)}\left[\frac{\sin(n+1)\frac{t}{2}}{\sin\frac{t}{2}}\right]^2.$$

This formula shows that the Fejér kernel is a positive even function. If, as usual, we put $K_n(2k\pi) = \frac{n+1}{2}$, then K_n is a continuous function on all of $\mathbb{R}$. A simple computation (see Exercise 2 at the end of this section) also shows that

$$\frac{2}{\pi} \int_0^{\pi} K_n(t)\, dt = 1. \tag{$\star$}$$

Now, let us take another look at the nth arithmetic mean of the partial sums of the Fourier series. By the preceding discussion, we have

$$\sigma_n(x) = \frac{1}{\pi} \int_0^{2\pi} f(t) K_n(x - t)\, dt = \frac{1}{\pi} \int_{x-\pi}^{x+\pi} f(x + u) K_n(u)\, du$$

$$= \frac{1}{\pi} \int_{-\pi}^{\pi} f(x + u) K_n(u)\, du.$$

Taking into account that $K_n(-u) = K_n(u)$, it easily follows that

$$\int_{-\pi}^{0} f(x + u) K_n(u)\, du = \int_0^{\pi} f(x - u) K_n(u)\, du,$$

---

[13]Leopold Fejér (1880–1959), a Hungarian mathematician. He worked mainly on Fourier series and their singularities.

and consequently,

$$\sigma_n(x) = \frac{2}{\pi} \int_0^\pi \left[\frac{f(x+u) + f(x-u)}{2} \right] K_n(u)\, du. \qquad (\star\star)$$

Next, we introduce a new function σ_f defined by the formula

$$\sigma_f(x) = \lim_{u \to 0^+} \frac{f(x+u) + f(x-u)}{2}.$$

The function σ_f is, of course, defined at those points for which the preceding limit exists. For instance, if f is continuous at some point x, then $\sigma_f(x) = f(x)$ and if f has a jump discontinuity at x_0, then $\sigma_f(x_0) = \frac{f(x_0+) + f(x_0-)}{2}$.

We are now ready to state and proof the following remarkable result of L. Fejér concerning the convergence of the sequence of arithmetic means $\{\sigma_n\}$.

Theorem 35.8 (Fejér). *Let $f: [0, 2\pi] \to \mathbf{C}$ be a periodic integrable function. Then, the sequence $\{\sigma_n\}$ of arithmetic means of the partial sums of the Fourier series of f satisfies*

$$\lim_{n \to \infty} \sigma_n(x) = \sigma_f(x) = \lim_{u \to 0^+} \frac{f(x+u) + f(x-u)}{2}$$

at every point x for which $\sigma_f(x)$ exists. Moreover, if f is continuous on some closed subinterval $[a, b]$ of $[0, 2\pi]$, then $\{\sigma_n\}$ converges uniformly to f on $[a, b]$.

Proof. Assume that $\sigma_f(x)$ exists at some point $x \in [0, 2\pi]$. Then there exists some $0 < \delta < \pi$ such that

$$\left| \frac{f(x+u) + f(x-u)}{2} - \sigma_f(x) \right| < \epsilon$$

for all $0 < u < \delta$. Notice that the δ depends upon the ϵ and the point x. If f is continuous on $[a, b]$, then notice that (in view of the uniform continuity of f) we can select the δ to be independent of $x \in [a, b]$. Now, from $(\star)$ and $(\star\star)$, it follows that

$$\begin{aligned}
|\sigma_n(x) - \sigma_f(x)| &= \frac{2}{\pi} \left| \int_0^\pi \left[\frac{f(x+u) + f(x-u)}{2} - \sigma_f(x) \right] K_n(u)\, du \right| \\
&\leq \frac{2}{\pi} \int_0^\delta \left| \frac{f(x+u) + f(x-u)}{2} - \sigma_f(x) \right| K_n(u)\, du \\
&\quad + \frac{2}{\pi} \int_\delta^\pi \left| \frac{f(x+u) + f(x-u)}{2} - \sigma_f(x) \right| K_n(u)\, du.
\end{aligned}$$

Now, observe that

$$\frac{2}{\pi} \int_0^\delta \left| \frac{f(x+u) + f(x-u)}{2} - \sigma_f(x) \right| K_n(u)\, du$$

$$\leq \frac{2}{\pi} \int_0^\delta \epsilon K_n(t)\, dt \leq \frac{2}{\pi} \int_0^\pi \epsilon K_n(t)\, dt = \epsilon.$$

Moreover, since $K_n(t) \leq \frac{1}{2(n+1)\sin^2 \frac{\delta}{2}}$ for each $\delta \leq x \leq \pi$, it follows that

$$\frac{2}{\pi} \int_\delta^\pi \left| \frac{f(x+u) + f(x-u)}{2} - \sigma_f(x) \right| K_n(u)\, du$$

$$\leq \frac{1}{(n+1)\pi \sin^2 \frac{\delta}{2}} \int_0^\pi \left| \frac{f(x+u) + f(x-u)}{2} - \sigma_f(x) \right| du \leq \frac{C}{n+1},$$

where the constant $C = \frac{1}{\pi \sin^2 \frac{\delta}{2}} \int_0^{2\pi} \left| \frac{f(x+u)+f(x-u)}{2} - \sigma_f(x) \right| du$ is independent of x.

Finally, notice that if we choose some n_0 such that $\frac{C}{n+1} < \epsilon$, then the preceding estimates show that

$$|\sigma_n(x) - \sigma_f(x)| < \epsilon + \epsilon = 2\epsilon$$

holds true for all $n \geq n_0$ (and all $x \in [a, b]$ if f is continuous on $[a, b]$). This completes the proof of the theorem. ∎

The following consequence of the preceding theorem plays an important role in applications.

Corollary 35.9. *Let $f : [0, 2\pi] \to \mathbf{C}$ be a periodic integrable function. Assume that the Fourier series of f converges at some point x and that $\sigma_f(x)$ exists, then*

$$\sigma_f(x) = \sum_{n=-\infty}^\infty c_n e^{-inx} = \frac{a_0}{2} + \sum_{n=1}^\infty (a_n \cos nx + b_n \sin nx).$$

In particular, if the Fourier series converges at some point of continuity x of f, then

$$f(x) = \sum_{n=-\infty}^\infty c_n e^{-inx} = \frac{a_0}{2} + \sum_{n=1}^\infty (a_n \cos nx + b_n \sin nx).$$

Proof. Assume that $s_n(x) \to s$ holds and that $\sigma_f(x)$ exists. By Theorem 35.8, we know that $\sigma_n(x) \to \sigma_f(x)$. Since $\{\sigma_n(x)\}$ is also convergent to s (see Exercise 11 of Section 4), it follows that $s = \sigma_f(x)$, and we are done. ∎

Here is an example that demonstrates the far-reaching implications of Corollary 35.9.

Example 35.10.   Consider the periodic function $f: [0, 2\pi] \to \mathbb{R}$ defined by

$$f(x) = \begin{cases} 1 & \text{if } 0 \leq x < \pi \\ 0 & \text{if } \pi \leq x < 2\pi \\ 1 & \text{if } x = 2\pi. \end{cases}$$

A direct computation shows that the Fourier coefficients of f are given by

$$a_0 = \frac{1}{\pi} \int_0^{2\pi} f(x)\, dx = \frac{1}{\pi} \int_0^{\pi} 1\, dx = 1,$$

$$a_n = \frac{1}{\pi} \int_0^{2\pi} f(x) \cos nx\, dx = \frac{1}{\pi} \int_0^{\pi} \cos nx\, dx = 0, \text{ and}$$

$$b_n = \frac{1}{\pi} \int_0^{2\pi} f(x) \sin nx\, dx = \frac{1}{\pi} \int_0^{\pi} \sin nx\, dx = \frac{1 - \cos n\pi}{n\pi} = \frac{1 - (-1)^n}{n\pi}.$$

Thus, the Fourier coefficients satisfy $a_0 = 1$, $a_n = 0$ for each n, $b_n = \frac{2}{n\pi}$ for n odd and $b_n = 0$ for n even. Therefore, the Fourier series of f is given by

$$\frac{1}{2} + \frac{2}{\pi} \left(\frac{\sin x}{1} + \frac{\sin 3x}{3} + \frac{\sin 5x}{5} + \frac{\sin 7x}{7} + \cdots \right).$$

This series converges at each x; see Example 9.7. By Corollary 35.9, this Fourier series converges to $f(x)$ at the points x of continuity of f. More precisely, we have

$$\frac{1}{2} + \frac{2}{\pi} \left(\frac{\sin x}{1} + \frac{\sin 3x}{3} + \frac{\sin 5x}{5} + \frac{\sin 7x}{7} + \cdots \right) = \begin{cases} f(x) & \text{if } x \in (0, \pi) \cup (\pi, 2\pi), \\ \frac{1}{2} & \text{if } x = 0, \pi, 2\pi. \end{cases}$$

This conclusion is typical and very powerful for applications.   ∎

Finally, we close the section with an example that guarantees the existence of a continuous periodic function whose Fourier series diverges on a given countable set.

Example 35.11.   This example demonstrates the existence of a continuous periodic function whose Fourier series diverges on a countable set of points. Consider the Banach space $C[0, 2\pi]$ of all continuous real-valued functions with the sup norm. Let X be the subspace of $C[0, 2\pi]$ consisting of all continuous periodic real-valued functions. Clearly, X is a closed subspace of $C[0, 2\pi]$, and hence, it is a Banach space in its own right.

Next, fix a point $x \in [0, 2\pi]$ and define the linear functional $S_n: X \to \mathbb{R}$ by the nth partial sum of the Fourier series of the given function evaluated at the point x. That is, for each $f \in X$ we let

$$S_n(f) = \sum_{k=-n}^{n} c_k e^{-ikx} = \frac{1}{\pi} \int_0^{2\pi} f(t) D_n(x - t) \, dt.$$

It should be clear that this formula defines a bounded linear functional on X whose norm (according to Exercise 3 at the end of the section) is given by the formula

$$\|S_n\| = \frac{1}{\pi} \int_0^{2\pi} |D_n(x - t)| \, dt.$$

Next, we estimate this norm. Since the Dirichlet kernel D_n is periodic, it follows that

$$\int_0^{2\pi} |D_n(x - t)| \, dt = \int_0^{2\pi} |D_n(t)| \, dt.$$

So, using the inequality $|\sin u| \leq |u|$ and changing the variable, we get

$$\int_0^{2\pi} |D_n(t)| \, dt = \int_0^{2\pi} \left| \frac{\sin \left(n + \frac{1}{2} \right) t}{2 \sin \frac{t}{2}} \right| \, dt = \int_0^{\pi} \frac{|\sin(2n + 1)u|}{|\sin u|} \, du$$

$$\geq \int_0^{\pi} \frac{|\sin(2n + 1)u|}{|u|} \, du = \int_0^{(2n+1)\pi} \frac{|\sin u|}{|u|} \, du$$

$$= \sum_{k=0}^{2n} \int_{k\pi}^{(k+1)\pi} \frac{|\sin u|}{|u|} \, du$$

$$\geq \sum_{k=0}^{2n} \left[\frac{1}{(k + 1)\pi} \int_{k\pi}^{(k+1)\pi} |\sin u| \, du \right] = \sum_{k=0}^{2n} \frac{2}{(k + 1)\pi}.$$

This implies

$$\|S_n\| \geq \frac{2}{\pi^2} \sum_{k=0}^{2n} \frac{1}{k + 1},$$

and consequently $\lim \|S_n\| = \infty$ independently of the point x.

Now, let $\{x_1, x_2, x_3, \ldots\}$ be a (possibly dense) countable subset of the interval $[0, 2\pi]$. Let $S_{n,m}$ be the bounded linear functional (defined as above) on X whose value at $f \in X$ is the nth partial sum of the Fourier series of f evaluated at the point x_m. From the above discussion, for each fixed m we have

$$\lim_{n \to \infty} \|S_{n,m}\| = \infty.$$

Therefore, by the Principle of Condensation of Singularities (Theorem 28.11), there exists some $f \in X$ (i.e., a continuous periodic real-valued function f on $[0, 2\pi]$) satisfying

$$\limsup_{n \to \infty} |S_{n,m}(f)| = \limsup_{n \to \infty} \left| \sum_{k=-n}^{n} c_k e^{-ikx_m} \right| = \infty$$

for each m. This implies that the Fourier series of f diverges at each point x_m. ∎

EXERCISES

1. Show that $\sin^n x$ is a linear combination of

    $$\{1, \sin x, \cos x, \sin 2x, \cos 2x, \sin 3x, \cos 3x, \ldots, \sin nx, \cos nx\}.$$

    Furthermore, show that the coefficients of the cosine terms are zero when n is an odd integer, and the coefficients of the sine terms are zero when n is an even integer.

2. Show that the Dirichlet kernel D_n and the Fejér kernel K_n satisfy

    $$\frac{1}{\pi} \int_{-\pi}^{\pi} D_n(t)\,dt = \frac{1}{\pi} \int_{-\pi}^{\pi} K_n(t)\,dt = 1.$$

3. Show that the norm of the linear functional S_n defined in Example 35 satisfies

    $$\|S_n\| = \frac{1}{\pi} \int_0^{2\pi} |D_n(x - t)|\,dt.$$

4. Show that the sequence of functions

    $$\left\{ \left(\frac{1}{\pi}\right)^{\frac{1}{2}}, \left(\frac{2}{\pi}\right)^{\frac{1}{2}} \cos x, \left(\frac{2}{\pi}\right)^{\frac{1}{2}} \cos 2x, \left(\frac{2}{\pi}\right)^{\frac{1}{2}} \cos 3x, \left(\frac{2}{\pi}\right)^{\frac{1}{2}} \cos 4x, \ldots \right\}$$

    is an orthonormal basis in $L_2[0, \pi]$. Also show that the preceding sequence is an orthogonal sequence of functions in $L_2[0, 2\pi]$ which is not complete.

5. Show that the sequence of functions

    $$\left\{ \left(\frac{2}{\pi}\right)^{\frac{1}{2}} \sin x, \left(\frac{2}{\pi}\right)^{\frac{1}{2}} \sin 2x, \left(\frac{2}{\pi}\right)^{\frac{1}{2}} \sin 3x, \left(\frac{2}{\pi}\right)^{\frac{1}{2}} \sin 4x, \ldots \right\}$$

    is an orthonormal basis of $L_2[0, \pi]$. Also prove that this set of functions is an orthogonal set of functions in $L_2[0, 2\pi]$ which is not complete.

6. The original Weierstrass approximation theorem showed that every continuous function of period 2π can be uniformly approximated by trigonometric polynomials. Establish this result.

7. Find the Fourier coefficients of the function

    $$f(x) = \begin{cases} 1 & \text{if } 0 \le x < \frac{\pi}{2} \\ 0 & \text{if } \frac{\pi}{2} \le x < 2\pi. \end{cases}$$

8. Find the Fourier series of the function

$$f(x) = \begin{cases} \sin x & \text{if } 0 \le x < \pi \\ -\sin x & \text{if } \pi \le x < 2\pi. \end{cases}$$

9. Show that for each $0 < x < 2\pi$ we have

$$x = \pi - 2\sum_{n=1}^{\infty} \frac{\sin nx}{n}.$$

10. Show that

$$\frac{x^2}{2} = \pi x - \frac{\pi^2}{3} + 2\sum_{n=1}^{\infty} \frac{\cos nx}{n^2}$$

holds for all $0 \le x \le 2\pi$. (The value $x = 0$ yields $\sum_{n=1}^{\infty} \frac{1}{n^2} = \frac{\pi^2}{6}$.)

11. Show that

$$x^2 = \frac{4}{3}\pi^2 + 4\sum_{n=1}^{\infty} \left(\frac{\cos nx}{n^2} - \frac{\pi \sin nx}{n} \right)$$

holds for each $0 < x < 2\pi$.

12. Consider the "integral" operator $T: L_2[0, \pi] \to L_2[0, \pi]$ defined by

$$Tf(x) = \int_0^{\pi} K(x, t) f(t)\, dt,$$

where the kernel $K: [0, \pi] \times [0, \pi] \to \mathbb{R}$ is given by

$$K(x, t) = \sum_{n=1}^{\infty} \frac{[\sin(n+1)x]\sin nt}{n^2}.$$

Show that the norm of the operator T satisfies $\|T\| = \pi/2$.
[HINT: Use the basis described in Exercise 5.]

SPECIAL TOPICS IN INTEGRATION

The powerful techniques of measure and integration theory are utilized in many scientific contexts. However, in a number of applications, set functions that are not measures appear naturally. For this reason a study of more general set functions promises to be very fruitful. In this chapter the set functions known as "signed measures" will be investigated.

Loosely speaking, a signed measure is an extended real-valued σ-additive function on a σ-algebra of sets. The first section of the chapter deals with the algebraic and lattice structures of signed measures, while the following adheres to comparison properties of signed measures. The most important comparison properties are those of "absolute continuity" and "singularity." Regarding absolute continuity, the far-reaching classical Radon–Nikodym theorem is proven here: If a finite signed measure ν on a σ-algebra Σ is absolutely continuous with respect to a σ-finite measure μ, then there exists a (unique) μ-integrable function f such that

$$\nu(A) = \int_A f \, d\mu$$

holds for all $A \in \Sigma$. This powerful theorem is used to show that $L_p^*(\mu) = L_q(\mu)$ holds for each $1 < p < \infty$.

After comparing signed measures, our attention is turned to regular Borel measures on a Hausdorff locally compact topological space X. Another classical result, known as the "Riesz Representation Theorem," is proved: If F is a positive linear functional on $C_c(X)$, then there exists a unique regular Borel measure μ such that

$$F(f) = \int f \, d\mu$$

holds for all $f \in C_c(X)$. As an application of this theorem, the norm dual of $C_c(X)$ will be characterized in terms of regular Borel measures.

The last two sections of this book deal with differentiation and integration in $\mathbb{R}^n$. First, the study focuses on the differentiation of Borel signed measures on $\mathbb{R}^n$. It will be shown that every Borel signed measure is differentiable almost everywhere. This basic theorem will then be used effectively to derive the classical results about ordinary derivatives of functions of bounded variation. Finally, a detailed proof of the familiar "change of variables formula" will be presented.

36. SIGNED MEASURES

Throughout this section, Σ will be a fixed σ-algebra of subsets of a set X, and the measures considered will be assumed defined on Σ. Since a variety of measures will be studied, it is a custom (for simplicity) to call the members of Σ the measurable subsets of X.

Let μ and ν be two measures, and let $\alpha \geq 0$. Then the two set functions $\mu + \nu$ and $\alpha\mu$ defined by

$$(\mu + \nu)(A) = \mu(A) + \nu(A),$$
$$(\alpha\mu)(A) = \alpha\mu(A)$$

for each $A \in \Sigma$ are obviously measures. That is, the collection of all measures on Σ is closed under addition and also by multiplication by nonnegative scalars. Clearly, this collection cannot be a vector space since multiplication by -1 yields negative-valued set functions.

An order relation $\leq$ can be introduced among measures by letting $\mu \leq \nu$ whenever $\mu(A) \leq \nu(A)$ holds for each $A \in \Sigma$. The reader should stop and check that $\leq$ is indeed an order relation. Remarkably, the collection of all measures under this ordering is a lattice. That is, for every pair μ and ν of measures, the least upper bound $\mu \vee \nu$ and the greatest lower bound $\mu \wedge \nu$ exist. The details are included in the next theorem.

Theorem 36.1. *The collection of all measures on Σ forms a lattice, where for each pair of measures μ and ν the lattice operations are given by*

$$\mu \vee \nu(A) = \sup\{\mu(B) + \nu(A \setminus B): B \in \Sigma \text{ and } B \subseteq A\}, \text{ and}$$
$$\mu \wedge \nu(A) = \inf\{\mu(B) + \nu(A \setminus B): B \in \Sigma \text{ and } B \subseteq A\}$$

for each $A \in \Sigma$. Moreover,

$$\mu \wedge \nu + \mu \vee \nu = \mu + \nu.$$

Proof. Let μ and ν be a pair of measures on Σ. For each $A \in \Sigma$ define

$$\omega(A) = \sup\{\mu(B) + \nu(A \setminus B): B \in \Sigma \text{ and } B \subseteq A\}.$$

First, we shall verify that ω is a measure, and then, that it is the least upper bound of μ and ν.

Clearly, $\omega(A) \geq 0$ holds for each $A \in \Sigma$ and $\omega(\emptyset) = 0$. It remains to be shown that ω is σ-additive. To this end, let $\{A_n\}$ be a disjoint sequence of Σ and put $A = \bigcup_{n=1}^{\infty} A_n$.

If $B \in \Sigma$ satisfies $B \subseteq A$, then

$$\mu(B) + \nu(A \setminus B) = \mu\left(\bigcup_{n=1}^{\infty} A_n \cap B\right) + \nu\left(\bigcup_{n=1}^{\infty}(A_n \setminus B)\right)$$

$$= \sum_{n=1}^{\infty}[\mu(A_n \cap B) + \nu(A_n \setminus (A_n \cap B))]$$

$$\leq \sum_{n=1}^{\infty} \omega(A_n),$$

and so, $\omega(A) \leq \sum_{n=1}^{\infty} \omega(A_n)$ holds.

For the reverse inequality, note that if $\omega(A) = \infty$, then $\omega(A) = \sum_{n=1}^{\infty} \omega(A_n) = \infty$ is clearly true. Hence, assume $\omega(A) < \infty$, and let $\epsilon > 0$; clearly, $\omega(A_n) \leq \omega(A) < \infty$ holds for all n. Thus, for each n there exists some $B_n \in \Sigma$ with $B_n \subseteq A_n$ and $\mu(B_n) + \nu(A_n \setminus B_n) > \omega(A_n) - \epsilon 2^{-n}$. Obviously, $\{B_n\}$ is a disjoint sequence of Σ, and if $B = \bigcup_{n=1}^{\infty} B_n \subseteq A$, then $\bigcup_{n=1}^{\infty}(A_n \setminus B_n) = A \setminus B$ holds. Moreover,

$$\omega(A) > \mu(B) + \nu(A \setminus B) = \mu\left(\bigcup_{n=1}^{\infty} B_n\right) + \nu\left(\bigcup_{n=1}^{\infty} A_n \setminus B_n\right)$$

$$= \sum_{n=1}^{\infty}[\mu(B_n) + \nu(A_n \setminus B_n)] \geq \sum_{n=1}^{\infty}[\omega(A_n) - \epsilon 2^{-n}]$$

$$= \sum_{n=1}^{\infty} \omega(A_n) - \epsilon$$

holds for each $\epsilon > 0$. That is, $\omega(A) \geq \sum_{n=1}^{\infty} \omega(A_n)$ and so $\omega(A) = \sum_{n=1}^{\infty} \omega(A_n)$, as required.

It should be clear that $\mu \leq \omega$ and $\nu \leq \omega$ both hold. Now, assume that another measure π satisfies $\mu \leq \pi$ and $\nu \leq \pi$. Let $A \in \Sigma$. If $B \in \Sigma$ satisfies $B \subseteq A$, then

$$\mu(B) + \nu(A \setminus B) \leq \pi(B) + \pi(A \setminus B) = \pi(B \cup (A \setminus B)) = \pi(A),$$

and so, $\omega(A) \leq \pi(A)$ holds for each $A \in \Sigma$. That is, $\omega \leq \pi$, and therefore, ω is the lest upper bound of μ and ν. That is, $\omega = \mu \vee \nu$. The proof of the infimum parallels the preceding one and is left for the reader.

To see that $\mu \wedge \nu + \mu \vee \nu = \mu + \nu$ holds, let $A \in \Sigma$. Then for every $B \in \Sigma$ with $B \subseteq A$ we have

$$
\begin{aligned}
\mu(B) + \nu(A \setminus B) + \mu \vee \nu(A) &\geq \mu(B) + \nu(A \setminus B) + \mu(A \setminus B) + \nu(B) \\
&= \mu(A) + \nu(A), \\
\mu \wedge \nu(A) + \mu(B) + \nu(A \setminus B) &\leq \mu(A \setminus B) + \nu(B) + \mu(B) + \nu(A \setminus B) \\
&= \mu(A) + \nu(A).
\end{aligned}
$$

By taking the inf and sup over all measurable subsets B contained in A, the preceding inequalities yield $\mu \wedge \nu(A) + \mu \vee \nu(A) = \mu(A) + \nu(A)$, and the proof is complete. ∎

The formula $\mu \wedge \nu + \mu \vee \nu = \mu + \nu$ is reminiscent of the familiar identity in vector lattices.

The next result describes an order completeness property of the lattice of all measures.

Theorem 36.2. *If a sequence $\{\mu_n\}$ of measures satisfies $\mu_n \uparrow$ (i.e., $\mu_n \leq \mu_{n+1}$ for each n), then the set function $\mu \colon \Sigma \to [0, \infty]$ defined by $\mu(A) = \lim \mu_n(A)$ for each $A \in \Sigma$ is a measure. Moreover, $\mu_n \uparrow \mu$ holds; that is, μ is the least upper bound of the sequence $\{\mu_n\}$.*

Proof. Clearly, $0 \leq \mu(A) \leq \infty$ for each $A \in \Sigma$ and $\mu(\emptyset) = 0$ hold. Also, if $A \subseteq B$, then $\mu(A) \leq \mu(B)$ is obviously true.

For the σ-additivity, let $\{A_n\}$ be a disjoint sequence of Σ. Let $A = \bigcup_{n=1}^{\infty} A_n$. In view of $\mu_k(A) = \sum_{n=1}^{\infty} \mu_k(A_n) \leq \sum_{n=1}^{\infty} \mu(A_n)$ for each k, it follows that $\mu(A) \leq \sum_{n=1}^{\infty} \mu(A_n)$. On the other hand, for each k we have

$$\sum_{i=1}^{k} \mu(A_i) = \lim_{n \to \infty} \sum_{i=1}^{k} \mu_n(A_i) = \lim_{n \to \infty} \mu_n \left(\bigcup_{i=1}^{k} A_i \right) \leq \lim_{n \to \infty} \mu_n(A) = \mu(A),$$

and so, $\sum_{i=1}^{\infty} \mu(A_i) \leq \mu(A)$ also holds. Thus, $\mu(A) = \sum_{n=1}^{\infty} \mu(A_n)$, as required. The verification of $\mu_n \uparrow \mu$ is straightforward. ∎

As mentioned before, multiplication of a measure by -1 yields a negative-valued set function. For this reason, it is desirable to consider σ-additive set functions that also assume extended negative values. However, if we do this, we run immediately into trouble. Suppose that a set function $\mu: \Sigma \to \mathbb{R}^*$ satisfies $\mu(A) = \infty$ and $\mu(B) = -\infty$, with $A \cap B = \emptyset$. If μ is to be additive, then $\mu(A \cup B) = \mu(A) + \mu(B) = \infty - \infty$ must hold, and we face the problem of having to give a meaning to the expression $\infty - \infty$.

The preceding difficulty can be avoided by excluding from the range of the set function at least one of the infinite values. If this is assumed, then the form $\infty - \infty$ does not appear, and the additivity property does not cause problems. To be distinguished from a measure, such a σ-additive set function is usually referred to as a *signed measure*. Its precise definition follows.

Definition 36.3. *A set function* $\mu: \Sigma \to \mathbb{R}^*$ *is said to be a* **signed measure** *if it satisfies the following properties*:

a. μ *assumes at most one of the values* ∞ *and* $-\infty$,

b. $\mu(\emptyset) = 0$, *and*

c. μ *is* σ-*additive, that is, if* $\{A_n\}$ *is a disjoint sequence of members of* Σ, *then* $\mu(\bigcup_{n=1}^{\infty} A_n) = \sum_{n=1}^{\infty} \mu(A_n)$ *holds*.

If μ is a signed measure, and a disjoint sequence $\{A_n\}$ of Σ satisfies $|\mu(\bigcup_{n=1}^{\infty} A_n)| < \infty$, then (since every permutation of $\{A_n\}$ has the same union as the original sequence) it follows from (c) of the preceding definition that $\sum_{n=1}^{\infty} \mu(A_n)$ is re-arrangement invariant. Thus, $\sum_{n=1}^{\infty} |\mu(A_n)|$ is convergent in $\mathbb{R}$ (see Exercise 7 of Section 5).

Clearly, every measure is a signed measure. Also, (b) and (c) of Definition 36 together show that every signed measure is finitely additive. The next few results will reveal that signed measures behave in a manner similar to measures. The first one informs us that a signed measure is always subtractive.

Theorem 36.4. *Let* μ *be a signed measure on* Σ, *and let* $A \in \Sigma$ *be such that* $|\mu(A)| < \infty$. *If* $B \in \Sigma$ *satisfies* $B \subseteq A$, *then* $|\mu(B)| < \infty$ *and*

$$\mu(A \setminus B) = \mu(A) - \mu(B).$$

Proof. The identity $A = (A \setminus B) \cup B$ combined with the additivity of μ imply $\mu(A) = \mu(A \setminus B) + \mu(B)$. Since $\mu(A)$ is a real number and μ assumes at most one of the values ∞ and $-\infty$, it follows that both $\mu(A \setminus B)$ and $\mu(B)$ are real numbers. Now, the identity $\mu(A \setminus B) = \mu(A) - \mu(B)$ should be obvious. ∎

An immediate and useful conclusion of the preceding theorem is the following: If a set $A \in \Sigma$ has a measurable subset of infinite signed measure, then A itself has infinite signed measure.

The usual continuity properties of measures are inherited by signed measures.

Theorem 36.5. *For a signed measure μ and a sequence $\{A_n\}$ of Σ, the following statements hold:*

a.   *If $A_n \uparrow A$, then $\lim \mu(A_n) = \mu(A)$.*
b.   *If $A_n \downarrow A$ and $\mu(A_k)$ is a real number for at least one k, then*

$$\lim_{n \to \infty} \mu(A_n) = \mu(A).$$

Proof. Repeat the proof of Theorem 15.4 taking into consideration the conclusion of Theorem 36.4. ∎

A measurable set A is said to be a **positive set** with respect to a signed measure μ, in symbols $A \geq 0$, whenever $\mu(E \cap A) \geq 0$ holds for each $E \in \Sigma$. Equivalently, A set $A \in \Sigma$ is a positive set whenever $\mu(E) \geq 0$ holds true for all $E \in \Sigma$ with $E \subseteq A$.

Obviously, the empty set is a positive set. Also, it should be clear that any measurable subset of a positive set is likewise a positive set. In addition, any countable union of positive sets is a positive set. To see this, let $\{A_n\}$ be a sequence of positive sets and let $A = \bigcup_{n=1}^{\infty} A_n$. Now, let $B_1 = A_1$, $B_{n+1} = A_{n+1} \setminus \bigcup_{i=1}^{n} A_i$ for $n \geq 1$, and note that $\{B_n\}$ is a disjoint sequence such that $A = \bigcup_{n=1}^{\infty} B_n$. Since $B_n \subseteq A_n$ holds for all n, it follows that each B_n is a positive set. Therefore, for each measurable set E we have $\mu(E \cap A) = \mu(\bigcup_{n=1}^{\infty} E \cap B_n) = \sum_{n=1}^{\infty} \mu(E \cap B_n) \geq 0$, so that A is a positive set.

Similarly, a set A is called a **negative set** for a signed measure μ, in symbols $A \leq 0$, whenever $\mu(A \cap E) \leq 0$ holds for each $E \in \Sigma$. As before, measurable subsets of negative sets are negative, and countable unions of negative sets are likewise negative sets.

The next lemma is a basic result for this section and guarantees the existence of nonempty positive sets.

Lemma 36.6. *Let μ be a signed measure on Σ, and let $E \in \Sigma$ with $\mu(E) > 0$. Then there exists a positive set A such that $A \subseteq E$ and $\mu(A) > 0$.*

Proof. If for every measurable subset B of E we have $\mu(B) \geq 0$, then E is itself a positive set and there is nothing to prove. Thus, assume that there exists some $B \in \Sigma$ with $B \subseteq E$ and $\mu(B) < 0$.

By Zorn's lemma (how?) there exists a maximal collection $\mathcal{C}$ of mutually disjoint measurable subsets of E such that $\mu(C) < 0$ holds for each $C \in \mathcal{C}$. We claim that $\mathcal{C}$ is at-most countable. To see this, note first that $\mathcal{C} = \bigcup_{n=1}^{\infty} \mathcal{C}_n$, where $\mathcal{C}_n = \{C \in \mathcal{C}: \mu(C) < -\frac{1}{n}\}$. On the other hand, if some $\mathcal{C}_n$ is not finite, then it must contain a countable subset of $\mathcal{C}$, say $\{C_1, C_2, \ldots\}$. But then, the measurable set $C = \bigcup_{i=1}^{\infty} C_i$ satisfies $C \subseteq E$, and $\mu(C) = \sum_{i=1}^{\infty} \mu(C_i) = -\infty$. This implies $\mu(E) = \mu(E \setminus C) + \mu(C) = -\infty$, which is a contradiction. Thus, each C_n is finite, and so C is at most countable.

Hence, the set $D = \bigcup_{C \in \mathcal{C}} C$ belongs to Σ, and we claim that $A = E \setminus D$ is a positive set satisfying $\mu(A) > 0$. Indeed, since $0 < \mu(E) = \mu(A) + \mu(D)$ and $\mu(D) < 0$, it is easy to see that $\mu(A) > 0$ holds. On the other hand, if $\mu(F) < 0$ holds for some measurable subset F of A, then we can incorporate F into C and violate the maximality property of C. Thus, A is a positive set, and the proof is finished. ∎

If μ is a signed measure on Σ and two disjoint sets A and B satisfy $A \geq 0, B \leq 0$, and $A \cup B = X$, then the pair (A, B) is referred to as a **Hahn decomposition** of X with respect to μ. Loosely speaking, a Hahn decomposition is a splitting of the space X into two pieces, where μ is positive on one of the pieces and negative on the other. Such a splitting always exists and is "essentially" unique, as the next theorem shows.

Theorem 36.7. *Let μ be a signed measure on Σ. Then X has a Hahn decomposition with respect to μ. That is, there exist a positive set A and a negative set B such that $X = A \cup B$ and $A \cap B = \emptyset$.*

Moreover, if (A, B) and (A_1, B_1) are two Hahn decompositions of X with respect to μ, then

a. $\mu(A \triangle A_1) = \mu(B \triangle B_1) = 0$,

b. $\mu(E \cap A) = \mu(E \cap A_1)$, and

c. $\mu(E \cap B) = \mu(E \cap B_1)$

hold for each $E \in \Sigma$.

Proof. We can assume without loss of generality that $\mu(E) \neq \infty$ holds for each $E \in \Sigma$ (otherwise, replace μ by $-\mu$).

Put $a = \sup\{\mu(E): E \geq 0\}$ and note that $a \geq 0$. Choose a sequence $\{A_n\}$ of positive sets such that $\lim \mu(A_n) = a$. Then $A = \bigcup_{n=1}^{\infty} A_n$ is a positive set, and since $\mu(A_n) \leq \mu(A) \leq a$, it follows that $a = \mu(A) < \infty$.

Now, we claim that $B = X \setminus A$ is a negative set for μ. To see this, assume by way of contradiction that there exists some measurable subset C of B with $\mu(C) > 0$. Then, by Lemma 36.5, there exists a positive set E with $E \subseteq C$ and $\mu(E) > 0$. It follows that $A \cup E \geq 0$ and

$$a + \mu(E) = \mu(A) + \mu(E) = \mu(A \cup E) \leq a < \infty,$$

which is impossible. Thus, (A, B) is a Hahn decomposition of X with respect to μ.

For the "uniqueness" of the Hahn decomposition, let (A_1, B_1) be another Hahn decomposition of X. Since $A \setminus A_1 \subseteq A$ and $A \setminus A_1 = A \cap A_1^c = A \cap B_1 \subseteq B_1$, it follows that $\mu(A \setminus A_1) \geq 0$ and $\mu(A \setminus A_1) \leq 0$, that is, $\mu(A \setminus A_1) = 0$. Similarly, $\mu(A_1 \setminus A) = 0$, and so

$$\mu(A \triangle A_1) = \mu((A \setminus A_1) \cup (A_1 \setminus A)) = \mu(A \setminus A_1) + \mu(A_1 \setminus A) = 0.$$

Now, if $E \in \Sigma$, then

$$\begin{aligned}
\mu(E \cap A) &= \mu(E \cap [(A \setminus A_1) \cup (A \cap A_1)]) \\
&= \mu(E \cap (A \setminus A_1)) + \mu(E \cap A \cap A_1) \\
&= \mu(E \cap A \cap A_1) = \mu(E \cap (A_1 \setminus A)) + \mu(E \cap A \cap A_1) \\
&= \mu(E \cap [(A_1 \setminus A) \cup (A_1 \cap A)]) = \mu(E \cap A_1).
\end{aligned}$$

By the symmetry of the situation, the corresponding formulas for B and B_1 are also true, and the proof is finished. ∎

If (A, B) is a Hahn decomposition of X with respect to a signed measure μ, then μ and $-\mu$ restricted to A and B, respectively, are in actuality measures. As we shall see, these two measures determine the structure of μ.

Definition 36.8. *Let μ be a signed measure on Σ, and let (A, B) be a Hahn decomposition of X with respect to μ. Then the three set functions*

$$\begin{aligned}
\mu^+(E) &= \mu(E \cap A), \\
\mu^-(E) &= -\mu(E \cap B), \text{ and} \\
|\mu|(E) &= \mu(E \cap A) - \mu(E \cap B) = \mu^+(E) + \mu^-(E),
\end{aligned}$$

*for each $E \in \Sigma$ are called the **positive variation**, the **negative variation**, and the **total variation** of μ, respectively.*

A glance at Theorem 36.7 guarantees that the values of μ^+, μ^-, and $|\mu|$ are independent of the chosen Hahn decomposition. Also, it should be clear that μ^+, μ^-, and $|\mu|$ are measures on Σ.

A signed measure μ is said to be a **finite signed measure** if $\mu(A) \in \mathbb{R}$ holds for each $A \in \Sigma$. Note that in view of Theorem 36.4, a signed measure μ is finite if and only if $|\mu(X)| < \infty$ holds. Similarly the expression "μ is a finite measure" means that $0 \leq \mu(A) \leq \mu(X) < \infty$ holds for all $A \in \Sigma$. Finally, let us say (as usual) that a signed measure is σ-**finite** if there exists a disjoint sequence $\{A_n\}$ of Σ with $X = \bigcup_{n=1}^{\infty} A_n$ and $\mu(A_n) \in \mathbb{R}$ for each n.

From Definition 36.8 it follows that

$$\mu = \mu^+ - \mu^-$$

holds. This identity is known as the **Jordan**[1] **decomposition** of the signed measure μ. Moreover, if (A, B) is a Hahn decomposition, then in view of the inequalities $\mu^+(E) \leq \mu(A)$ and $\mu^-(E) \leq -\mu(B)$ for each $E \in \Sigma$, it follows that at least one of the measures μ^+ and μ^- is a finite measure. Thus, after all, the Jordan decomposition shows that each signed measure is the difference of two measures, at least one of which is a finite measure.

Some other useful expressions for the different variations of a signed measure are presented next.

Theorem 36.9. *Let μ be a signed measure on Σ. Then for every $E \in \Sigma$ the following formulas hold:*

1. $\mu^+(E) = \sup\{\mu(F) \colon F \in \Sigma \text{ and } F \subseteq E\}$.
2. $\mu^-(E) = \sup\{-\mu(F) \colon F \in \Sigma \text{ and } F \subseteq E\}$.
3. $|\mu|(E) = \sup\{\sum |\mu(F_i)| \colon \{F_i\} \subseteq \Sigma \text{ is finite and disjoint with } \bigcup F_i \subseteq E\}$.

Proof. Let (A, B) be a Hahn decomposition of X with respect to μ, and let $E \in \Sigma$.

(1) Clearly, $\mu^+(E) = \mu(E \cap A) \leq \sup\{\mu(F) \colon F \in \Sigma \text{ and } F \subseteq E\}$ holds. On the other hand, if $F \in \Sigma$ satisfies $F \subseteq E$, then

$$\mu(F) = \mu(F \cap A) + \mu(F \cap B) \leq \mu(F \cap A) \leq \mu(E \cap A) = \mu^+(E),$$

so that $\sup\{\mu(F) \colon F \in \Sigma \text{ and } F \subseteq E\} \leq \mu^+(E)$. Therefore, the identity in (1) holds.

(2) This follows from (1) by observing that $\mu^- = (-\mu)^+$.

(3) Let

$$\nu(E) =$$
$$\sup\left\{\sum |\mu(F_i)| \colon \{F_i\} \text{ is a finite disjoint collection of } \Sigma \text{ with } \bigcup F_i \subseteq E\right\}.$$

Then,

$$|\mu|(E) = \mu^+(E) + \mu^-(E) = \mu(E \cap A) - \mu(E \cap B)$$
$$= |\mu(E \cap A)| + |\mu(E \cap B)| \leq \nu(E).$$

---

[1]Camille Jordan (1838–1921), an eminent French mathematician. He was considered a "universal" mathematician because he published papers in virtually all areas of mathematics of his time. His main contributions were in group theory.

Now, let $\{F_1, \ldots, F_n\}$ be a finite disjoint collection of Σ such that $\bigcup_{i=1}^{n} F_i \subseteq E$. Then

$$\sum_{i=1}^{n} |\mu(F_i)| = \sum_{i=1}^{n} |\mu^+(F_i) - \mu^-(F_i)| \le \sum_{i=1}^{n} [\mu^+(F_i) + \mu^-(F_i)]$$

$$= \sum_{i=1}^{n} |\mu|(F_i) = |\mu| \left(\bigcup_{i=1}^{n} F_i \right) \le |\mu|(E).$$

This implies $\nu(E) \le |\mu|(E)$. Thus, $|\mu|(E) = \nu(E)$ holds, and the proof of the theorem is complete. ∎

It is easy to establish, but important to observe, the following inequality regarding the total variation:

$$|\mu(A)| \le |\mu|(A) \quad \text{for each} \quad A \in \Sigma.$$

Summarizing, the different variations of a signed measure μ satisfy the following identities:

a. $\mu = \mu^+ - \mu^-$ and $|\mu| = \mu^+ + \mu^-$.
b. $\mu^+ \wedge \mu^- = 0$.
c. $\mu^- = (-\mu)^+$.

They are, of course, reminiscent of the usual identities of vector lattices. Furthermore, they suggest that the set of all signed measures forms a vector lattice. Unfortunately, this is not the case. Although the (pointwise) sum of two measures is always defined, the sum of two signed measures may fail to exist for the very simple reason that by adding the signed measures pointwise one might encounter the expression $\infty - \infty$.

However, if we restrict ourselves to the collection $M(\Sigma)$ of all finite signed measures, then we obtain a vector lattice. Note first that a signed measure μ satisfies $\mu \in M(\Sigma)$ if and only if $|\mu|(X) < \infty$. Indeed, if $|\mu|(X) < \infty$, then clearly, $\mu \in M(\Sigma)$. On the other hand, if $\mu \in M(\Sigma)$, then the identity $\mu = \mu^+ - \mu^-$ with either μ^+ or μ^- finite shows that μ^+ and μ^- are both finite, and so $|\mu|(X) = \mu^+(X) + \mu^-(X) < \infty$. (Because of the last statement, the finite signed measures are also known as the **signed measures of finite total variation**.)

Clearly, if $\mu, \nu \in M(\Sigma)$, then $\mu + \nu$ and $\alpha\mu$ belong to $M(\Sigma)$, where, of course, $(\mu + \nu)(A) = \mu(A) + \nu(A)$ and $(\alpha\mu)(A) = \alpha\mu(A)$ for each $A \in \Sigma$ and $\alpha \in \mathbb{R}$. That is, $M(\Sigma)$ is a vector space. Now, if $\mu \le \nu$ means $\mu(A) \le \nu(A)$ for each $A \in \Sigma$, then $\le$ is an order relation under which $M(\Sigma)$ is a vector lattice. Its lattice operations are given by

$$\mu \vee \nu(A) = \sup\{\mu(B) + \nu(A \setminus B) \colon B \in \Sigma \text{ and } B \subseteq A\}, \text{ and}$$

$$\mu \wedge \nu(A) = \inf\{\mu(B) + \nu(A \setminus B) \colon B \in \Sigma \text{ and } B \subseteq A\}.$$

The preceding formulas, combined with Theorem 36.9, show that $\mu^+ = \mu \vee 0$, $\mu^- = (-\mu) \vee 0$, and $|\mu| = \mu \vee (-\mu)$ in $M(\Sigma)$ are precisely the positive, negative, and total variations of μ as given by Definition 36.8.

The next thing to observe is that $\|\mu\| = |\mu|(X)$ defines a norm on $M(\Sigma)$. Indeed,

  a. $|\mu|(X) \geq 0$ holds, and since $|\mu(A)| \leq |\mu|(A) \leq |\mu|(X) = \|\mu\|$ for each $A \in \Sigma$, we have $\|\mu\| = 0$ if and only if $\mu = 0$.
  b. $\|\alpha\mu\| = |\alpha\mu|(X) = |\alpha| \cdot |\mu|(X) = |\alpha| \cdot (|\mu|(X)) = |\alpha| \cdot \|\mu\|$.
  c. $\|\mu + \nu\| = |\mu + \nu|(X) \leq (|\mu| + |\nu|)(X) = |\mu|(X) + |\nu|(X) = \|\mu\| + \|\nu\|$.

It is also true that $\|\cdot\|$ is a lattice norm on $M(\Sigma)$. Indeed, if $|\mu| \leq |\nu|$ holds, then $\|\mu\| = |\mu|(X) \leq |\nu|(X) = \|\nu\|$. Therefore, $M(\Sigma)$ is a normed vector lattice which is actually a Banach lattice, as the next result shows.

Theorem 36.10. *The collection of all finite signed measures on a σ-algebra is a Banach lattice.*

Proof. According to the preceding discussion, it remains to be shown that $M(\Sigma)$ is a Banach space. To this end, let $\{\mu_n\}$ be a Cauchy sequence of $M(\Sigma)$. We have to show that there exists some $\mu \in M(\Sigma)$ such that

$$\lim_{n \to \infty} \|\mu_n - \mu\| = 0.$$

Let $\epsilon > 0$. Choose some k such that $\|\mu_n - \mu_m\| < \epsilon$ for all $m, n \geq k$. The inequalities

$$|\mu_n(A) - \mu_m(A)| \leq |\mu_n - \mu_m|(A) \leq |\mu_n - \mu_m|(X) = \|\mu_n - \mu_m\| \qquad (\star)$$

show that $\{\mu_n(A)\}$ is a Cauchy sequence of real numbers for each $A \in \Sigma$. Let $\mu(A) = \lim \mu_n(A)$, so that μ is a real-valued set function. From $(\star)$ it follows that

$$|\mu_n(A) - \mu(A)| \leq \epsilon \quad \text{for each } A \in \Sigma \text{ and } n \geq k. \qquad (\star\star)$$

Clearly, $\mu(\emptyset) = \lim \mu_n(\emptyset) = 0$. For the σ-additivity of μ, let $\{A_n\}$ be a disjoint sequence of Σ, and let $A = \bigcup_{n=1}^{\infty} A_n$. Then

$$\left| \sum_{i=1}^{p} [\mu_k(A_i) - \mu_n(A_i)] \right| \leq \sum_{i=1}^{p} |\mu_k(A_i) - \mu_n(A_i)| \leq |\mu_k - \mu_n|(A)$$

$$\leq \|\mu_k - \mu_n\| < \epsilon$$

holds for all p and $n \geq k$. Hence, $|\sum_{i=1}^{p} [\mu_k(A_i) - \mu(A_i)]| \leq \epsilon$ holds for each p. Now, choose n_0 such that $|\mu_k(A) - \sum_{i=1}^{p} \mu_k(A_i)| < \epsilon$ for all $p > n_0$, and note that

$$\left| \mu(A) - \sum_{i=1}^{p} \mu(A_i) \right| \leq |\mu(A) - \mu_k(A)| + \left| \mu_k(A) - \sum_{i=1}^{p} \mu_k(A_i) \right|$$

$$+ \left| \sum_{i=1}^{p} [\mu_k(A_i) - \mu(A_i)] \right| < 3\epsilon$$

holds for all $p \geq n_0$. Therefore, $\mu(A) = \sum_{n=1}^{\infty} \mu(A_n)$, and so $\mu \in M(\Sigma)$.

Finally, combining $(\star\star)$ with Theorem 36.9, we get

$$(\mu_n - \mu)^+(X) = \sup\{\mu_n(A) - \mu(A): A \in \Sigma\} \leq \epsilon$$

and $(\mu_n - \mu)^-(X) \leq \epsilon$ for all $n \geq k$. Hence,

$$\|\mu_n - \mu\| = (\mu_n - \mu)^+(X) + (\mu_n - \mu)^-(X) \leq 2\epsilon$$

holds for all $n \geq k$. That is, $\lim \|\mu_n - \mu\| = 0$, as required.   ∎

EXERCISES

1. Give an example of a signed measure and two Hahn decompositions (A, B) and (A_1, B_1) of X with respect to the signed measure such that $A \neq A_1$ and $B \neq B_1$.
2. If μ is a signed measure, then show that $\mu^+ \wedge \mu^- = 0$.
3. If μ is a signed measure, then show that for each $A \in \Sigma$ we have

$$|\mu|(A) = \sup \left\{ \sum_{n=1}^{\infty} |\mu(A_n)| \colon \{A_n\} \text{ is a disjoint sequence of } \Sigma \text{ with } \bigcup_{n=1}^{\infty} A_n = A \right\}.$$

4. Verify that if μ and ν are two finite signed measures, then the least upper bound $\mu \vee \nu$ and the greatest lower bound $\mu \wedge \nu$ in $M(\Sigma)$ are given by

$$\mu \vee \nu(A) = \sup\{\mu(B) + \nu(A \setminus B): B \in \Sigma \text{ and } B \subseteq A\}, \text{ and}$$

$$\mu \wedge \nu(A) = \inf\{\mu(B) + \nu(A \setminus B): B \in \Sigma \text{ and } B \subseteq A\}$$

for each $A \in \Sigma$.

5. Let λ be the Lebesgue measure on the Lebesgue measurable subsets of $\mathbb{R}$. If μ is the Dirac measure, defined by $\mu(A) = 0$ if $0 \notin A$ and $\mu(A) = 1$ if $0 \in A$, describe $\lambda \vee \mu$ and $\lambda \wedge \mu$.

6. Show that the collection of all σ-finite measures forms a distributive lattice. That is, show that if μ, ν, and ω are three σ-finite measures, then

$$(\mu \vee \nu) \wedge \omega = (\mu \wedge \omega) \vee (\nu \wedge \omega) \quad \text{and} \quad (\mu \wedge \nu) \vee \omega = (\mu \vee \omega) \wedge (\nu \vee \omega).$$

[HINT: Every vector lattice is a distributive lattice.]

7. If Σ is a σ-algebra of subsets of a set X and $\mu\colon \Sigma \to \mathbb{R}^*$ is a signed measure, then show that

$$\Lambda_{\mu^+} \cap \Lambda_{\mu^-} = \Lambda_{|\mu|}.$$

8. Let μ and ν be two measures on a σ-algebra Σ with at least one of them finite. Assume also that S is a semiring such that $S \subseteq \Sigma$, $X \in S$, and that the σ-algebra generated by S equals Σ. Then show that $\mu = \nu$ on Σ if and only if $\mu = \nu$ on S.
   [HINT: See Theorem 15.10; see also Exercise 11 of Section 20.]

9. Let (X, S, μ) be a measure space and let $f \in L_1(\mu)$. Then show that

$$\nu(A) = \int_A f \, d\mu$$

for each $A \in \Lambda_\mu$ defines a finite signed measure on Λ_μ. Also, show that

$$\nu^+(A) = \int_A f^+ \, d\mu, \quad \nu^-(A) = \int_A f^- \, d\mu \quad \text{and} \quad |\nu|(A) = \int_A |f| \, d\mu$$

hold for each $A \in \Lambda_\mu$.
   [HINT: Use Theorem 15.11.]

10. Let ν be a signed measure on Σ. A function $f\colon X \to \mathbb{R}$ is said to be ν-integrable if f is simultaneously ν^+- and ν^--integrable (in this case, we write $\int f \, d\nu = \int f \, d\nu^+ - \int f \, d\nu^-$). Show that a function f is ν-integrable if and only if $f \in L_1(|\nu|)$.

11. Show that the Jordan decomposition is unique in the following sense: If ν is a signed measure, and μ_1 and μ_2 are two measures such that $\nu = \mu_1 - \mu_2$ and $\mu_1 \wedge \mu_2 = 0$, then $\mu_1 = \nu^+$ and $\mu_2 = \nu^-$.

12. In a vector lattice $x_n \downarrow x$ means that $x_{n+1} \leq x_n$ for each n and that x is the greatest lower bound of the sequence $\{x_n\}$. A normed vector lattice is said to have σ-**order continuous norm** if $x_n \downarrow 0$ implies $\lim \|x_n\| = 0$.

   a. Show that every $L_p(\mu)$ with $1 \leq p < \infty$ has σ-order continuous norm.
   b. Show that $L_\infty([0, 1])$ does not have σ-order continuous norm.
   c. Let Σ be a σ-algebra of sets, and let $\{\mu_n\}$ be a sequence of $M(\Sigma)$ such that $\mu_n \downarrow \mu$. Show that $\lim \mu_n(A) = \mu(A)$ holds for all $A \in \Sigma$.
   d. Show that the Banach lattice $M(\Sigma)$ has σ-order continuous norm.

13. Prove the following additivity property of the Banach lattice $M(\Sigma)$: If $\mu, \nu \in M(\Sigma)$ are disjoint (i.e., $|\mu| \wedge |\nu| = 0$), then $\|\mu + \nu\| = \|\mu\| + \|\nu\|$ holds.
   [HINT: In a vector lattice $|x| \wedge |y| = 0$ implies $|x + y| = |x| + |y|$. Reason:

$$|x + y| \geq |\, |x| - |y|\, | = |x| \vee |y| - |x| \wedge |y| = |x| \vee |y| = |x| + |y| \geq |x + y|.]$$

14. Let Σ be a σ-algebra of subsets of a set X and let $\{\mu_n\}$ be a disjoint sequence of $M(\Sigma)$. If the sequence of signed measures $\{\mu_n\}$ is order bounded, then show that $\lim \|\mu_n\| = 0$.

37. COMPARING MEASURES AND THE
RADON–NIKODYM THEOREM

Again, in this section, Σ will be a fixed σ-algebra of subsets of a nonempty set X and all set functions will be assumed defined on Σ.

Two important comparison notions for signed measures will be introduced here. Both notions will be defined in terms of measure properties. The first one is referred to as the "concept of absolute continuity."

> **Definition 37.1.**   *A signed measure ν is said to be **absolutely continuous** with respect to another signed measure μ, in symbols $\nu \ll \mu$ or $\mu \gg \nu$, whenever $A \in \Sigma$ and $|\mu|(A) = 0$ imply $\nu(A) = 0$.*

Some characterizations of absolute continuity in terms of variations are included in the next result.

> **Theorem 37.2.**   *For a pair of signed measures μ and ν, the following statements are equivalent:*
> a. $\nu \ll \mu$.
> b. $\nu^+ \ll |\mu|$ and $\nu^- \ll |\mu|$.
> c. $|\nu| \ll |\mu|$.

Proof.   (a) $\Longrightarrow$ (b) Let $A \in \Sigma$ satisfy $|\mu|(A) = 0$. If $B \in \Sigma$ satisfies $B \subseteq A$, then $|\mu|(B) = 0$, and so by our hypothesis $\nu(B) = 0$. Thus,

$$\nu^+(A) = \sup\{\nu(B): B \in \Sigma \text{ and } B \subseteq A\} = 0.$$

Similarly, $\nu^-(A) = 0$, and so, both $\nu^+ \ll |\mu|$ and $\nu^- \ll |\mu|$ hold.

(b) $\Longrightarrow$ (c) If $|\mu|(A) = 0$, then by hypothesis $\nu^+(A) = \nu^-(A) = 0$. Since $|\nu| = \nu^+ + \nu^-$, it follows that $|\nu|(A) = 0$, and thus $|\nu| \ll |\mu|$.

(c) $\Longrightarrow$ (a) It follows from the inequality $|\nu(A)| \le |\nu|(A)$. ∎

If a measure ν is absolutely continuous with respect to another measure μ, then it is natural to expect some relationship between the μ-measurable and ν-measurable sets. The following theorem tells us the precise relationship:

> **Theorem 37.3.**   *For two measures μ and ν on a σ-algebra Σ we have the following:*
> a. *If $\nu \ll \mu$ and a subset E of X satisfies $\mu^*(E) = 0$, then $\nu^*(E) = 0$.*
> b. *If $\nu \ll \mu$ and μ is σ-finite, then $\Lambda_\mu \subseteq \Lambda_\nu$ holds. In particular, in this case every μ-measurable function is also ν-measurable.*

Proof. (a) Assume $\mu^*(E) = 0$. By Theorem 15.11, there exists some $A \in \Sigma$ such that $E \subseteq A$ and $\mu(A) = \mu^*(E)$. From $\nu \ll \mu$ it follows that $\nu(A) = 0$. Therefore, $0 \leq \nu^*(E) \leq \nu(A) = 0$, so that $\nu^*(E) = 0$.

(b) Assume E is μ-measurable such that $\mu^*(E) < \infty$. By Theorem 15.11, there exists some $A \in \Sigma$ with $E \subseteq A$ and $\mu(A) = \mu^*(E) < \infty$. But then $\mu^*(A \setminus E) = 0$, and by (a) we have $\nu^*(A \setminus E) = 0$. Therefore, $A \setminus E$ is ν-measurable. The ν-measurability of E now follows from the identity $E = A \setminus (A \setminus E)$. Finally, the preceding conclusion, coupled with the σ-finiteness of μ, easily implies that $\Lambda_\mu \subseteq \Lambda_\nu$ holds. ∎

Now, let μ and ν be two measures. If $\nu \leq \mu$, then it should be clear that $\nu \ll \mu$ also holds. The converse of the latter is false. For instance, $2\mu \ll \mu$ holds for each measure μ, and $2\mu \leq \mu$ is true only if $\mu = 0$. However, if $\nu \ll \mu$ holds, then (up to a multiplication factor) μ and ν are "locally" comparable in the order sense. The exact details follow.

Theorem 37.4. *Let ν be a finite nonzero measure, and let μ be a σ-finite measure. If ν is absolutely continuous with respect to μ, then there exist some $\epsilon > 0$ and some $A \in \Sigma$ with $0 < \mu(A) < \infty$ such that*

$$\epsilon\mu(B) \leq \nu(B)$$

holds for all $B \in \Sigma$ with $B \subseteq A$.

Proof. Let $\{E_n\}$ be a disjoint sequence of Σ such that $X = \bigcup_{n=1}^\infty E_n$ and $\mu(E_n) < \infty$ for each n. Since ν is a nonzero measure, there exists some k with $\nu(E_k) > 0$. Choose some $\epsilon > 0$ so that

$$\nu(E_k) - \epsilon\mu(E_k) = (\nu - \epsilon\mu)(E_k) > 0.$$

By Lemma 36.6 there exists a measurable subset A of E_k that is a positive set for the signed measure $\nu - \epsilon\mu$ satisfying $(\nu - \epsilon\mu)(A) > 0$. Clearly, $\mu(A) < \infty$ holds. Now observe that $\mu(A) > 0$. Indeed, if $\mu(A) = 0$ holds, then the last inequality yields $\nu(A) > 0$, contradicting the absolute continuity of ν with respect to μ. On the other hand, the inequality $(\nu - \epsilon\mu)(B) \geq 0$ for each $B \in \Sigma$ with $B \subseteq A$ implies $\epsilon\mu(B) \leq \nu(B)$, and the proof of the theorem is finished. ∎

The opposite notion to absolute continuity is that of singularity. Two signed measures μ and ν are said to be **singular** (or **orthogonal**), in symbols $\mu \perp \nu$, if there exist two disjoint sets A and B of Σ with $A \cup B = X$ and $|\mu|(A) = |\nu|(B) = 0$.

If μ is a signed measure, then Theorem 36.7 shows that μ^+ and μ^- are two singular measures. The singularity concept is characterized in terms of lattice properties as follows.

Theorem 37.5. *Two signed measures μ and ν are singular if and only if $|\mu| \wedge |\nu| = 0$.*

Proof. Assume that $\mu \perp \nu$. Choose two disjoint sets A and B of Σ with $A \cup B = X$ and $|\mu|(A) = |\nu|(B) = 0$. Then

$$0 \le |\mu| \wedge |\nu|(X) = |\mu| \wedge |\nu|(A) + |\mu| \wedge |\nu|(B) \le |\mu|(A) + |\nu|(B) = 0,$$

so that $|\mu| \wedge |\nu| = 0$.

Conversely, assume that $|\mu| \wedge |\nu| = 0$. By Theorem 36.1, for each n there exists some $E_n \in \Sigma$ such that $|\mu|(E_n) + |\nu|(E_n^c) \le 2^{-n}$. Let $A_n = \bigcup_{i=n}^{\infty} E_i$, and then let $A = \bigcap_{n=1}^{\infty} A_n$ and $B = A^c$. Note that

$$|\mu|(A) \le |\mu|(A_n) = |\mu| \left(\bigcup_{i=n}^{\infty} E_i \right) \le \sum_{i=n}^{\infty} |\mu|(E_i) \le \sum_{i=n}^{\infty} 2^{-i} = 2^{1-n}$$

holds for all n, and thus, $|\mu|(A) = 0$.

Now, observe that $|\nu|(A_n^c) = |\nu|(\bigcap_{i=n}^{\infty} E_i^c) \le 2^{-i}$ for all $i \ge n$ implies $|\nu|(A_n^c) = 0$ for each n. Therefore, since $B = A^c = \bigcup_{n=1}^{\infty} A_n^c$, it follows that $|\nu|(B) = 0$. Hence, $X = A \cup B$ and $|\mu|(A) = |\nu|(B) = 0$ hold, proving that $\mu \perp \nu$. ∎

The following list of statements presents a number of useful relationships between absolute continuity and singularity. The set functions μ, ν, and ω are signed measures on a common σ-algebra Σ.

1. If $\mu \ll \omega$ and $\nu \ll \omega$, then $|\mu| + |\nu| \ll \omega$.
2. If $\mu \perp \omega$ and $\nu \perp \omega$, then $|\mu| + |\nu| \perp \omega$.
3. If $\mu \ll \omega$ and $|\nu| \le |\mu|$, then $\nu \ll \omega$.
4. If $\mu \perp \omega$ and $|\nu| \le |\mu|$, then $\nu \perp \omega$.
5. If $\nu \ll \mu$ and $\nu \perp \mu$, then $\nu = 0$.

Their proofs are straightforward applications of the definitions and Theorems 37.2 and 37.5. For instance, the proof of (5) goes as follows: Since $\nu \perp \mu$, there exists some $A \in \Sigma$ such that $|\nu|(A) = |\mu|(A^c) = 0$. From $\nu \ll \mu$ and Theorem 37.2, $|\nu|(A^c) = 0$, and so, $|\nu|(X) = |\nu|(A) + |\nu|(A^c) = 0$. That is, $|\nu| = 0$, so that $\nu = 0$.

Now, let μ be a σ-finite measure. A classical result of H. Lebesgue asserts that any other σ-finite measure ν can be decomposed as a sum of two measures ν_1 and ν_2 such that $\nu_1 \ll \mu$ and $\nu_2 \perp \mu$. Its proof will be based upon the following simple property of vector lattices:

Lemma 37.6. *In a vector lattice $x_n \uparrow x$ implies $x_n \wedge y \uparrow x \wedge y$ for each y.*

Proof. Note first that $x_n \wedge y \uparrow \leq x \wedge y$ holds. On the other hand, if $x_n \wedge y \leq z$ holds for each n and some z, then Theorem 30.1(5) shows that

$$(x \wedge y - z)^+ \leq (x \wedge y - x_n \wedge y)^+ \leq |x \wedge y - x_n \wedge y| \leq x - x_n.$$

Hence, $x_n \leq x - (x \wedge y - z)^+ \leq x$ holds for each n. But then $x_n \uparrow x$ implies $(x \wedge y - z)^+ = 0$, and so, $x \wedge y \leq z$. That is, $x \wedge y$ is the least upper bound of $\{x_n \wedge y\}$, so that $x_n \wedge y \uparrow x \wedge y$ holds, as claimed. ∎

We are now ready to state and prove the Lebesgue decomposition theorem.

Theorem 37.7 (Lebesgue). *Let μ and v be two σ-finite measures on Σ. Then there exist two unique measures v_1 and v_2 such that*

$$v = v_1 + v_2, \quad where \; v_1 \ll \mu \quad and \quad v_2 \perp \mu.$$

Proof. Assume first that both μ and v are finite measures. Then the formula

$$v_1(A) = \sup\{(v \wedge n\mu)(A): \; n = 1, 2, \ldots\} \leq v(A)$$

defines a finite measure on Σ (Theorem 36.2). Moreover, if $\mu(A) = 0$, then clearly $(v \wedge n\mu)(A) = 0$ for each n. Thus, $v_1(A) = 0$, that is, $v_1 \ll \mu$. Now, let $v_2 = v - v_1$, and note that v_2 is a finite measure such that $v_1 + v_2 = v$. Next, observe that $v \wedge n\mu \uparrow v_1$ implies $(\mu + v) \wedge (n + 1)\mu = \mu + v \wedge n\mu \uparrow \mu + v_1$, and so, by Lemma 37.6,

$$v \wedge (n + 1)\mu = v \wedge (\mu + v) \wedge (n + 1)\mu \uparrow v \wedge (\mu + v_1).$$

On the other hand, $v \wedge (n + 1)\mu \uparrow v_1$ implies $v_1 = v \wedge (\mu + v_1)$. Thus,

$$0 \leq v_2 \wedge \mu = (v - v_1) \wedge \mu = v \wedge (\mu + v_1) - v_1 = v_1 - v_1 = 0,$$

and hence, by Theorem 37.5, $v_2 \perp \mu$.

For the uniqueness of the decomposition, note first that $v_1 \perp v_2$. Indeed, since $v_2 \perp \mu$, there exist two disjoint measurable sets A and B with $X = A \cup B$ and $v_2(A) = \mu(B) = 0$. Then $v_1 \ll \mu$ implies $v_1(B) = 0$, so that $v_1 \perp v_2$. Now, assume that another pair of measures ω_1 and ω_2 satisfies $\omega_1 \ll \mu$, $\omega_2 \perp \mu$, and $v = \omega_1 + \omega_2$. Clearly, $v_1 - \omega_1 = \omega_2 - v_2$ holds, and on one hand $v_1 - \omega_1 \ll \mu$, and on the other $\omega_2 - v_2 \perp \mu$. It follows that $v_1 - \omega_1 = \omega_2 - v_2 = 0$, so that $v_1 = \omega_1$ and $v_2 = \omega_2$.

For the general case, choose a disjoint sequence $\{A_n\}$ of Σ with $v(A_n) < \infty$, $\mu(A_n) < \infty$, and $X = \bigcup_{n=1}^{\infty} A_n$. Let $v_n(A) = v(A \cap A_n)$ and $\mu_n(A) = \mu(A \cap A_n)$ for each $A \in \Sigma$. Then v_n and μ_n are finite measures on Σ. By the previous case, for each n there exists a unique pair of measures (ν_1^n, ν_2^n) such that $\nu_1^n \ll \mu_n$, $\nu_1^n \perp \mu_n$, and $\nu_n = \nu_1^n + \nu_2^n$. Now, define

$$\nu_1(A) = \sum_{n=1}^{\infty} \nu_1^n(A) \quad \text{and} \quad \nu_2(A) = \sum_{n=1}^{\infty} \nu_2^n(A)$$

for each $A \in \Sigma$. It is a routine matter to verify that $\nu_1 \ll \mu$, $\nu_2 \perp \mu$, and $\nu = \nu_1 + \nu_2$. The uniqueness of ν_1 and ν_2 follows easily from the uniqueness of the decomposition of each ν_n.   ∎

The identity $\nu = \nu_1 + \nu_2$ provided by the preceding theorem (where $\nu_1 \ll \mu$ and $\nu_2 \perp \mu$), is called the **Lebesgue decomposition** of ν with respect to μ.

Now, assume that μ is a measure and $f \in L_1(\mu)$. Then the formula

$$\nu(A) = \int_A f \, d\mu \tag{$\star$}$$

for each $A \in \Sigma$ defines a (typical) finite signed measure that is absolutely continuous with respect to μ. Sometimes formula ($\star$) is referred to as the **indefinite integral** of f. A direct verification shows that

$$\nu^+(A) = \int_A f^+ \, d\mu, \quad \nu^-(A) = \int_A f^- \, d\mu \quad \text{and} \quad |\nu|(A) = \int_A |f| \, d\mu$$

hold for each $A \in \Sigma$.

It is natural to ask whether the converse of the preceding statement is true. That is to say, whenever ν is absolutely continuous with respect to μ, does there exist some $f \in L_1(\mu)$ such that ν is given by ($\star$)? In general, the answer is negative; see Exercise 7 at the end of the section. On the other hand, whenever μ is σ-finite and ν is finite, the answer is yes. This result is of great importance because of its wide range of applications. It is known as the Radon–Nikodym theorem. It was established first by J. Radon[2] (in 1913) for the Euclidean spaces with the Lebesgue measure, and later (in 1930) O. Nikodym[3] extended it to the general case.

Theorem 37.8 (Radon–Nikodym). *Let ν be a finite signed measure which is absolutely continuous with respect to a σ-finite measure μ. Then there exists a*

---

[2]Johann Radon (1887–1956), an Austrian mathematician. He is well known for his contributions to analysis and especially to the calculus of variations.

[3]Otton Martin Nikodym (1889–1974), a Polish mathematician. He contributed to Boolean algebras, measure theory, and mathematical physics.

unique function f in $L_1(\mu)$ such that

$$\nu(A) = \int_A f\, d\mu$$

holds for all $A \in \Sigma$.

Proof. (**Uniqueness.**) This part is straightforward and is left for the reader; see Exercise 6 at the end of this section. (The function f is, of course, unique μ-a.e.)

(**Existence.**) Clearly, ν^+ and ν^- are finite measures, both are absolutely continuous with respect to μ, and $\nu = \nu^+ - \nu^-$ holds. This shows that we can assume without loss of generality that ν is itself a finite measure. By Theorem 37.3, we have $\Sigma \subseteq \Lambda_\mu \subseteq \Lambda_\nu$.

Now, let

$$C = \left\{ g \in L_1(\mu)\colon\ g \geq 0\ \mu\text{-a.e. and } \int_A g\, d\mu \leq \nu^*(A) \text{ for all } A \in \Lambda_\mu \right\}.$$

Observe that C is closed under finite suprema. Indeed, if $f,\ g \in C$ and $A \in \Lambda_\mu$, then the two sets

$$E = \{x \in A\colon\ f(x) \geq g(x)\} \quad \text{and} \quad F = \{x \in A\colon\ g(x) > f(x)\}$$

are μ-measurable (and hence, ν-measurable) and disjoint, and $E \cup F = A$. Therefore,

$$\int_A f \vee g\, d\mu = \int_E f \vee g\, d\mu + \int_F f \vee g\, d\mu = \int_E f\, d\mu + \int_F g\, d\mu$$
$$\leq \nu^*(E) + \nu^*(F) = \nu^*(E \cup F) = \nu^*(A),$$

so that $f \vee g \in C$.

Since $0 \in C$, the set C is nonempty. Let

$$a = \sup\left\{ \int g\, d\mu\colon\ g \in C \right\} \leq \nu(X) < \infty.$$

Choose a sequence $\{g_n\}$ of C with $\lim \int g_n\, d\mu = a$. Define $f_n = g_1 \vee \cdots \vee g_n$ for each n, and note that $\{f_n\}$ is a sequence of C with $0 \leq f_n \uparrow$ and $\lim \int f_n\, d\mu = a < \infty$. By Levi's theorem (Theorem 22.8) there exists some $f \in L_1(\mu)$ such that $f_n \uparrow f$ and $\int f\, d\mu = a$. In view of $0 \leq f_n \chi_A \uparrow f \chi_A$, $\int f_n \chi_A\, d\mu = \int_A f_n\, d\mu \leq \nu^*(A)$, and the Lebesgue dominated convergence theorem, it follows that $\int_A f \, d\mu \leq \nu^*(A)$ for each $A \in \Lambda_\mu$; that is, $f \in C$. To finish the proof, we show next that $\nu(A) = \int_A f \, d\mu$ holds for all $A \in \Sigma$.

Clearly, $\omega(A) = \nu^*(A) - \int_A f \, d\mu$ for $A \in \Lambda_\mu$ defines a finite measure on Λ_μ which is absolutely continuous with respect to μ^*. Assume by way of contradiction that $\omega \neq 0$. Then, by Theorem 37.4, there exists some $\epsilon > 0$ and some $B \in \Lambda_\mu$ such that $0 < \mu^*(B) < \infty$ and $\epsilon \mu^*(E) \leq \omega(E)$ for each $E \in \Lambda_\mu$ with $E \subseteq B$. Moreover, the function $g = f + \epsilon \chi_B \geq 0$ belongs to $L_1(\mu)$, and since $a = \int f \, d\mu < \int g \, d\mu$, the function g does not belong to C. On the other hand, if $A \in \Lambda_\mu$, then

$$
\int_A g \, d\mu = \int_A (f + \epsilon \chi_B) \, d\mu = \int_A f \, d\mu + \epsilon \mu^*(B \cap A) \leq \int_A f \, d\mu + \omega(B \cap A)
$$

$$
= \int_A f \, d\mu + \nu^*(B \cap A) - \int_{B \cap A} f \, d\mu = \int_{A \setminus B} f \, d\mu + \nu^*(B \cap A)
$$

$$
\leq \nu^*(A \setminus B) + \nu^*(B \cap A) = \nu^*(A),
$$

and therefore $g \in C$, which is a contradiction. The proof of the theorem is now complete. ∎

The unique (μ-a.e.) function f determined by the Radon–Nikodym theorem is called the **Radon–Nikodym derivative** of ν with respect to μ, in symbols

$$
\frac{d\nu}{d\mu} = f \quad \text{or} \quad d\nu = f \, d\mu.
$$

The function f is also referred to as the **density function** of ν with respect to μ.

In Section 31 it was shown that if $(X, \mathcal{S}, \mu)$ is a measure space and $1 < p < \infty$, then for each $g \in L_q(\mu)$, where $\frac{1}{p} + \frac{1}{q} = 1$, the function

$$
F_g(f) = \int fg \, d\mu \tag{$\star\star$}
$$

for $f \in L_p(\mu)$ defines a continuous linear functional on $L_p(\mu)$ such that $\|F_g\| = \|g\|_q$. Also, in the same section we proved that the mapping $g \mapsto F_g$ from $L_q(\mu)$ to $L_p^*(\mu)$ was a lattice isometry and promised the reader to establish that this lattice isometry is also onto. That is (according to F. Riesz), every member of $L_p^*(\mu)$ can be obtained from a function of $L_q(\mu)$ as in ($\star\star$). With the help of the Radon–Nikodym theorem, we are now in the position to establish this claim.

Theorem 37.9 (F. Riesz). *Let $(X, \mathcal{S}, \mu)$ be a measure space, $1 < p < \infty$, and let F be a continuous linear functional on $L_p(\mu)$. Then there exists a unique*

$g \in L_q(\mu)$, where $\frac{1}{p} + \frac{1}{q} = 1$, such that

$$F(f) = \int fg \, d\mu$$

holds for all $f \in L_p(\mu)$. Moreover, $\|F\| = \|g\|_q$.

Proof. The uniqueness of g has been discussed before. The proof of the "existence" of g has two steps.

Step I. Assume that $(X, \mathcal{S}, \mu)$ is a finite measure space.

Define $\nu \colon \Lambda_\mu \to \mathbb{R}$ by $\nu(A) = F(\chi_A)$ for each $A \in \Lambda_\mu$. Clearly, $\nu(\emptyset) = 0$. Also, if $\{A_n\}$ is a disjoint sequence of measurable sets and $A = \bigcup_{n=1}^\infty A_n$, then

$$\left\| \sum_{i=1}^n \chi_{A_i} - \chi_A \right\|_p = \left[\mu^*(A) - \sum_{i=1}^n \mu^*(A_i) \right]^{\frac{1}{p}} \to 0.$$

Therefore, by the continuity of F, we have

$$\sum_{i=1}^n \nu(A_i) = \sum_{i=1}^n F(\chi_{A_i}) = F\left(\sum_{i=1}^n \chi_{A_i} \right) \to F(\chi_A) = \nu(A),$$

That is, $\nu(A) = \sum_{n=1}^\infty \nu(A_n)$ holds, and hence, ν is a finite signed measure on Λ_μ. Moreover, if $A \in \Lambda_\mu$ satisfies $\mu^*(A) = 0$, then $\chi_A = 0$; therefore, $\nu(A) = F(\chi_A) = F(0) = 0$, so that ν is absolutely continuous with respect to μ^*.

By the Radon–Nikodym theorem, there exists a unique function g in $L_1(\mu)$ such that $\nu(A) = \int_A g \, d\mu$ holds for all $A \in \Lambda_\mu$. Thus, for every $A \in \Lambda_\mu$ we have $F(\chi_A) = \int \chi_A g \, d\mu$. By linearity, it follows that $F(\phi) = \int \phi g \, d\mu$ for every step function ϕ. Now let

$$A = \{x \in X \colon g(x) \geq 0\} \quad \text{and} \quad B = \{x \in X \colon g(x) < 0\},$$

and let $0 \leq f \in L_p(\mu)$. Choose a sequence of step functions $\{\phi_n\}$ with $0 \leq \phi_n \uparrow f$. Then $\phi_n \chi_A g = \phi_n g^+ \uparrow fg^+$ and $\lim \|\phi_n \chi_A - f \chi_A\|_p = 0$. By the continuity of F we have $\lim \int \phi_n g^+ \, d\mu = \lim F(\phi_n \chi_A) = F(f \chi_A) < \infty$, and so, by Levi's theorem $fg^+ \in L_1(\mu)$ and $F(f \chi_A) = \int fg^+ \, d\mu$. Similarly, $fg^- \in L_1(\mu)$ and $F(f \chi_B) = -\int fg^- \, d\mu$. Thus, for every $f \in L_p(\mu)$ we have $fg \in L_1(\mu)$ and $F(f) = \int fg \, d\mu$. To complete the proof of this case, it remains to be shown that $g \in L_q(\mu)$.

To this end, let $E_n = \{x \in X \colon |g(x)| \leq n\}$, and then define $f_n = |g|^{q-1} \chi_{E_n} \operatorname{Sgn} g$. Since each f_n is essentially bounded, $\{f_n\} \subseteq L_p(\mu)$ holds, and moreover, we have

$f_n g = |f_n|^p = |g|^q \chi_{E_n} \uparrow |g|^q$. Therefore,

$$\int |g|^q \chi_{E_n}\, d\mu = \int f_n g\, d\mu = F(f_n) \le \|F\| \cdot \left(\int |f_n|^p\, d\mu \right)^{\frac{1}{p}}$$

$$= \|F\| \cdot \left(\int |g|^q \chi_{E_n}\, d\mu \right)^{\frac{1}{p}},$$

from which it follows that $(\int |g|^q \chi_{E_n}\, d\mu)^{\frac{1}{q}} \le \|F\| < \infty$ for each n. Levi's theorem now implies that $g \in L_q(\mu)$.

Step II.   The general case.

For each $A \in \Lambda_\mu$ write $L_p(A) = \{f \in L_p(\mu): f = 0 \text{ on } A^c\}$, and note that $L_p(A) = \{f \chi_A: f \in L_p(\mu)\}$. Also, let

$$C = \{A \in \Lambda_\mu: \mu^*(A) < \infty\}.$$

By the preceding case, it should be obvious that for each $A \in C$ there exists a unique $g_A \in L_q(A)$ such that

$$F(f \chi_A) = \int f g_A\, d\mu$$

holds for all $f \in L_p(\mu)$. Moreover,

$$\|g_A\|_q = \sup\{|F(f \chi_A)|: \|f\|_p \le 1\} \le \|F\| < \infty$$

holds. Let $a = \sup\{\|g_A\|_q: A \in C\} \le \|F\| < \infty$.

Now, observe that if $A, B \in C$ satisfy $A \subseteq B$, then $g_A = g_B$ on A; therefore, $|g_A| \le |g_B|$, and so, $\|g_A\|_q \le \|g_B\|_q$. It follows that there exists an increasing sequence $\{A_n\}$ of C such that $\|g_{A_n}\|_q \uparrow a < \infty$. The latter, combined with Levi's theorem, shows that the function g defined by $g(x) = \lim g_{A_n}(x)$ for each $x \in X$ belongs to $L_q(\mu)$. Clearly, g vanishes outside of the set $A = \bigcup_{n=1}^{\infty} A_n$.

Next, we claim that F vanishes on $L_p(A^c)$. Indeed, if F is not zero on $L_p(A^c)$, then in view of the norm denseness of the step functions there exists some $B \in C$ with $B \subseteq A^c$, so that F does not vanish on $L_p(B)$. Thus, $g_B \ne 0$, and since $B \cap A_n = \emptyset$ for all n, it follows that for each n

$$a^q \le (\|g_{B \cup A_n}\|_q)^q = (\|g_B\|_q)^q + (\|g_{A_n}\|_q)^q$$

holds. Since $\lim \|g_{A_n}\|_q = a$, it follows that $\|g_B\|_q = 0$, which is impossible. Hence, F vanishes on $L_p(A^c)$.

Now, let $f \in L_p(\mu)$. Note first that since $g \in L_q(\mu)$, we have $fg \in L_1(\mu)$. Thus, from $fg_{A_n} \to fg$, $|fg_{A_n}| \le |fg|$, and the Lebesgue dominated convergence theorem, it follows that $\lim \int fg_{A_n} \, d\mu = \int fg \, d\mu$. Similarly, it follows that $\lim \| f\chi_{A_n} - f\chi_A \|_p = 0$, and so, by the continuity of F we must have $\lim F(f\chi_{A_n}) = F(f\chi_A)$. Therefore,

$$F(f) = F(f\chi_A + f\chi_{A^c}) = F(f\chi_A) + F(f\chi_{A^c}) = F(f\chi_A)$$

$$= \lim_{n\to\infty} F(f\chi_{A_n}) = \lim_{n\to\infty} \int fg_{A_n} \, d\mu = \int fg \, d\mu,$$

as required.

The equality $\|F\| = \|g\|_q$ was established in Theorem 31.16. The proof of the theorem is now complete. ∎

The conjugate space of $L_1(\mu)$ will be considered next. If we assume that (X, S, μ) is a σ-finite measure space, then the norm dual of $L_1(\mu)$ is lattice isomorphic to $L_\infty(\mu)$. For a general measure space this may not be the case (see Exercise 16 of Section 31).

Theorem 37.10. Let (X, S, μ) be a σ-finite measure space, and let F be a continuous linear functional on $L_1(\mu)$. Then there exists a unique $g \in L_\infty(\mu)$ such that

$$F(f) = \int fg \, d\mu$$

holds for all $f \in L_1(\mu)$. Moreover, $\|F\| = \|g\|_\infty$. In other words,

$$L_1^*(\mu) = L_\infty(\mu).$$

Proof. First consider a finite measure space (X, S, μ). Then the arguments of Step I in the proof of Theorem 37.9 also apply here and show that there exists a unique function $g \in L_1(\mu)$ such that $F(f) = \int fg \, d\mu$ holds for all $f \in L_1(\mu)$. Next, we show $g \in L_\infty(\mu)$ and $\|F\| = \|g\|_\infty$.

Indeed, if $\epsilon > 0$, and $A = \{x \in X : \|F\| + \epsilon < |g(x)|\}$, then

$$(\|F\| + \epsilon)\mu^*(A) \le \int_A |g| \, d\mu = \int_A g \operatorname{Sgn} g \, d\mu = F(\chi_A \operatorname{Sgn} g)$$

$$\le \|F\| \cdot \|\chi_A \operatorname{Sgn} g\|_1 = \|F\| \cdot \mu^*(A)$$

holds. Thus, $\mu^*(A) = 0$, and so, $|g(x)| \le \|F\| + \epsilon$ for almost all x. This shows that $g \in L_\infty(\mu)$ and that $\|g\|_\infty \le \|F\| + \epsilon$. Since $\epsilon > 0$ is arbitrary, $\|g\|_\infty \le \|F\|$ holds. On the other hand, $\|F\| \le \|g\|_\infty$ holds trivially, and thus, $\|F\| = \|g\|_\infty$.

To extend the theorem to the σ-finite case, let $\{X_n\}$ be a disjoint sequence of measurable satisfying $X = \bigcup_{n=1}^{\infty} X_n$ and $\mu^*(X_n) < \infty$ for each n. By the preceding case, for each n there exists a unique $g_n \in L_\infty(\mu)$ which vanishes off X_n, with $\|g_n\|_\infty \leq \|F\|$ and satisfying $F(f\chi_{X_n}) = \int f g_n \, d\mu$ for each $f \in L_1(\mu)$. Let g be the function that equals g_n on each X_n. Clearly, g is measurable, and since $\|g\|_\infty \leq \|F\|$, it follows that $g \in L_\infty(\mu)$.

We leave now the details for the reader to verify that $F(f) = \int f g \, d\mu$ holds for all $f \in L_1(\mu)$. ∎

As an application of Theorem 37.9 we shall establish that every L_p-space with $1 < p < \infty$ is a reflexive Banach space. Recall that a Banach space is said to be reflexive if the natural embedding of the space to its second dual is onto.

Theorem 37.11. *Let (X, S, μ) be a measure space. If $1 < p < \infty$, then $L_p(\mu)$ is a reflexive Banach lattice.*

Proof. Let $\phi \in L_p^{**}(\mu)$. Then $\psi(g) = \phi(F_g)$ for $g \in L_q(\mu)$ defines a bounded linear functional on $L_q(\mu)$, since

$$|\psi(g)| = |\phi(F_g)| \leq \|\phi\| \cdot \|F_g\| = \|\phi\| \cdot \|g\|_q.$$

By Theorem 37.9, there exists some $f \in L_p(\mu)$ such that $\psi(g) = \int g f \, d\mu$ holds for all $g \in L_q(\mu)$. On the other hand, $\hat{f}(F_g) = F_g(f) = \int g f \, d\mu = \psi(g) = \phi(F_g)$ holds for all $g \in L_q(\mu)$, and this (by Theorem 37.9) guarantees that $\phi = \hat{f}$. That is, the natural embedding of $L_p(\mu)$ to its second dual $L_p^{**}(\mu)$ is onto, and hence, $L_p(\mu)$ is a reflexive Banach lattice. ∎

EXERCISES

1. Verify the following properties of signed measures:

   a. $\mu \ll \mu$.
   b. $\nu \ll \mu$ and $\mu \ll \omega$ imply $\nu \ll \omega$.
   c. If $0 \leq \nu \leq \mu$, then $\nu \ll \mu$.
   d. If $\mu \ll 0$, then $\mu = 0$.

2. Verify statements (1) through (5) following Theorem 37.5.

3. Let μ and ν be two measures on a σ-algebra Σ. If ν is a finite measure, then show that the following statements are equivalent:

   a. $\nu \ll \mu$ holds.
   b. For each sequence $\{A_n\}$ of Σ with $\lim \mu(A_n) = 0$, we have $\lim \nu(A_n) = 0$.
   c. For each $\epsilon > 0$ there exists some $\delta > 0$ (depending on ϵ) such that whenever $A \in \Sigma$ satisfies $\mu(A) < \delta$, then $\nu(A) < \epsilon$ holds.

[HINT: If (a) does not imply (b), then there exists some $\epsilon > 0$ and a sequence $\{A_n\}$ of Σ such that $\mu(A_n) < 2^{-n}$ and $\nu(A_n) > \epsilon$. Let $A = \bigcap_{n=1}^{\infty} \bigcup_{i=n}^{\infty} A_i$, and note that $\mu(A) = 0$, while $\nu(A) \geq \epsilon$.]

4. Let μ be a finite measure and let $\{\nu_n\}$ be a sequence of finite measures (all on Σ) such that $\nu_n \ll \mu$ holds for each n. Furthermore, assume that $\lim \nu_n(A)$ exists in $\mathbb{R}$ for each $A \in \Sigma$. Then show that:

   a. For each $\epsilon > 0$ there exists some $\delta > 0$ such that whenever $A \in \Sigma$ satisfies $\mu(A) < \delta$, then $\nu_n(A) < \epsilon$ holds for each n.
   b. The set function $\nu: \Sigma \to [0, \infty]$, defined by $\nu(A) = \lim \nu_n(A)$ for each $A \in \Sigma$, is a measure such that $\nu \ll \mu$.

   [HINT: Consider Σ equipped with the distance $d(A, B) = \mu(A \triangle B)$, where A and B are identified if $\mu(A \triangle B) = 0$. Then (Σ, d) is a complete metric space; see Exercise 3 of Section 31. Next, use $\nu_n \ll \mu$ to verify that each ν_n is a well-defined continuous function on Σ. Furthermore, if $\epsilon > 0$, then each $C_k = \{A \in \Sigma: |\nu_n(A) - \nu_m(A)| \leq \epsilon$ for $n, m \geq k\}$ is a closed set, and $\Sigma = \bigcup_{k=1}^{\infty} C_k$ holds. By Theorem 6.18, some C_k has an interior point. Use this and the preceding exercise to establish (a). Statement (b) follows from (a).]

5. Let $\{\nu_n\}$ be a sequence of nonzero finite measures such that $\lim \nu_n(A)$ exists in $\mathbb{R}$ for each $A \in \Sigma$. Show that $\nu(A) = \lim \nu_n(A)$ for $A \in \Sigma$ is a finite measure.
   [HINT: Consider the finite measure $\mu(A) = \sum_{n=1}^{\infty} 2^{-n}[\nu_n(A)/\nu_n(X)]$, and then apply the conclusion of the preceding exercise.]

6. Verify the uniqueness of the Radon–Nikodym derivative by proving the following statement: If (X, S, μ) is a measure space and $f \in L_1(\mu)$ satisfies $\int_A f \, d\mu = 0$ for all $A \in S$, then $f = 0$ a.e.

7. This exercise shows that the hypothesis of σ-finiteness of μ in the Radon–Nikodym theorem cannot be omitted. Consider $X = [0, 1]$, Σ the σ-algebra of all Lebesgue measurable subsets of $[0, 1]$, ν the Lebesgue measure on Σ and μ the measure defined by $\mu(\emptyset) = 0$ and $\mu(A) = \infty$ if $A \neq \emptyset$. (Incidentally, μ is the largest measure on Σ.) Show that:

   a. ν is a finite measure, μ is not σ-finite, and $\nu \ll \mu$.
   b. There is no function $f \in L_1(\mu)$ such that $\nu(A) = \int_A f \, d\mu$ holds for all $A \in \Sigma$.

8. Let μ be a finite signed measure on Σ. Show that there exists a unique function $f \in L_1(|\mu|)$ such that

$$\mu(A) = \int_A f \, d|\mu|$$

holds for all $A \in \Sigma$.

9. Assume that ν is a finite measure and μ is a σ-finite measure such that $\nu \ll \mu$. Let $g = d\nu/d\mu \in L_1(\mu)$ be the Radon–Nikodym derivative of ν with respect to μ. Then show that:

   a. If $Y = \{x \in X: g(x) > 0\}$, then $Y \cap A$ is a μ-measurable set for each ν-measurable set A.
   b. If $f \in L_1(\nu)$, then $fg \in L_1(\mu)$ and $\int f \, d\nu = \int fg \, d\mu$ holds.

[HINT: Note first that $\Sigma \subseteq \Lambda_\mu \subseteq \Lambda_\nu$ holds, and then (by Theorem 15.10) that $\nu^*(A) = \int_{A} g \, d\mu$ must hold for each $A \in \Lambda_\mu$; hence, $\nu^*(Y^c) = 0$. For (a) use Theorem 37.3 and the fact that Σ is a σ-algebra. Now, if $\phi = \sum_{i=1}^{n} a_i \chi_{A_i}$ is a ν-step function, then

$$\int \phi \, d\nu = \sum_{i=1}^{n} a_i \nu^*(A_i) = \sum_{i=1}^{n} a_i \nu^*(A_i \cap Y) = \sum_{i=1}^{n} a_i \int_{A_i \cap Y} g \, d\mu = \int \phi g \, d\mu,$$

and so, $\phi g \in L_1(\mu)$. Now, if $0 \le f \in L_1(\nu)$, then pick a sequence $\{\phi_n\}$ of ν-step functions such that $0 \le \phi_n(x) \uparrow f(x)$ for all $x \in X$. To finish the proof, observe that $0 \le \phi_n g \uparrow fg$ and $\int \phi_n g \, d\nu = \int \phi_n \, d\nu \uparrow \int f \, d\nu < \infty$.]

10. Establish the **chain rule** for Radon–Nikodym derivatives: If ω is a σ-finite measure and ν and μ are two finite measures (all on Σ) such that $\nu \ll \mu$ and $\mu \ll \omega$, then $\nu \ll \omega$ and

$$\frac{d\nu}{d\omega} = \frac{d\nu}{d\mu} \cdot \frac{d\mu}{d\omega}, \quad \omega\text{-a.e.,}$$

holds.

[HINT: Use the preceding exercise.]

11. All measures considered here will be assumed defined on a fixed σ-algebra Σ.

   a. Call two measures μ and ν equivalent (in symbols, $\mu \equiv \nu$) if $\mu \ll \nu$ and $\nu \ll \mu$ both hold. Show that $\equiv$ is an equivalence relation among the measures on Σ.
   b. If μ and ν are two equivalent σ-finite measures, then show that $\Lambda_\mu = \Lambda_\nu$.
   c. Show that if μ and ν are two equivalent finite measures, then

$$\frac{d\mu}{d\nu} \cdot \frac{d\nu}{d\mu} = 1 \quad \text{a.e. holds.}$$

   d. If μ and ν are two equivalent finite measures, then show that $f \mapsto f \cdot \frac{d\mu}{d\nu}$, from $L_1(\mu)$ to $L_1(\nu)$, is an onto lattice isometry. Thus, under this identification, $L_1(\mu) = L_1(\nu)$ holds.
   e. Generalize (d) to equivalent σ-finite measures. That is, if μ and ν are two equivalent σ-finite measures, then show that the Banach lattices $L_1(\mu)$ and $L_1(\nu)$ are lattice isometric.
   f. Show that if μ and ν are two equivalent σ-finite measures, then the Banach lattices $L_p(\mu)$ and $L_p(\nu)$ are lattice isometric for each $1 \le p \le \infty$.

12. Let μ be a σ-finite measure, and let $AC(\mu)$ be the collection of all finite signed measures that are absolutely continuous with respect to μ; that is,

$$AC(\mu) = \{\nu \in M(\Sigma): \nu \ll \mu\}.$$

   a. Show that $AC(\mu)$ is a norm closed ideal of $M(\Sigma)$ (and hence, $AC(\mu)$ with the norm $\|\nu\| = |\nu|(X)$, is a Banach lattice in its own right).
   b. For each $f \in L_1(\mu)$, let μ_f be the finite signed measure defined by $\mu_f(A) = \int_A f \, d\mu$ for each $A \in \Sigma$. Then show that $f \mapsto \mu_f$ is a lattice isometry from $L_1(\mu)$ onto $AC(\mu)$.

13. Let Σ be a σ-algebra of subsets of a set X and μ a measure on Σ. Assume also that Σ^* is a σ-algebra of subsets of a set Y and that $T: X \to Y$ has the property that $T^{-1}(A) \in \Sigma$ for each $A \in \Sigma^*$.

a. Show that $v(A) = \mu(T^{-1}(A))$ for each $A \in \Sigma^*$ is a measure on Σ^*.

b. If $f \in L_1(v)$, then show that $f \circ T \in L_1(\mu)$ and

$$\int_Y f \, dv = \int_X f \circ T \, d\mu.$$

c. If μ is finite and ω is a σ-finite measure on Σ^* such that $v \ll \omega$, then show that there exists a function $g \in L_1(\omega)$ such that

$$\int_X f \circ T \, d\mu = \int_Y fg \, d\omega$$

holds for each $f \in L_1(v)$.

14. Let (X, S, μ) be a σ-finite measure space, and let g be a measurable function. Show that if for some $1 \leq p < \infty$ we have $fg \in L_1(\mu)$ for all $f \in L_p(\mu)$, then $g \in L_q(\mu)$, where $\frac{1}{p} + \frac{1}{q} = 1$. Also, show by a counterexample that for $1 < p < \infty$, the σ-finiteness of μ cannot be dropped.
[HINT: We can assume that $g \geq 0$ (why?). Then $F(f) = \int fg \, d\mu$ for $f \in L_p(\mu)$ defines a positive linear functional on $L_p(\mu)$. By Theorem 30.10, F is continuous. Now, by Theorems 37.9 and 37.10 there exists some $h \in L_q(\mu)$ such that $\int fg \, d\mu = \int fh \, d\mu$ for each $f \in L_p(\mu)$. Show that $g = h$ a.e. holds.]

15. Let (X, S, μ) be a σ-finite measure space, g a measurable function, and $1 \leq p < \infty$. Assume that there exists some real number $M > 0$ such that $\phi g \in L_1(\mu)$ and $\int \phi g \, d\mu \leq M \|\phi\|_p$ holds for every step function ϕ. Then show that:

a. $g \in L_q(\mu)$, where $\frac{1}{p} + \frac{1}{q} = 1$, and

b. $\int fg \, d\mu \leq M \|f\|_q$ holds for all $f \in L_p(\mu)$.

16. Let μ be a Borel measure on $\mathbf{R}^k$ and suppose that there exists a constant $c > 0$ such that whenever a Borel set E satisfies $\lambda(E) = c$, then $\mu(E) = c$. Show that μ coincides with λ, i.e., show that $\mu = \lambda$.

17. Let μ and v be two σ-finite measures on a σ-algebra Σ of subsets of a set X such that $v \ll \mu$ and $v \neq 0$. Show that there exist a set $E \in \Sigma$ and an integer n such that

a. $v(E) > 0$; and

b. $A \in \Sigma$ and $A \subseteq E$ imply $\frac{1}{n}\mu(A) \leq v(A) \leq n\mu(A)$.

18. Let μ be a finite Borel measure on $[1, \infty)$ such that

a. $\mu \ll \lambda$, and

b. $\mu(B) = a\mu(aB)$ for each $a \geq 1$ and each Borel subset B of $[1, \infty)$, where $aB = \{ab: b \in B\}$.

If the Radon–Nikodym derivative $d\mu/d\lambda$ is a continuous function, then show that there exists a constant $c \geq 0$ such that $[d\mu/d\lambda](x) = c/x^2$ for each $x \geq 1$.

19. Let μ be a finite Borel measure on $(0, \infty)$ such that

a. $\mu \ll \lambda$, and

b. $\mu(aB) = \mu(B)$ for each $a > 0$ and each Borel subset B of $(0, \infty)$.

If the Radon–Nikodym derivative is a continuous function, then show that there exists a constant $c \geq 0$ such that $[d\mu/d\lambda](x) = c/x$ for each $x > 0$.

38. THE RIESZ REPRESENTATION THEOREM

In this section, X will denote a Hausdorff locally compact topological space. Recall that $C_c(X)$ denotes the vector lattice of all real-valued continuous functions defined on X having compact support. That is, $f: X \to \mathbb{R}$ belongs to $C_c(X)$ if and only if f is continuous and vanishes off a compact set. The main purpose of our discussion here is to characterize the positive linear functionals on $C_c(X)$. As usual, a linear functional F on $C_c(X)$ is said to be positive if $0 \le F(f)$ holds whenever $0 \le f \in C_c(X)$.

Clearly, every continuous function on X is Borel measurable. Thus, if μ is a Borel measure on X (i.e., μ is a measure on the σ-algebra $\mathcal{B}$ generated by the open sets and satisfies $\mu(K) < \infty$ for each compact set K), then the formula

$$F(f) = \int f \, d\mu, \quad f \in C_c(X),$$

defines a positive linear functional on $C_c(X)$. In general, if F is a positive linear functional on $C_c(X)$ and a Borel measure μ on X satisfies $F(f) = \int f \, d\mu$ for each $f \in C_c(X)$, then μ is called a **representing measure** for F (or that μ represents F). The following basic result (known as the Riesz representation theorem) will be obtained: *Every positive linear functional on $C_c(X)$ is represented by a unique regular Borel measure.*

Recall that a Borel measure μ on X is called a **regular Borel measure** (see Definition 18.4) if μ satisfies the following two extra properties:

    a. $\mu(B) = \inf\{\mu(V): V \text{ open and } B \subseteq V\}$ for each Borel set B; and
    b. $\mu(V) = \sup\{\mu(K): K \text{ compact and } K \subseteq V\}$ for each open set V.

Of course, the preceding two conditions are enough to ensure that whenever a Borel set B has finite measure, then

$$\mu(B) = \sup\{\mu(K): K \text{ compact and } K \subseteq B\}$$

holds; see Lemma 18.5.

Let us start by introducing some standard notation. Let V be an open set, and let $f \in C_c(X)$. The symbol $f \prec V$ means that $0 \le f(x) \le 1$ holds for each $x \in X$ and Supp $f \subseteq V$. Similarly, for a compact set K and a function $f \in C_c(X)$, the notation $K \prec f$ means that $0 \le f(x) \le 1$ holds for all $x \in X$, and $f(x) = 1$ for all $x \in K$. The notation $K \prec f \prec V$ simply means that $K \prec f$ and $f \prec V$ both hold. In this terminology, for instance, Theorem 10.9 can be stated as follows: *If $K \subseteq \bigcup_{i=1}^n V_i$ holds with K compact and each V_i open, then there exist functions $f_1, \ldots, f_n \in C_c(X)$ such that $f_i \prec V_i$ for each i and $\sum_{i=1}^n f_i = \mathbf{1}$ on K.*

Now, let F be a positive linear functional on $C_c(X)$. For each open subset V of X, define

$$\mu(V) = \sup\{F(f): f \prec V\}.$$

Clearly, $0 \leq \mu(V) \leq \infty$ holds for each open set V. It is easily seen that $\mu(V) \leq \mu(W)$ holds for every pair of open sets with $V \subseteq W$ (since $f \prec V$ implies $f \prec W$). This observation allows us to extend the set function μ from the open sets to all subsets A of X by defining

$$\mu(A) = \inf\{\mu(V): V \text{ open and } A \subseteq V\}.$$

As expected, the set function μ is an outer measure.

Theorem 38.1. *Let F be a positive linear functional on $C_c(X)$. Then the set function μ (previously defined) is an outer measure on X.*

Proof. Clearly, $0 \leq \mu(A) \leq \infty$ holds for every subset A of X. Since $f \prec \emptyset$ if and only if $f = 0$, it follows that $\mu(\emptyset) = 0$. Also, it should be obvious that if $A \subseteq B$, then $\mu(A) \leq \mu(B)$ holds. To complete the proof, it remains to be shown that μ is σ-subadditive.

To this end, let $\{A_n\}$ be a sequence of subsets of X. Let $A = \bigcup_{n=1}^{\infty} A_n$. If $\sum_{n=1}^{\infty} \mu(A_n) = \infty$, then $\mu(A) \leq \sum_{n=1}^{\infty} \mu(A_n)$ is obvious. Hence, assume $\sum_{n=1}^{\infty} \mu(A_n) < \infty$. Let $\epsilon > 0$. For each n, choose an open set V_n such that $A_n \subseteq V_n$ and $\mu(V_n) < \mu(A_n) + \epsilon 2^{-n}$. Put $V = \bigcup_{n=1}^{\infty} V_n$. Now, if $f \prec V$ holds, then $K = \text{Supp } f \subseteq \bigcup_{n=1}^{\infty} V_n$, and in view of the compactness of K there exists some m such that $K \subseteq \bigcup_{n=1}^{m} V_n$. By Theorem 10.9, there exist functions $f_1, \ldots, f_m \in C_c(X)$ such that $f_n \prec V_n$ for $n = 1, \ldots, m$ and $\sum_{n=1}^{m} f_n = 1$ on K. Clearly, $f \leq \sum_{n=1}^{m} f_n$ holds, and by the positivity of F it follows that

$$F(f) \leq \sum_{n=1}^{m} F(f_n) \leq \sum_{n=1}^{m} \mu(V_n) \leq \sum_{n=1}^{\infty} \mu(V_n) \leq \sum_{n=1}^{\infty} \mu(A_n) + \epsilon.$$

Thus, since the latter holds for each $f \prec V$, it follows that $\mu(V) \leq \sum_{n=1}^{\infty} \mu(A_n) + \epsilon$. Now, notice that $A \subseteq V$ implies $\mu(A) \leq \mu(V) \leq \sum_{n=1}^{\infty} \mu(A_n) + \epsilon$ for all $\epsilon > 0$, from which it follows that

$$\mu(A) = \mu\left(\bigcup_{n=1}^{\infty} A_n\right) \leq \sum_{n=1}^{\infty} \mu(A_n).$$

The proof of the theorem is now complete. ∎

The outer measure μ determined by the positive linear functional F is called **the outer measure induced** by F (or the outer measure associated with F). The next goal is to show that μ is a regular Borel measure on $\mathcal{B}$.

Theorem 38.2. *Let F be a positive linear functional on $C_c(X)$, and let μ be its induced outer measure. Then every Borel set is μ-measurable (i.e., $\mathcal{B} \subseteq \Lambda_{\mu}$), and μ restricted to $\mathcal{B}$ is a regular Borel measure.*

Proof.   The proof goes by steps.

Step I.   For each compact set K, we have $\mu(K) < \infty$.

Let K be a compact set. Choose an open set V with compact closure such that $K \subseteq V$. By Theorem 10.9, there exists a function $g \in C_c(X)$ such that $\overline{V} \prec g$. Now, if $f \prec V$ holds, then $f \leq g$, and so $F(f) \leq F(g)$. Hence,

$$\mu(K) \leq \mu(V) = \sup\{F(f)\colon\ f \prec V\} \leq F(g) < \infty.$$

Step II.   If K_1 and K_2 are disjoint compact sets, then $\mu(K_1 \cup K_2) = \mu(K_1) + \mu(K_2)$.

In view of the σ-subadditivity of μ, it suffices to show $\mu(K_1) + \mu(K_2) \leq \mu(K_1 \cup K_2)$.

Since $K_1 \subseteq K_2^c$ holds, there exists (by Lemma 10.7) an open set V_1 with compact closure such that $K_1 \subseteq V_1 \subseteq \overline{V_1} \subseteq K_2^c$. Thus, $K_2 \subseteq (\overline{V_1})^c$ holds. Let $V_2 = (\overline{V_1})^c$, and note that V_1 and V_2 are two disjoint open sets.

Now, let $\epsilon > 0$. Choose an open set V such that $K_1 \cup K_2 \subseteq V$ and $\mu(V) < \mu(K_1 \cup K_2) + \epsilon$; clearly, $K_1 \subseteq V \cap V_1$ and $K_2 \subseteq V \cap V_2$. Next, select two continuous functions f and g such that $f \prec V \cap V_1$, $g \prec V \cap V_2$, $\mu(V \cap V_1) < F(f) + \epsilon$, and $\mu(V \cap V_2) < F(g) + \epsilon$. From $V_1 \cap V_2 = \emptyset$, it follows that $f + g \prec V$. Now, note that

$$\mu(K_1) + \mu(K_2) \leq \mu(V \cap V_1) + \mu(V \cap V_2) < F(f) + F(g) + 2\epsilon$$
$$= F(f + g) + 2\epsilon \leq \mu(V) + 2\epsilon < \mu(K_1 \cup K_2) + 3\epsilon,$$

holds for all $\epsilon > 0$. Therefore, $\mu(K_1) + \mu(K_2) \leq \mu(K_1 \cup K_2)$, as claimed.

Step III.   For every subset A of X we have $\mu(A) = \inf\{\mu(V)\colon\ V$ open and $A \subseteq V\}$.

This is precisely the definition of μ.

Step IV.   For every open set V, $\mu(V) = \sup\{\mu(K)\colon\ K$ compact and $K \subseteq V\}$.

To see this, let V be an open set and let $r \in \mathbb{R}$ satisfy $r < \mu(V)$. Choose a continuous function $f \prec V$ with $r < F(f)$, and let $K = \operatorname{Supp} f$. Now if W is an open set such that $K \subseteq W$, then $f \prec W$ holds, and so $r < F(f) \leq \mu(W)$. Therefore,

$$\mu(V) \geq \mu(K) = \inf\{\mu(W)\colon\ W \text{ open and } K \subseteq W\} \geq F(f) > r$$

holds for every $r \in \mathbb{R}$ with $r < \mu(V)$, and the desired identity follows.

Step V.   Every Borel set is μ-measurable, that is, $\mathcal{B} \subseteq \Lambda_\mu$.

Note first that if K is a compact set and V is an open set disjoint from K, then $\mu(K) + \mu(V) = \mu(K \cup V)$ holds. Indeed, by Step II, for every compact subset K_1

of V we have

$$\mu(K) + \mu(K_1) = \mu(K \cup K_1) \leq \mu(K \cup V) \leq \mu(K) + \mu(V),$$

and the claim follows from the identity of Step IV.

Since Λ_μ is a σ-algebra, it is enough to show that Λ_μ contains every open set. To establish this, let V be an open set. We must show that $\mu(A \cap V) + \mu(A \cap V^c) \leq \mu(A)$ holds for each $A \subseteq X$. If $\mu(A) = \infty$, then the inequality is obvious. Thus, we can consider only the case $\mu(A) < \infty$.

Assume at the beginning that A is also an open set. Let K be a compact set such that $K \subseteq A \cap V$. Then the open set $W = A \setminus K$ satisfies $K \cap W = \emptyset$ and $A \cap V^c \subseteq W \subseteq A$. Thus, by what was observed previously,

$$\mu(K) + \mu(A \cap V^c) \leq \mu(K) + \mu(W) = \mu(K \cup W) \leq \mu(A)$$

holds for all compact sets K with $K \subseteq A \cap V$. From the identity of Step IV the desired inequality follows easily.

Assume now that A is an arbitrary set. If W is an open set such that $A \subseteq W$, then by the preceding case

$$\mu(A \cap V) + \mu(A \cap V^c) \leq \mu(W \cap V) + \mu(W \cap V^c) \leq \mu(W)$$

holds. Thus, by the identity of Step III, we get

$$\mu(A \cap V) + \mu(A \cap V^c) \leq \mu(A),$$

and the proof is finished. ∎

Now, we come to the main result of this section. Namely: The regular Borel measure, induced by a positive linear functional F on $C_c(X)$, is the only regular Borel measure that represents F.

Theorem 38.3 (The Riesz Representation Theorem). *For every positive linear functional F on $C_c(X)$, there exists a unique regular Borel measure μ such that*

$$F(f) = \int f \, d\mu$$

holds for every $f \in C_c(X)$. Moreover, the representing measure is the regular Borel measure induced by F on X.

Proof. (**Uniqueness.**) Let μ and ν be two regular Borel measures on X such that $\int f \, d\mu = \int f \, d\nu$ holds for each $f \in C_c(X)$. We shall show that $\mu(A) = \nu(A)$

holds for each Borel set A. In view of the regularity properties of μ and ν, it is enough to establish that $\mu(K) = \nu(K)$ holds for each compact set K.

To this end, let K be a compact set. Given $\epsilon > 0$, choose an open set V such that $K \subseteq V$ and $\mu(V) < \mu(K) + \epsilon$. By Theorem 10.9, there exists a function $f \in C_c(X)$ such that $K \prec f \prec V$. But then, since $\chi_K \leq f \leq \chi_V$, we have

$$\nu(K) = \int \chi_K \, d\nu \leq \int f \, d\nu = \int f \, d\mu \leq \int \chi_V \, d\mu = \mu(V) < \mu(K) + \epsilon$$

for all $\epsilon > 0$. That is, $\nu(K) \leq \mu(K)$. By the symmetry of the situation, $\mu(K) \leq \nu(K)$, and so $\mu(K) = \nu(K)$, as desired.

(**Existence.**) Let F be a positive linear functional on $C_c(X)$, and let μ be its induced outer measure. By Theorem 38.2, μ restricted to the Borel sets of X is a regular Borel measure. We shall show that $F(f) = \int f \, d\mu$ holds for each $f \in C_c(X)$.

To this end, let $f \in C_c(X)$. Fix an open set V such that $K = \text{Supp} f \subseteq V$ and $\mu(V) < \infty$. Also, choose $c > 0$ satisfying $|f(x)| < c$ for all $x \in X$.

Given $\epsilon > 0$, pick n such that $2c/n < \epsilon$, and let $y_i = -c + i(2c/n)$ for $i = 0, 1, \ldots, n$; that is, $\{y_0, y_1, \ldots, y_n\}$ is the partition of $[-c, c]$ with $y_i - y_{i-1} = 2c/n$ for $i = 1, \ldots, n$. For each $1 \leq i \leq n$ let $A_i = \{x \in K : y_{i-1} < f(x) \leq y_i\}$, and note that the open set $W_i = \{x \in V : y_i - \epsilon < f(x) < y_i + \epsilon\}$ satisfies $A_i \subseteq W_i$. Notice that the Borel sets $A_1, \ldots, A_n$ are pairwise disjoint and $\bigcup_{i=1}^{n} A_i = K$.

By the regularity of μ, for each $1 \leq i \leq n$ there exists an open set V_i satisfying $A_i \subseteq V_i \subseteq W_i$ and $\mu(V_i) - \mu(A_i) < \epsilon/n$; clearly, $K \subseteq \bigcup_{i=1}^{n} V_i \subseteq V$ holds. By Theorem 10.9 there exist functions $g_1, \ldots, g_n \in C_c(X)$ such that $g_i \prec V_i$ for $i = 1, \ldots, n$ and $\sum_{i=1}^{n} g_i(x) = 1$ for each $x \in K$. Note that $fg_i \leq (y_i + \epsilon)g_i$ holds for each i and $f = \sum_{i=1}^{n} fg_i$. Therefore,

$$F(f) - \int f \, d\mu = \sum_{i=1}^{n} F(fg_i) - \sum_{i=1}^{n} \int_{A_i} f \, d\mu$$

$$\leq \sum_{i=1}^{n} (y_i + \epsilon)F(g_i) - \sum_{i=1}^{n} (y_i - \epsilon)\mu(A_i)$$

$$\leq \sum_{i=1}^{n} (y_i + \epsilon)\mu(V_i) - \sum_{i=1}^{n} (y_i - \epsilon)\mu(A_i)$$

$$= \sum_{i=1}^{n} (y_i + \epsilon)[\mu(V_i) - \mu(A_i)] + 2\epsilon \sum_{i=1}^{n} \mu(A_i)$$

$$\leq \sum_{i=1}^{n} (c + \epsilon)\frac{\epsilon}{n} + 2\epsilon\mu(K) = \epsilon[c + \epsilon + 2\mu(K)].$$

Since $\epsilon > 0$ is arbitrary, $F(f) - \int f \, d\mu \leq 0$ holds for all $f \in C_c(X)$. Replacing f by $-f$, we get $F(f) - \int f \, d\mu \geq 0$. Thus, $F(f) = \int f \, d\mu$ holds for each $f \in C_c(X)$, and the proof of the theorem is complete. ∎

Any Borel measure that fails to be a regular Borel measure can be considered as a "pathological" case. The next result reveals that nonregular Borel measures cannot appear with "good" topological spaces. Recall that a subset of a topological space is called σ-**compact** if it is the union of a countable collection of compact sets.

Theorem 38.4. *Let X be a Hausdorff locally compact topological space whose open sets are σ-compact (for instance, each Euclidean space $\mathbb{R}^n$ has this property). Then every Borel measure on X is a regular Borel measure.*

Proof. Let ν be a Borel measure on X, i.e., $\nu(K) < \infty$ holds for each compact set K. Then every continuous function with compact support is ν-integrable, and so, $F(f) = \int f \, d\nu$, $f \in C_c(X)$, defines a positive linear functional on $C_c(X)$. By the Riesz representation theorem, there exists a regular Borel measure μ such that $F(f) = \int f \, d\mu$ holds for each $f \in C_c(X)$. To complete the proof, we shall show that $\nu = \mu$.

If K is compact, and V open with $K \subseteq V$, then there exists a function $f \in C_c(X)$ with $K \prec f \prec V$. Therefore,

$$\nu(K) = \int \chi_K \, d\nu \leq \int f \, d\nu = \int f \, d\mu \leq \int \chi_V \, d\mu = \mu(V),$$

and so, by the regularity of μ it follows that $\nu(K) \leq \mu(K)$ holds for each compact set K.

Now, let $\mathcal{O}$ be an open set. Since (by hypothesis) $\mathcal{O}$ is σ-compact, there exists a sequence $\{K_n\}$ of compact sets with $K_n \uparrow \mathcal{O}$. But then the inequalities $\nu(K_n) \leq \mu(K_n)$ imply $\nu(\mathcal{O}) = \lim \nu(K_n) \leq \lim \mu(K_n) = \mu(\mathcal{O})$. On the other hand, for each n there exists some $f_n \in C_c(X)$ with $K_n \prec f_n \prec \mathcal{O}$, and so,

$$\mu(K_n) = \int \chi_{K_n} \, d\mu \leq \int f_n \, d\mu = \int f_n \, d\nu \leq \nu(\mathcal{O})$$

holds for each n. It follows that $\mu(\mathcal{O}) \leq \nu(\mathcal{O})$, and thus, $\nu(\mathcal{O}) = \mu(\mathcal{O})$ for each open set $\mathcal{O}$.

Next, let K be a compact set. Choose an open set $\mathcal{O}$ with $K \subseteq \mathcal{O}$ and $\nu(\mathcal{O}) = \mu(\mathcal{O}) < \infty$. Then $\mu(\mathcal{O}) - \mu(K) = \mu(\mathcal{O} \setminus K) = \nu(\mathcal{O} \setminus K) = \nu(\mathcal{O}) - \nu(K)$, and so, $\nu(K) = \mu(K)$ for each compact set K.

Finally, if B is an arbitrary Borel set, K a compact set, and $\mathcal{O}$ an open set such that $K \subseteq B \subseteq \mathcal{O}$, then the inequalities

$$\mu(K) = \nu(K) \le \nu(B) \le \nu(\mathcal{O}) = \mu(\mathcal{O}),$$

coupled with the regularity of μ, imply $\nu(B) = \mu(B)$. The proof of the theorem is now complete.   ∎

With the usual algebraic and lattice operations and the sup norm, $C_c(X)$ is a normed vector lattice. Our next objective is to describe the norm dual of $C_c(X)$. To do this, we need some algebraic and lattice properties of the regular Borel measures.

Theorem 38.5. *For two regular Borel measures μ and ν on X, we have the following:*

1. *$\mu + \nu$ and $\alpha\mu$ for $\alpha \ge 0$ are regular Borel measures.*
2. *$\mu \vee \nu$ is a regular Borel measure.*
3. *If μ and ν are both σ-finite, then $\mu \wedge \nu$ is also a regular Borel measure.*

Proof.   (1) Obvious.
(2) We start the proof of this part by recalling that

$$\mu \vee \nu(A) = \sup\{\mu(B) + \nu(A \setminus B) \colon B \in \mathcal{B} \text{ and } B \subseteq A\}$$

holds for each $A \in \mathcal{B}$. The proof now goes by steps.
(a) Let K be a compact subset of X. Since $\mu \vee \nu(K) \le \mu(K) + \nu(K) < \infty$ holds, it follows that $\mu \vee \nu$ is a Borel measure.
(b) Let $A \in \mathcal{B}$ satisfy $\mu \vee \nu(A) < \infty$. Then $\mu(A) < \infty$ and $\nu(A) < \infty$ hold. Let $a = \sup\{\mu \vee \nu(K) \colon K \text{ compact and } K \subseteq A\}$.
If $B \in \mathcal{B}$ satisfies $B \subseteq A$, then by the regularity of μ and ν there exist two compact sets K_1 and K_2 such that $K_1 \subseteq B$, $K_2 \subseteq A \setminus B$, $\mu(K_1) > \mu(B) - \epsilon$, and $\nu(K_2) > \nu(A \setminus B) - \epsilon$, where, of course, $\epsilon > 0$ is given. Then, $K_1 \cap K_2 = \emptyset$ and

$$\mu \vee \nu(A) \ge a \ge \mu \vee \nu(K_1 \cup K_2) \ge \mu(K_1) + \nu(K_2)$$
$$> \mu(B) - \epsilon + \nu(A \setminus B) - \epsilon = \mu(B) + \nu(A \setminus B) - 2\epsilon.$$

This implies $\mu \vee \nu(A) \ge a \ge \mu \vee \nu(A) - 2\epsilon$ for each $\epsilon > 0$. Hence, $a = \mu \vee \nu(A)$.
Now, let A be an open set such that $\mu \vee \nu(A) = \infty$. Since $\mu \vee \nu \le \mu + \nu$ holds, it follows that either $\mu(A) = \infty$ or $\nu(A) = \infty$. We can assume that $\mu(A) = \infty$. Since $\mu \le \mu \vee \nu$ and $\sup\{\mu(K) \colon K \text{ compact and } K \subseteq A\} = \mu(A) = \infty$, it easily

follows that

$$\sup\{\mu \vee v(K): \ K \text{ compact and } K \subseteq A\} = \infty = \mu \vee v(A).$$

(c) Let $A \in \mathcal{B}$ and put $b = \inf\{\mu\vee v(V): V \text{ open and } A \subseteq V\}$. If $\mu\vee v(A) = \infty$, then $b = \mu\vee v(A) = \infty$ is obvious. Hence, we can assume $\mu\vee v(A) < \infty$. Clearly, $\mu(A) < \infty$ and $v(A) < \infty$ both hold. Since μ and v are regular Borel measures, there exists an open set $\mathcal{O}$ such that $A \subseteq \mathcal{O}$, $\mu(\mathcal{O}) < \infty$, and $v(\mathcal{O}) < \infty$.

Now, let $\epsilon > 0$. By (a) there exists a compact set K such that $K \subseteq \mathcal{O} \setminus A$ and $\mu \vee v(K) > \mu \vee v(\mathcal{O} \setminus A) - \epsilon$; and so, $\mu \vee v(K) > \mu \vee v(\mathcal{O}) - \mu \vee v(A) - \epsilon$. Thus, the open set $V = \mathcal{O} \setminus K$ satisfies $A \subseteq V$ and

$$\mu \vee v(A) + \epsilon > \mu \vee v(\mathcal{O}) - \mu \vee v(K) = \mu \vee v(\mathcal{O} \setminus K) = \mu \vee v(V) \geq \mu \vee v(A).$$

This implies that $b = \mu \vee v(A)$ holds, and thus, $\mu \vee v$ is a regular Borel measure.

(3) Since $\mu \wedge v \leq \mu$, it is easy to see that $\mu \wedge v$ is a Borel measure. Also, recall that

$$\mu \wedge v(A) = \inf\{\mu(B) + v(A \setminus B): B \in \mathcal{B} \text{ and } B \subseteq A\}$$

for each $A \in \mathcal{B}$.

(α) Let $A \in \mathcal{B}$, put $c = \inf\{\mu \wedge v(V): V \text{ open and } A \subseteq V\}$, and let $\epsilon > 0$. Given $B \in \mathcal{B}$ with $B \subseteq A$, choose two open sets V_1 and V_2 such that $B \subseteq V_1$, $A \setminus B \subseteq V_2$, $\mu(V_1) \leq \mu(B) + \epsilon$, and $v(V_2) \leq v(A \setminus B) + \epsilon$. Then

$$\mu \wedge v(A) \leq c \leq \mu \wedge v(V_1 \cup V_2) \leq \mu \wedge v(V_1) + \mu \wedge v(V_2) \leq \mu(V_1) + v(V_2)$$
$$\leq \mu(B) + \epsilon + v(A \setminus B) + \epsilon = \mu(B) + v(A \setminus B) + 2\epsilon.$$

Thus, $\mu \wedge v(A) \leq c \leq \mu \wedge v(A) + 2\epsilon$ holds for each $\epsilon > 0$, and so, $c = \mu \wedge v(A)$.

(β) Let $A \in \mathcal{B}$ satisfy $\mu \vee v(A) < \infty$. Then the relation

$$\mu \wedge v(K) = \mu(K) + v(K) - \mu \vee v(K)$$

for each compact set $K \subseteq A$ and the regularity of μ, v, and $\mu \vee v$ easily imply that

$$\mu \wedge v(A) = \sup\{\mu \wedge v(K): K \text{ compact and } K \subseteq A\}. \qquad (\star)$$

On the other hand, using the preceding conclusion and that μ and v are both σ-finite, it is easy to see that $(\star)$ holds for each open set A. (Note that this is the only place where the σ-finiteness of μ and v is used.) Thus, $\mu \wedge v$ is a regular Borel measure, and the proof of the theorem is complete. ∎

Observe that if μ and ν are regular Borel measures, then $\mu + \nu$ and $\alpha\mu$ (for $\alpha \geq 0$) satisfy

$$\int f \, d(\mu + \nu) = \int f \, d\mu + \int f \, d\nu \quad \text{and} \quad \int f \, d(\alpha\mu) = \alpha \int f \, d\mu$$

for each $f \in C_c(X)$.

In identifying the lattice isomorphisms among positive operators between two vector lattices, the following lemma is often very useful:

Lemma 38.6. *Let $T: X \to Y$ be a one-to-one, onto linear operator between two vector lattices. If T and T^{-1} are both positive, then T is a lattice isomorphism.*

Proof. Let $x, y \in X$. Since $x \leq x \vee y$ holds, it follows from the positivity of T that $T(x) \leq T(x \vee y)$. Similarly, $T(y) \leq T(x \vee y)$, and therefore,

$$T(x) \vee T(y) \leq T(x \vee y). \tag{$\star\star$}$$

Reversing the roles of T, x, and y to those of T^{-1}, $T(x)$, and $T(y)$, respectively, it follows from $(\star\star)$ that

$$x \vee y = T^{-1}(T(x)) \vee T^{-1}(T(y)) \leq T^{-1}(T(x) \vee T(y)).$$

By applying T to the last inequality, we get $T(x \vee y) \leq T(x) \vee T(y)$. That is, $T(x \vee y) = T(x) \vee T(y)$ holds, as required. ∎

Next, the norm dual of $C_c(X)$ will be identified. We already know that $C_c^*(X)$ is a Banach lattice (Theorem 30.9), and so, every continuous linear functional on $C_c(X)$ can be written as a difference of two positive continuous linear functionals.

Let us denote by $M_b(X)$ the collection of all finite signed measures that can be written as a difference of two positive regular Borel measures. That is,

$$M_b(X) = \{\mu \in M(\mathcal{B}) : \exists \text{ regular Borel measures } \mu_1, \mu_2 \text{ with } \mu = \mu_1 - \mu_2\}.$$

Clearly, $M_b(X)$ is a vector subspace of $M(\mathcal{B})$, the Banach lattice of all finite signed measures on $\mathcal{B}$ (see Theorem 36.10). Moreover, since $\mu^+ = (\mu_1 - \mu_2)^+ = \mu_1 \vee \mu_2 - \mu_2$ and (by Theorem 38.5) $\mu_1 \vee \mu_2$ is a finite regular Borel measure, it follows that μ^+ belongs to $M_b(X)$. Therefore, $M_b(X)$ is a vector sublattice of $M(\mathcal{B})$. Thus, $M_b(X)$ is a normed vector lattice having, of course, the total variation as its norm, that is, $\|\mu\| = |\mu|(X)$ for each $\mu \in M_b(X)$.

Also, it is easy to see that if μ_1 and μ_2 are two finite regular Borel measures, then $(\mu_1 - \mu_2)^+ = \mu_1 \vee \mu_2 - \mu_2$ is likewise a finite regular Borel measure. This implies that $M_b(X)$ could be defined as follows:

$$M_b(X) = \{\mu \in M(\mathcal{B})\colon \mu^+ \text{ and } \mu^- \text{ are both finite regular Borel measures}\}.$$

Now, assume that $\mu = \mu_1 - \mu_2 = \nu_1 - \nu_2$ holds with μ_1, μ_2 and ν_1, ν_2 finite regular Borel measures. Then $\mu_1 + \nu_2 = \nu_1 + \mu_2$, and so, $\int f\, d\mu_1 + \int f\, d\nu_2 = \int f\, d\nu_1 + \int f\, d\mu_2$ holds for each $f \in C_c(X)$; therefore,

$$\int f\, d\mu_1 - \int f\, d\mu_2 = \int f\, d\nu_1 - \int f\, d\nu_2$$

holds for each $f \in C_c(X)$. This shows that if $\mu = \mu_1 - \mu_2 \in M_b(X)$, then

$$F_\mu(f) = \int f\, d\mu_1 - \int f\, d\mu_2$$

defines a linear functional on $C_c(X)$ which is independent of the representation of μ as a difference of two finite regular Borel measures. [It is also a custom to write $\int f\, d\mu$ instead of $F_\mu(f)$]. Moreover, the estimate

$$|F_\mu(f)| \le \left| \int f\, d\mu_1 \right| + \left| \int f\, d\mu_2 \right| \le [\mu_1(X) + \mu_2(X)] \cdot \|f\|_\infty$$

shows that F_μ is a continuous linear functional, that is, $F_\mu \in C_c^*(X)$.

Thus, a mapping $\mu \mapsto F_\mu$ from $M_b(X)$ to $C_c^*(X)$ can be established. Clearly, this mapping is linear, and as the next theorem shows it is onto and a lattice isometry.

Theorem 38.7. *The mapping $\mu \mapsto F_\mu$ (defined above) is a lattice isometry from $M_b(X)$ onto $C_c^*(X)$. That is, $C_c^*(X) = M_b(X)$ holds.*

Proof. Note first that $\mu \mapsto F_\mu$ is one-to-one. Indeed, if $\mu = \mu_1 - \mu_2$ and $F_\mu = 0$, then $\int f\, d\mu_1 = \int f\, d\mu_2$ holds for each $f \in C_c(X)$, and so, by the Riesz representation theorem, $\mu_1 = \mu_2$. That is, $\mu = \mu_1 - \mu_2 = 0$.

Now, let F be a positive linear functional on $C_c(X)$ and μ its representing regular Borel measure [i.e., $F(f) = \int f\, d\mu$ for each $f \in C_c(X)$]. Since $0 \le f \in C_c(X)$ satisfies $\|f\|_\infty \le 1$ if and only if $f \prec X$, it follows that

$$\mu(X) = \sup\{F(f)\colon f \prec X\} = \sup\{F(f)\colon \|f\|_\infty \le 1\} = \|F\|. \qquad (\dagger)$$

Thus, F is continuous if and only if $\mu(X) < \infty$, i.e., if and only if $\mu \in M_b(X)$. From this and the fact that every continuous linear functional can be written as a difference of two positive continuous linear functionals, it follows that $\mu \mapsto F_\mu$ is onto. Moreover, it is easy to see that $\mu \geq 0$ holds in $M_b(X)$ if and only if $F_\mu \geq 0$. Thus, by Lemma 38.6, $\mu \mapsto F_\mu$ is a lattice isomorphism from $M_b(X)$ onto $C_c^*(X)$.

Finally, since both $C_c^*(X)$ and $M_b(X)$ are normed vector lattices, it follows from (†) that

$$\|F_\mu\| = \| \, |F_\mu| \, \| = \|F_{|\mu|}\| = |\mu|(X) = \|\mu\|,$$

so that $\mu \mapsto F_\mu$ is a lattice isometry.   ∎

It should be noticed that Theorem 38.7 gives an indirect proof that $M_b(X)$ is a Banach lattice. The members of $M_b(X)$ are also referred to as (finite) **regular Borel signed measures** on X, and they are characterized as follows: A finite signed measure μ on $\mathcal{B}$ belongs to $M_b(X)$ if and only if for each $A \in \mathcal{B}$ and $\epsilon > 0$ there exists a compact set K and an open set V with $K \subseteq A \subseteq V$, so that $|\mu(B)| < \epsilon$ holds for all $B \in \mathcal{B}$ with $B \subseteq V \setminus K$; see Exercise 10 at the end of this section.

Finally, it is a custom to call a signed measure μ on $\mathcal{B}$ a **Borel signed measure** if $\mu(K) \in \mathbb{R}$ holds for every compact set K. This is, of course equivalent to saying that μ^+ and μ^- are both Borel measures. Indeed, if this is the case, then in view of $\mu = \mu^+ - \mu^-$ we have $\mu(K) \in \mathbb{R}$ for each compact set K. On the other hand, if μ is a Borel signed measure, then since at least one of the measures μ^+ and μ^- is a finite measure, it follows that both μ^+ and μ^- are Borel measures.

We shall close the section with a brief introduction to the Haar[4] integral which is associated with Hausdorff locally compact groups. To understand the Haar integral, we need some background.

Recall that a **group** is a nonempty set G together with a function $(x, y) \mapsto x \star y$ (from $G \times G$ to G) which is associative (i.e., $x \star (y \star z) = (x \star y) \star z$ for all $x, y, z \in G$), has an identity element e (i.e., $x \star e = e \star x = x$ for all $x \in G$) and every element has an inverse (i.e., for each $x \in G$ there exists some x^{-1} such that $x \star x^{-1} = x^{-1} \star x = e$). It is customary to write xy instead of $x \star y$. If $xy = yx$ holds for all $x, y \in G$, then G is called a **commutative group**. A **topological group** is a group G equipped with a topology under which the function $(x, y) \mapsto xy^{-1}$ (from $G \times G$ to G) is continuous. The Euclidean spaces $\mathbb{R}^n$ under addition are examples of commutative Hausdorff locally compact topological groups.

Now, let G be a Hausdorff locally compact topological group. If $f \in C_c(G)$, then the **left** and **right translates** of f by an element $a \in G$ are the functions f_a

---

[4] Alfred Haar (1885–1933), a Hungarian mathematician. He studied orthogonal systems of functions and partial differential equations and is best remembered for his introduction of the integral that bears his name on Hausdorff locally compact topological groups.

and f^a defined respectively by

$$f_a(x) = f(ax) \quad \text{and} \quad f^a(x) = f(xa)$$

for each $x \in G$. Note that f_a and f^a both belong to $C_c(G)$ and that $f_a = f^a$ holds if G is commutative. A linear functional $F: C_c(G) \to \mathbb{R}$ is said to be **left-invariant** (resp. **right-invariant**) if $F(f) = F(f_a)$ (resp. $F(f) = F(f^a)$) holds for all $a \in G$.

Here is the fundamental result regarding left-invariant linear functionals.

Theorem 38.8 (Haar). *If G is a Hausdorff locally compact topological group, then there exists a unique (aside of scalar factors) non-zero positive linear functional $H: C_c(G) \to \mathbb{R}$ which is left-invariant.*

By the symmetry of the situation, there exists, of course, also a unique non-zero positive linear functional on $C_c(G)$ which is right-invariant. The unique non-zero left-invariant positive linear functional $H: C_c(G) \to \mathbb{R}$ is called the **left-Haar integral** of G. If G is commutative, then the left- and right-Haar integrals coincide and the common integral is called the **Haar integral** of G.

Now, let G be a Hausdorff locally compact topological group and let $H: C_c(G) \to \mathbb{R}$ be its unique left-Haar integral. By the Riesz Representation Theorem 38.3, there exists a unique (aside of scalar factors) regular Borel measure μ_G satisfying

$$H(f) = \int_G f(g) \, d\mu_G(g).$$

It is easy to prove that the left-Haar measure satisfies $\mu_G(aA) = \mu_G(A)$ for each Borel set A. The measure μ_G is called the **left Haar measure** of G. Here are two examples of the Haar integral and the Haar measure on two familiar Hausdorff locally compact topological groups.

1. Consider the additive Hausdorff locally compact topological group $G = \mathbb{R}^n$. Its Haar integral coincides with the Lebesgue integral and its Haar measure is the Lebesgue measure.

2. Consider $G = (0, \infty)$ as a commutative group under the ordinary multiplication. Then the Haar integral $H: G \to \mathbb{R}$ is given by $H(f) = \int_0^\infty \frac{f(x)}{x} \, dx$ for each $f \in C_c(G)$ and the Haar measure satisfies $\mu_G(A) = \int_A \frac{dx}{x}$ for each Borel subset A of G.

For proofs and details regarding the Haar measure and Haar integral we refer the reader to [14, Chapter 6].

EXERCISES

Unless otherwise specified, in the exercises below X will denote a Hausdorff locally compact topological space.

1. If X is a compact topological space, then show that a continuous linear functional F on $C(X)$ is positive if and only if $F(1) = \|F\|$ holds.

2. Let X be a compact topological space, and let F and G be two positive linear functionals on $C(X)$. If $F(1) + G(1) \leq \|F - G\|$, then show that $F \wedge G = 0$.
   [HINT: Show that $F \vee G(1) = \|F - G\|$.]

3. Let $c_0(X) = \{f \in C(X): \forall \epsilon > 0 \exists K$ compact with $|f(x)| < \epsilon \ \forall x \notin K\}$. Show that:

   a. $c_0(X)$ equipped with the sup norm is a Banach lattice.
   b. The norm completion of $C_c(X)$ is the Banach lattice $c_0(X)$.

4. Let F be a positive linear functional on $C_c(X)$, and let μ be the outer measure induced by F on X. Show that if μ^* is the outer measure generated by the measure space $(X, \mathcal{B}, \mu)$, then $\mu^*(A) = \mu(A)$ holds for every subset A of X.

5. Let μ and ν be two regular Borel measures on X. Then show that $\mu \geq \nu$ holds if and only if $\int f \, d\mu \geq \int f \, d\nu$ for each $0 \leq f \in C_c(X)$.

6. Fix a point $x \in X$, and define $F(f) = f(x)$ for each $f \in C_c(X)$. Show that F is a positive linear functional on $C_c(X)$ and then describe the unique regular Borel measure μ that satisfies $F(f) = \int f \, d\mu$ for each $f \in C_c(X)$. What is the support of μ?

7. Let X be a compact Hausdorff topological space. If μ and ν are regular Borel measures, then show that the regular Borel measures $\mu \vee \nu$ and $\mu \wedge \nu$ satisfy

   a. $\text{Supp}(\mu \vee \nu) = \text{Supp} \, \mu \cup \text{Supp} \, \nu$, and
   b. $\text{Supp}(\mu \wedge \nu) \subseteq \text{Supp} \, \mu \cap \text{Supp} \, \nu$.

   Use (b) to show that if $\text{Supp} \, \mu \cap \text{Supp} \, \nu = \emptyset$, then $\mu \perp \nu$ holds. Also, give an example for which $\text{Supp}(\mu \wedge \nu) \neq \text{Supp} \, \mu \cap \text{Supp} \, \nu$.

8. Characterize the positive linear functionals F on $C_c(X)$ that are also lattice homomorphisms, that is, $F(f \vee g) = \max\{F(f), F(g)\}$ holds for each pair $f, g \in C_c(X)$.
   [HINT: Let μ be the representing regular Borel measure for F. Show that $\text{Supp} \, \mu$ contains at most one point.]

9. If X is an uncountable set, then show that

   a. $C_c^*(X)$ is not separable, and
   b. $C[0, 1]$ (with the sup norm) is not a reflexive Banach space.

   [HINT: For every $x \in X$ define the positive linear functional $F_x(f) = f(x)$. Show that $\|F_x - F_y\| = 2$ if $x \neq y$. For (b) combine (a) with Exercise 8 of Section 27 and Exercise 12 of Section 12.]

10. For a finite signed measure μ on $\mathcal{B}$, show that the following statements are equivalent:

    a. μ belongs to $M_b(X)$.
    b. μ^+ and μ^- are both finite regular Borel measures.
    c. For each $A \in \mathcal{B}$ and $\epsilon > 0$, there exist a compact set K and an open set V with $K \subseteq A \subseteq V$ such that $|\mu(B)| < \epsilon$ holds for all $B \in \mathcal{B}$ with $B \subseteq V \setminus K$.

11. A sequence $\{x_n\}$ in a normed space is said to **converge weakly** to some vector x if $\lim f(x_n) = f(x)$ holds for every continuous linear functional f.

    a. Show that a sequence in a normed space can have at most one weak limit.
    b. Let X be a Hausdorff compact topological space. Then show that a sequence $\{f_n\}$ of $C(X)$ converges weakly to some function $f \in C(X)$ if and only if $\{f_n\}$ is norm bounded and $\lim f_n(x) = f(x)$ holds for each $x \in X$.

    [HINT: For (b) use Theorem 28.8 and the Riesz representation theorem.]

12. Let μ be a regular Borel measure on X, and let $f \in L_1(\mu)$. Show that the finite signed measure ν, defined by

$$\nu(E) = \int_E f \, d\mu$$

    for each Borel set E, is a (finite) regular Borel signed measure. In other words, show that $\nu \in M_b(X)$.
    [HINT: Use the fact that $\{x \in X : f(x) \neq 0\}$ is a σ-finite set.]

13. Generalize part (3) of Theorem 38.5 as follows: If μ and ν are two regular Borel measures on a Hausdorff locally compact topological space and one of them is σ-finite, then show that $\mu \wedge \nu$ is also a regular Borel measure.

14. Show that every finite Borel measure on a complete separable metric space is a regular Borel measure. Use this conclusion to present an alternate proof of the fact that the Lebesgue measure is a regular Borel measure.

15. Let X be a Hausdorff compact topological space. If $\phi: X \to X$ is a continuous function, then show that there exists a regular Borel measure on X such that

$$\int f \circ \phi \, d\mu = \int f \, d\mu$$

    holds for each $f \in C(X)$.
    [HINT: If $\mathcal{L}im$ is a Banach–Mazur limit (see Exercise 7 of Section 29) and $\omega \in X$ is a fixed point, then consider the positive linear functional $F: C(X) \to \mathbb{R}$ defined by

$$F(f) = \mathcal{L}im(f(\phi(\omega)), f(\phi^2(\omega)), f(\phi^3(\omega)), \ldots).]$$

16. This exercise gives an identification of the order dual $C_c^{\sim}(X)$ of $C_c(X)$. Consider the collection $\mathcal{M}(X)$ of all formal expressions $\mu_1 - \mu_2$ with μ_1 and μ_2 regular Borel measures. That is,

$$\mathcal{M}(X) = \{\mu_1 - \mu_2 : \mu_1 \text{ and } \mu_2 \text{ are regular Borel measures on } X\}.$$

    a. Define $\mu_1 - \mu_2 \equiv \nu_1 - \nu_2$ in $\mathcal{M}(X)$ to mean $\mu_1(A) + \nu_2(A) = \nu_1(A) + \mu_2(A)$ for all $A \in \mathcal{B}$. Show that $\equiv$ is an equivalence relation.
    b. Denote the collection of all equivalence classes by $\mathcal{M}(X)$ again. That is, $\mu_1 - \mu_2$ and $\nu_1 - \nu_2$ are considered to be identical if $\mu_1 + \nu_2 = \nu_1 + \mu_2$ holds. In $\mathcal{M}(X)$ define the algebraic operations

$$(\mu_1 - \mu_2) + (\nu_1 - \nu_2) = (\mu_1 + \nu_1) - (\mu_2 + \nu_2),$$

$$\alpha(\mu_1 - \mu_2) = \begin{cases} \alpha\mu_1 - \alpha\mu_2 & \text{if } \alpha \geq 0 \\ (-\alpha)\mu_2 - (-\alpha)\mu_1 & \text{if } \alpha < 0. \end{cases}$$

Show that these operations are well-defined (i.e., show that they depend only upon the equivalence classes) and that they make $\mathcal{M}(X)$ a vector space.

c.   Define an ordering in $\mathcal{M}(X)$ by $\mu_1 - \mu_2 \geq \nu_1 - \nu_2$ whenever $\mu_1(A) + \nu_2(A) \geq \nu_1(A) + \mu_2(A)$ holds for each $A \in \mathcal{B}$. Show that $\geq$ is well defined and that it is an order relation on $\mathcal{M}(X)$ under which $\mathcal{M}(X)$ is a vector lattice.

d.   Consider the mapping $\mu = \mu_1 - \mu_2 \mapsto F_\mu$, from $\mathcal{M}(X)$ to $C_c^\sim(X)$, defined by $F_\mu(f) = \int f \, d\mu_1 - \int f \, d\mu_2$ for each $f \in C_c(X)$. Show that F_μ is well-defined and that $\mu \mapsto F_\mu$ is a lattice isomorphism (Lemma 38.6 may be helpful here) from $\mathcal{M}(X)$ onto $C_c^\sim(X)$. That is, show that $C_c^\sim(X) = \mathcal{M}(X)$ holds.

17.   This exercise shows that for a noncompact space X, in general $C_c^*(X)$ is a proper ideal of $C_c^\sim(X)$. Let X be a Hausdorff locally compact topological space having a sequence $\{\mathcal{O}_n\}$ of open sets such that $\mathcal{O}_n \subseteq \mathcal{O}_{n+1}$ and $\mathcal{O}_n \neq \mathcal{O}_{n+1}$ for each n, and with $X = \bigcup_{n=1}^\infty \mathcal{O}_n$.

a.   Show that if X is σ-compact but not a compact space, then X admits a sequence $\{\mathcal{O}_n\}$ of open sets with the above properties.

b.   Choose $x_1 \in \mathcal{O}_1$ and $x_n \in \mathcal{O}_n \setminus \mathcal{O}_{n-1}$ for $n \geq 2$. Then show that

$$F(f) = \sum_{n=1}^\infty f(x_n), \quad \text{for } f \in C_c(X),$$

defines a positive linear functional on $C_c(X)$ that is not continuous.

c.   Determine the (unique) regular Borel measure μ on X that represents F. What is the support of μ?

39.   DIFFERENTIATION AND INTEGRATION

In this section the differentiability properties of a Borel signed measure on $\mathbb{R}^k$ will be studied. The derivative of a measure will always be taken with respect to the Lebesgue measure. For a differentiation theory of arbitrary set functions in a more general context, the interested reader is referred to Chapter 8 of [34]. The results obtained about differentiation of signed measures will be used to derive the classical properties of functions of bounded variation.

Throughout this section, the domain of all signed measures will be the σ-algebra $\mathcal{B}$ of all Borel sets of $\mathbb{R}^k$. Also, unless otherwise specified, the expression "almost everywhere" will be synonymous with "almost everywhere with respect to the Lebesgue measure." The main Borel sets considered for the purposes of differentiation will be the open balls. Recall that the **open ball** with center at x and radius r is the subset of $\mathbb{R}^k$ defined by $\{y \in \mathbb{R}^k : \|x - y\| < r\}$, where $\|x - y\| = [\sum_{i=1}^k (x_i - y_i)^2]^{\frac{1}{2}}$.

Occasionally, measures defined on the Borel subsets of an open set V of $\mathbb{R}^k$ appear. By assigning the value zero to the Borel subsets of V^c, it is easy to see that these measures can be assumed defined on all Borel sets of $\mathbb{R}^k$. Therefore, theorems about Borel measures on $\mathbb{R}^k$ can be applied equally well to such a situation.

Let μ be a signed measure on the σ-algebra of all Borel sets of $\mathbb{R}^k$. For each $x \in \mathbb{R}^k$ and $r > 0$, we define the following two extended real numbers:

$$\Delta_r^*(x) = \sup \left\{ \frac{\mu(B)}{\lambda(B)} : B \text{ is an open ball of radius } \leq r \text{ and } x \in B \right\}, \text{ and}$$

$$\Delta_r(x) = \inf \left\{ \frac{\mu(B)}{\lambda(B)} : B \text{ is an open ball of radius } \leq r \text{ and } x \in B \right\}.$$

Observe that x is not required to be the center of the open balls B where the sup and inf are taken. Clearly,

$$-\infty \leq \Delta_r(x) \leq \Delta_r^*(x) \leq \infty$$

holds for each $x \in \mathbb{R}^k$ and all $r > 0$.

Also, it should be obvious that if $0 < r < s$, then $\Delta_r^*(x) \leq \Delta_s^*(x)$, and $\Delta_s(x) \leq \Delta_r(x)$ both hold for each x. Therefore, the limits

$$D^*\mu(x) = \lim_{r \downarrow 0} \Delta_r^*(x) \quad \text{and} \quad D_*\mu(x) = \lim_{r \downarrow 0} \Delta_r(x)$$

exist in $\mathbb{R}^*$, and they satisfy $-\infty \leq D_*\mu(x) \leq D^*\mu(x) \leq \infty$ for each $x \in \mathbb{R}^k$. The extended real numbers $D_*\mu(x)$ and $D^*\mu(x)$ are called the **lower** and **upper derivatives** of μ (with respect to λ) at the point x. If they are real and equal, then this common value is called the derivative of μ at x.

Definition 39.1. *Let μ be a signed measure on $\mathcal{B}$ and $x \in \mathbb{R}^k$. If*

$$-\infty < D_*\mu(x) = D^*\mu(x) < \infty$$

*holds, then μ is said to be **differentiable** at the point x. The common value is called the **derivative** of μ at x and is denoted by $D\mu(x)$, that is,*

$$D\mu(x) = D_*\mu(x) = D^*\mu(x).$$

Here is a rephrasing of the preceding definition: A real number m satisfies $m = D\mu(x)$ if for each $\epsilon > 0$, there exists some $\delta > 0$ such that $|\frac{\mu(B)}{\lambda(B)} - m| < \epsilon$ holds for each open ball B with $x \in B$ and with a radius less than δ.

An alternative definition of the derivative using sequences is this: A signed measure μ on $\mathcal{B}$ is differentiable at some point $x \in \mathbb{R}^k$ if and only if there exists a real number m such that $\lim \frac{\mu(B_n)}{\lambda(B_n)} = m$ holds for every sequence $\{B_n\}$ of open balls containing x whose radii tend to zero. (The number m is, of course, the derivative $D\mu(x)$.)

The expression "$D\mu(x)$ exists" is synonymous (as usual) with "μ is differentiable at x." It is easy to see that if two signed measures μ and ν are differentiable at some point x and their sum $\mu + \nu$ defines a signed measure, then $\mu + \nu$ is also differentiable at x and

$$D(\mu + \nu)(x) = D\mu(x) + D\nu(x)$$

holds. Similarly, if $\mu - \nu$ defines a signed measure, then

$$D(\mu - \nu)(x) = D\mu(x) - D\nu(x).$$

Our first objective is to show that a finite signed Borel measure μ that is absolutely continuous with respect to the Lebesgue measure is differentiable almost everywhere and that the derivative $D\mu$ coincides with the Radon–Nikodym derivative $d\mu/d\lambda$. This will be the key differentiation result for this section. To prove this theorem, we need some preliminary discussion.

The reader can verify easily the following property of the Lebesgue measure: If A is a subset of $\mathbb{R}^k$ and $r > 0$, then the set $rA = \{rx: x \in A\}$ satisfies $\lambda(rA) = r^k\lambda(A)$.

This, combined with the fact that λ is translation invariant, shows that if B is an open ball with radius r and B^* is another open ball with radius αr, then $\lambda(B^*) = \alpha^k\lambda(B)$ holds. The latter will be used in the proof of the next lemma.

Lemma 39.2. *Let $B_1, \ldots, B_n$ be open balls in $\mathbb{R}^k$. Then there exist pairwise disjoint open balls $B_{k_1}, \ldots, B_{k_m}$ among the $B_1, \ldots, B_n$ such that*

$$\lambda\left(\bigcup_{i=1}^n B_i\right) \le 3^k \sum_{j=1}^m \lambda\left(B_{k_j}\right).$$

Proof. Let r_i be the radius of B_i. Rearranging the balls, we can assume without loss of generality that $r_1 \ge r_2 \ge \cdots \ge r_n$.

Put $k_1 = 1$, and let k_2 be the smallest integer (if there is any) such that B_{k_2} is disjoint from B_{k_1}. Let k_3 be the smallest integer such that B_{k_3} is disjoint from B_{k_1} and B_{k_2}. Continue this process until the finite set $\{B_1, \ldots, B_n\}$ has been exhausted. This gives us the open balls $B_{k_1}, \ldots, B_{k_m}$, and we claim that they satisfy the property of the lemma.

To see this, let A_{k_i} be the open ball with the same center as B_{k_i} and with radius three times that of B_{k_i}. Now, note that each B_i must intersect some B_{k_j}. It follows that $B_i \subseteq A_{k_j}$ must hold for that j. Thus, $\bigcup_{i=1}^n B_i \subseteq \bigcup_{j=1}^m A_{k_j}$, and so,

$$\lambda\left(\bigcup_{i=1}^n B_i\right) \le \lambda\left(\bigcup_{j=1}^m A_{k_j}\right) \le \sum_{j=1}^m \lambda(A_{k_j}) = 3^k \sum_{j=1}^m \lambda\left(B_{k_j}\right),$$

as claimed. ∎

Now, let μ be a signed measure on $\mathbb{R}^k$. From the definition of $\Delta_r^*(x)$, it is easy to see that for each $a \in \mathbb{R}$ and $r > 0$ the set $\{x \in \mathbb{R}^k \colon \Delta_r^*(x) > a\}$ is an open set. Thus, if $\{r_n\}$ is a sequence of positive real numbers with $r_n \downarrow 0$, then the identity

$$\{x \in \mathbb{R}^k \colon D^*\mu(x) > a\} = \bigcup_{m=1}^{\infty} \bigcap_{n=1}^{\infty} \left\{ x \in \mathbb{R}^k \colon \Delta_{r_n}^*(x) > a + \frac{1}{m} \right\}$$

shows that $\{x \in \mathbb{R}^k \colon D^*\mu(x) > a\}$ is a Borel set for each $a \in \mathbb{R}$. This observation will be used in the next key lemma.

Lemma 39.3. *Let μ be a Borel measure on $\mathbb{R}^k$, and let A be a Borel set such that $\mu(A) = 0$. Then $D\mu(x) = 0$ holds for almost all points x in A.*

Proof. Let μ be a Borel measure on $\mathbb{R}^k$, and let A be a Borel set such that $\mu(A) = 0$. Since μ is a measure, $0 \le D_*\mu(x) \le D^*\mu(x)$ holds for each x. To establish the lemma it must be shown that $\lambda(\{x \in A \colon D^*\mu(x) > 0\}) = 0$. For this, it is enough to show that $\lambda(\{x \in A \colon D^*\mu(x) > \epsilon\}) = 0$ for each $\epsilon > 0$.

To this end, fix $\epsilon > 0$. Let $E = \{x \in A \colon D^*\mu(x) > \epsilon\}$, and note that E is a Borel set. Let K be an arbitrary compact subset of E and V an arbitrary open set such that $A \subseteq V$. Now, if $x \in K$, then $D^*\mu(x) > \epsilon$ holds, and so, there exists an open ball B containing x with $B \subseteq V$ and $\mu(B) > \epsilon\lambda(B)$. In view of the compactness of K, there exists a finite number of such open balls $B_1, \ldots, B_n$ that cover K. Let $B_{k_1}, \ldots, B_{k_m}$ be the disjoint open balls among $B_1, \ldots, B_n$ that satisfy Lemma 39.2. Then

$$\lambda(K) \le \lambda\left(\bigcup_{i=1}^{n} B_i \right) \le 3^k \sum_{j=1}^{m} \lambda(B_{k_j}) \le \frac{3^k}{\epsilon} \sum_{j=1}^{m} \mu(B_{k_j})$$

$$= \frac{3^k}{\epsilon} \mu\left(\bigcup_{j=1}^{m} B_{k_j} \right) \le \frac{3^k}{\epsilon} \mu(V).$$

Since μ is also regular Borel measure (Theorem 38.4), it follows that $\lambda(K) = 0$ for every compact subset K of E. But then the regularity of λ implies $\lambda(E) = 0$, and we are done. ∎

Now, we come to the main differentiation theorem regarding Borel measures.

Theorem 39.4. *Every finite signed Borel measure μ on $\mathbb{R}^k$ that is absolutely continuous with respect to the Lebesgue measure (i.e., $\mu \ll \lambda$), is differentiable almost everywhere.*

Moreover, its derivative $D\mu$ coincides (a.e., of course) with the Radon–Nikodym derivative $d\mu/d\lambda$; that is, $D\mu = d\mu/d\lambda$ holds.

Proof. Let $f = d\mu/d\lambda \in L_1(\lambda)$ be the Radon–Nikodym derivative provided by Theorem 37.8. That is, $\mu(E) = \int_E f \, d\lambda$ holds for each Borel set E.

Fix a real number r, and let $A = \{x \in \mathbb{R}^k : f(x) \geq r\}$. Clearly, A is Lebesgue measurable, and by Theorem 22.5, $\lambda(A) < \infty$. Also, by Theorem 15.11, there exists a Borel set F such that $A \subseteq F$ and $\lambda(F) = \lambda(A)$; clearly, $\lambda(F \setminus A) = 0$.

Next, consider the Borel measure ν on $\mathbb{R}^k$ defined by

$$\nu(E) = \int_{F \cap E} [f(x) - r] \, d\lambda(x) = \int_{A \cap E} [f(x) - r] \, d\lambda(x)$$

for each Borel set E. Note that if B is an open ball, then

$$\mu(B) - r\lambda(B) = \int_B [f(x) - r] \, d\lambda(x) \leq \int_{B \cap A} [f(x) - r] \, d\lambda(x) = \nu(B)$$

holds. Thus,

$$\frac{\mu(B)}{\lambda(B)} \leq r + \frac{\nu(B)}{\lambda(B)}$$

holds for each open ball B, from which it follows that

$$D^*\mu(x) \leq r + D^*\nu(x) \quad \text{for each} \quad x \in \mathbb{R}^k. \tag{$\star$}$$

Since $\nu(F^c) = 0$, it follows from Lemma 39.3 that $D\nu(x) = 0$ holds for almost all $x \in F^c$. Thus, $(\star)$ implies $D^*\mu(x) \leq r$ for almost all $x \in F^c$, and consequently, $D^*\mu(x) \leq r$ for almost all $x \in A^c$. Since $f(x) < r$ implies $x \in A^c$, it follows that the set

$$E_r = \{x \in \mathbb{R}^k : f(x) < r < D^*\mu(x)\}$$

satisfies $E_r \subseteq A^c$. Therefore, by the last conclusion, each E_r has Lebesgue measure zero. But then $\{x \in \mathbb{R}^k : D^*\mu(x) > f(x)\} \subseteq \bigcup_{r \in \mathcal{Q}} E_r$, where $\mathcal{Q}$ is the set of all rational numbers, shows that $\{x \in \mathbb{R}^k : D^*\mu(x) > f(x)\}$ has Lebesgue measure zero.

Applying the preceding arguments to $-\mu$, and taking into consideration that $d(-\mu)/d\lambda = -f$ and $D^*(-\mu) = -D_*\mu$, we easily infer that the set

$$\{x \in \mathbb{R}^k : D_*\mu(x) < f(x)\}$$

has Lebesgue measure zero. In other words,

$$D^*\mu(x) \leq f(x) \leq D_*\mu(x)$$

holds for almost all $x \in \mathbb{R}^k$. That is, $D\mu(x)$ exists for almost all x and $D\mu = f$ a.e. holds, as desired. ∎

As a first application of the preceding theorem, let us establish a classical result.

Theorem 39.5. *If E is a Lebesgue measurable subset of $\mathbb{R}$, then*

1. $\lim_{\epsilon \to 0^+} \frac{\lambda(E \cap (x-\epsilon, x+\epsilon))}{2\epsilon} = 1$ *for almost all x in E, and*
2. $\lim_{\epsilon \to 0^+} \frac{\lambda(E \cap (x-\epsilon, x+\epsilon))}{2\epsilon} = 0$ *for almost all x in E^c.*

Proof. Without loss of generality, we can assume that $\lambda(E) < \infty$. Consider the finite Borel measure μ on $\mathbb{R}$ defined by

$$\mu(A) = \lambda(E \cap A) = \int_A \chi_E \, d\lambda.$$

Clearly, $\mu \ll \lambda$ and $d\mu/d\lambda = \chi_E$. By Theorem 39.4, $D\mu = \chi_E$ a.e. holds, and the formulas in (1) and (2) follow. ∎

A point x of a Lebesgue measurable subset E of $\mathbb{R}$ for which (1) in the preceding theorem holds is referred to as a **density point** of E. In this terminology, Theorem 39.5(1) is usually stated as follows: *If E is a Lebesgue measurable subset of $\mathbb{R}$, then almost every point of E is a density point.*

Now, let μ be a signed Borel measure. If μ is singular with respect to the Lebesgue measure λ, then we shall show that its derivative equals zero almost everywhere. This, coupled with Theorem 39.4, will yield the following important differentiation result.

Theorem 39.6. *Every signed Borel measure μ on $\mathbb{R}^k$ is differentiable almost everywhere. Moreover, if $\mu = \mu_1 + \mu_2$ is the Lebesgue decomposition of μ (i.e., $\mu_1 \ll \lambda$ and $\mu_2 \perp \lambda$), then*

$$D\mu_2 = 0 \ \text{a.e.} \quad \text{and} \quad D\mu = D\mu_1 = \frac{d\mu}{d\lambda} \ \text{a.e.}$$

Proof. Let μ be a Borel signed measure on $\mathbb{R}^k$. By considering separately the positive and negative parts of μ, we can assume without loss of generality that μ is a Borel measure. Clearly, μ is σ-finite. Let $\mu = \mu_1 + \mu_2$ be the Lebesgue decomposition of μ (see Theorem 37.7), where $\mu_1 \ll \lambda$ and $\mu_2 \perp \lambda$.

If $B_n = \{x \in \mathbb{R}^k : \|x\| < n\}$, then the set function defined by $\nu_n(E) = \mu_1(E \cap B_n)$ for each Borel set E is a finite Borel measure satisfying $\nu_n \ll \lambda$. Thus, by Theorem 39.4, ν_n is differentiable for almost all points of B_n, and as a routine verification shows, $D\mu_1(x) = D\nu_n(x)$ holds for almost all $x \in B_n$. Since $B_n \uparrow \mathbb{R}^k$, it follows that μ_1 is differentiable almost everywhere.

On the other hand, $\mu_2 \perp \lambda$ implies the existence of a Borel set A with $\mu_2(A) = \lambda(A^c) = 0$. But by Lemma 39.3, $D\mu_2(x) = 0$ holds for almost all points x of A, and hence (since $\lambda(A^c) = 0$), $D\mu_2(x) = 0$ for almost all points x of $\mathbb{R}^k$. Using Theorem 39.4, we see that

$$D\mu(x) = D\mu_1(x) + D\mu_2(x) = D\mu_1(x) = \frac{d\mu}{d\lambda}(x)$$

holds for almost all $x \in \mathbb{R}^k$, and the proof is finished.	∎

Having established the basic differentiation properties of measures, our attention is now turned to the relationship between measures and real valued functions defined on an interval. In the following discussion some of the deepest classical results of ordinary derivatives will be obtained.

Recall that a function $f: I \to \mathbb{R}$ (where I is an interval) that satisfies $f(x) \leq f(y)$ whenever $x \leq y$ is called an **increasing function** (and if $f(x) \geq f(y)$ whenever $x \leq y$, then f is called **decreasing**). An increasing or a decreasing function is referred to as a **monotone function**. Likewise, if $x > y$ implies $f(x) > f(y)$, then f is called a **strictly increasing function** (and if $f(x) < f(y)$ whenever $x > y$, then f is called **strictly decreasing**). A strictly increasing or a strictly decreasing function is known as a **strictly monotone function**.

Every monotone function f has the property that both limits

$$f(x-) = \lim_{t \uparrow x} f(t) = \lim_{t \to x^-} f(t) \quad \text{and} \quad f(x+) = \lim_{t \downarrow x} f(t) = \lim_{t \to x^+} f(t)$$

exist in $\mathbb{R}$ for each x. Moreover, the **oscillation** of f at any point x is given by

$$\omega_f(x) = |f(x+) - f(x-)| < \infty.$$

Therefore, the discontinuities of a monotone function are jump discontinuities, and as the next result shows, they are at-most countably many.

Theorem 39.7.	*The set of all discontinuities of a monotone function is at-most countable.*

Proof.	Let $f: I \to \mathbb{R}$ be monotone. Replacing f by $-f$ (if necessary), we can assume that f is increasing. Since the interior of the interval can be written as a countable union of open finite intervals, it suffices to show that f has at-most countably many discontinuities in each finite open interval. To this end, let (a, b) be a finite open subinterval of I.

If A is the set of all discontinuities of f in (a, b), then $A = \bigcup_{n=1}^{\infty} A_n$, where $A_n = \{x \in (a, b): f(x+) - f(x-) \geq \frac{1}{n}\}$. Now, if $a < x_1 < \cdots < x_k < b$ belong to some A_n, then it is easy to see that

$$\frac{k}{n} \leq \sum_{i=1}^{k} [f(x_i+) - f(x_i-)] \leq f(b) - f(a) < \infty$$

holds, which implies that each A_n is finite. Hence, A is at-most countable, and the proof is finished. ∎

An immediate and very useful consequence of the preceding theorem is that a monotone function has a point of continuity in every open interval. In applications this observation is translated as follows: If f is a monotone function, then for every point x there exist sequences $\{x_n\}$ and $\{y_n\}$ such that $x_n \downarrow x$, $y_n \uparrow x$ and with f continuous at each x_n and y_n.

Recall that if a function $f: \mathbb{R} \to \mathbb{R}$ is increasing and left continuous (i.e., satisfying $\lim_{t \uparrow x} f(t) = f(x)$ for each x), then a set function μ_f can be defined on the semiring $S = \{[a, b): a \leq b\}$ by $\mu_f([a, b)) = f(b) - f(a)$. In Example 13.6 we verified that μ_f is indeed a measure. Clearly, every Borel set is μ_f-measurable; therefore, the outer measure μ_f^* restricted to $\mathcal{B}$ is a Borel measure. For simplicity, the restriction of μ_f^* to $\mathcal{B}$ will be denoted by μ_f again, and it will be called the **Borel measure induced** by f. It is a routine matter to verify that $\mu_f((a, b)) = f(b) - f(a+)$ holds for each open interval (a, b). On the other hand, for $a < b$, the relation

$$f(b) - f(a) = \mu_f([a, b)) = \mu_f(\{a\}) + \mu_f((a, b))$$

shows that f is continuous at some $a \in \mathbb{R}$ if and only if $\mu_f(\{a\}) = 0$ holds.

The first "differentiation" relation between f and μ_f is stated in the next important theorem. Keep in mind that the open balls of $\mathbb{R}$ are precisely the finite open intervals.

Theorem 39.8. *Let $f: \mathbb{R} \to \mathbb{R}$ be an increasing left continuous function, and let x_0 be a real number. Then the Borel measure μ_f is differentiable at x_0 if and only if f is differentiable at x_0. Moreover, in this case*

$$D\mu_f(x_0) = f'(x_0).$$

Proof. Assume first that f is differentiable at x_0 and let $m = f'(x_0)$. Given $\epsilon > 0$, choose $\delta > 0$ such that

$$m - \epsilon < \frac{f(x) - f(x_0)}{x - x_0} < m + \epsilon \quad \text{whenever} \quad 0 < |x - x_0| < \delta. \quad (\star)$$

If an open interval (a, b) satisfies $x_0 \in (a, b)$ and $b - a < \delta$, then it is easy to see from $(\star)$ that $m - \epsilon \leq \frac{f(b) - f(a+)}{b - a} \leq m + \epsilon$. This implies

$$\left| \frac{\mu_f((a, b))}{\lambda((a, b))} - m \right| = \left| \frac{f(b) - f(a+)}{b - a} - m \right| \leq \epsilon$$

for each open interval (a, b) with $x_0 \in (a, b)$ and $b - a < \delta$. Hence, $D\mu_f(x_0) = m$ holds.

For the converse, assume that $m = D\mu_f(x_0)$ exists in $\mathbb{R}$ and let $\epsilon > 0$. Choose some $\delta > 0$ such that whenever $x_0 \in (a, b)$ and $b - a < \delta$, then

$$\left| \frac{\mu_f((a, b))}{b - a} - m \right| < \epsilon.$$

Let $b > x_0$ satisfy $b - x_0 < \delta$. Since f has at most countably many discontinuities (Theorem 39.7), there exists a sequence $\{a_n\}$ such that $a_n < x_0$, $\lim a_n = x_0$, $b - a_n < \delta$, and f is continuous at each a_n. Thus, $\mu_f((a_n, b)) = f(b) - f(a_n+) = f(b) - f(a_n)$, and so

$$\left| \frac{f(b) - f(a_n)}{b - a_n} - m \right| < \epsilon$$

for each n. By the left continuity of f, it follows that

$$\left| \frac{f(b) - f(x_0)}{b - x_0} - m \right| \leq \epsilon$$

for each $b > x_0$ with $b - x_0 < \delta$. This means that f is differentiable from the right at x_0 and thus, it is right continuous at x_0. Reason:

$$\lim_{b \downarrow x_0} [f(b) - f(x_0)] = \lim_{b \downarrow x_0} \left[\frac{f(b) - f(x_0)}{b - x_0} \cdot (b - x_0) \right] = m \cdot 0 = 0.$$

Now, by the symmetry of the situation, we have

$$\left| \frac{f(a) - f(x_0)}{a - x_0} - m \right| \leq \epsilon$$

for all $a < x_0$ with $x_0 - a < \delta$. Thus, $f'(x_0)$ exists, and $f'(x_0) = m$. The proof of the theorem is now complete. ∎

By Theorem 39.7 we know that a monotone function has at-most countably many discontinuities. Now, we are ready to prove that a monotone function also has a derivative at almost every point.

Theorem 39.9 (Lebesgue). *Every monotone function is differentiable almost everywhere.*

Proof. Let f be a monotone function. Replacing f by $-f$ (if necessary), we can assume that f is increasing. Define a new function $f_*: \mathbb{R} \to \mathbb{R}$ by

$$f_*(x) = f(x-) = \lim_{t \uparrow x} f(t) = \lim_{t \to x^+} f(t).$$

Clearly, $f_*(x) \leq f(x)$ holds for all x. Moreover, f_* is increasing and left continuous. (Indeed, for each x, there exists a sequence $\{x_n\}$ such that $x_n \uparrow x$ and with f continuous at each x_n; therefore, $f_*(x-) = \lim f_*(x_n) = \lim f(x_n) = f(x-) = f_*(x)$.) Also, a similar argument shows that $\omega_{f_*}(x) = \omega_f(x) = f(x+) - f(x-)$ holds for each x. In particular, this shows that f and f_* have precisely the same points of continuity.

Now, let μ_{f_*} be the Borel measure induced by f_*. By Theorem 39.6, μ_{f_*} is differentiable almost everywhere, and hence, by Theorem 39.8, f_* is also differentiable almost everywhere.

Assume that $m = f_*'(a)$ exists at some point a. Then f_* is continuous at a, and so $f_*(a) = f(a)$. Given $\epsilon > 0$, choose some $\delta > 0$ such that

$$m - \epsilon < \frac{f_*(x) - f(a)}{x - a} < m + \epsilon \quad \text{holds whenever} \quad 0 < |x - a| < \delta.$$

Now, fix some x with $0 < |x - a| < \delta$, and then choose a sequence $\{x_n\}$ such that $x_n \downarrow x, 0 < |x_n - a| < \delta$, and with f continuous at each x_n. It follows that

$$m - \epsilon < \frac{f_*(x) - f(a)}{x - a} \leq \frac{f(x) - f(a)}{x - a} \leq \frac{f(x+) - f(a)}{x - a}$$

$$= \lim_{n \to \infty} \frac{f(x_n) - f(a)}{x_n - a} \leq m + \epsilon.$$

Therefore,

$$\left| \frac{f(x) - f(a)}{x - a} - m \right| \leq \epsilon$$

holds whenever $0 < |x - a| < \delta$. Hence, $f'(a)$ exists and $f'(a) = f_*'(a) = m$ holds. That is, at every point where f_* is differentiable, so is f. Therefore, f is differentiable almost everywhere. ∎

For the rest of the section, only real-valued functions defined on a (finite) closed interval $[a, b]$ will be considered. To simplify matters (when the circumstances require it), a function defined on $[a, b]$ will also be tacitly assumed defined on all of $\mathbb{R}$ by $f(x) = f(a)$ if $x < a$ and $f(x) = f(b)$ if $x > b$.

We start by reviewing some basic properties of functions of bounded variation. Recall that a collection of points $P = \{t_0, \ldots, t_n\}$ is a **partition** of an interval $[a, b]$ if $a = t_0 < t_1 < \cdots < t_n = b$ holds.

Let $f: [a, b] \to \mathbb{R}$ be a function. Then the **(total) variation** V_f of f over $[a, b]$ is defined to be

$$V_f = \sup\left\{ \sum_{i=1}^{n} |f(t_i) - f(t_{i-1})|: \ P = \{t_0, \ldots, t_n\} \text{ is a partition of } [a, b] \right\}.$$

If $V_f < \infty$ holds, then f is said to be a function of **bounded variation**.

A function of bounded variation is necessarily a bounded function. [Reason: If $a < x < b$, then

$$(|f(x)| - |f(a)|) + (|f(x)| - |f(b)|) \le |f(x) - f(a)| + |f(b) - f(x)| \le V_f.]$$

Also, it is a routine matter to verify that if f and g are functions of bounded variation (on $[a, b]$) and $\alpha \in \mathbb{R}$, then $f + g$, αf, fg, and $|f|$ are all functions of bounded variation. Thus, if $BV[a, b]$ denotes the collection of all real-valued functions of bounded variation on $[a, b]$, then the latter shows that $BV[a, b]$ is a function space and an algebra of functions.

Every monotone function f is of bounded variation, and $V_f = |f(b) - f(a)|$ holds. [Indeed, if f is increasing, and $a = t_0 < t_1 < \cdots < t_n = b$, then

$$\sum_{i=1}^{n} |f(t_i) - f(t_{i-1})| = \sum_{i=1}^{n} [f(t_i) - f(t_{i-1})] = f(b) - f(a).]$$

Therefore, every function that can be written as a difference of two monotone functions is of bounded variation. In actuality, these are the only types of functions that are of bounded variation. The following discussion will clarify the situation.

Now, let $f: [a, b] \to \mathbb{R}$ be a function of bounded variation. Then f restricted to any closed subinterval $[c, d]$ of $[a, b]$ is of bounded variation there. As a matter of fact, if for any closed subinterval $[c, d]$ of $[a, b]$ we denote the (total) variation of f over $[c, d]$ by $\mathrm{Var}_f(c, d)$, i.e.,

$$\mathrm{Var}_f(c, d) =$$
$$\sup\left\{ \sum_{i=1}^{n} |f(t_i) - f(t_{i-1})|: \ P = \{t_0, \ldots, t_n\} \text{ is a partition of } [c, d] \right\},$$

then it is easy to verify (and the reader should do so) that

$$\text{Var}_f(c, d) = \text{Var}_f(c, e) + \text{Var}_f(e, d).$$

for all $a \leq c < e < d \leq b$. Therefore, we can define a new real function $V_f(\cdot)$ by $V_f(a) = 0$ and $V_f(x) = $ the variation of f over $[a, x]$ for $a < x \leq b$. Note that $V_f(b) = V_f$. The function $V_f(\cdot)$ is called the **variation function** of f, and its basic properties are included in the next theorem.

Theorem 39.10. *For a function $f: [a, b] \to \mathbb{R}$ of bounded variation, the following statements hold:*

1. *The total variation function $V_f(\cdot)$ of f is increasing.*
2. $|f(y) - f(x)| \leq V_f(y) - V_f(x)$ *holds for all $a \leq x < y \leq b$.*
3. *The function $g: [a, b] \to \mathbb{R}$ defined by $g(x) = V_f(x) - f(x)$ is increasing.*
4. *The function f is continuous at some $x_0 \in [a, b]$ if and only if its total variation function V_f is continuous at x_0.*

Proof. (1) Assume $a < x < y \leq b$. If $a = t_0 < t_1 < \cdots < t_n = x < y$ holds, then

$$\sum_{i=1}^{n} |f(t_i) - f(t_{i-1})| + |f(y) - f(x)| \leq V_f(y).$$

Hence, $V_f(x) \leq V_f(x) + |f(y) - f(x)| \leq V_f(y)$ holds.
(2) The preceding inequality immediately gives (2).
(3) If $a \leq x < y \leq b$ holds, then (2) implies

$$f(y) - f(x) \leq |f(y) - f(x)| \leq V_f(y) - V_f(x),$$

so that $g(y) - g(x) = [V_f(y) - f(y)] - [V_f(x) - f(x)] \geq 0$.
(4) If V_f is continuous at x_0, then it should be obvious from (2) that f is also continuous at x_0. For the converse, assume that f is continuous at $x_0 \in [a, b]$. We shall establish that V_f is right-continuous at x_0, and we shall leave the identical arguments for the left continuity of V_f at x_0 for the reader.

So, assume $a \leq x_0 < b$ and let $\epsilon > 0$. Choose some $\delta > 0$ such that $x \in [a, b]$ and $|x - x_0| < \delta$ imply $|f(x) - f(x_0)| < \epsilon$. Next, fix some $x_0 < s < b$ such that $s - x_0 < \delta$, and then select a partition $P = \{a = t_0 < t_1 < \cdots < t_n = s\}$ of $[a, s]$ such that

$$V_f(s) - \epsilon < \sum_{i=1}^{n} |f(t_i) - f(t_{i-1})|.$$

By adding the point x_0 to the partition P, we can assume without loss of generality that

$$P = \{a = t_0 < t_1 < \cdots < t_j = x_0 < t_{j+1} < \cdots < t_n = s\}.$$

Now, assume that $x_0 < x < t_{j+1}$. Then, we have

$$V_f(x) + \mathrm{Var}_f(x, s) = V_f(s) < \sum_{i=1}^{n} |f(t_i) - f(t_{i-1})| + \epsilon$$

$$\leq \sum_{i=1}^{j} |f(t_i) - f(t_{i-1})| + |f(x) - f(x_0)| + |f(x) - f(t_{j+1})|$$

$$+ \sum_{i=j+2}^{n} |f(t_i) - f(t_{i-1})| + \epsilon$$

$$\leq V_f(x_0) + |f(x) - f(x_0)| + \mathrm{Var}_f(x, s) + \epsilon$$

$$< V_f(x_0) + \mathrm{Var}_f(x, s) + 2\epsilon.$$

This implies $0 \leq V_f(x) - V_f(x_0) < 2\epsilon$ for all $x_0 < x < t_{j+1}$, and so V_f is right-continuous at x_0, as desired. ∎

An immediate application of the preceding theorem is the following:

Theorem 39.11. *If* $f: [a, b] \to \mathbb{R}$ *is of bounded variation, then*

1. *f is the difference of two increasing functions (which, in addition, can be taken to be continuous if f is also continuous), and*
2. *f is differentiable almost everywhere.*

Proof. For (1) apply Theorem 39.10 to $f(x) = V_f(x) - [V_f(x) - f(x)]$, and for (2) use Theorem 39.9. ∎

Let $f: [a, b] \to \mathbb{R}$ be an increasing function and consider it defined on all of $\mathbb{R}$ by $f(x) = f(b)$ if $x > b$ and $f(x) = f(a)$ if $x < a$. Then the set function

$$\mu_f([c, d)) = f(d-) - f(c-)$$

defines a measure on the semiring $\{[c, d) : c, d \in \mathbb{R} \text{ and } c \leq d\}$. Clearly, the measurable sets of μ_f contain the Borel sets of $\mathbb{R}$, and hence, μ_f is a Borel measure. Also, μ_f vanishes off $[a, b]$. That is, $\mathrm{Supp}\, \mu_f \subseteq [a, b]$, and therefore, μ_f is a finite Borel measure.

Now, consider a function $f: [a, b] \to \mathbb{R}$ of bounded variation. Again, f is considered defined on all of $\mathbb{R}$ by $f(x) = f(b)$ if $x > b$ and $f(x) = f(a)$ if

$x < a$. By Theorem 39.11, there exist two increasing functions g and h defined on $[a, b]$ such that $f = g - h$ on $[a, b]$. Then the set function

$$\mu_f(E) = \mu_g(E) - \mu_h(E)$$

for each Borel subset E of $\mathbb{R}$ defines a finite Borel signed measure. It is easy to see (by Theorem 15.10) that the value $\mu_f(E)$ does not depend upon the particular representation of f as a difference of two increasing functions. In addition, note that μ_f vanishes off $[a, b]$ and that

$$\mu_f([c, d)) = f(d-) - f(c-) \quad \text{and} \quad \mu_f(\{x\}) = f(x+) - f(x-).$$

The finite Borel signed measure μ_f is referred to as the **Lebesgue–Stieltjes**[5] **measure** generated by the function of bounded variation f.

Which Lebesgue–Stieltjes measures are absolutely continuous with respect to the Lebesgue measure? The answer is provided by the next theorem.

Theorem 39.12. *For a continuous function $f : [a, b] \to \mathbb{R}$ of bounded variation, the following statements are equivalent:*

1. μ_f *is absolutely continuous with respect to the Lebesgue measure.*
2. *For each $\epsilon > 0$, there exists some $\delta > 0$ such that if $(a_1, b_1), \dots, (a_n, b_n)$ are disjoint open subintervals of $[a, b]$ satisfying $\sum_{i=1}^{n}(b_i - a_i) < \delta$, then $\sum_{i=1}^{n}|f(b_i) - f(a_i)| < \epsilon$ holds.*

Proof. (1) $\implies$ (2) Assume (2) is false. Then there exists some $\epsilon > 0$ so that given any δ there exist disjoint open subintervals $(a_1, b_1), \dots, (a_n, b_n)$ of $[a, b]$ such that

$$\sum_{i=1}^{n}(b_i - a_i) < \delta \quad \text{and} \quad \sum_{i=1}^{n}|f(b_i) - f(a_i)| \geq \epsilon.$$

Note that the open set $\mathcal{O} = \bigcup_{i=1}^{n}(a_i, b_i)$ satisfies $\lambda(\mathcal{O}) = \sum_{i=1}^{n}(b_i - a_i) < \delta$ and

$$\epsilon \leq \sum_{i=1}^{n}|f(b_i) - f(a_i)| = \sum_{i=1}^{n}|\mu_f((a_i, b_i))| \leq \sum_{i=1}^{n}|\mu_f|((a_i, b_i)) = |\mu_f|(\mathcal{O}).$$

Thus, for each n there exists an open set $\mathcal{O}_n \subseteq (a, b)$ such that $\lambda(\mathcal{O}_n) < 2^{-n}$ and $|\mu_f|(\mathcal{O}_n) \geq \epsilon$. Let $A = \bigcap_{n=1}^{\infty} \bigcup_{i=n}^{\infty} \mathcal{O}_i$, and note that A is a Borel set. Since

$$\lambda\left(\bigcup_{i=n}^{\infty} \mathcal{O}_i\right) \leq \sum_{i=n}^{\infty} \lambda(\mathcal{O}_i) < 2^{1-n},$$

---

[5]Thomas Jan Stieltjes (1856–1894), a Dutch mathematician. He published extensively in all parts of analysis in his time. He is remembered today mainly for his generalization of the Riemann integral.

it follows that $\lambda(A) = 0$. On the other hand,

$$|\mu_f|\left(\bigcup_{i=n}^{\infty} \mathcal{O}_i\right) \geq |\mu_f|(\mathcal{O}_n) \geq \epsilon$$

holds for each n, and so, $|\mu_f|(A) = \lim |\mu_f|(\bigcup_{i=n}^{\infty} \mathcal{O}_i) \geq \epsilon$, contrary to $|\mu_f| \ll \lambda$ (recall that $|\mu_f| \ll \lambda$ if and only if $\mu_f \ll \lambda$). Hence, (2) must be true.

$(2) \implies (1)$ Let A be a Borel set with $\lambda(A) = 0$. We can suppose that $A \subseteq (a, b)$. Let $\epsilon > 0$. Choose some $\delta > 0$ so that statement (2) is satisfied.

Since μ_f is a finite regular Borel signed measure and $\lambda(A) = 0$ holds, there exists an open set $\mathcal{O}$ such that $A \subseteq \mathcal{O} \subseteq (a, b)$, $|\mu_f(\mathcal{O}) - \mu_f(A)| < \epsilon$, and $\lambda(\mathcal{O}) < \delta$. Let $\mathcal{O} = \bigcup(a_i, b_i)$ be written as an (at most countable) union of disjoint open intervals. Then, it is easy to see that

$$|\mu_f(\mathcal{O})| = \left| \sum_i \mu_f((a_i, b_i)) \right| = \left| \sum_i [f(b_i) - f(a_i)] \right| \leq \sum_i |f(b_i) - f(a_i)| \leq \epsilon.$$

Hence, $|\mu_f(A)| \leq |\mu_f(A) - \mu_f(\mathcal{O})| + |\mu_f(\mathcal{O})| < 2\epsilon$ holds for each $\epsilon > 0$. That is, $\mu_f(A) = 0$, so that $\mu_f \ll \lambda$, and the proof of the theorem is complete. ∎

Statement (2) of the preceding theorem is taken as the definition of absolute continuity of functions.

Definition 39.13. *A function $f: [a, b] \to \mathbb{R}$ is said to be **absolutely continuous** if for every $\epsilon > 0$ there exists some $\delta > 0$ such that whenever $(a_1, b_1), \ldots, (a_n, b_n)$ are disjoint open subintervals of $[a, b]$, then*

$$\sum_{i=1}^{n}(b_i - a_i) < \delta \quad implies \quad \sum_{i=1}^{n} |f(b_i) - f(a_i)| < \epsilon.$$

It should be clear that an absolutely continuous function is necessarily continuous. The converse is false; see Exercises 7 and 8 at the end of this section. Also, it is easy to verify that if f and g are absolutely continuous functions on $[a, b]$ and $\alpha \in \mathbb{R}$, then $f + g$, αf, fg, and $|f|$ are all absolutely continuous functions. Thus, the collection $AC[a, b]$ of all absolutely continuous functions on $[a, b]$ is a function space and an algebra of functions.

Theorem 39.14. *For an absolutely continuous function $f: [a, b] \to \mathbb{R}$, the following statements hold:*

1. *f is of bounded variation, and hence, $AC[a, b]$ is a vector sublattice of $BV[a, b]$.*

2. *The variation function $V_f(\cdot)$ is absolutely continuous, and hence, f is the difference of two increasing absolutely continuous functions.*

Proof. (1) For any closed subinterval $[c, d]$ of $[a, b]$, we denote (as usual) by $\text{Var}_f(c, d)$ the total variation of f over $[c, d]$. Keep in mind that if $c < e < d$, then

$$\text{Var}_f(c, d) = \text{Var}_f(c, e) + \text{Var}_f(e, d).$$

Next, choose some $\delta > 0$ so that the definition of absolute continuity is satisfied with $\epsilon = 1$. Now, fix some natural number n with $\frac{b-a}{n} < \delta$, and let $a = t_0 < t_1 < \cdots < t_n = b$ be the partition of $[a, b]$ with $t_i - t_{i-1} = \frac{b-a}{n}$ for each i. Clearly, $\text{Var}_f(t_{i-1}, t_i) \leq 1$ holds for each i, and thus,

$$V_f = \text{Var}_f(a, b) = \sum_{i=1}^{n} \text{Var}_f(t_{i-1}, t_i) \leq n < \infty.$$

(2) If $[c, d]$ is a closed subinterval of $[a, b]$ and $P = \{t_0, t_1, \ldots, t_n\}$ is a partition of $[c, d]$, then write $v(c, d, P) = \sum_{i=1}^{n} |f(t_i) - f(t_{i-1})|$. Note that

$$V_f(d) - V_f(c) = \text{Var}_f(c, d) = \sup\{v(c, d, P) \colon P \text{ is a partition of } [c, d]\}.$$

Now, let $\epsilon > 0$. Choose some $\delta > 0$ for which the definition of absolute continuity of f is satisfied. Assume that $(a_1, b_1), \ldots, (a_n, b_n)$ are pairwise disjoint open subintervals of $[a, b]$ such that $\sum_{i=1}^{n}(b_i - a_i) < \delta$. For each $1 \leq i \leq n$, let P_i be an arbitrary partition of $[a_i, b_i]$. Each P_i subdivides (a_i, b_i) into a finite number of open subintervals. Clearly, these open subintervals taken together are pairwise disjoint, and the sum of their lengths equals $\sum_{i=1}^{n}(b_i - a_i) < \delta$. The absolute continuity of f applied to the open subintervals yields

$$v(a_1, b_1, P_1) + \cdots + v(a_n, b_n, P_n) < \epsilon.$$

It follows that

$$\sum_{i=1}^{n} [V_f(b_i) - V_f(a_i)] \leq \epsilon,$$

so that $V_f(x)$ is absolutely continuous. ■

The final theorem of this section is a classical result. It asserts that for the Lebesgue integration, the fundamental theorem of calculus holds true precisely for the absolutely continuous functions. (As is customary, the Lebesgue measure on $\mathbb{R}$ is denoted by dt.)

Theorem 39.15.   *A function $f: [a, b] \to \mathbb{R}$ is absolutely continuous if and only if $f' \in L_1([a, b])$ and*

$$f(x) - f(a) = \int_a^x f'(t)\, dt$$

holds for each $x \in [a, b]$.

Proof.   Assume that f is absolutely continuous. By Theorem 39.14, we can suppose that f is increasing. Also, by Theorem 39.12, $\mu_f \ll \lambda$ holds, and so, by the Radon–Nikodym theorem $\mu_f(E) = \int_E g\, d\lambda$ holds for some $g \in L_1([a, b])$. Now, combine Theorems 39.4 and 39.8 to obtain that $g = f'$ a.e. To get the desired identity let $E = [a, x)$.

The converse is straightforward and is left as an exercise for the reader. See also Exercise 3 of Section 37.   ∎

EXERCISES

1. If μ is a Borel measure on $\mathbb{R}^k$, then show that $\mu \perp \lambda$ holds if and only if $D\mu(x) = 0$ for almost all x.

2. Generalize Theorem 39.5 to Lebesgue measurable subsets of $\mathbb{R}^k$. That is, show that if E is a Lebesgue measurable subset of $\mathbb{R}^k$, then almost all points of E are density points.

3. Write $B_r(a)$ for the open ball with center at $a \in \mathbb{R}^k$ and radius r. If f is a Lebesgue integrable function on $\mathbb{R}^k$, then a point $a \in \mathbb{R}^k$ is called a **Lebesgue point** for f if

$$\lim_{r \to 0^+} \frac{1}{\lambda(B_r(a))} \int_{B_r(a)} |f(x) - f(a)|\, d\lambda(x) = 0.$$

Show that if f is a Lebesgue integrable function on $\mathbb{R}^k$, then almost all points of $\mathbb{R}^k$ are Lebesgue points.

[HINT: Let Q be the set of all rational numbers in $\mathbb{R}$. Apply Theorem 39.4 to conclude that for each $a \in Q$ there exists a null set E_a such that

$$\lim_{r \to 0^+} \frac{1}{\lambda(B_r(x))} \int_{B_r(x)} |f(t) - a|\, d\lambda(t) = |f(x) - a|$$

holds for all $x \notin E_a$. Let $E = \bigcup_{a \in Q} E_a$, and then show that every point of E^c is a Lebesgue point.]

4. Let $f: \mathbb{R} \to \mathbb{R}$ be an increasing, left continuous function. Show directly (i.e., without using Theorem 38.4) that the Lebesgue–Stieltjes measure μ_f is a regular Borel measure.

5. **(Fubini)** Let $\{f_n\}$ be a sequence of increasing functions defined on $[a, b]$ such that $\sum_{n=1}^{\infty} f_n(x) = f(x)$ converges in $\mathbb{R}$ for each $x \in [a, b]$. Then show that f is differentiable almost everywhere and that $f'(x) = \sum_{n=1}^{\infty} f_n'(x)$ holds for almost all x.

[HINT: Replacing each f_n by $f_n - f_n(a)$, we can assume that $f_n \geq 0$ holds for each n. Let $s_n = f_1 + \cdots + f_n$, and note that each s_n is increasing and $s_n(x) \uparrow f(x)$ for each x. By Theorem 39.9, f and all the f_n and s_n are differentiable almost everywhere. Since $s_{n+1} - s_n = f_{n+1}$ (an increasing function), $s'_{n+1}(x) \geq s'_n(x)$ must hold for almost all x; similarly, $f'(x) \geq s'_n(x)$ for almost all x. Now, choose a subsequence $\{s_{k_n}\}$ of $\{s_n\}$ such that $\sum_{n=1}^{\infty}[f(x) - s_{k_n}(x)] \leq \sum_{n=1}^{\infty}[f(b) - s_{k_n}(b)] < \infty$. Observe that $\{f - s_{k_n}\}$ is a sequence of increasing functions. Use the preceding arguments to obtain that $s'_{k_n} \to f'$ a.e., and then conclude that $\sum_{n=1}^{\infty} f'_n = f'$ a.e. holds.]

6. Suppose $\{f_n\}$ is a sequence of increasing functions on $[a, b]$ and that f is an increasing function on $[a, b]$ such that $\mu_{f_n} \uparrow \mu_f$. Establish that $f'(x) = \lim f'_n(x)$ holds for almost all x.

7. This exercise presents some basic properties of functions of bounded variation on an interval $[a, b]$.

   a. If f is differentiable at every point and $|f'(x)| \leq M < \infty$ holds for all $x \in [a, b]$, then show that f is absolutely continuous (and hence, of bounded variation).

   b. Show that the function $f: [0, 1] \to \mathbb{R}$ defined by $f(0) = 0$ and $f(x) = x^2 \cos(x^{-2})$ for $0 < x \leq 1$ is differentiable at each x, but not of bounded variation (and hence, f is continuous but not absolutely continuous).

   c. If f is a function of bounded variation and $|f(x)| \geq M > 0$ holds for each $x \in [a, b]$, then show that $\frac{1}{f}$ is a function of bounded variation.

   d. If a function $f: [a, b] \to \mathbb{R}$ satisfies a Lipschitz condition (i.e., if there exists a constant $M > 0$ such that $|f(x) - f(y)| \leq M|x - y|$ holds for all $x, y \subset [a, b]$), then show that f is absolutely continuous.

8. This exercise presents an example of a continuous increasing function (and hence, of bounded variation) that is not absolutely continuous.

   Consider the Cantor set C as constructed in Example 6.15. Recall that C was obtained from $[0, 1]$ by removing certain open intervals by steps. In the first step we removed the open middle third interval. At the nth step there were 2^{n-1} closed intervals, all of the same length, and we removed the open middle third interval from each one of them. Let us denote by $I_1^n, \ldots, I_{2^{n-1}}^n$ (counted from left to right) the removed open intervals at the nth step. Now, define the function $f: [0, 1] \to [0, 1]$ as follows:

   i. $f(0) = 0$;
   ii. if $x \in I_i^n$ for some $1 \leq i \leq 2^{n-1}$, then $f(x) = (2i - 1)/2^n$; and
   iii. if $x \in C$ with $x \neq 0$, then $f(x) = \sup\{f(t): t < x$ and $t \in [0, 1] \setminus C\}$.

   Part of the graph of f is shown in Figure 7.1.

   a. Show that f is an increasing continuous function from $[0, 1]$ to $[0, 1]$.
   b. Show that $f'(x) = 0$ for almost all x.
   c. Show that f is not absolutely continuous.
   d. Show that $\mu_f \perp \lambda$ holds.

9. Let $f: [a, b] \to \mathbb{R}$ be an absolutely continuous function. Then show that f is a constant function if and only if $f'(x) = 0$ holds for almost all x.

10. Let f and g be two left continuous functions (on $\mathbb{R}$). Show that $\mu_f = \mu_g$ holds if and only if $f - g$ is a constant function.

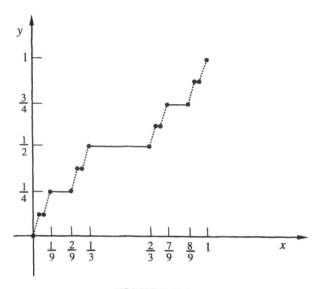

FIGURE 7.1.

11. This exercise presents another characterization of the norm dual of $C[a, b]$. Start by letting L denote the collection of all functions of bounded variation on $[a, b]$ that are left continuous and vanish at a.

   a. Show that L under the usual algebraic operations is a vector space, and that $f \mapsto \mu_f$, from L to $M_b([a, b])$, is linear, one-to-one, and onto.

   b. Define $f \succeq g$ to mean that $f - g$ is an increasing function. (Note that $f \succeq g$ does not imply $f \geq g$.) Show that L under $\succeq$ is a partially ordered vector space such that $f \succeq g$ holds in L if and only if $\mu_f \geq \mu_g$ in $M_b([a, b])$.

   c. Establish that L with the norm $\|f\| = V_{|f|}$ is a Banach lattice.

   d. Show, with an appropriate interpretation, that $C^*[a, b] = L$.

12. If $f:[a, b] \to \mathbb{R}$ is an increasing function, then show that $f' \in L_1([a, b])$ and that $\int_a^b f'(x)\, dx \leq f(b) - f(a)$ holds. Give an example for which $\int_a^b f'(x)\, dx < f(b) - f(a)$ holds.

   [HINT: Let $g_n(x) = n[f(x + \frac{1}{n}) - f(x)]$ for each n and $x \in [a, b]$. Note that $g_n \to f'$ a.e. holds. Now, apply Fatou's lemma to the sequence $\{g_n\}$.]

13. If $f:[a, b] \to \mathbb{R}$ is an absolutely continuous function, then show that

$$V_f = \int_a^b |f'(x)|\, dx$$

holds.

14. For a continuously differentiable function $f:[a, b] \to \mathbb{R}$, establish the following properties:

   a. The signed measure μ_f is absolutely continuous with respect to the Lebesgue measure and $d\mu_f / d\lambda = f'$ a.e.

b. If $g:[a, b] \to \mathbb{R}$ is Riemann integrable, then gf' is also Riemann integrable and

$$\int g \, d\mu_f = \int_a^b g(x)f'(x) \, dx.$$

15. For each n, consider the increasing continuous function $f_n: \mathbb{R} \to \mathbb{R}$ defined by

$$f_n(x) = \begin{cases} 1 & \text{if } x > 0, \\ n(x-1)+1 & \text{if } 1-\frac{1}{n} < x < 1, \\ 0 & \text{if } x \leq 1-\frac{1}{n}. \end{cases}$$

If $f: \mathbb{R} \to \mathbb{R}$ is a continuous function, then show that

a. f is μ_{f_n}-integrable for each n, and
b. $\lim \int f \, d\mu_{f_n} = f(1)$.

16. Let $f: \mathbb{R} \to \mathbb{R}$ be a bounded function and let

$$E = \{x \in \mathbb{R}: \ f'(x) \text{ exists in } \mathbb{R}\}.$$

If $\lambda(E) = 0$, then show that $\lambda(f(E)) = 0$.

17. This exercise presents an example of a continuous function $f: \mathbb{R} \to \mathbb{R}$ which is nowhere differentiable. Consider the function $\phi: [0, 2] \to \mathbb{R}$ defined by $\phi(x) = x$ if $0 \leq x \leq 1$ and $\phi(x) = 2 - x$ if $1 < x \leq 2$. Extend ϕ to all of $\mathbb{R}$ (periodically) so that $\phi(x) = \phi(x+2)$ holds for all $x \in \mathbb{R}$. Now define the function $f: \mathbb{R} \to \mathbb{R}$ by

$$f(x) = \sum_{n=0}^{\infty} \left(\frac{3}{4}\right)^n \phi(4^n x).$$

Show that f is a continuous nowhere differentiable function. (Compare this conclusion with Exercise 28 of Section 9.)

40. THE CHANGE OF VARIABLES FORMULA

The objective of this section is to establish the familiar formula known as the "change of variables formula." To do this, we need some preliminary discussion. For the rest of the section, V and W will be two fixed open sets of some Euclidean space $\mathbb{R}^k$. Also, $\|\cdot\|$ will always denote the Euclidean norm on $\mathbb{R}^k$.

We remind the reader about the representation of a linear mapping on $\mathbb{R}^k$ by a matrix. Let $T: \mathbb{R}^k \to \mathbb{R}^k$ be a linear mapping, and let $\{e_1, \ldots, e_k\}$ be the standard basis of $\mathbb{R}^k$ (that is, the ith coordinate of e_j is 1 if $i = j$ and 0 if $i \neq j$). For each j, there exist constants $a_{1j}, \ldots, a_{kj}$ (uniquely determined) such that $T(e_j) = \sum_{i=1}^k a_{ij}e_i$. If we define the matrix

$$A = \begin{bmatrix} a_{11} & a_{12} & \cdots & a_{1k} \\ a_{21} & a_{22} & \cdots & a_{2k} \\ \vdots & \vdots & \ddots & \vdots \\ a_{k1} & a_{k2} & \cdots & a_{kk} \end{bmatrix},$$

then

$$T(x) = Ax$$

holds for each $x \in \mathbb{R}^k$. The matrix A is called the **matrix representation** of the linear mapping T (with respect to the standard basis). Conversely, any $k \times k$ matrix $A = [a_{ij}]$ defines a linear operator $T \colon \mathbb{R}^k \to \mathbb{R}^k$ by the formula $T(x) = Ax$ for each $x \in \mathbb{R}^k$. Also, it is easy to see that if $M = k \cdot \max\{|a_{ij}| \colon i, j = 1, \ldots, k\}$, then $\|Ax\| \le M\|x\|$ holds so that each matrix defines a continuous linear operator. The **determinant** of T is simply the determinant of the matrix A; that is, $\det T = \det A$.

A function $T \colon V \to \mathbb{R}^k$ is said to be **differentiable** at some point a of (the open set) V if there exists a linear operator $A \colon \mathbb{R}^k \to \mathbb{R}^k$ and some $r > 0$ such that

$$T(x) = T(a) + A(x - a) + o(x - a)$$

holds for all $x \in V$ with $\|x - a\| < r$. Here (as usual) $o(x - a)$ is a function from V to $\mathbb{R}^k$ such that

$$\lim_{x \to a} \frac{o(x - a)}{\|x - a\|} = 0.$$

The linear operator A is denoted by $T'(a)$ and is called the **derivative** of T at the point a. It should be clear that if T is differentiable at a, then T must be continuous at a.

A function $T \colon V \to \mathbb{R}^k$ is said to be **differentiable** if $T'(x)$ exists at each point $x \in V$. Moreover, if $T = (T_1, \ldots, T_n)$ is differentiable, then it follows that each partial derivative

$$\frac{\partial T_j}{\partial x_i}(x) = \lim_{h \to 0} \frac{T_j(x + he_i) - T_j(x)}{h}$$

exists at every point $x \in V$. In addition, the matrix representation of $T'(x)$ is

$$\begin{bmatrix} \frac{\partial T_1}{\partial x_1}(x) & \cdots & \frac{\partial T_1}{\partial x_k}(x) \\ \vdots & \ddots & \vdots \\ \frac{\partial T_k}{\partial x_1}(x) & \cdots & \frac{\partial T_k}{\partial x_k}(x) \end{bmatrix}.$$

This matrix is called the **Jacobian**[6] **matrix**, and its determinant is called the **Jacobian**. The Jacobian determinant is denoted by $J_T(x)$; that is, $J_T(x) = \det T'(x) = \det[(\partial T_i / \partial x_j)(x)]$.

---

[6] Carl Gustav Jacob Jacobi (1804–1851), a German mathematician. He was a prolific writer who contributed to many diverse mathematical disciplines—including number theory, mathematical physics, mechanics, and the history of mathematics.

A function $T: V \rightarrow \mathbb{R}^k$ is said to be C^1-**differentiable** if T is differentiable and all of its partial derivatives are continuous functions on V. Observe that if T is C^1-differentiable, then the Jacobian $J_T(\cdot)$ is a continuous real valued function on V.

It is important to know that C^1-differentiable mappings carry null sets onto null sets. The details follow.

Lemma 40.1. *Let $V \subseteq \mathbb{R}^k$ be open and let $T: V \rightarrow \mathbb{R}^k$ be a C^1-differentiable function. If $A \subseteq V$ satisfies $\lambda(A) = 0$, then $\lambda(T(A)) = 0$ also holds.*

Proof. Assume first that there exists some $M > 0$ such that $|(\partial T_i / \partial x_j)(x)| \leq M$ holds for each $x \in V$ and $i, j = 1, \ldots, k$. Let $C = kM + 1$. If $a \in V$, then by the differentiability of T there exists some open ball $B_a \subseteq V$ with center at a such that

$$\|T(x) - T(a)\| < kM \|x - a\| + \|x - a\| = C \|x - a\|$$

holds for each $x \in B_a$.

Let $\epsilon > 0$. Choose an open set $\mathcal{O}$ with $A \subseteq \mathcal{O} \subseteq V$ and $\lambda(\mathcal{O}) < \epsilon$, and then select a sequence $\{K_n\}$ of compact sets such that $\mathcal{O} = \bigcup_{n=1}^{\infty} K_n$. Since T is continuous, each $T(K_n)$ is compact, and the identity $T(\mathcal{O}) = \bigcup_{n=1}^{\infty} T(K_n)$ shows that $T(\mathcal{O})$ is a Borel set.

Now, let K be an arbitrary compact subset of $T(\mathcal{O})$. For each $y \in K$ fix some $z \in \mathcal{O}$ with $y = T(z)$. By the above, there exists an open ball $B_y \subseteq \mathcal{O}$ with center at z such that

$$\|T(x) - y\| < C \|x - z\|$$

holds for each $x \in B_y$. Let B_y^* be the open ball with center at y, and radius C times that of B_y; clearly, $T(B_y) \subseteq B_y^*$. Since K is compact, there exist $y_1, \ldots, y_n \in K$ such that $K \subseteq \bigcup_{i=1}^{n} B_{y_i}^*$.

By Lemma 39.2, there exist pairwise disjoint balls $B_1^*, \ldots, B_m^*$ among the $B_{y_1}^*, \ldots, B_{y_n}^*$ such that

$$\lambda \left(\bigcup_{i=1}^{n} B_{y_i}^* \right) \leq 3^k \sum_{j=1}^{m} \lambda(B_j^*) = (3C)^k \sum_{j=1}^{m} \lambda(B_j).$$

Note that the corresponding $B_1, \ldots, B_m$ are necessarily pairwise disjoint open balls. Therefore,

$$\lambda(K) \leq \lambda \left(\bigcup_{i=1}^{n} B_{y_i}^* \right) \leq (3C)^k \sum_{j=1}^{m} \lambda(B_j) = (3C)^k \lambda \left(\bigcup_{j=1}^{m} B_j \right)$$

$$\leq (3C)^k \lambda(\mathcal{O}) < (3C)^k \epsilon$$

holds for each compact subset K of $T(\mathcal{O})$. Since $T(\mathcal{O})$ is a Borel set, the regularity of λ implies $\lambda(T(\mathcal{O})) \leq (3C)^k \epsilon$, and hence, $\lambda(T(A)) \leq \lambda(T(\mathcal{O})) \leq (3C)^k \epsilon$ for each $\epsilon > 0$. That is, $\lambda(T(A)) = 0$.

For the general case, let $B_1, B_2, \ldots$ be an enumeration of the open balls with "rational" centers and rational radii whose closures lie entirely in V. Clearly, $V = \bigcup_{n=1}^{\infty} B_n$. Since each $\overline{B_n}$ is compact and T is C^1-differentiable, it follows easily that $T: B_n \to \mathbb{R}^k$ satisfies the hypotheses of the preceding case. Thus, $\lambda(T(A \cap B_n)) = 0$ holds for each n, and consequently, we have $\lambda(T(A)) = \lambda(\bigcup_{n=1}^{\infty} T(A \cap B_n)) = 0$, as required.   ■

The notion of a diffeomorphism plays an important role for this section. Its definition follows.

Definition 40.2. *Let V and W be two open sets of $\mathbb{R}^k$. A function $T: V \to W$ is said to be a **diffeomorphism** if*

a.   *T is one-to-one and onto,*
b.   *T is C^1-differentiable,*
c.   *$J_T(x) \neq 0$ holds for all $x \in V$, and*
d.   *T is a homeomorphism (from V onto W).*

We remark that (a), (b), and (c) are enough to ensure that T is an open mapping and thus, a homeomorphism. However, property (d) has been added to the definition of a diffeomorphism in order to emphasize its importance. Also, it should be noted that the inverse mapping $T^{-1}: W \to V$ is necessarily a diffeomorphism. For details about this and other properties of differentiable functions see [2].

The next theorem tells us that a diffeomorphism preserves the Borel and the Lebesgue measurable sets.

Theorem 40.3. *Let $T: V \to W$ be a diffeomorphism between two open subsets of $\mathbb{R}^k$, and let E be a subset of V. Then:*

a.   *$T(E)$ is a Borel set if and only if E is a Borel set.*
b.   *$T(E)$ is Lebesgue measurable if and only if E is Lebesgue measurable.*

Proof. (a) Let $\mathcal{B}_V$ and $\mathcal{B}_W$ denote the Borel sets of V and W, respectively. Clearly, $T(\mathcal{B}_V) = \{T(E): E \in \mathcal{B}_V\}$ is a σ-algebra of subsets of W containing the open sets of W; hence, $\mathcal{B}_W \subseteq T(\mathcal{B}_V)$. By the symmetry of the situation, $\mathcal{B}_V \subseteq T^{-1}(\mathcal{B}_W)$ holds, and from this we get $T(\mathcal{B}_V) = \mathcal{B}_W$. The conclusion now follows easily from the last identity.

(b) Let E be Lebesgue measurable. We can suppose $\lambda(E) < \infty$ (why?). Choose a Borel set B with $E \subseteq B \subseteq V$ and $\lambda(B \setminus E) = 0$. By Lemma 40.1, $\lambda(T(B \setminus E)) = 0$, so that $T(B \setminus E)$ is Lebesgue measurable. Now from (a) the set $T(B)$ is a Borel set, and hence, the relation

$$T(E) = T(B) \setminus T(B \setminus E)$$

shows that $T(E)$ is Lebesgue measurable. The converse should be obvious from the symmetry of the situation.   ∎

Now, assume that $T: V \to W$ is a diffeomorphism. From the preceding theorem it is easy to see that the set function

$$\mu(E) = \lambda(T(E))$$

defined for each Lebesgue measurable subset E of V is a measure. On the other hand, it follows from Lemma 40.1 that $\mu \ll \lambda$ holds. The proof of the "change of variables formula" rests upon the fact that the Radon–Nikodym derivative $d\mu/d\lambda$ satisfies $(d\mu/d\lambda)(x) = |J_T(x)|$ for each $x \in V$.

To establish this identity, we need to know how a linear operator alters the volume of the unit cube of $\mathbb{R}^k$.

Lemma 40.4.   *If $A: \mathbb{R}^k \to \mathbb{R}^k$ is a linear operator (which we identify with a matrix), then*

$$\lambda(A(E)) = |\det A| \cdot \lambda(E)$$

holds for all Lebesgue measurable subsets E of $\mathbb{R}^k$.

Proof.   If A is not invertible, then A maps $\mathbb{R}^k$ onto a linear subspace of lower dimension, and hence, onto a set of measure zero (why?). Hence, $\lambda(A(E)) = |\det A| \cdot \lambda(E) = 0$ holds trivially.

Now, assume that A is invertible; clearly, A is a diffeomorphism. Let $\mu(E) = \lambda(A(E))$ for each Borel set E. Then μ is a Borel measure, and by Theorem 15.10 it is enough to establish $\mu = |\det A| \cdot \lambda$ on the Borel sets. Start by observing that μ is a translation invariant Borel measure on $\mathbb{R}^k$, and hence, by Lemma 18.7, there exists a constant C such that $\mu = C\lambda$ holds. In particular, if $U = [0, 1] \times \cdots \times [0, 1]$, then $C = \mu(U)$. To complete the proof, we must show that $C = |\det A|$.

To this end, note first that if $A = A_1 \circ A_2$ and $C, C_1,$ and C_2 are the constants associated with $A, A_1,$ and A_2, respectively, then

$$C = \lambda(A(U)) = \lambda(A_1(A_2(U))) = C_1\lambda(A_2(U)) = C_1 C_2 \lambda(U) = C_1 C_2.$$

If we denote the matrix representing A by A again, then the matrix A can be written as a product of matrices $A = A_1 A_2 \cdots A_n$, where each A_i is either

1.   a matrix obtained from the $(k \times k)$ identity matrix by multiplying one of its rows by a nonzero constant,
2.   a matrix obtained from the identity matrix by interchanging two of its rows, or
3.   a matrix obtained from the identity matrix by adding to one of its rows a nonzero multiple of some other row of the identity.

(The three types of matrices described previously are known as **elementary matrices**.) Thus, since $C = C_1 C_2 \cdots C_n$ [where $C_i = \lambda(A_i(U))$] and $\det A = \det A_1 \cdot \det A_2 \cdots \det A_n$, in order to complete the proof it suffices to verify the formula for the above three types of matrices.

Type 1.   Without loss of generality we can assume that the matrix A is obtained from the identity matrix by multiplying its first row by a nonzero number α. Note that $\det A = |\alpha|$. Here U is mapped onto $[0, \alpha] \times [0, 1] \times \cdots \times [0, 1]$ if $\alpha > 0$ and $[\alpha, 0] \times [0, 1] \times \cdots \times [0, 1]$ if $\alpha < 0$. In either case, $C = \lambda(A(U)) = |\alpha| = |\det A|$.

Type 2.   In this case, $\det A = -1$. It is easy to see that U is mapped under A onto U. Hence, $C = \lambda(A(U)) = \lambda(U) = 1 = |\det A|$.

Type 3.   It is easy to see that this case is reduced to the matrix A obtained from the identity matrix by adding to its first row the second one. Clearly, $\det A = 1$.

Now, observe that if $x = (x_1, x_2, \ldots, x_k)$, then $Ax = (x_1 + x_2, x_2, x_3, \ldots, x_k)$. In particular, when $k = 2$ the unit square in $\mathbb{R}^2$ is mapped under A onto the parallelogram P as shown in Figure 7.2. Obviously, $\lambda(P) = 1$. On the other hand, if we consider the Lebesgue measure on $\mathbb{R}^k$ (where $k > 2$) as the product measure of the Lebesgue measure on $\mathbb{R}^2$ and the Lebesgue measure on $\mathbb{R}^{k-2}$, then U is mapped under A onto the set $P \times U_1$, where U_1 is the $(k - 2)$-dimensional unit cube. By Theorem 26.2, $\lambda(A(U)) = \lambda(P) \cdot \lambda(U_1) = 1 = |\det A|$, and the proof of the lemma is complete.   ∎

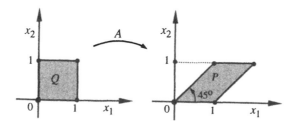

FIGURE 7.2.   The Transformation of the Unit Square

Now, we are in the position to establish the identity $d\mu/d\lambda = |J_T|$.

Theorem 40.5. *Let $T: V \to W$ be a diffeomorphism between two open subsets of $\mathbb{R}^k$. If $\mu(E) = \lambda(T(E))$ for each Borel set E of V, then μ is a Borel measure on V whose derivative satisfies*

$$D\mu(x) = |J_T(x)|$$

for all $x \in V$.

Proof. Let $a \in V$. Replacing T by $S(x) = T(a + x) - T(a)$, we can assume without loss of generality that $a = 0$ and $T(0) = 0$. We shall prove first the result for the case when the Jacobian matrix equals the $k \times k$ identity matrix I, and so $J_T(0) = 1$.

Let $\epsilon > 0$. Fix $0 < \alpha < \frac{1}{4}$ such that $1 - \epsilon < (1 - 2\alpha)^k < (1 + 2\alpha)^k < 1 + \epsilon$, and then select some $\delta > 0$ so that

$$\|T(x) - x\| < \alpha\|x\|$$

holds for each $x \in V$ with $\|x\| < \delta$. (Since $T(x) = T'(x)x + o(x) = x + o(x)$ such a $\delta > 0$ always exists.)

Now, let $B \subseteq V$ be an open ball containing zero with center at x_0 and radius $r < \frac{\delta}{2}$. Let B_1 and B_2 be the open balls both with center at x_0 and radii $(1 - 2\alpha)r$ and $(1 + 2\alpha)r$, respectively. We claim that

$$B_1 \subseteq T(B) \subseteq B_2.$$

Indeed, note first that if $x \in B$, then $\|x\| \leq \|x - x_0\| + \|x_0\| < 2r < \delta$, and so

$$\|T(x) - x_0\| \leq \|T(x) - x\| + \|x - x_0\| < \alpha\|x\| + r < (1 + 2\alpha)r$$

for each $x \in B$, which shows that $T(B) \subseteq B_2$. Now, write

$$B_1 = [B_1 \cap T(B)] \cup [B_1 \setminus T(B)],$$

and note that the two sets of the union are disjoint. Since T is a diffeomorphism, $T(B)$ is an open set, and hence, $B_1 \cap T(B)$ is also open. Since

$$\|T(x_0) - x_0\| < \alpha\|x_0\| < 2\alpha r < (1 - 2\alpha)r,$$

it follows that $T(x_0) \in B_1 \cap T(B)$, and so $B_1 \cap T(B) \neq \emptyset$. Also, if $\|x - x_0\| = r$ holds, then $\|x\| < 2r < \delta$ implies

$$r = \|x - x_0\| \leq \|x - T(x)\| + \|T(x) - x_0\| < 2\alpha r + \|T(x) - x_0\|,$$

and so $\|T(x) - x_0\| > (1 - 2\alpha)r$. This shows that no boundary point of B is mapped under T into B_1. In other words, $B_1 \setminus T(B) = B_1 \setminus T(\overline{B})$ holds. Since T is continuous, $T(\overline{B})$ is compact, and hence, $B_1 \setminus T(B)$ is also an open set. But then, it follows that $B_1 \setminus T(B) = \emptyset$ (see Exercise 1 at the end of this section), and so, $B_1 = B_1 \cap T(B) \subseteq T(B)$ holds.

Now, if c is the Lebesgue measure of the open ball with center at zero and radius one, then

$$(1 - \epsilon)\lambda(B) = c(1 - \epsilon)r^k < c(1 - 2\alpha)^k r^k = \lambda(B_1) \leq \lambda(T(B))$$
$$= \mu(B) \leq \lambda(B_2) = c(1 + 2\alpha)^k r^k$$
$$< (1 + \epsilon)cr^k = (1 + \epsilon)\lambda(B).$$

Thus, $1 - \epsilon < \mu(B)/\lambda(B) < 1 + \epsilon$ holds for all open balls $B \subseteq V$ containing zero and having radius less than $\delta/2$. That is, $D\mu(0) = 1$ holds.

For the general case, let A be the Jacobian matrix at zero. Since $\det A = J_T(0) \neq 0$, the matrix A is invertible. Then $S(x) = A^{-1}(T(x))$ defines a diffeomorphism between V and $A^{-1}(W)$ whose Jacobian matrix at zero equals I. By the preceding case, it follows that the Borel measure $\nu(E) = \lambda(S(E))$ satisfies $D\nu(0) = 1$. Now, by Lemma 40.4, we have

$$\nu(E) = |\det A^{-1}| \cdot \lambda(T(E)) = |\det A|^{-1} \cdot \mu(E).$$

This implies $D\mu(0) = |\det A| = |J_T(0)|$ and the proof is finished. ∎

A particular case of the "change of variables formula" is stated next.

Theorem 40.6. *Let $T: V \to W$ be a diffeomorphism between two open subsets of $\mathbf{R}^k$ such that $\lambda(W) < \infty$. Then for each Lebesgue measurable subset E of V we have*

$$\lambda(T(E)) = \int_E |J_T| \, d\lambda.$$

Proof. Let $\mu(E) = \lambda(T(E))$ for each Lebesgue measurable subset E of V. Then, by Lemma 40.1, $\mu \ll \lambda$ holds, and hence, by the Radon–Nikodym theorem $\mu(E) = \int_E (d\mu/d\lambda) \, d\lambda$ holds for each Lebesgue measurable set E. Now, combine Theorem 39.4 and the preceding theorem to obtain $d\mu/d\lambda = D\mu = |J_T|$.

Therefore,

$$\lambda(T(E)) = \int_E |J_T| \, d\lambda.$$

holds for each Lebesgue measurable subset E of V.   ∎

We now come to the main result of this section.

Theorem 40.7 (The Change of Variables Formula).   *Let* $T:V \to W$ *be a diffeomorphism between two open sets of* $\mathbb{R}^k$. *Then for every function* $f \in L_1(W)$, *the function* $(f \circ T) \cdot |J_T|$ *belongs to* $L_1(V)$ *and*

$$\int_W f \, d\lambda = \int_V (f \circ T) \cdot |J_T| \, d\lambda$$

holds.

Proof.   Assume first that $\lambda(W) < \infty$. Let $f = \chi_E$, where E is a Lebesgue measurable subset of W. By Theorem 40.6, we have

$$\int_W f \, d\lambda = \lambda(E) = \lambda(T(T^{-1}(E))) = \int_{T^{-1}(E)} |J_T| \, d\lambda$$

$$= \int_V (\chi_E \circ T) \cdot |J_T| \, d\lambda = \int_V (f \circ T) \cdot |J_T| \, d\lambda.$$

Thus, the conclusion is valid for characteristic functions of Lebesgue measurable subsets of W. It follows that the formula is true for step functions, and then by a simple continuity argument for each $f \in L_1(W)$.

Now, if $\lambda(W) = \infty$, then let $W_n = \{x \in W : \|x\| < n\}$ and $V_n = T^{-1}(W_n)$. Clearly, $\lambda(W_n) < \infty$ for each n, and so, if $0 \le f \in L_1(W)$, then by the preceding case (and Levi's Theorem 22.8)

$$\int_W f \, d\lambda = \lim_{n \to \infty} \int_{W_n} f \, d\lambda = \lim_{n \to \infty} \int_{V_n} (f \circ T) \cdot |J_T| \, d\lambda = \int_V (f \circ T) \cdot |J_T| \, d\lambda,$$

and the conclusion follows.   ∎

The formula appearing in the preceding theorem is referred to as the **change of variables formula** and is commonly written as follows:

$$\int_{T(V)} f(y) \, dy = \int_V f(T(x)) \cdot |J_T(x)| \, dx.$$

The following version of Theorem 40.7 is most often used in applications, and its proof follows immediately from the preceding theorem.

Theorem 40.8. *Let $T: A \to B$ be a a mapping between two Lebesgue measurable subsets of $\mathbf{R}^k$. Assume that there exist two open sets $V \subseteq A$ and $W \subseteq B$ such that $T(V) = W$, $T: V \to W$ is a diffeomorphism, and $\lambda(A \setminus V) = \lambda(B \setminus W) = 0$. Then for each $f \in L_1(B)$, the function $(f \circ T) \cdot |J_T|$ (defined a.e. on A) belongs to $L_1(A)$ and*

$$\int_B f \, d\lambda = \int_A (f \circ T) \cdot |J_T| \, d\lambda$$

holds.

EXERCISES

1. Show that an open ball in a Banach space is a connected set. That is, show that if B is an open ball in a Banach space such that $B = \mathcal{O}_1 \cup \mathcal{O}_2$ holds with both $\mathcal{O}_1$ and $\mathcal{O}_2$ open and disjoint, then either $\mathcal{O}_1 = \emptyset$ or $\mathcal{O}_2 = \emptyset$.
   [HINT: If $a \in \mathcal{O}_1$ and $b \in \mathcal{O}_2$, then let $\alpha = \inf\{t \in [0, 1]: ta + (1 - t)b \in \mathcal{O}_1\}$, and note that $c = \alpha a + (1 - \alpha)b \in B$. To obtain a contradiction, show that $c \notin \mathcal{O}_1$ and $c \notin \mathcal{O}_2$.]
2. Let $T: V \to \mathbf{R}^k$ be C^1-differentiable. Show that the mapping $x \mapsto T'(x)$ from V into $L(\mathbf{R}^k, \mathbf{R}^k)$ is a continuous function.
3. Show that the Lebesgue measure on $\mathbf{R}^2$ is "rotation" invariant.
4. **(Polar Coordinates)** Let $E = \{(r, \theta) \in \mathbf{R}^2: r \geq 0 \text{ and } 0 \leq \theta \leq 2\pi\}$. The transformation $T: E \to \mathbf{R}^2$ defined by $T(r, \theta) = (r \cos\theta, r \sin\theta)$, or as it is usually written

$$x = r \cos\theta \quad \text{and} \quad y = r \sin\theta,$$

is called the **polar coordinate transformation** on $\mathbf{R}^2$, shown graphically in Figure 7.3.

   a. Show that $\lambda(E \setminus E^\circ) = 0$.
   b. If $A = \{(x, 0): x \geq 0\}$, then show that A is a closed subset of $\mathbf{R}^2$ whose (two-dimensional) Lebesgue measure is zero.
   c. Show that $T: E^\circ \to \mathbf{R}^2 \setminus A$ is a diffeomorphism whose Jacobian determinant satisfies $J_T(r, \theta) = r$ for each $(r, \theta) \in E^\circ$.
   d. Show that if G is a Lebesgue measurable subset of E with $\lambda(G \setminus G^\circ) = 0$, then $T(G)$ is a Lebesgue measurable subset of $\mathbf{R}^2$. Moreover, show that if $f \in L_1(T(G))$, then

$$\int_{T(G)} f \, d\lambda = \iint_G f(r \cos\theta, r \sin\theta) r \, dr \, d\theta$$

   holds.

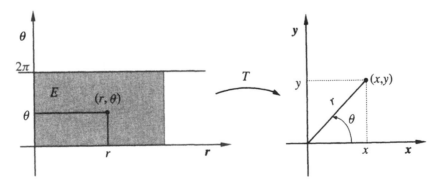

FIGURE 7.3.   The Polar Coordinate Transformation

5. This exercise uses polar coordinates (introduced in the preceding exercise) to present an alternate proof of Euler's formula $\int_0^\infty e^{-x^2}\,dx = \sqrt{\pi}/2$.

   a. For each $r > 0$, let $C_r = \{(x, y) \in \mathbf{R}^2: x^2 + y^2 \leq r^2, x \geq 0, y \geq 0\}$ and $S_r = [0, r] \times [0, r]$. Show that $C_r \subseteq S_r \subseteq C_{r\sqrt{2}}$.

   b. If $f(x, y) = e^{-(x^2+y^2)}$, then show that

$$\int_{C_r} f\,d\lambda \leq \int_{S_r} f\,d\lambda \leq \int_{C_{r\sqrt{2}}} f\,d\lambda,$$

   where λ is the two-dimensional Lebesgue measure.

   c. Use the change of variables to polar coordinates and Fubini's theorem to show that

$$\int_{C_r} f\,d\lambda = \int_0^{\frac{\pi}{2}} \int_0^r e^{-t^2} t\,dt\,d\theta = \frac{\pi}{4}\left(1 - e^{-r^2}\right).$$

   d. Use (b) to establish that

$$\frac{\pi}{4}\left(1 - e^{-r^2}\right) \leq \left(\int_0^r e^{-x^2}\,dx\right)^2 \leq \frac{\pi}{4}\left(1 - e^{-2r^2}\right),$$

   and then let $r \to \infty$ to obtain the desired formula.

6. In $\mathbf{R}^4$, "double" polar coordinates are defined by

$$x = r\cos\theta, \quad y = r\sin\theta, \quad z = \rho\cos\phi, \quad w = \rho\sin\phi.$$

   State the change of variables formula for this transformation, and use it to show that the "volume" of the open ball in $\mathbf{R}^4$ with center at zero and radius a is $\frac{1}{2}\pi^2 a^4$.

7. **(Cylindrical Coordinates)** Let $E = \{(r, \theta, z) \in \mathbf{R}^3: r \geq 0, 0 \leq \theta \leq 2\pi, z \in \mathbf{R}\}$. The transformation $T: E \to \mathbf{R}^3$ defined by $T(r, \theta, z) = (r\cos\theta, r\sin\theta, z)$ or as it is usually written

$$x = r\cos\theta, \quad y = r\sin\theta, \quad z = z,$$

   is called the **cylindrical coordinate transformation**, shown graphically in Figure 7.4.

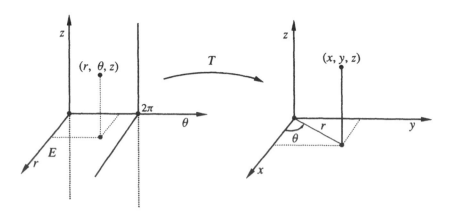

FIGURE 7.4. The Cylindrical Coordinate Transformation

a. Show that $\lambda(E \setminus E^\circ) = 0$.
b. If $A = \{(x, 0, z) \in \mathbf{R}^3 : x \geq 0,\ z \in \mathbf{R}\}$, then show that A is a closed subset of $\mathbf{R}^3$ whose (three-dimensional) Lebesgue measure is zero.
c. Show that $T : E^\circ \to \mathbf{R}^3 \setminus A$ is a diffeomorphism whose Jacobian determinant satisfies $J_T(r, \theta, z) = r$ for each $(r, \theta, z) \in E^\circ$.
d. Show that if G is a Lebesgue measurable subset of E with $\lambda(G \setminus G^\circ) = 0$, then $T(G)$ is a Lebesgue measurable subset of $\mathbf{R}^3$. Moreover, show that if $f \in L_1(T(G))$, then

$$\int_{T(G)} f\, d\lambda = \iiint_G f(r\cos\theta, r\sin\theta, z) r\, dr\, d\theta\, dz$$

holds.

8. **(Spherical Coordinates)** Let

$$E = \{(r, \theta, \phi) \in \mathbf{R}^3 : r \geq 0,\ 0 \leq \theta \leq 2\pi,\ 0 \leq \phi \leq \pi\}.$$

The transformation $T : E \to \mathbf{R}^3$ defined by

$$T(r, \theta, \phi) = (r\cos\theta\sin\phi, r\sin\theta\sin\phi, r\cos\phi),$$

or as it is usually written

$$x = r\cos\theta\sin\phi, \quad y = r\sin\theta\sin\phi, \quad z = r\cos\phi,$$

is called the **spherical coordinate transformation**, shown graphically in Figure 7.5.

a. Show that $\lambda(E \setminus E^\circ) = 0$.
b. If $A = \{(x, 0, z) : x \geq 0 \text{ and } z \in \mathbf{R}\}$, then show that A is a closed subset of $\mathbf{R}^3$ whose (three-dimensional) Lebesgue measure is zero.
c. Show that $T : E^\circ \to \mathbf{R}^3 \setminus A$ is a diffeomorphism whose Jacobian determinant satisfies $J_T(r, \theta, \phi) = -r^2 \sin\phi$.

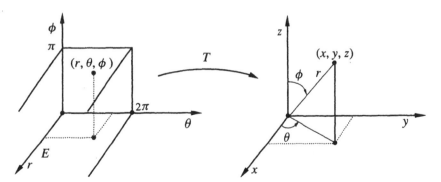

FIGURE 7.5. The Spherical Coordinate Transformation

d. Show that if G is a Lebesgue measurable subset of E with $\lambda(G \setminus G^{\circ}) = 0$, then $T(G)$ is a measurable subset of $\mathbf{R}^3$. In addition, show that if $f \in L_1(T(G))$, then

$$\int_{T(G)} f \, d\lambda = \int \int \int_G f(r \cos \theta \sin \phi, r \sin \theta \sin \phi, r \cos \phi) r^2 \sin \phi \, dr \, d\theta \, d\phi$$

holds.

BIBLIOGRAPHY

1. C. D. Aliprantis and K. C. Border, *Infinite Dimensional Analysis*, Studies in Economic Theory, #4, Springer–Verlag, New York & Heidelberg, 1994.
2. T. M. Apostol, *Mathematical Analysis*, Second Edition, Addison-Wesley, Reading, MA, 1974.
3. G. Birkhoff and S. MacLane, *A Survey of Modern Algebra*, Third Edition, Macmillan, New York, 1965.
4. E. Borel, *Leçons sur la Théorie des Fonctions*, Gauthier-Villars, Paris, 1898. (Third Edition, 1928.)
5. P. J. Daniell, A general form of integral, *Annals of Mathematics* **19** (1917), 279 294.
6. M. M. Day, *Normed Linear Spaces*, Third Edition, Springer–Verlag, Heidelberg & New York, 1973.
7. J. Dieudoné, *Foundations of Modern Analysis*, Academic Press, New York & London, 1969.
8. A. A. Fraenkel, *Abstract Set Theory*, Fourth Edition, North-Holland, Amsterdam, 1976.
9. M. Frantz, On Sierpiński's nonmeasurable set, *Fundamenta Mathematicae* **139** (1991), 17–22.
10. B. R. Gelbaum and J. M. H. Olmsted, *Theorems and Counterexamples in Mathematics*, Springer–Verlag, Heidelberg & New York, 1990.
11. C. Goffman, *Real Functions*, Prindle, Weber & Schmidt, New York, 1953.
12. C. Goffman and G. Pedrick, *First Course in Functional Analysis*, Prentice-Hall, Englewood Cliffs, NJ, 1965.
13. P. R. Halmos, *Naive Set Theory*, Springer–Verlag, Heidelberg & New York, 1974.
14. P. R. Halmos, *Measure Theory*, Van Nostrand, New York, 1950.
15. E. Hewitt and K. Stromberg, *Real and Abstract Analysis*, Third Edition, Springer–Verlag, Heidelberg & New York, 1975.
16. T. Jech, *Set Theory*, Academic Press, New York, 1978.
17. I. Kaplansky, *Set Theory and Metric Spaces*, Allyn and Bacon, Boston, 1972.
18. J. L. Kelley, *General Topology*, Graduate Texts in Mathematics, #27, Springer–Verlag, Heidelberg & New York, 1975.
19. A. N. Kolmogorov and S. V. Fomin, *Measure, Lebesgue Integrals, and Hilbert Space*, Academic Press, New York & London, 1961.
20. K. Kuratowski and A. Mostowski, *Set Theory*, North-Holland, Amsterdam, 1976.
21. H. Lebesgue, Intégrale, longueur, aire, Annali Mat. Pura Appl., Ser. 3, **7** (1902), 231–359.

22. J. R. Munkres, *Topology: a First Course*, Prentice-Hall, Englewood Cliffs, NJ, 1975.
23. F. Riesz and B. Sz.-Nagy, *Functional Analysis*, L. Boron translator, F. Ungar, New York, 1955.
24. H. L. Royden, *Real Analysis*, Third Edition, Macmillan, New York & London, 1988.
25. W. Rudin, *Principles of Mathematical Analysis*, Third Edition, McGraw-Hill, New York, 1976.
26. W. Rudin, *Real and Complex Analysis*, Third Edition, McGraw-Hill, New York, 1987.
27. W. Sierpiński, Sur un problème concernant les ensembles mesurables superficiellement, *Fundamenta Mathematicae* **1** (1920), 112–115.
28. G. F. Simmons, *Introduction to Topology and Modern Analysis*, McGraw-Hill, Macmillan, New York, 1963.
29. G. Strang, *Linear Algebra and its Applications*, Second Edition, Academic Press, New York, 1980.
30. A. E. Taylor, *General Theory of Functions and Integration*, Blaisdell, Waltham, MA, 1965.
31. A. E. Taylor and D. C. Lay, *Introduction to Functional Analysis*, Second Edition, Wiley, New York, 1980.
32. A. Torchinsky, *Real Variables*, Addison Wesley, New York, 1988.
33. R. L. Wheeden and A. Zygmund, *Measure and Integral*, Marcel Dekker, Inc., New York & Basel, 1977.
34. A. C. Zaanen, *Integration*, North–Holland, Amsterdam, 1967.

LIST OF SYMBOLS

INDEX